International Conference on Mechanical, Electronics, Computer, and Industrial Technology (MECnIT 2017)

Journal of Physics: Conference Series Volume 1007

Prima, Indonesia
6 – 8 December 2017

ISBN: 978-1-5108-6239-5
ISSN: 1742-6588

Printed from e-media with permission by:

Curran Associates, Inc.
57 Morehouse Lane
Red Hook, NY 12571

CURRAN ASSOCIATES INC. proceedings.com

Some format issues inherent in the e-media version may also appear in this print version.

Printed by Curran Associates, Inc. (2018)

For permission requests, please contact the Institute of Physics
at the address below.

Institute of Physics
Dirac House, Temple Back
Bristol BS1 6BE UK

Phone: 44 1 17 929 7481
Fax: 44 1 17 920 0979

techtracking@iop.org

Additional copies of this publication are available from:

Curran Associates, Inc.
57 Morehouse Lane
Red Hook, NY 12571 USA
Phone: 845-758-0400
Fax: 845-758-2633
Email: curran@proceedings.com
Web: www.proceedings.com

Table of contents

Volume 1007

International Conference on Mechanical, Electronics, Computer, and Industrial Technology

6–8 December 2017, Prima, Indonesia

Accepted papers received: 9 April 2018
Published online: 30 April 2018

Preface

Papers

Computer Science

Electrical Engineering

IOP Conf. Series: Journal of Physics: Conf. Series **1007** (2018) 011001 doi:10.1088/1742-6596/1007/1/011001

2017 International Conference on

Mechanical, Electronics, Computer, and Industrial Technology (MECnIT 2017)

December 6th-8th, 2017 | Medan, Indonesia

Organized By:

Co-Organized By:

Supported By:

IOP Conf. Series: Journal of Physics: Conf. Series **1007** (2018) 011001 doi:10.1088/1742-6596/1007/1/011001

Preface

International Conference on Mechanical, Electronics, Computers, and Industrial Technology (MECnIT 2017) was held in Medan, Indonesia. MECnIT 2017 was hosted by Faculty of Technology and Computer Sciences, Universitas Prima Indonesia. This conference is co-organized by TIMERS (Technology, Informatics, Management, and Engineering Research Support), Universiti Utara Malaysia, National Taiwan University of Science and Technology, Universitas Sumatera Utara, Institut Teknologi Bandung, Universitas Padjajaran, and Asosiasi Perguruan Tinggi Informatika dan Ilmu Komputer (APTIKOM).

The MECnIT 2017 have attracted many academicians, scientists, engineers, postgraduates, and other professionals from many countries. These conferences aim to promote interaction among engineers, researchers, and scientists active in the related areas. The events are intended to provide a high-level international forum to present, to exchange, and to discuss recent advances, new techniques, and applications in the field of knowledge discussed in this conference.

Our special thank also goes to all individuals and organizations such as the international program committees (IPC), the conference organizers, the reviewers, and the authors, for their contribution in making MECnIT 2017 not only a successful international conference but also as a memorable gathering event. We are also grateful for the support of the publication service of IOP. We hope that it should give you a beautiful memory to bring home in addition to new insights and friends gathered during the conference. We are truly grateful for your contribution and interest. We hope that you will get pleasure from MECnIT 2017 in this beautiful city, Medan, Indonesia.

Best regards,
Abdi Dharma (General Chair of MECnIT 2017)

IOP Conf. Series: Journal of Physics: Conf. Series **1007** (2018) 011001 doi:10.1088/1742-6596/1007/1/011001

List of Committee

Conference General Chair:
Abdi Dharma, S. Kom, M. Kom

Secretary Chair 1
Mardi Turnip (Universitas Prima Indonesia, Indonesia)
Secretary Chair 2
Yonata Laia (Universitas Prima Indonesia, Indonesia)
Technical Program Chair
Poltak Sihombing (Universitas Sumatera Utara, Indonesia)
Publication Chair
Muhammad Zarlis (Universitas Sumatera Utara, Indonesia)
Publicity Chair
Hairus Abdullah (National Taiwan University of Science and Technology, Taiwan)
IPC Chair
Nooraini Binti Yusoff (University Utara Malaysia, Malaysia)

International Program Committee (IPC)
Dessy Novita, Universitas Padjadjaran, Indonesia
Diah Chaerani, Universitas Padjadjaran, Indonesia
Gea O.F. Parikesit, Universitas Gadjah Mada, Indonesia
Irwan Purnama, Indonesian Institute of Sciences, Indonesia
Ismoyo Haryanto, Universitas Diponegoro, Indonesia
Mauridhi H. Purnomo, Institut Teknologi Sepuluh Nopember, Indonesia
Mohammad Taufik, Universitas Padjajaran, Indonesia
Salmah, Universitas Gadjah Mada, Indonesia
Juni Astel Rajagukguk, Universtas Negeri Medan, Indonesia
Augie Widyotriatmo, Institut Teknologi Bandung, Indonesia
Meilita Tryana Sembiring, Universitas Sumatera Utara, Indonesia
Juliza Hidayati, Universitas Sumatera Utara, Indonesia
Daniel Sutopo, Politeknik Batam, Indonesia
Joni Welman Simatupang, Universitas President, Indonesia
Nutthita Chuankrerkkul, Chulalongkom University, Thailand
Surapong Chatpun, Institute of Biomedical Engineering, Prince of Songkla University, Thailand
Ching Yern Chee, University of Malaya, Malaysia
Leong Loong Kong, University Tunku Abdul Rahman, Malaysia
Rini Akmeliawati, International Islamic University Malaysia, Malaysia
Wayan Suparta, University Kebangsaan Malaysia, Malaysia
Lo Chin Kim, TNB Research Sdn, Malaysia
Le Hoa Nguyen, Hanoi University of Science and Technology, Vietnam
Nguyen Van Cuong, Can Tho University, Vietnam

IOP Conf. Series: Journal of Physics: Conf. Series **1007** (2018) 011001 doi:10.1088/1742-6596/1007/1/011001

IOP Conf. Series: Journal of Physics: Conf. Series **1007** (2018) 011001 doi:10.1088/1742-6596/1007/1/011001

CONFERENCES PHOTOGRAPHS

IOP Conf. Series: Journal of Physics: Conf. Series **1007** (2018) 011001 doi:10.1088/1742-6596/1007/1/011001

IOP Conf. Series: Journal of Physics: Conf. Series **1007** (2018) 011001 doi:10.1088/1742-6596/1007/1/011001

IOP Conf. Series: Journal of Physics: Conf. Series **1007** (2018) 011001 doi:10.1088/1742-6596/1007/1/011001

IOP Conf. Series: Journal of Physics: Conf. Series **1007** (2018) 011001 doi:10.1088/1742-6596/1007/1/011001

IOP Conf. Series: Journal of Physics: Conf. Series **1007** (2018) 011002 doi:10.1088/1742-6596/1007/1/011002

Peer review statement

All papers published in this volume of *Journal of Physics: Conference Series* have been peer reviewed through processes administered by the proceedings Editors. Reviews were conducted by expert referees to the professional and scientific standards expected of a proceedings journal published by IOP Publishing.

IOP Conf. Series: Journal of Physics: Conf. Series **1007** (2018) 012001 doi:10.1088/1742-6596/1007/1/012001

Overview Electrotactile Feedback for Enhancing Human Computer Interface

Daniel S Pamungkas[1*], Wahyu Caesarendra[2]

[1]Electrical Engineering Department, Politeknik Negeri Batam, Batam, Indonesia
[2]Department of Mechanical Engineering, Diponegoro University, Semarang 50275, Indonesia

E-mail: *daniel@polibatam.ac.id

Abstract. To achieve effective interaction between a human and a computing device or machine, adequate feedback from the computing device or machine is required. Recently, haptic feedback is increasingly being utilised to improve the interactivity of the Human Computer Interface (HCI). Most existing haptic feedback enhancements aim at producing forces or vibrations to enrich the user's interactive experience. However, these force and/or vibration actuated haptic feedback systems can be bulky and uncomfortable to wear and only capable of delivering a limited amount of information to the user which can limit both their effectiveness and the applications they can be applied to. To address this deficiency, electrotactile feedback is used. This involves delivering haptic sensations to the user by electrically stimulating nerves in the skin via electrodes placed on the surface of the skin. This paper presents a review and explores the capability of electrotactile feedback for HCI applications. In addition, a description of the sensory receptors within the skin for sensing tactile stimulus and electric currents alsoseveral factors which influenced electric signal to transmit to the brain via human skinare explained.

1. Introduction

New and innovative Human Computer Interaction (HCI) technologies have recently attracted increased interest. This is mainly due to the proliferation of various interactive computing devices and applications including gaming, virtual reality, augmented reality, teleoperation, telepresence, sports and skill training systems, etc. (see [1] [2], [3]). To be effective, HCI requires the input device to be intuitive and user friendly and the output device to provide appropriate information or feedback to the user. Normally, the user input is delivered to the computer via a keyboard and mouse and information or feedback is received via a visual display and generated sounds. Recently, gesture based input devices that involve tracking the body, hands or finger movements, have been gaining attention over the conventional keyboard and mouse for certain interactive applications [4]. Similarly, 3D stereo head mounted displays and haptic feedback devices involving vibrating actuators or electro-mechanical forces have been developed to improve the interactivity of many applications (e.g. [5-6]).

The word of haptic comes from the Greek word 'hapto' which means to touch and handle objects. Touch sensing relates to stimuli on the skin and is what gives both humans and animals the ability to detect and classify objects that come in contact with the skin. Touch sensing is often referred to as tactile sensing by researchers who attempt to implement, enhance or facilitate touch sensing via electronic or mechanical means [7].

IOP Conf. Series: Journal of Physics: Conf. Series **1007** (2018) 012001 doi:10.1088/1742-6596/1007/1/012001

Adding 3D stereo perception and haptic feedback to the interface can help the user to gain a sense of being immersed in a remote or virtual environment and result to improved interactivity [8]. Most haptic feedback systems aim at enriching the user's sense of touch making it possible to interpret or interact with the application, or a remote or virtual environment, more effectively.

Haptic interfaces use artificially induced haptic sensations to deliver information or feedback signals from a computer or a machine to the user. Haptic feedback can make it possible to simulate the tactile characteristics of a remote or virtual object (such as weight, texture, force, movement, etc.) via electro-mechanical or electronic means. These physical characteristics are difficult, if not impossible, to perceive visually. Since touch sensing is important for manipulating objects, haptic feedback can enable the user to operate a machine or a computer program with more agility and dexterity. Applications that make use of haptic interface are wide and expanding. Some prominent examples are: surgery training simulators [9], teleoperation of robots in hazardous environments [10] and VR applications [11]. These applications have haptic interfaces that can range from simple force feedback joystick devices with a single Degree of Freedom (DOF) [12], to more complex interfaces with multiple DOF [10].

Most haptic feedback systems are aimed at replicating certain natural tactile sensations and consist of force and/or vibrating electro-mechanical actuators coupled to the user with mechanical linkages or bulky garments. These systems are often difficult to setup and bulky to wear and can compromise the user's movements and comfort. Furthermore, electro-mechanical haptic feedback systems are often custom built and configured for specific applications which can make it difficult to use the same haptic feedback system for different applications [13].To overcome these drawbacks, the electrical signal was introduced to deliver information to the user. This paper reviews several methods and for transmitting information signals and its applications by the electrical signal.

Electrotactile feedback involves producing haptic sensations by electrically stimulating nerves in the skin via electrodes placed on the surface of the skin. They were initially developed for providing substitute visual perception to the blind [14]. Recently, electrotactile feedback has become recognised by some researchers as being simpler and more versatile than electro-mechanical feedback systems. The main benefits of electrotactile feedback is that it is inexpensive, it has no mechanical or moving parts and it can deliver a wide variety of sensations to the user by simply varying the electrical signal delivered to the electrodes [14].

To explain this in more detail and provide some understanding on the factors that influence the delivery of electrotactile feedback information to the brain, via neurons in the skin, the following section provides some background information on the human tactile sensory system.

2. The Human Tactile Sensory System

Compared to other sensory organs, human skin has the largest number of sensory receptors. Touch sensing actually combines several sub-modalities (e.g. pressure, temperature, texture, movement, pain, etc.) These tactile sensations from the skin are mediated by the somatosensory system which also processes sensory information from the epithelial tissues, skeletal muscles, bones, internal organs and the cardiovascular system. To understand the possibilities of any tactile feedback device, it helps to know the structure and functionality of skin, including the sensory nerves and receptors within the skin.

The two types of human skin are hairy skin and glabrous (or hairless) skin. Hairy skin has hair follicles and comprises almost 95 percent of human skin. Glabrous skin covers about 5 percent of the human body and has a thinner dermal layer, for increased touch sensitivity, and thicker epidermis without hair follicles. Glabrous skin also has a ridged appearance with circular patterns (like fingerprints), e.g. finger pads, lips, palms and the soles of feet [15].

Human skin is sensitive to several sensations, e.g. temperature, vibration, pressure or electric current/voltage. To accommodate these sensations, skin has a number of sensory receptors, namely, mechanoreceptors, thermoreceptors, nociceptors and proprioceptors. Furthermore, mechanoreceptors can be divided into seven classes and thermoreceptors have two classes. In addition, there are four and three classes for nociceptors and proprioceptors, respectively [16]. Electrotactile feedback works mainly on the mechanoreceptors. Table 1 lists the four major mechanoreceptors and their respective sensory modalities.

Table 1. Skin Mechanoreceptors (adapted from [16]).

	Meissner's Corpuscles	Merkel's Cells	Pacinian Corpuscle	Ruffini Endings
Class	Rapid Adapting Type I	Slow Adapting Type I	Rapid Adapting Type II	Slow Adapting Type II
Sensory Correlate	Stroking, Fluttering, Low-frequency vibration	Pressure, Texture	High vibration	Skin Stretch
Frequency range	10-200 Hz	0.4-100 Hz	40-800 Hz	7 Hz

The following section presents certain related factors that can influence the design and control of electrotactile interfaces.

3. Electrotactile Sensory Substitution Methods

Electrotactile feedback has been shown to be capable of delivering a variety of tactile sensations [17]. Although it is not possible to replicate all the possible tactile sensations perceivable through the skin, electrotactile stimulation can provide a reasonable substitute in certain applications. There are a number of factors that can influence effective information transmission through the skin while minimising feelings of discomfort. In general, these factors are mainly related to the characteristics of the electro-neural stimulation parameters, the conductivity and location of the electrodes on the skin, and the properties of the skin.

3.1 Electrodes

Chemical compounds can be produced by electrodes in an electrochemical process. For example, any electrolysis reactions in water might damage skin cells. Electrodes made of noble metals or conductive polymers may reduce or eliminate any electrochemical reaction [18]. The size of the electrodes can also be important. Small electrodes can generate higher density currents which can cause discomfort to the user. Electrodes with an area of $10mm^2$ or greater have been shown to perform better than smaller electrodes [19].

3.2 Waveform Stimulus

Square wave electrical pulses are more commonly used for electro-neural stimulation than sinusoidal waves due to their improved efficiency and ease of implementation. Most comparisons find no significant differences between the performances of these two waveforms [16,20]. However, biphasic stimulation may help prevent half-cell reactions and aid in preventing ions being absorbed from electrodes and conductive gels [21].

3.3 Polarity

Some researchers have described the use of bipolar and unipolar signals for electrotactile stimulation (e.g. [22]). These results suggest different sensations can be produced by changing the polarity of the electrodes. A bipolar signal has an alternating current phase that is positive and negative. A unipolar signal only has a positive or a negative phase. Generally, bipolar signals are preferable in order to prevent skin irritation caused by the transfer of ions within axon membranes. Generating a negative pulse shortly after a positive pulse, or visa-versa, can also reduce polarisation caused by the first stimulation. Studies suggest a delay of between 50-500µs can produce delayed depolarisation [23]. For unipolar signals, positive pulses tend to produce a lower sensory threshold and are more comfortable than negative pulses [23].

3.4 Frequency
Experiments conducted by Saunders [24] suggest that a frequency range between 2-100 Hz is the optimal for electrotactile stimulation. These frequencies are within the detectable range of all rapid adapting Type-I receptors and some slow adapting Type-I receptors [25]. Investigations by Collin [26], indicate that the perceptual response to a continuous pulse rate at frequencies higher than 60hz results in fast adaptation, whereas, frequencies around 10hz produce slower adaptation [28,29]. However, high frequency pulses in bursts can also produce slower adaptation.

3.5 Pulse Width
It has been shown that pulse widths between 50-150µs provide effective receptor stimulus [21,27]. Furthermore, Mortimer [20]have shown that there is an inverse relationship between the pulse width and the threshold current. Generally, to reduce the power consumption and to prevent potential harm to the user's skin, it is preferable to keep the pulse width to a minimum. Experiments by Rollman [29] and Mortimer [20] suggest a pulse width around 100µs is optimal in terms of power consumption and receptor stimulation.

3.6 Burst Rate
Some researchers, e.g. [30], have shown that long continuous pulse trains, particularly at low power and higher frequencies, can cause rapid adapting Type-I receptors to become adapted to the stimulus. To prevent receptors from becoming adapted to stimulus, some researchers have opted for grouping pulses into bursts. For example, Kaczmarek has shown that short repeated bursts at a low rate (<15hz) could produce good results at avoiding adaptation [27].

3.7 Pulses per Burst
The dynamic range and sensitivity of rapid adapting Type-I receptors can also be affected by number of pulses per burst, particularly at higher frequencies. For example, Kaczmarek [22], showed that increasing the number of pulses per burst beyond six can fail to provide any additional performance improvement. Kaczmarek's results show that improved performance could be achieved with 200Hz pulses grouped into bursts of 6 pulses with a delay of 37ms per burst.

3.8 Modulation Method
The adaptation of receptors can also be avoided by modulating the frequency and/or the intensity of the electrical stimulus. For example, Szeto [31]conducted experiments on test subjects where various stimulus parameters were modulated including the polarity, frequency, intensity and the number of active electrodes. Apparently, identifying the position and/or the number of activated electrodes performed the best. Furthermore, test results with lower frequencies were better than results with higher frequencies [28].

3.9 Skin Features
Skin impedance and sensitivity can vary depending on its moisture content and location on the body. Generally, glabrous skin performs worse than hairy skin for electrotactile stimulation because of its thickness. This can make electrical conductivity difficult and individual receptors hard to target [20]. Glabrous skin is also found only on small inconvenient regions of the body, e.g. palms and the soles of feet. Moistening the skin with conductive gels prior to applying electrodes can also improve the electrical conductivity and reduce the possibility of painful sensations occurring as a result of current flowing directly through sweat glands [20]. The following section provides further examples of electrotactile haptic interfaces and a more detailed explanation of their implementation and operation.

4. Electrotactile Feedback Devices

Electrotactile feedback involves producing haptic sensations by electrically stimulating nerves in the skin via electrodes placed on the surface of the skin. The main benefit of electrotactile feedback, over vibrotactile or force feedback systems, is that there are no mechanical or moving parts and it can deliver a wide variety of sensations to the user that are not available with existing vibrotactile or force feedback systems.

With electrotactile feedback, the electrodes can also be placed on various body locations where more cutaneous sensory neurons are available, e.g. tongue [32], hands/arms [13], or abdomen [33]. Since the hands and fingertips have high neural density, they are often studied for delivering haptic information to the brain via electro-neural stimulation [34].

Electro-neural stimulation has been intensively studied for providing Tactile Visual Substitution (TVS) for the blind. For example, in the late 1990s, Kazmarek et al [32], proposed a Tongue Display Unit (TDU) for the blind. The tongue was used because it has high tactile sensitivity and spatial acuity. Additionally, the tongue is always wet and therefore provides good electrical conductivity and stable electrical properties.

Electro-neural stimulation is not commonly used for developing information feedback systems. Most haptic researchers traditionally use force and/or vibrotactile haptic interfaces, probably due to fears that electro-neural stimulation might cause pain and/or discomfort to the user. However, some researchers have recently attempted to develop haptic feedback systems using electro-neural stimulation mainly because of its success at providing substitute visual perception to the blind as well as the advantages it has over electromechanical interfaces, as explained in [14]; e.g. no mechanical components, low cost and increased bandwidth for interpreting information.

Another advantage of electro-neural stimulation is that the electrodes can be configured into compact arrays and used to implement electrotactile displays. These are mostly used for TVS systems, like the TDU. They have also been used for helping patients to recover from posture and balance-disorders such as Meniere's disease and as a navigation aid [35]. Electrotactile arrays, like that used for the TDU, are typically comprised of 144 electrodes configured into a 12x12 matrix. These electrodes are generally made from gold-plated copper pads fixed to a flexible printed circuit polymer sheet.

Other researchers, e.g. Kajimoto et al, developed an electrotactile display system called the SmartTouch [36]. SmartTouch is a 4x4 matrix array made from 1mm diameter concentric stainless steel electrodes that can generate both pressure and vibration sensations. These sensations occur because the positive inner electrodes are designed to stimulate the Meissner's corpuscles (near the surface) which are responsible for sensing pressure. The negative outer electrodes are meant for stimulating the deeper Pacini corpuscles, which are involved in sensing vibrations.

To explore the potential of electrotactile feedback, Daniel and Koren use hand gesture based input devices (such as Leap Motion or P5 data glove [4]) combined with electrotactile feedback for achieving more effective control and improved interaction of various computer controlled machines and applications for example teleoperation[37], VR [38], training [39], and controlled prosthesis hand [40]. This generally involves using hand and/or finger movements to operate a remotely located machine, or an agent in Virtual Reality (VR), while receiving haptic sensations from the electrotactile feedback system. The electrotactile stimulus is usually derived from real or virtual sensors mounted on a remote machine, or a VR environment. When used in conjunction with a 3D head mounted display, electrotactile feedback can help the user to feel more immersed within the remote or virtual environment and achieve more dexterous control of the machine or virtual agent.

5. Conclusions

The main benefits of electrotactile feedback is that it has no mechanical or moving parts, requires no-invasive surgery and can deliver haptic information to the user via a wide variety of sensations that are not available with existing vibrotactile or force feedback systems. These reason makes electrotactile feedback has a potential to enrich the HCI systems, to make the user more engaged with the system.

IOP Conf. Series: Journal of Physics: Conf. Series **1007** (2018) 012001 doi:10.1088/1742-6596/1007/1/012001

References

[1] Bednarz, T.P., C. Caris, and O. Dranga, Human-computer interaction experiments in an immersive virtual reality environment for e-learning applications. Annual Conference for the Australasian Assosiation for Engineering Education.. 2010, Adelaide p. 834-839.

[2] Do, T.N., Tjahjowidodo. T., Lau, M.W.S., Phee, S.J., (2013). Nonlinear Modeling and Parameter Identification of Dynamic Friction Model in Tendon Sheath for Flexible Endoscopic Systems, Proceedings of the 10th International Conference on Informatics in Control, Automation and Robotics, Reykjavik, Iceland, pp. 5-10.

[3] Wu, C.-I., HCI and Eye Tracking Technology for Learning Effect. Procedia - Social and Behavioral Sciences, 2012. 64: p. 626-632.

[4] Nielsen, M., et al., Gesture Interfaces, in HCI Beyond the GUI: Design for Haptic, Speech, Olfactory, and Other Nontraditional Interfaces, P. Kortum, Editor. 2008, Elsevier Inc. p. 75-106.

[5] Pawar, V.M. and A. Steed, Profiling the behaviour of 3D selection tasks on movement time when using natural haptic pointing gestures, in Proceedings of the 16th ACM Symposium on Virtual Reality Software and Technology. 2009, ACM: Kyoto, Japan. p. 79-82.

[6] Diego, J.P.S., et al., PHANTOM prototype: exploring the potential for learning with multimodal features in dentistry, in Proceedings of the 10th international conference on Multimodal interfaces. 2008, ACM: Chania, Crete, Greece. p. 201-202.

[7] Mihelj, M., D. Novak, and S. Beguš, Virtual reality technology and applications. Intelligent systems, control and automation--science and engineering: volume 68. 2014: Dordrecht Springer.

[8] Boll, S., N. Magnenat-Thalmann, and U. Bonanni, Haptics in Virtual Reality and Multimedia. IEEE Multimedia, 2006. 13(3): p. 6-11.

[9] Delp, S.L. and P. Loan, Surgical simulation: An emerging technology for training in emergency medicine. Presence: Teleoperators & Virtual Environments, 1997. 6(2): p. 147-159.

[10] Yoon Sang, K., et al., A force reflected exoskeleton-type masterarm for human-robot interaction. Systems, Man and Cybernetics, Part A: Systems and Humans, IEEE Transactions on, 2005. 35(2): p. 198-212.

[11] Sadihov, D., et al. Prototype of a VR upper-limb rehabilitation system enhanced with motion-based tactile feedback. in World Haptics Conference (WHC), 2013. Daejon, p. 449-454.

[12] Lawrence, D.A. and J.D. Chapel. Performance trade-offs for hand controller design. in Robotics and Automation, 1994. Proceedings., 1994 IEEE International Conference on.,1994. San Diego, p. 3211-3216

[13] Pamungkas, D., Ward, Koren. Tele-operation of a Robot Arm with Electro Tactile Fedback. in 2013 IEEE/ASME International Conference on Advanced Intelegent Mechatronics. 2013. Wollongong, Australia. p. 704-709.

[14] Meers, S. and K. Ward. A vision system for providing 3D perception of the environment via transcutaneous electro-neural stimulation. in Information Visualisation, 2004. IV 2004. Proceedings. Eighth International Conference on. 2004 :Los Alamitos. p. 546-552.

[15] Darian-Smith, I., The sense of touch: performance and peripheral neural processes in Supplement 3: Handbook of Physiology, The Nervous System, Sensory Processes. 1984: Bethesda, MD. p. 739-788.

[16] Toney, A., et al. A shoulder pad insert vibrotactile display. in Wearable Computers, 2003. Proceedings. Seventh IEEE International Symposium on. 2003. Washington : p. 35-44.

[17] Kajimoto, H., et al. Electro-Tactile Display with Tactile Primary Color Approach. in Proceedings of International Conference on Intelligent Robots and Systems. 2004.

[18] Brummer, S.B. and M.J. Turner, Electrical stimulation of the nervous system: The principle of safe charge injection with noble metal electrodes. Bioelectrochemistry and Bioenergetics, 1975. 2: p. 13-25.

[19] Kaczmarek, K.A., et al., A 16-channel 8-parameter waveform electrotactile stimulation system.

IEEE Transactions On Bio-Medical Engineering, 1991. 38(10): p. 933-943.
[20] Mortimer, J.T., Motor Prostheses, in Supplement 2: Handbook of Physiology, The Nervous System, Motor Control. 1981: Bethesda, MD. p. 155-187.
[21] Kajimoto, H., N. Kawakami, and S. Tachi, Optimal design method for selective nerve stimulation and its application to electrocutaneous display. Proceedings 10th Symposium on Haptic Interfaces for Virtual Environment & Teleoperator Systems. HAPTICS 2002, Orlando: p. 303-310.
[22] Kaczmarek, K.A., J.G. Webster, and R.G. Radwin, Maximal dynamic range electrotactile stimulation waveforms. IEEE Transactions On Bio-Medical Engineering, 1992. 39(7): p. 701-715.
[23] Lori A Birder and E.R. Perl, Cutaneous Sensory Receptors. Journal of Clinical Neurophysiology, 1994. 11(6): p. 534-552.
[24] Saunders, F.A., Recommended procedures for electrocutaneous displays, in Functional Electrical Stimulation, F.T. Hambrecht and J.B. Reswick, Editors. 1977, Marcel Dekker: New York. p. 303–309.
[25] Knibestöl, M. and A.B. Vallbo, Intensity of sensation related to activity of slowly adapting mechanoreceptive units in the human hand. The Journal Of Physiology, 1980. 300: p. 251-267.
[26] Collins, C., Tactile Television - Mechanical and Electrical Image Projection. IEEE Transactions on Man-Machine Systems, 1970. 11(1): p. 65-71.
[27] Kaczmarek, K.A., Electrotactile adaptation on the abdomen: preliminary results. IEEE Transactions On Rehabilitation Engineering: A Publication Of The IEEE Engineering In Medicine And Biology Society, 2000. 8(4): p. 499-505.
[28] Kaczmarek, K.A. and S.J. Haase, Pattern Identification and Perceived Stimulus Quality as a Function of Stimulation Waveform on a Fingertip-Scanned Electrotactile Display. IEEE Transactions on Neural Systems & Rehabilitation Engineering, 2003. 11(1): p. 9-16.
[29] Rollman, G., Electrocutaneous stimulation: Psychometric functions and temporal integration. Perception & Psychophysics, 1969. 5(5): p. 289-293.
[30] Lindblom, Y. and D.N. Tapper, Integration of impulse activity in a peripheral sensory unit. Experimental Neurology, 1966. 15(1): p. 63-69.
[31] Szeto, A.Y., Comparison of codes for sensory feedback using electrocutaneous tracking. Annals Of Biomedical Engineering, 1977. 5(4): p. 367-383.
[32] Kaczmarek, K.A., The tongue display unit (TDU) for electrotactile spatiotemporal pattern presentation. Scientia Iranica, 2011. 18(6): p. 1476-1485.
[33] Haase, S.J. and K.A. Kaczmarek, Electrotactile perception of scatterplots on the fingertips and abdomen. Medical & Biological Engineering & Computing, 2005. 43(2): p. 283-289.
[34] Igmo, K., et al., Wearable Fingertip Tactile Display. 2006 SICE-ICASE International Joint Conference, 2006: Busan, p. 1911-1916.
[35] Segond, H., D. Weiss, and E. Sampaio, Human spatial navigation via a visuo-tactile sensory substitution system. Perception, 2005. 34(10): p. 1231-1249
[36] Kajimoto, H., et al., SmartTouch: electric skin to touch the untouchable. Computer Graphics and Applications, IEEE, 2004. 24(1): p. 36-43
[37] Pamungkas, D. S. & Ward, K. Immersive Teleoperation of a Robot Arm Using Electrotactile Feedback. The 6th International Conference on Automation, Robotics and Applications (ICARA 2015). February 2015, Queenstown, New Zealand. p. 300-305
[38] Pamungkas, D. S. & Ward, K. Electrotactile Feedback System to Enhance Virtual Reality Experience. To appear in: International Journal of Computer Theory and Engineering. (IJCTE). pp. 465-470. Vol.8, No.6, Dec. 2016, ISSN: 1793-820.
[39] Pamungkas, D. S. & Ward, K. Using Hand Tracking and Electrotactile Stimulation to Facilitate Learning Touch Typing. 5th Annual International Conference on Education & e-Learning (EeL2015) Sept.2015. Bangkok, Thailand.

[40] Pamungkas, D. S. & Ward, K. Electrotactile Feedback System for a Prosthetic Hand. International Conference on Machine Vision and Mechatronics in Practice. (M2VIP). Dec 2015, Toowoomba, Australia.

IOP Conf. Series: Journal of Physics: Conf. Series **1007** (2018) 012002 doi:10.1088/1742-6596/1007/1/012002

An online condition monitoring system implemented an internet connectivity and FTP for low speed slew bearing

W Caesarendra[1,2*], B Kosasih[2], T Tjahjowidodo[3], M Ariyanto[1], LWQ Daryl[3], D Pamungkas[4]

[1]Department of Mechanical Engineering, Diponegoro University, Semarang, Indonesia

[2]School of Mechanical, Materials, Mechatronic and Biomedical Engineering, University of Wollongong, Wollongong NSW, Australia

[3]School of Mechanical and Aerospace Engineering, Nanyang Technological University, Singapore

[4]Electrical Engineering Department, Politeknik Negeri Batam, Indonesia

E-mail: *wc026@uowmail.edu.au

Abstract. Rapid and reliable information in slew bearing maintenance is not trivial issue. This paper presents the online monitoring system to assist maintenance engineer in order to monitor the bearing condition of low speed slew bearing in sheet metal company. The system is able to pass the vibration information from the place where the bearing and accelerometer sensors are attached to the data center; and from the data center it can be access by opening the online monitoring website from any place and by any person. The online monitoring system is built using some programming languages such as C language, MATLAB, PHP, HTML and CSS. Generally, the flow process is start with the automatic vibration data acquisition; then features are calculated from the acquired vibration data. These features are then sent to the data center; and form the data center, the vibration features can be seen through the online monitoring website. This online monitoring system has been successfully applied in School of Mechanical, Materials and Mechatronic Engineering, University of Wollongong.

1. Introduction

Slew bearings have been developed as new machine parts in the past 30 years. They were first used in excavators and cranes, and then applied to others machines such as loading and unloading machines, transporting machines and material processing machines. Slew bearings are almost the same as ball and roller bearings which have roller and rolling ring with the raceway. But compared with common roller bearings, they have many special features. They can be summarized in the following points: The size of slew bearing are great, diameter is generally between 0.4-10m. Slew bearings usually bear loads from different parts. They not only bear the axial and radial force but also endure tipping simultaneously. The speed of slew bearings is fairly low; usually ranging from 0.5 to 15 rpm. Slew bearing maintenance is relatively complicated, failure go unnoticed will result to breakdown maintenance and costly. Therefore the online condition monitoring of slew bearings has become necessary to provide the immediate information of bearing condition.

The industrial revolution has growth from Industry 3.0 which uses the automation system in the manufacturing process to Industry 4.0 which uses the advantages of cyber world into physical world of manufacturing which in return allows transmission and reception of data from and to anywhere on earth that connects to the internet.

2. A brief review on vibration online monitoring

There were many published articles used the word ‘online’ for their method [1-3]. Before going further, the definition of ‘online’ should be determined first. This paper defines the word ‘online’ as “the system that can provide the rapid and reliable information which convert the raw data into the useful information to support the decision making in maintenance work without any human interference”. The example of recent article which put the word ‘online’ on their title is Loutas et.al [4]. The paper presented the on-line monitoring for three monitoring techniques: vibration, acoustic emission and oil debris using National Instrument. Even though the paper said the three monitoring are conducted by online but the post processing signal analysis methods such as PCA, DWT and ICA are calculated after the whole data from beginning until the catastrophic failure occurred are acquired. It can be said that this is an off-line analysis. Another paper was presented by Eric et.al. [5], the Authors developed the online remote monitoring for vibration and acoustic emission using National Instrument and LabVIEW. The monitoring system consists of data acquisition (DAQ) control, real-time analysis, off-line analysis, feature extraction and monitoring trending features. The signals are sampled at the DAQ card and saved in the PXI system. The PXI is connected to internet via local network with security function allowing access for user through internet.

Robust and complicated online monitoring systems have been developed by some vibration monitoring companies. Those systems are complicated and expensive but the quality is superior. In this project simple and low cost online monitoring system based C language, MATLAB executable, File Transfer Protocol (FTP) and Web design is proposed. This project is partially developed for PhD project in University of Wollongong.

3. Experimental setup

The slew bearing test-rig is operated continuously under controlled conditions. The bearing under test was run from the new condition to failure. Figure 1 shows the schematic of a slew bearing test-rig. The test-rig can be operated at speeds of 1 to 12 rpm. The test-rig was designed to simulate the actual working conditions of a steel manufacturing company. In this study, the test-rig is expected to run in reversible rotation at a speed of 1 rpm. The slew bearing type is an axial/radial bearing YRT260 supplied by China bearing manufacture with an inner and outer diameter of 260mm and 385mm. The constant load via hydraulic pressure was applied to bearing of 30 tonne.

Three types of sensors were installed on the bearing: accelerometer, temperature sensor and acoustic emissions (AE) sensor. We used four accelerometer and temperature sensors to maintain 4 symmetrical measurement area. These sensors were placed on the inner radial surface at 90 deg to each other as can be seen in figure 2. Two AE sensors are attached on the top axial surface at 180 deg to each other as presented in figure 2. The IMI608 A11 ICP type sensor is used in this study. The accelerometers were connected to high speed DAQ (PS3424) from Pico scope. To avoid noise interruption from laboratory environment, the vibration data was collected at midnight outside of working hours with 4880Hz sampling rates.

Figure 1. Photo of slew bearing lab test rig [6].

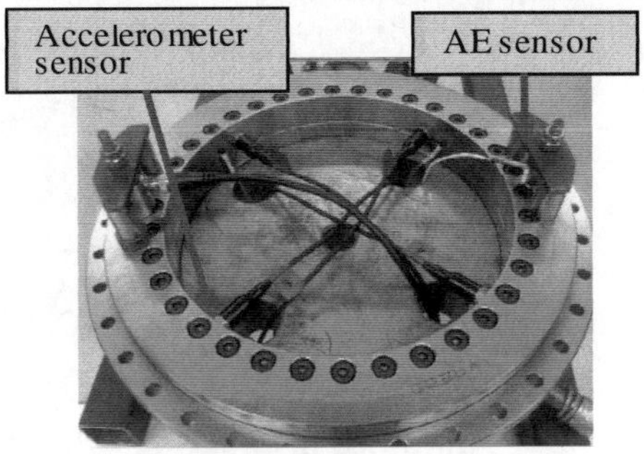

Figure 2. Sensors location.

4. Online monitoring based feature extraction

The objective of this system design is to facilitate online monitoring to the currently used slew experiment. The resulted software is expected to be able to send the data automatically from one location, to the data centre and then accessed from any other site using web browser. The locations involved in this online monitoring system are divided into 3 main locations: WORKSHOP, DATA CENTRE/SERVER and CLIENT. The illustration is shown in figure 3.

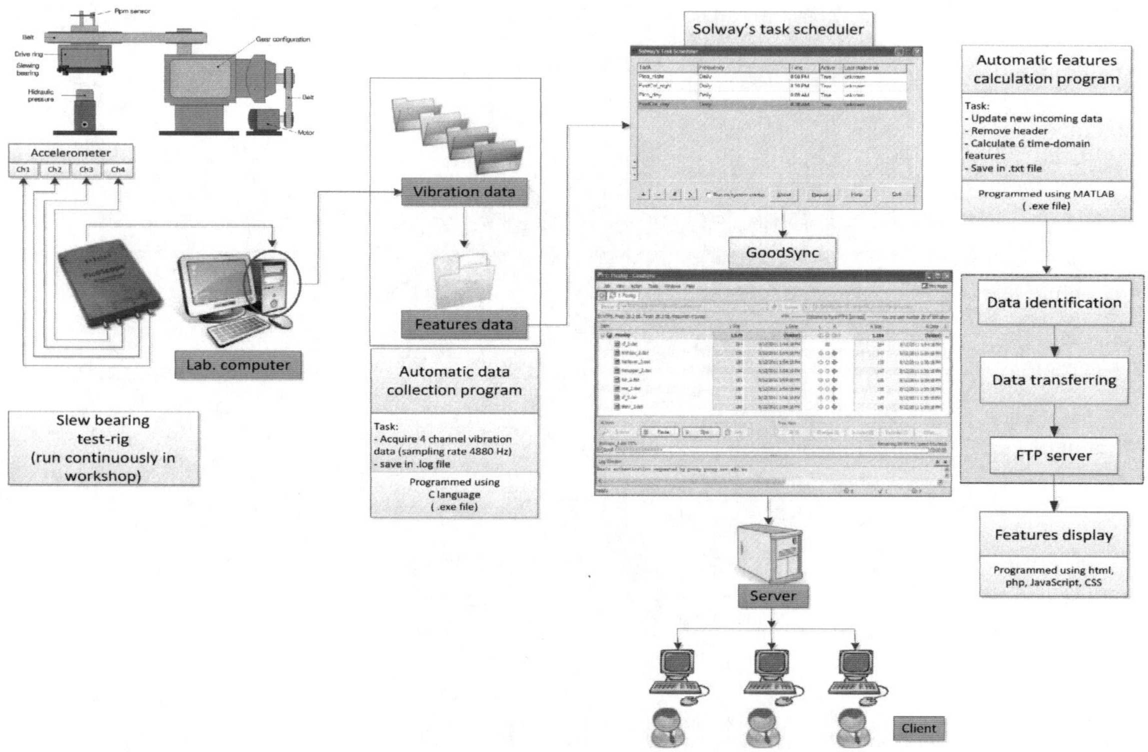

Figure 3. Online monitoring framework based FTP protocol.

4.1. Workshop

Since the vibration data is generated in the machine located in workshop, the control of data format should be done here. Disregarding the process of how the result is yielded, the end resulted data should be in form of simple human readable file, to be precise is TXT or DAT format. Each file will be produced every particular period of time, and each file will contain the readings of the bearing conditions by the machine at that time. Two automatic data acquisition will be conducted at day and night. This is will be done by doing two things. First, the executable data acquisition program is written in C language and saves in the EXE format. This is the advantage of PICO DAQ that it can be used not only for human interface but also for automatic way. Second, the automatic data acquisition will be used by running the executable file using Solway's task scheduler shown in figure 4. The vibration data are saving in desktop computer where the filename contain the information about date, month and year when the readings are taken.

Initial online monitoring process used free hosting. There is limited data transfer for free hosting. Therefore the huge file of vibration data should be transform into less capacity by calculating the useful information. This information is usually called features. In this work, six time domain features are used: root mean square (RMS), shape factor (SF), skewness, kurtosis, crest factor (CF), entropy, histogram lower and histogram upper [7]. The executable MATLAB program is created to do the six feature calculation. This program also runs automatically using Solway's task scheduler approximately every 30 minutes after the automatic data acquisition as shown in figure 4. The illustration of automatic data acquisition, automatic feature calculation and online monitoring process are shown in figure 3.

The readings as had been mentioned before are divided into four different channels (channel 1 to channel 4). In other words, each file will contain four values that represent the condition of bearing from four different accelerometer locations, separated by space. As automatic data acquisition will be carried out at two different times (day and night), the feature calculation filename will be consisting of two data name as well. For example the kurtosis feature will be saved as 'kur_1.txt' for day and 'kur_2.txt' for night. Each file will consist of five column where the first column will be the date, month and year when the feature are calculated. Second columns to fifth column are the accelerometer channel 1 to accelerometer channel 4. The six different feature calculation data will be stored temporarily in the connected desktop computer. The illustration of data format is shown in figure 5. Once the data on figure 5 saved and collected in the main PC, the data is then transferred and sent to the web hosting through GoodSync as depicted in figure 6.

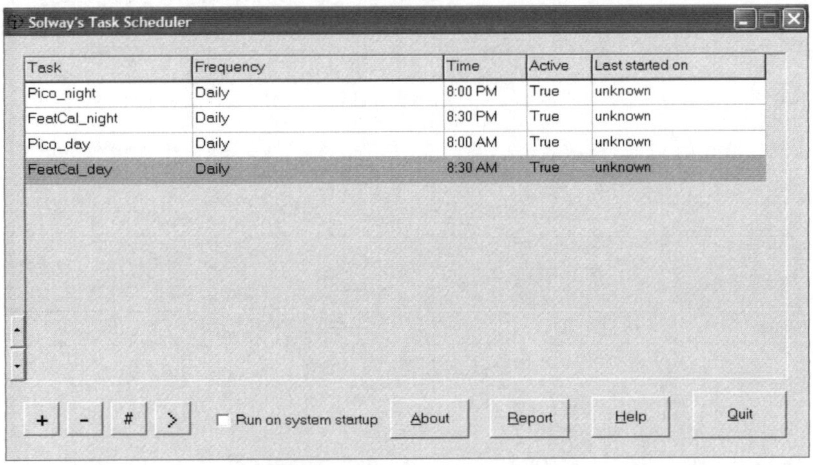

Figure 4. Solway's task scheduler.

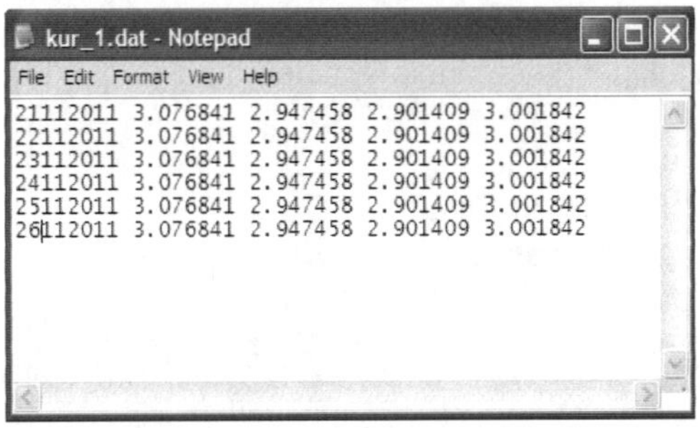

Figure 5. Example of feature file format.

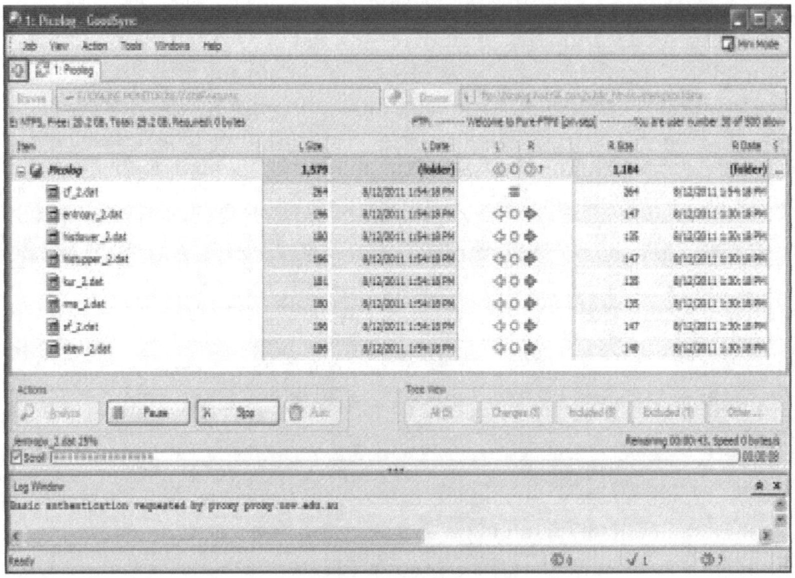

Figure 6. GoodSync.

4.2. Server

The server is expected to handle a lot of storage and bandwidth size to access this site must be reasonable enough (if use paid hosting). Since the transaction including downloading the data from any client are done regularly, it is very important to choose a competent hosting site. Most likely, the best choice for this project is the unlimited storage and unlimited bandwidth. The costs are ranged between approximately 4 AUD per month, and greater than that. Moreover, the hosting site should satisfy the following needs:

a) Allow FTP connection to their site
b) Support PHP and MySQL just in Case
c) Again, unlimited storage and bandwidth is favourable.

Since the needed data for the client site is as the following:

a) Root mean square (RMS)
b) Shape factor (SF)
c) Skewness
d) Kurtosis
e) Crest factor (CF)
f) Entropy
g) Histogram lower and
h) Histogram upper.

Then the data must be prepared before fetched by the client site. The calculation for each variable will be done in 2 different triggers, automatically and an update button pressed. The first option should check the last time the webpage is accessed. As long as current timestamp and the previous timestamp fall under the same timeframe (frames are like sessions, depend on the FTP frequency) no pre-processor are executed, because definitely no additional raw data. In case the timeframe are different, then the calculations are needed, since we already have new data to be displayed. This way, we can achieve optimization as we do not waste unnecessary process. In case of manual update, this button will be provided, if and only if the user wishes to do so and not necessarily to do it. Just to reassure the user that the data can be updated at user's willed.

4.3. Client

Any computer with internet browser is considered as client. After simple verification the calculated data then displayed accordingly. The users are given several configurations option to choose what data preferred to be displayed by default. Nonetheless, as the user may see fit, users are allowed to download the data. May it be individually or in several at once. Furthermore, the data displayed is powered by JavaScript to make it more interactive and easy to deal with.

The expected configurations are:

a) Display all data on start-up
b) Download latest data on start-up
c) Only display last 3 days data

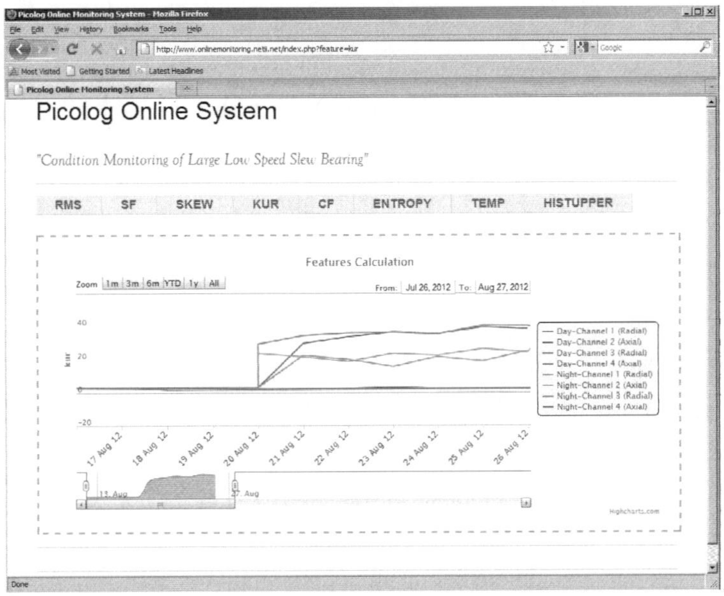

Figure 7. Kurtosis feature from online monitoring website.

5. Conclusions

Low cost, simple and easy to implement online monitoring system for low speed condition monitoring is presented. There are undergoing work to improve this system such as find the proper feature which indicates the bearing condition and embed the intelligent condition monitoring and prognosis algorithm to the online monitoring system.

References

[1] Naranjo, A.R. and Otero, M.E.M. 2005 A method for the correlation dimension estimation for on-line condition monitoring of large rotating machinery *Mech. Syst. Signal Process.* **19** p 939-954

[2] Oberholster, A.J. and Heyns, P.S. 2009 Online condition monitoring of axial-flow turbomachinery blades using rotor-axial Eulerian laser doppler vibromerty *Mech. Syst. Signal Process.* **23** p 1634-1643

[3] Wang, W., Tavner, P.J. and Court, R. 2013 An online technique for condition monitoring the induction generators used in wind and marine turbines *Mech. Syst. Signal Process.* **38** p 103-112

[4] Loutas, T,H., Roulias, D., Pauly, E. and Kostopoulos, V. 2011 The combine use of vibration, acoustic emission and oil debris on-line monitoring towards a more effective condition monitoring of rotation machinery *Mech. Syst. Signal Process.* **25** p 1339-1352

[5] Kim, E.Y., Tan, A.C.C., Mathew, J. and Yang, B.S. 2010 Development of an online condition monitoring system for low speed machinery *Proc of the 4th World Congress of Engineering Asset Management* p 1-9

[6] Caesarendra, W. 2015 Vibration and acoustic emission-based condition monitoring and prognostic methods for very low speed slew bearing *PhD Thesis, School of Mechanical, Materials and Mechatronic Engineering, University of Wollongong*

[7] Caesarendra, W., Kosasih, B., Tieu, K. and Moodie, C.A.S. An application of nonlinear feature extraction – a case study for low speed slewing bearing condition monitoring and prognosis *Proc. of IEEE/ASME Int. Conf. on Advanced Intelligent Mechatronics* p 1713-1718

Combination Base64 Algorithm and EOF Technique for Steganography

Robbi Rahim[1*], Heri Nurdiyanto[2], Rahmat Hidayat[3], Ansari Saleh Ahmar[4], Dodi Siregar[5], Andysah Putera Utama Siahaan[3], Ilham Faisal[5], Sayuti Rahman[5], Diana Suita[6], Ahmad Zamsuri[7], Dahlan Abdullah[8], Darmawan Napitupulu[9], Muhammad Ikhsan Setiawan[10] and S Sriadhi[11]

[1]School of Computer and Communication Engineering, Universiti Malaysia Perlis, Malaysia
[2]Department of Informatics, STMIK Dharma Wacana, Metro Lampung, Indonesia
[3]Department of Information Technology, Politeknik Negeri Padang, Indonesia
[4]Departement of Statistics, Universitas Negeri Makasar, Indonesia
[5]Department of Informatics Engineering, Universitas Harapan Medan, Indonesia
[6]Department of Informatics, Universitas Harapan Medan, Medan, Indonesia
[7]Department of Informatics, Universitas Lancang Kuning, Pekanbaru, Indonesia
[8]Department of Informatics, Universitas Malikussaleh, Aceh, Indonesia
[9]Research Center for Quality System and Testing Technology, Indonesian Institute of Sciences, Indonesia
[10]Department of Civil Engineering, Narotama University, Indonesia
[11]Department of Electrical Engineering, Universitas Negeri Medan, Indonesia

E-mail: *usurobbi85@zoho.com

Abstract. The steganography process combines mathematics and computer science. Steganography consists of a set of methods and techniques to embed the data into another media so that the contents are unreadable to anyone who does not have the authority to read these data. The main objective of the use of base64 method is to convert any file in order to achieve privacy. This paper discusses a steganography and encoding method using base64, which is a set of encoding schemes that convert the same binary data to the form of a series of ASCII code. Also, the EoF technique is used to embed encoding text performed by Base64. As an example, for the mechanisms a file is used to represent the texts, and by using the two methods together will increase the security level for protecting the data, this research aims to secure many types of files in a particular media with a good security and not to damage the stored files and coverage media that used.

1. Introduction

Data[1] is crucial for maintaining the confidentiality of information, especially vital information[2] and can only be known by certain parties only and the information are transmitted through the public network[1], [3], [4], if the data is not secured enough it will be easily tapped by others and the content easily known by users who do not have the authority[1], [5], [6]. One way to secure data is using a steganography technique[7], [8]. Steganography is the art and science to hide secret messages in other media so that the existence of such secret messages is unknowable, steganography is concealed by covering up or hiding messages[9]–[11], hidden messages can be read or extract from a file by using a certain method in this paper EoF were a technique using for steganography and Base64 a method to convert any file to printable text before hidden to media.

Base64 is a coding schemes that encode binary data and translate into a base64 representation[12], [13]. End of File (EOF) technique is a method that works by inserting data at the end of the file[14]. Singh and Supriya[12] explains the encoding and decoding process with the base64 algorithm can be performed for all types of files by converting into printable text, M. Ramalingam[15] conducts research for steganography on video types using 2 different methods for doing steganography and cryptography process with good result, another research was perform by Nurdiyanto[16] by using Pixel Value Differencing a message will be hidden in pixel image but the result from this made a distortion in pixel, it can happen when insert a large file or a lot of message to be hidden. Hidden message in pixel in maybe a good one but it can distort pixel, one of reason why combine Base64 and EoF for hidden information.

2. Methodology

Base64 algorithm[17], [18] and EOF technique are implemented by using Visual Basic.Net 2015 programming language, the experiment process is done step by step for each file to embedded into various media, the first step is to change the binary of each file into the form of text encoding using the Base64 algorithm, the result of text encoding is inserted into the media used as cover by using EOF technique, for base64 procedure encoding and decoding can be seen in the following function:

```
FileToBase64(ByVal xobject As object)
  Using ms As New MemoryStream()
   xobject.Save(ms, format)
   Dim objBytes As Byte() = ms.ToArray()
   Dim base64String As String = Convert.ToBase64String(objBytes)
   Return base64String
  End Using

Base64ToFile(ByVal base64Code As String)
Dim objBytes As Byte() = Convert.FromBase64String(base64Code)
Dim ms As New MemoryStream(objbytes, 0, objBytes.Length)
Dim tmpObject As Object= objbytes.FromStream(ms, True)
Return tmpObject
```

The above function is a process snippet code to get the base64 value of each object to be embedded, the procedure Base64[12], [17] are like below:

1. Look for the binary code of each object
2. Convert binary number 8 bits to the ASCII code
3. Combine the last 8 bits to 24 bits.
4. Then, broke a 24 bit earlier to 6 bits. It will produce four fractions.
5. Each fragment is converted into a decimal value.
6. Lastly, make value the decimal value to an index to choose a character constituent of base64 and the maximum is 63 or 64 to the index.

The index value of the base64 algorithm can be seen in the table below:

Table 1. base64 index value.

Index	Value	Index	Value	Index	value
0	A	28	C	56	4
1	B	29	D	57	5
2	C	30	E	58	6
3	D	31	F	59	7

4	E	31	G	60	8
5	F	33	H	61	9
6	G	34	I	62	+
7	H	35	J	63	-
8	I	36	K		
9	J	37	L		
10	K	38	M		
11	L	39	N		
12	M	40	O		
13	N	41	P		
14	O	42	Q		
15	P	43	R		
16	Q	44	S		
17	R	45	T		
18	S	46	U		
19	T	47	V		
20	U	48	W		
21	V	49	X		
22	W	50	Y		
23	X	51	Z		
24	Y	52	0		
25	Z	53	1		
26	a	54	2		
27	b	55	3		

For example, Base64 algorithm encoding process performed on binary = 0100100101101110 0111010001100101011100100110111001100001011101000110100101101111011011100110000101 101100 using the above procedure obtained the following encoding results in table 2 below:

Table 2. base64 decimal value.

No	Binary 6 Bit	Decimal	Base64 Value	No	Binary 6 Bit	Decimal	Base64 Value
1	010010	18	S	**10**	010111	23	X
2	010110	22	W	**11**	010001	17	R
3	111001	57	5	**12**	101001	41	p
4	110100	52	0	**13**	011011	27	b
5	011001	25	Z	**14**	110110	54	2
6	010111	23	X	**15**	111001	57	5
7	001001	9	J	**16**	100001	33	h
8	101110	46	u	**17**	011011	27	b
9	011000	24	Y	**18**	000000	0	A

From table 2 above obtained the encoding results SW50ZXJuYXRpb25hbA, then the encoding results are embedded into file by using EOF technique with the following functions:

```
FileOpen(1, filename, OpenMode.Binary)
Seek(1, LOF(1) + 1)
FilePut(1, StartData)
FilePut(1, Data)
FilePut(1, EndData)
FileClose(1)
```

IOP Conf. Series: Journal of Physics: Conf. Series **1007** (2018) 012003 doi:10.1088/1742-6596/1007/1/012003

The above functions make it possible to embed base64 encoding text into any file, so there is no limit how big the file size to be embedded to cover media.

3. Result and Discussion

Experiments with Base64 algorithms and EOF techniques are perform with various types of files to be embeded, table 3 below is the file information in the experiment.

Table 3. file information.

No	Media to be Secure	Size in Kb	Cover Media	Size in Kb
1	IC_1.jpg	109	Lena.bmp	117
2	IC_2.jpg	115	Lena.bmp	117
3	IPS7-1.jpg	831	Lena.bmp	117
4	Sertifikat IC Robbi.jpg	112	Lena.bmp	117
5	WhatsApp Image 2017-07-26 at 20.41.00.png	299	Sample.avi	539
6	IJSRST.doc	571	Proceeding.pdf	1326
7	IJRTER.doc	753	Proceeding.pdf	1326
8	IJSRSET.PDF	412	Proceeding.pdf	1326

Table 3 shows the file name and size of the file to be inserted and also the size of media as the cover, the next process is to perform steganography the existing file, and the results as in table 4:

Table 4. steganography results.

No	Media to be Secure	Cover Media	Result Steganography (Kb)
1	IC_1.jpg	Lena.bmp	262
2	IC_2.jpg	Lena.bmp	270
3	IPS7-1.jpg	Lena.bmp	1190
4	Sertifikat IC Robbi.jpg	Lena.bmp	271
5	WhatsApp Image 2017-07-26 at 20.41.00.png	Sample.avi	937
6	IJSRST.doc	Proceeding.pdf	2167
7	IJRTER.doc	Proceeding.pdf	2273
8	IJSRSET.PDF	Proceeding.pdf	1960

Each of file will convert into printable text using base64 method and then insert into a cover media with EOF method, the size of cover media will increase when text are embeded, based on experiment process any file type can be well hide using the EoF method.

4. Conclusion

An experiments for all types can be done as well, before EoF process a file will be convert into printable text by using Base64 method, the steganography process with EoF techniques can perform will with Base64 encoding text. Next improvement a Base64 and EoF method can combine with compression algorithm, so the encoding base64 result compress by an algortithm and then embed to file by using EoF.

References

[1] R. Rahim and A. Ikhwan, "Study of Three Pass Protocol on Data Security," Int. J. Sci. Res., vol. 5, no. 11, pp. 102–104, Nov. 2016.

[2] R. Rahim, M. Dahria, M. Syahril, and B. Anwar, "Combination of the Blowfish and Lempel-Ziv-Welch algorithms for text compression," World Trans. Eng. Technol. Educ., vol. 15, no. 3, pp. 292–297, 2017.
[3] R. Rahim and A. Ikhwan, "Cryptography Technique with Modular Multiplication Block Cipher and Playfair Cipher," Int. J. Sci. Res. Sci. Technol., vol. 2, no. 6, pp. 71–78, 2016.
[4] H. Nurdiyanto, R. Rahim, and N. Wulan, "Symmetric Stream Cipher using Triple Transposition Key Method and Base64 Algorithm for Security Improvement," J. Phys. Conf. Ser., vol. 930, no. 1, p. 12005, Dec. 2017.
[5] E. Hariyanto and R. Rahim, "Arnold's Cat Map Algorithm in Digital Image Encryption," Int. J. Sci. Res., vol. 5, no. 10, pp. 1363–1365, Oct. 2016.
[6] A. Putera, U. Siahaan, and R. Rahim, "Dynamic Key Matrix of Hill Cipher Using Genetic Algorithm," Int. J. Secur. Its Appl., vol. 10, no. 8, pp. 173–180, Aug. 2016.
[7] S. Islam, M. R. Modi, and P. Gupta, "Edge-based image steganography," Eurasip J. Inf. Secur., vol. 2014, 2014.
[8] M. H. and M. Hussain, "A survey of image steganography techniques," Int. J. Adv. Sci. Technol., vol. 54, no. 3, pp. 113–124, 2013.
[9] E. Zielińska, W. Mazurczyk, and K. Szczypiorski, "Trends in steganography," Commun. ACM, vol. 57, no. 3, pp. 86–95, 2014.
[10] A. Binny and M. Koilakuntla, "Hiding secret information using LSB based audio Steganography," in Proceedings - 2014 International Conference on Soft Computing and Machine Intelligence, ISCMI 2014, 2014, pp. 56–59.
[11] A. Cheddad, J. Condell, K. Curran, and P. Mc Kevitt, "Digital image steganography: Survey and analysis of current methods," Signal Processing, vol. 90, no. 3. pp. 727–752, 2010.
[12] G. Singh and Supriya, "Modified vigenere encryption algorithm and its hybrid implementation with Base64 and AES," in Proceedings - 2nd International Conference on Advanced Computing, Networking and Security, ADCONS 2013, 2013, pp. 232–237.
[13] L. Yu, Z. Wang, and W. Wang, "The Application of Hybrid Encryption Algorithm in Software Security," Fourth Int. Conf. Comput. Intell. Commun. Networks, pp. 762–765, 2012.
[14] Y. Anggraini and D. V. S. Y. Sakti, "Penerapan Steganografi Metode End of File (Eof) Dan Enkripsi Metode Data Encryption Standard (Des) Pada Aplikasi Pengamanan Data Gambar Berbasis Java," Konf. Nas. Sist. Informasi, STMIK Dipanegara Makassar, no. September 2016, pp. 1743–1753, 2014.
[15] M. Ramalingam and N. A. M. Isa, "A steganography approach over video images to improve security," Indian J. Sci. Technol., vol. 8, no. 1, pp. 79–86, 2015.
[16] H. Nurdiyanto and R. Rahim, "Enhanced pixel value differencing steganography with government standard algorithm," in 2017 3rd International Conference on Science in Information Technology (ICSITech), 2017, pp. 366–371.
[17] K. Fiscus and D. Shinburg, "Base64 Can Get You Pwned," 2011.
[18] Z. Liu, L. Liu, R. Hill, and Y. Zhan, "Base62x: An alternative approach to Base64 for non-alphanumeric characters," in Proceedings - 2011 8th International Conference on Fuzzy Systems and Knowledge Discovery, FSKD 2011, 2011, vol. 4, pp. 2667–2670.

Searching Process with Raita Algorithm and its Application

Robbi Rahim*[1], Ansari Saleh Ahmar[2], Dahlan Abdullah[3], Dedy Hartama[4], Darmawan Napitupulu[5], Andysah Putera Utama Siahaan[1], Muhammad Noor Hasan Siregar[6], Nurliana Nasution[7], Siti Sundari[8] and S Sriadhi[9]

[1]Schoolof Computer Engineering and Communication, Universiti Malaysia Perlis, Malaysia
[2]Departement of Statistics, Universitas Negeri Makassar, Indonesia
[3]Department of Informatics, Universitas Malikussaleh, Aceh, Indonesia
[4]Department of Information System, STIKOM Tunas Bangsa, Pematang Siantar, Indonesia
[5]Research Center for Quality System and Testing Technology, Indonesian Institute of Sciences, Indonesia
[6]Department of Management, Universitas Graha Nusantara, Indonesia
[7]Department of Informatics, Universitas Lancang Kuning, Pekanbaru, Indonesia
[8]Department of Informatics, Universitas Harapan Medan, Medan, Indonesia
[9]Department of Electrical Engineering, Universitas Negeri Medan, Indonesia

E-mail: *usurobbi85@zoho.com

Abstract. Searching is a common process performed by many computer users, Raita algorithm is one algorithm that can be used to match and find information in accordance with the patterns entered. Raita algorithm applied to the file search application using java programming language and the results obtained from the testing process of the file search quickly and with accurate results and support many data types.

1. Introduction

Searching[1 - 4] on the computer is a common process performed by computer users, before computer technology develops like this searching process is perform on the archive organized with the number documents that much and also searching process not easy to do and require a lot of time[5], with the presence of computer technology[5 - 7] right now it makes easier for human to search files because it's already in digital formand also indexed by the computer[8 - 9]. A result from searching process will appear in a few seconds due to the speed of the computer and amount of data.The searching process on the computer can be achievedusing available tools on the operating system[6], [10], the length of time searching process on the operating system depends on the process of indexing from operating system to speed up searching process, Raita algorithm [11 – 12] were applied for searching a file using a keyword that represents name of the file from operating system.

The results of this paper is applied Raita algorithm in a fast file search application using java language.R. Rahim, et.al[1], [2], [13]explains the search process can be done using many algorithms such as Knuth Morris Pratt to search for a particular word or file, or also by using double hashing techniques to make the search process faster by creating the main table and overflow table, based on

IOP Conf. Series: Journal of Physics: Conf. Series **1007** (2018) 012004 doi:10.1088/1742-6596/1007/1/012004

previous experiment from other researcher the searching process can be done for various type of data you want. Raita algorithm is not a new algorithm,Patel [11] with his finding describe a different result of many searching algorithm with advantage and weakness, based of his result this paper is to prove that Raita algorithm is much more fastest than usual algorithm by implement in searching file. The Raita algorithm implemented in the file search application is supposed to be able to search file much faster than using ordinary file searching, one of the reasons that allows the search process faster because Raita algorithm has a procedureof matching and searching simultaneously and this makes the process time more faster.

2. Methodology

The RAITA algorithm has a good pattern in searching [11], [12], especially in mid-word character checks from given pattern by comparing the last character of both patterns (m-1) with the second character from length of the character. The middle character of the pattern is compared twice for the corresponding character of the text. This procedure is repeated until does not reach n-m + 1.
An analyzing RAITA algorithm for searching a file can be assumption as text searching, see an example process below.

P(Text) = ROBBIRAHIMKEREN
K (Pattern) = RAHIM

Table 1. Index K position.

i	0	1	2	3	4
a	R	A	H	I	M

Based on the words to be tested made table BmBc by using equation formula: $m-2$ which serves as pattern matching boundary$m-1-i$.

$m = K = 5$
$m = Pattern\ Length$
$m = P$
$m = 5 - 2 = 3$

BmBc (a) = m - 1 - i
1) 5 - 1 - i (5 - 1 - 0) = 4, the value is put on index to 0 with character that is R
2) 5 - 1 - i (5 - 1 - 1) = 3, the value is placed on index to 1 with character that is A
3) 5 - 1 - i (5 - 1 - 2) = 2, the value is placed on the second index with the character of H
4) 5 - 1 - i (5 - 1 - 3) = 1, the value is placed on index to 3 with character that is I
5) 5 - 1 - i (5 - 1 - 4) = 0, the value is placed on index to 4 with character that is M

Based on the calculation of BmBc above obtained the following results:

Table 2. BmBc position.

i	0	1	2	3	4	
a	R	A	H	I	M	*
BmBc	4	3	2	1	0	5

The value of m is 5 in accordance with the length of the pattern, because the alphabet that is not in the table then is initialized with the sign (*) and then the value corresponds to the length of the pattern. The searching process can be viewed step by step in tables 3 and table 4 by similar the character pattern, table 3 are the first table form Raita process.

IOP Conf. Series: Journal of Physics: Conf. Series **1007** (2018) 012004 doi:10.1088/1742-6596/1007/1/012004

Table 3. Text and sentence pattern phase 1.

Text	R	O	B	B	I	R	A	H	I	M	K	E	R	E	N
Pattern	R	A	H	I	M										

In the first stage the character I pattern does not match the M text, since the character M does not have BmBc, and then the pattern will advance as much as 5 characters. Next phase can see in table 4 below:

Table 4. Text and sentence pattern phase 1.

Text	R	O	B	B	I	R	A	H	I	M	K	E	R	E	N
Pattern						R	A	H	I	M					

In the second stage the character M pattern corresponds to the M text, since the M character exists in the BmBc table with the shift, then the process stops performed and is tested, the test is performed for all characters ranging from 0 - 4 according to the character position values in the BmBc table and all corresponding characters then the search process stops because the word has been found. Raita algorithm process above then applied to the application created by using Java programming language with Windows 8.1 platform and also can be applied to any operating system that provides java library, below is pseudo code from Raita algorithm applied to application.

```
if (pattern.length == 1) {
      final int nLimit = Math.min(text.length, textEnd);
      for (int n = textStart; n < nLimit; n++) {
        if (text[n] == pattern[0])
          return n;
      }
      return -1;
    }
    else if (pattern.length == 2) {
      final int nLimit = Math.min(text.length, textEnd) - 1;
      for (int n = textStart; n < nLimit; n++) {
        if (text[n] == pattern[0]) {
          if (text[n + 1] == pattern[1])
            return n;
        }
      }
      return -1;
    }

    CharIntMap m = (CharIntMap) processed;

    int i, j, k, mMinusOne;
    char last, first;

    i = pattern.length - 1;
    mMinusOne = i - 1;

    last = pattern[i];
    first = pattern[0];
```

```
i += textStart;

while (i < textEnd) {

   if (text[i] == last && text[i - (pattern.length - 1)] == first) {

      k = i - 1;
      j = mMinusOne;

      while (k > -1 && j > -1 && text[k] == pattern[j]) {
         --k;
         --j;
      }
      if (j == -1) {
         return k + 1;
      }

   }
   i += m.get(text[i]);
}

return -1;
```

3. Result and Discussion

Searching application with Raita algorithm designed by using java language, and the trial of searching file are using keyword that represent a file name, figure 1 are the application interface when searching process already finish.

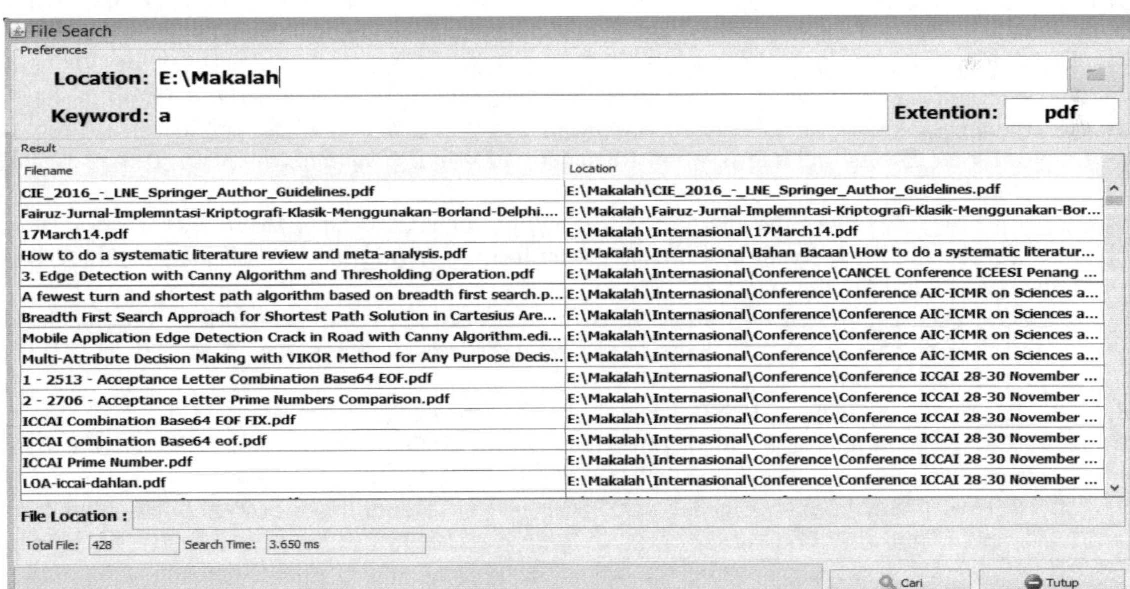

Figure 1. Search result.

Searching on a folder with name *makalah* containing 8.132 files with size 5.32 GB, searchingprocess using Raita algorithm with character ***a***and PDF file typeobtained total files as 423

IOP Conf. Series: Journal of Physics: Conf. Series **1007** (2018) 012004 doi:10.1088/1742-6596/1007/1/012004

pdf files and it takes 3,650 millisecond. Implementation Raita algorithm for searching file were perform as much 5 times with same type and different keywords, Table 5 below is the result of the testing performed.

Table 5. Experiment searching process.

No	Number of Files	Raita Application (Second)
1	432	3.650 ms
2	519	4.121 ms
3	812	7.712 ms
4	1012	14.812 ms
5	1200	56.139 ms

Based on testing done with varying results, the search time is quite fast and with more files found will require a lot of time than usual process.

4. Conclusion

File search application by applying Raita algorithm to find file on operating system got good result with fast enough time, application only done limited to matching and searching process by using Raita algorithm and result obtained not compared with search by using windows search or also with the use of other algorithms, future development makes it possible to implement multiple algorithms in one search process and compare the results.

References

[1] R. Rahim, I. Zulkarnain, and H. Jaya, "A review: search visualization with Knuth Morris Pratt algorithm," in *IOP Conference Series: Materials Science and Engineering*, 2017, vol. 237, no. 1, p. 12026.

[2] R. Rahim, I. Zulkarnain, and H. Jaya, "Double hashing technique in closed hashing search process," *IOP Conf. Ser. Mater. Sci. Eng.*, vol. 237, no. 1, p. 12027, Sep. 2017.

[3] S. J. Putra and I. Khalil, "Context for the intelligent search of information," in *2017 5th International Conference on Cyber and IT Service Management (CITSM)*, 2017, pp. 1–4.

[4] R. Rahim, S. Nurarif, M. Ramadhan, S. Aisyah, and W. Purba, "Comparison Searching Process of Linear, Binary and Interpolation Algorithm," *J. Phys. Conf. Ser.*, vol. 930, no. 1, p. 12007, Dec. 2017.

[5] M. Andrews, "Searching the internet," *IEEE Softw.*, vol. 29, no. 2, pp. 13–16, 2012.

[6] J. G. Reed and P. M. Baxter, "Computer searching.," *Library use: Handbook for psychology (3rd ed.).* pp. 27–37, 2003.

[7] G. G. S. Fiates, J. E. A. Fiates, F. A. R. Serra, and M. P. Ferreira, "Innovation environment in small technology-based companies," *J. Technol. Manag. Innov.*, vol. 5, no. 3, pp. 81–95, 2010.

[8] R. Rahim, D. Hartama, H. Nurdiyanto, A. S. Ahmar, D. Abdullah, and D. Napitupulu, "Keylogger Application to Monitoring Users Activity with Exact String Matching Algorithm," *J. Phys. Conf. Ser.*, vol. 954, no. 1, p. 12008, 2018.

[9] R. Rahim *et al.*, "Block Architecture Problem with Depth First Search Solution and Its Application," *J. Phys. Conf. Ser.*, vol. 954, no. 1, p. 12006, 2018.

[10] R. Rahim, A. S. Ahmar, A. P. Ardyanti, and D. Nofriansyah, "Visual Approach of Searching Process using Boyer-Moore Algorithm," *J. Phys. Conf. Ser.*, vol. 930, no. 1, p. 12001, Dec. 2017.

[11] U. Patel and M. Thakkar, "An Efficient Exact Single Pattern Matching Algorithm," vol. 3, no. 5, pp. 1955–1958, 2014.

[12] H. N. Verma and R. Singh, "A fast string matching algorithm," vol. 2, no. 6, pp. 1877–1883,

2011.
[13] R. Rahim, Nurjamiyah, and A. R. Dewi, "Data Collision Prevention with Overflow Hashing Technique in Closed Hash Searching Process," *J. Phys. Conf. Ser.*, vol. 930, no. 1, p. 12012, Dec. 2017.

IOP Conf. Series: Journal of Physics: Conf. Series **1007** (2018) 012005 doi:10.1088/1742-6596/1007/1/012005

EMG finger movement classification based on ANFIS

W Caesarendra*[1,2],T Tjahjowidodo[*2],Y Nico[2],S Wahyudati[3] and L Nurhasanah[3]

[1]Department of Mechanical Engineering, Diponegoro University, Semarang, Indonesia

[2]Schoolof Mechanical and Aerospace Engineering, Nanyang TechnologicalUniversity,Singapore

[3]Medical Specialty Programs Physical Medicine and Rehabilitation, Medical Faculty, Diponegoro University, Semarang,Indonesia

E-mail:*wc026@uowmail.edu.au

Abstract. An increase number of people suffering from stroke has impact to the rapid development of finger hand exoskeleton to enable an automatic physical therapy. Prior to the development of finger exoskeleton, a research topic yet important i.e. machine learning of finger gestures classification is conducted. This paper presents a study on EMG signal classification of 5 finger gestures as a preliminary study toward the finger exoskeleton design and development in Indonesia. The EMG signals of 5 finger gestures were acquired using Myo EMG sensor. The EMG signal features were extracted and reduced using PCA. The ANFIS based learning is used to classify reduced features of 5 finger gestures. The result shows that the classification of finger gestures is less than the classification of 7 hand gestures.

1. Introduction

Cerebral vascular accident, well-known as stroke is a major cause of long-term disability in the world, with over 775,000 cases in United States alone in 2014 [1]. The main cause of stroke is the damage in the brain cells as a result of an interruption in the blood flow that supplies nutrient and oxygen to the brain, causes paralysis on one side of the body (hemiparesis). Relearning the basic movement is one way to treat this disability. Repetitive training is highly suggested as to speed up the recovery of the paralyzed limbs. However, the conventional treatment method is labour intensive as it requires at least one assistant to guide the patient while using the treatment equipment.Many research has been done to create an independent rehabilitation equipment as a method to increase the treatment frequency and reduce number of assistant needed. One solution that has current been developed is by creatingan exoskeleton to guide and control the movement of the paralyzed limb[2].

The previous research that introduced the exoskeleton equipment was an actuated finger exoskeleton (AFX) that provided treatment specifically only to the joints of the index finger and thumb [3]. It was designed to provide independent control of pinch and reach-to-pinch of each joint of both fingers. Another exoskeleton device that provide control to all five fingers was Hand EXOskeleton Rehabilitation Robot (HEXORR), which used electric motors and a low-friction gear-trains[4].

This paper is the preliminary study of finger exoskeleton design that allows the non-paralyzed fingers to guide, control and enable self-therapy the paralyzed fingers in order to recover the motor function. The first stage of study is identifying the characteristics of EMG signal acquired from 5 finger gestures. Previous studies has analysed the EMG signals for the purpose of the classification of several hand movements[5-8]. Methods such as ANN[5], SVM[6] and ANFIS[7] has been employed and examined. The result show that ANFIS is over performed than other methods. Therefore this paper used ANFIS as a classification method for EMG signals.

2. Methods

2.1. Features extraction

This paper employed sixteen feature extraction methods as presented in table 1.

Table 1. EMG and time-domain feature extraction methods.

No	Method	Equation
1.	Integrated EMG Integrated EMG (IEMG) is used as an onset detection index in clinical application [9, 10].	$IEMG = \sum_{i=1}^{N} \lvert x_i \rvert$
2.	Mean absolute value Mean absolute value (MAV) is also used as an onset index, especially in detection of the surface EMG signal[9, 10].	$MAV = \frac{1}{N}\sum_{i=1}^{N} \lvert x_i \rvert$
3.	Modified mean absolute value type 1 Modified mean absolute value type 1 (MAV1) is an extension of MAV feature [9, 10].	$MAV1 = \frac{1}{N}\sum_{i=1}^{N} w_i \lvert x_i \rvert$
4.	Modified mean absolute value type 2 Modified mean absolute value type 2 (MAV2) is an expansion of MAV feature which is similar to MAV1 [9, 10].	$MAV2 = \frac{1}{N}\sum_{i=1}^{N} w_i \lvert x_i \rvert$
5.	Simple square integral Simple square integral (SSI) or integral square uses energy of the EMG signal as feature [9, 10].	$SSI = \sum_{i=1}^{N} x_i^2$
6.	Variance of EMG Variance of EMG (VAR) is defined as an average of square values of the deviation of the EMG signal[9, 10].	$VAR = \frac{1}{N-1}\sum_{i=1}^{N} x_i^2$
7.	Root mean square Root mean square (RMS) is a measure that relate to constant force and non–fatiguing contraction [9, 10].	$RMS = \sqrt{\frac{1}{N}\sum_{i=1}^{N} x_i^2}$
8.	Waveform length Waveform length (WL) is a measure of complexity of the EMG signal[9, 10].	$WL = \sum_{i=1}^{N-1} \lvert x_{i+1} - x_i \rvert$
9.	Difference absolute standard deviation value (DASDV) DASDV is a standard deviation of the wavelength [9, 10].	$DASDV = \sqrt{\frac{1}{N-1}\sum_{i=1}^{N-1} (x_{i+1} + x_i)^2}$
10.	Autoregressive Autoregressive (AR) is a common approach for modeling univariate time series as AR model follows [9, 10].n= 4.	$y_t = a_1 y_{t-1} + a_2 y_{t-2} + \ldots + a_n y_{1-n} + \varepsilon_t$ $= \sum_{i=1}^{n} a_i y_{t-1} + \varepsilon_t$
11.	Hjorth activity Hjorth activity (H1) can indicate the surface of power spectrum in the frequency domain [11].	$HI = \sigma_x^2 = \frac{1}{N}\sum_{i=1}^{N}\left(x_i - \bar{x}\right)^2$

12.	Hjorth mobility Hjorth mobility (H2) parameter represents the proportion of standard deviation of the power spectrum [11].	$H2 = \frac{\sigma_{x'}}{\sigma_x}$
13.	Hjorth complexity Hjorth complexity (H3) parameter represents the change in frequency[11].	$H3 = \frac{\sigma_{x''}/\sigma_{x'}}{\sigma_{x''}/\sigma_x}$

2.2. Dimensional features reduction and feature classification method

This paper employed principal component analysis (PCA) for dimensional features reduction and adaptive neuro-fuzzy inference system (ANFIS) for reduced features classification. The PCA is a technique using sophisticated underlying mathematical principles to transforms a number of possibly correlated variables into a significant smaller number of uncorrelated variables called principal components[12]. The visualisation of dimensional features reduction using PCA used in this paper is presented in figure 1(a). Where F_1, F_2, F_3, ... F_{16} is the number of features and a subscript index a, b, ..., e is the number of finger gestures (there are 5 finger gestures in this paper). It can be seen from figure 1(a) that the PCA reduced 16 features into 3 new features namely principal components (PCs). We can select how many new features are required, in this paper three PCs is obtained which are PC1, PC2 and PC3.

The adaptive neuro-fuzzy inference system (ANFIS) [13] is a specific kind of neuro-fuzzy classifier approach which integrates the ANNs adaptive capability and the fuzzy logic qualitative approach.The ANFIS architecture consists of five layers as shown in figure 1(b).A detail of PCA and ANFIS algorithm is presented in previous study [12]. Both PCA and ANFIS method is used for pattern recognition method of seven hand gestures [12].

$$\begin{bmatrix} F_{1a} & F_{1b} & \cdots & F_{1e} \\ F_{2a} & F_{1b} & \cdots & F_{2e} \\ F_{3a} & F_{1b} & \cdots & F_{3e} \\ \cdot & \cdot & \cdot & \cdot \\ \cdot & \cdot & \cdot & \cdot \\ \cdot & \cdot & \cdot & \cdot \\ F_{16a} & F_{16b} & \cdots & F_{16e} \end{bmatrix} \Rightarrow \begin{bmatrix} PC_{1a} & PC_{1b} & \cdots & PC_{1e} \\ PC_{2a} & PC_{2b} & \cdots & PC_{2e} \\ PC_{3a} & PC_{3b} & \cdots & PC_{3e} \end{bmatrix}$$

(a)

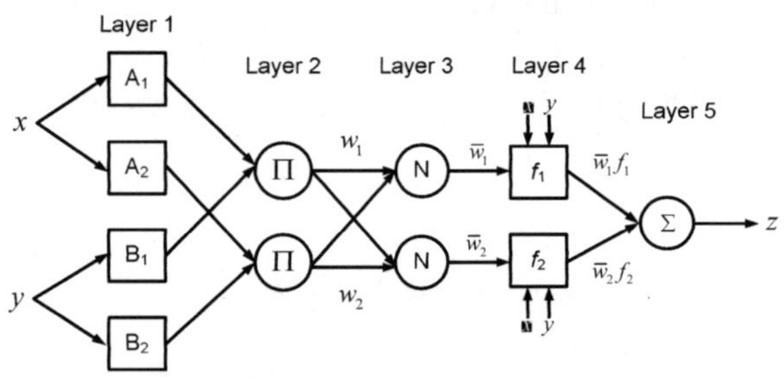

(b)

Figure1. (a) Dimensional feature reduction using PCA; (b) Two-input ANFIS architecture with first-order Sugeno fuzzy model and two rules.

2.3 Experimental setup

The EMG datasets were collected from each individual finger. The subject is person with no neurological or muscular disorders. The participant was seated on an armchair with his arm supported on the armchair, and levelled with the heart to avoid the variation generated by the arm position on the EMG signal.EMG signals were acquired by using Myo gesture controlled armband with 200 Hz sampling rate. The EMG electrodes were attached near to the flexor digitorum superficialis. The photograph of Myo device attached on the arm during the measurement is presented in figure 2. The Myo device consist of 8 EMG sensors. This paper discussed the result of EMG sensor 1. The participant was instructed to elicit a contraction (flexor) from the rest and hold on that position for the period of 5 seconds, continued with the relaxation (extensor) for about the same 5 seconds. The data collection was repeated ten times, until we collected 10 flexor and 10 extensor data. The example of EMG signals and the 5 fingers gesture are shown in figure 3.

IOP Conf. Series: Journal of Physics: Conf. Series **1007** (2018) 012005 doi:10.1088/1742-6596/1007/1/012005

(a)
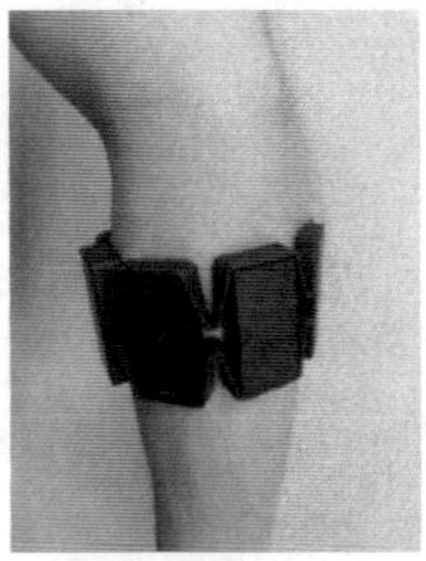
(b)

Figure 2. (a) The position of the armband on the arm;(b) The EMG channel assignments.

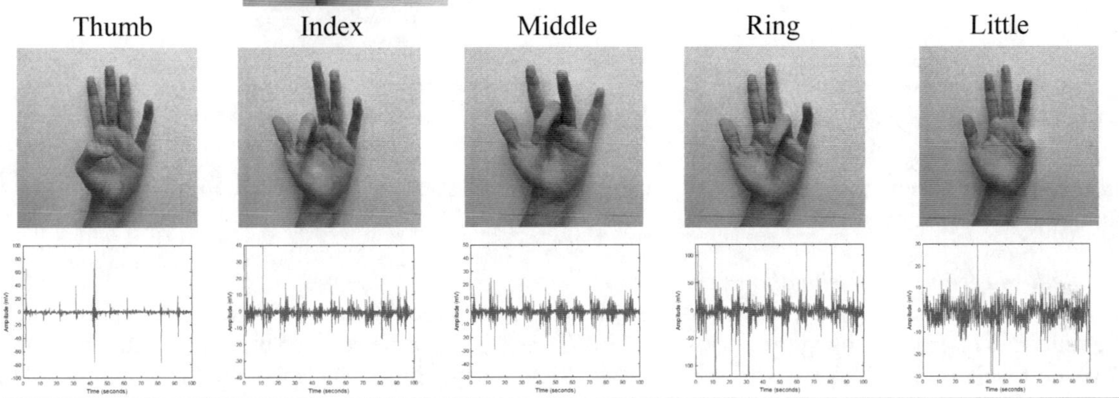

Figure 3. Finger gestures during EMG signal data acquisition.

3. Results and discussion

3.1. Dim1ensional features reduction using PCA

This paper employed As mentioned inSection 3 that 10 EMG signals were acquired for each finger gesture. We used 5 signals for training and another 5 signals for testing. Sixteen features were extracted from each EMG signal. Both training and testing feature dataset from features extraction process is 16 by 25. Where 16 indicate the number of features and 25 is the number of EMG data resulted from 5 finger gestures times 5 measurement. Prior to the feature classification, the feature datasets were reduced using PCA to obtain more sensitive features for each finger gesture. The three new features are computationally calculated from 16 features namely principal component1 (PC1), principal component2 (PC2) and principal component3 (PC3). The feature dimensional reduction using PCA for trainingand testing data is presented in figure 4 and 5. Compare to previous study in [11], the PCA plot for 7 hand gestures are easier to identify rather than the PCA plot of 5 finger gestures. The features of 7 hand gestures are groups in certain area and there is no overlapping between one feature group to other feature groups. In contrast, the PCA of five finger gestures are grouped in close distance betweenone group and others,for instance index finger group, middle finger group and little finger group. Only the ring fingergroup that located in far away from other groups.In addition there is one obvious scattered group i.e. thumb finger. This is because the muscle of one finger to other fingersin the arm are anatomically closed, therefore, once the EMG signal is acquired, there are influenced each other.

IOP Conf. Series: Journal of Physics: Conf. Series **1007** (2018) 012005 doi:10.1088/1742-6596/1007/1/012005

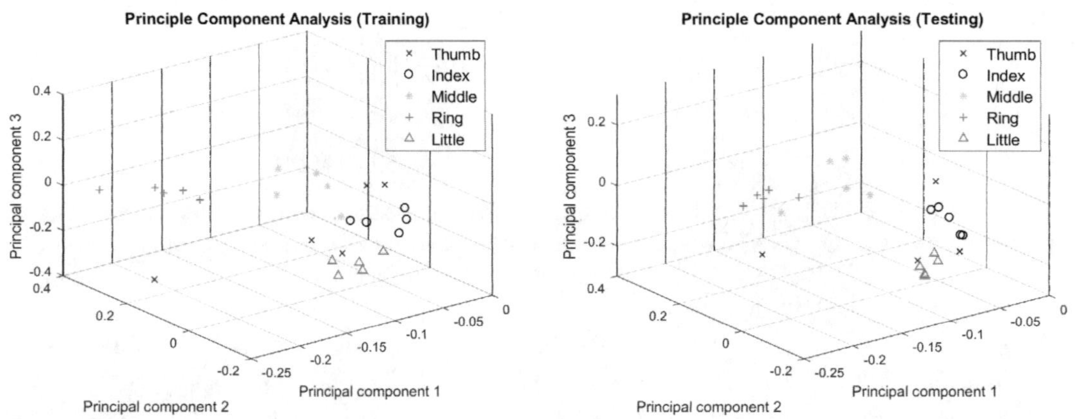

(a) (b)

Figure 4. Three principal components plot resulted from PCA feature reduction method: (a) for training process; (b) from testing process.

3.2. Reduced features classification using ANFIS

The first-order Sugeno type ANFIS was trained with new three features obtained from PCA result. The reduce dataset for training process is 3 x 25. Where 3 is the dimension of principal component and 25 is the result from 5 finger gestures by 5 EMG signal acquisition. The membership function (MF) result from training process is presented in figure 5(a). Three input was indicate the number of PCs and the five membership indicate the finger gestures (classes). Once the ANFIS was trained, the ANFIS model was then tested using another datasets (testing datasets) and the MF is presented in figure 5(b). Input1, input2 and input 3 in figure 5 represent the PC1, PC2 and PC3 respectively.It can be seen from figure 5 that some MFs of finger gestures were obviously changed,particularly in PC1 i.e. class #1 to #4 and PC2 i.e. class #1

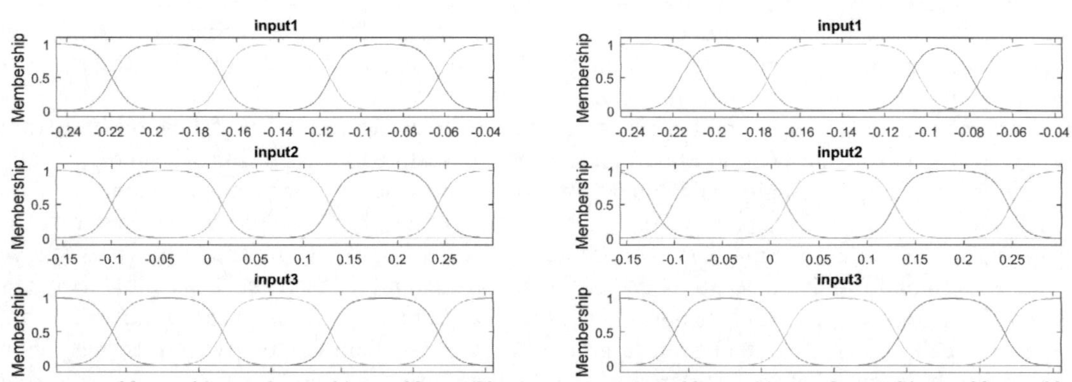

(a) (b)

Figure 5. ANFIS MFs for 3 PCs and 5 finger gestures: (a) training process; (b) testing process.

The classification result of five finger gestures is presented in figure 6. The ANFIS training parameters used in this paper such as number of training and testing set, membership function, learning rules and number of epoch are similar to the parameters used in [11].The x-axis and y-axis of figure 6 represents the number of features for each finger gesture and the number of class (finger gestures) respectively. Where 1, 2, 3, 4 and 5 is thumb finger, index finger, middle finger, ring finger and little finger.The classification accuracy is presented on each finger gesture. Where the thumb finger has lowest accuracy result of 20% and ring and little fingers achieved 100% accuracy. The average classification accuracy of 72%. The missed classification are occurred one in index and middle finger and most of missed classification are occurred in thumb finger. This result is corresponding to the MF result where the class #1 to class #4 in PC1 were changed from training MF and testing MF. In addition, the class #1 in PC2 was also changed.

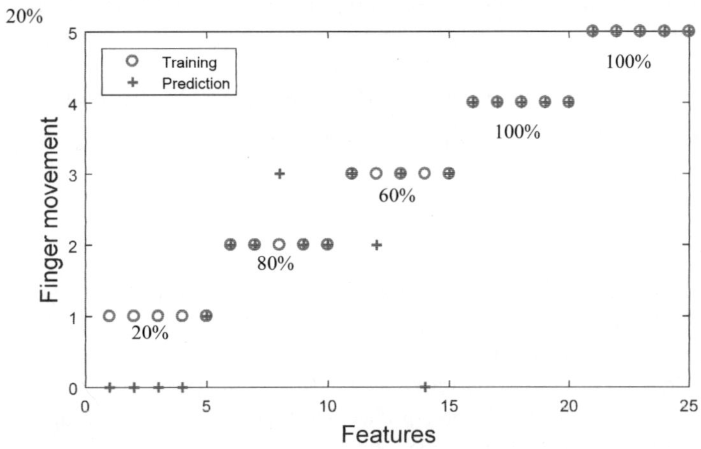

Figure 6. ANFIS classification result for 5 finger gestures.

4. Conclusions

A study of pattern recognition on 5 finger gestures based on EMG signal analysis has been presented. Finger therapy is the most difficult part in post-stroke rehabilitation and physical therapy. The EMG signal of finger gestures is more difficult to be analysed than the EMG signal from hand gestures that conducted in previous study. This is because the position of muscles that connect to certain finger are close each other. Once a certain finger was moved, the other muscle may also be moved. Therefore the accuracy result of classification of finger gestures especially for thumb finger is less than the classification accuracy of hand gestures. However, there is a future work that will be done in the future that is to identify the proper muscle position for each finger movement through the Myo EMG sensor. Through the study to find the appropriate sensor corresponding to certain finger, the classification accuracy will be improve.

References

[1] A. S. Go, D. Mozaffarian, V. L. Roger, E. J. Benjamin, J. D. Berry, M. J. Blaha, et al. (2014, Heart disease and stroke statistics—2014 update: a report from the American Heart Association. Circulation129(3), e28–e292. Available: https://www.ncbi.nlm.nih.gov/pmc/articles/PMC5408159/

[2] P. Heo, G. M. Gu, S.-j. Lee, K. Rhee, and J. Kim, "Current hand exoskeleton technologies for rehabilitation and assistive engineering," International Journal of Precision Engineering and Manufacturing, vol. 13, pp. 807-824, 2012

[3] T. Worsnopp, M. Peshkin, J. Colgate, and D. Kamper, "An actuated finger exoskeleton for hand rehabilitation following stroke," in Rehabilitation Robotics, 2007. ICORR 2007. IEEE 10th International Conference on, 2007, pp. 896-901

[4] C. N. Schabowsky, S. B. Godfrey, R. J. Holley, and P. S. Lum, "Development and pilot testing of HEXORR: hand EXOskeleton rehabilitation robot," Journal of neuroengineering and rehabilitation, vol. 7, p. 36, 2010

[5] W. Caesarendra, S. U. Lekson, K. A. Mustaqim, A. R. Winoto, and A. Widyotriatmo, "A classification method of hand EMG signals based on principal component analysis and artificial neural network," in Instrumentation, Control and Automation (ICA), 2016 International Conference on, 2016, pp. 22-27

IOP Conf. Series: Journal of Physics: Conf. Series **1007** (2018) 012005 doi:10.1088/1742-6596/1007/1/012005

[6] W. Caesarendra and M. Irfan, "Classification Method of Hand Gestures based on Support Vector Machine," presented at the 1st International Conference on Informatics, Robotics, Network, Control and Systems (IRONCONS), 2016
[7] W. Caesarendra, T. Tjahjowidodo, and D. Pamungkas, "EMG based Classification of Hand Gestures using PCA and ANFIS," presented at the 2nd International Conference on Robotics, Biomimetics, & Intelligent Computational Systems (Robionetics), Bali, Indonesia, 2017
[8] M. Ariyanto, W. Caesarendra, K. A. Mustaqim, M. Irfan, J. A. Pakpahan, J. D. Setiawan, et al., "Finger movement pattern recognition method using artificial neural network based on electromyography (EMG) sensor," in Automation, Cognitive Science, Optics, Micro Electro-Mechanical System, and Information Technology (ICACOMIT), 2015 International Conference on, 2015, pp. 12-17
[9] A. Phinyomark, P. Phukpattaranont, and C. Limsakul, "Feature reduction and selection for EMG signal classification," Expert Systems with Applications, vol. 39, pp. 7420-7431, 2012
[10] B.-S. Yang and A. Widodo, Introduction of intelligent machine fault diagnosis and prognosis: Nova Science Publishers, 2009
[11] R. N. Khushaba, S. Kodagoda, M. Takruri, and G. Dissanayake, "Toward improved control of prosthetic fingers using surface electromyogram (EMG) signals," Expert Systems with Applications, vol. 39, pp. 10731-10738, 2012
[12] S. Theodoridis, A. Pikrakis, K. Koutroumbas, and D. Cavouras, Introduction to pattern recognition: a matlab approach: Academic Press, 2010.
[13] J.-S. R. Jang, C.-T. Sun, and E. Mizutani, "Neuro-fuzzy and soft computing; a computational approach to learning and machine intelligence," 1997

Acknowledgement
Authors thank toMedical Specialty Programs Physical Medicine and Rehabilitation, Medical Faculty, Diponegoro University, Semarang, Indonesia.

A Comparison of One Time Pad Random Key Generation using Linear Congruential Generator and Quadratic Congruential Generator

D Apdilah[1], M K Harahap[2], and N Khairina[2], A M Husein[3], M Harahap[3]

[1]Lecturer of Amik INTeL Com GLOBAL INDO, Medan, Indonesia
[2] Lecturer of Politeknik Ganesha Medan, Medan, Indonesia
[3]Lecturer of Universitas Prima Indonesia, Medan, Indonesia

E-mails: dicky@nusa.net.id, choir.harahap@yahoo.com, nurulkhairina27@gmail.com, amirmahmud@unprimdn.ac.id, mawaddah@unprimdn.ac.id

Abstract. One Time Pad algorithm always requires a pairing of the key for plaintext. If the length of keys less than a length of the plaintext, the key will be repeated until the length of the plaintext same with the length of the key. In this research, we use Linear Congruential Generator and Quadratic Congruential Generator for generating a random number. One Time Pad use a random number as a key for encryption and decryption process. Key will generate the first letter from the plaintext, we compare these two algorithms in terms of time speed encryption, and the result is a combination of OTP with LCG faster than the combination of OTP with QCG.

1. Introduction

Research on data security is very interesting. This study is also always discussed to improve the convenience of using confidential data. Cryptography depends on key strengths, by using a strong key on the process of encryption and decryption, we can improve the security of confidential data [1][2].
The One Time Pad (OTP) algorithm is characterized, in that the key is used only once in every encryption process [3][4]. The OTP algorithm uses random numbers that can be constructed from different types of random number generator algorithms. An OTP algorithm key will be established along the plaintext to be encrypted. If the plaintext is longer than the key, the key will be repeated throughout the plaintext [5][6][7]. Because of this unique key, the OTP algorithm can be considered as the most powerful algorithm in cryptography [8].

Muhammad Iqbal et al [5] implements the OTP algorithm in the SMS app to keep the security of confidential messages. XOR operators are used in encryption and decryption processes. Key selection is done manually by selecting some characters that will be converted into ASCII code. Messages generated from this application are messages that can be guaranteed confidentiality because the message can only be read by the sender and the recipient. The OTP algorithm is the latest version of the Vernam Cipher algorithm [9]. The keys used in the OTP Algorithm can also be generated through a random number generator algorithm. The first random number to be referred to is called Seed [10]. Jhessica Clawdia et al [11] uses the Linear Congruential Generator (LCG) algorithm to generate

IOP Conf. Series: Journal of Physics: Conf. Series **1007** (2018) 012006 doi:10.1088/1742-6596/1007/1/012006

random numbers. This algorithm is used to avoid generating repetitive keys. From the key generation process can be seen that the longer the range of m values the less likely the key will be repeated.

Using the LCG algorithm to generate keys, can also generate dynamic keys. As has been done by Zeenat Mahmood et al [12] it appears that the cryptographic algorithm is already difficult to solve by cryptanalysis. Random number generator algorithms have an important role in cryptography because they have resistance to cryptanalysis attacks. Mina Mishra et al [13] he did some research on testing some random number generator algorithms, one of them is LCG. The LCG algorithm has resistance to cryptanalysis attacks and has a good key sensitivity.

Various combinations of algorithms are also performed to generate random numbers. Massoud Sokouti et al [14] he generates random numbers with a genetic algorithm. The key that has been formed will be used by the OTP algorithm. The results of this study, OTP algorithm has a strong resistance and able to maintain data security from cryptanalyst attacks.

In addition to the LCG random number generator algorithm, there is also a Quadratic Congruential Generator (QCG) algorithm. This algorithm generates random numbers with mathematical formulas and quadratic ranks as written by Jurgen Eichenauer et al [15].

From some previous research, researchers are still conducting trials to generate random numbers that aim to form a strong cryptographic key. In this study we propose and compare the combination of OTP with LCG algorithm and OTP with QCG algorithm. And we also compare these two algorithms in terms of time speed encryption. The author will also modify the OTP algorithm. Differences exist in the encryption and decryption processes that do not involve the first plaintext characters.

2. Literature Review

2.1 One Time Pad Algorithm

One Time Pad Algorithm (OTP) is a symmetric key cryptography algorithm with encryption and decryption process using the same key [1]. In this research, we use 256 characters from the table of ASCII so we will do encryption and decryption process in mod 256.

OTP encryption formula is [16]:

$$C_i = (P_i + K_i) \bmod 256 \tag{1}$$

AndOTP decryption formula is [16]:

$$P_i = (C_i - K_i) \bmod 256 \tag{2}$$

Description of formula :

C_i = Ciphertext
P_i = Plaintext
K_i = Key
Mod 255 is mean total of ASCII numbers

2.2 Linear Congruential Generator Algorithm

This is the formula of Linear Congruential Generator [1]:

$$X_n = (a * X_{n-1} + b) \bmod m \tag{3}$$

Description of formula :

X_n = Random number to n of the series
X_{n-1} = previous random number
a = multiplier

IOP Conf. Series: Journal of Physics: Conf. Series **1007** (2018) 012006 doi:10.1088/1742-6596/1007/1/012006

c = increment
m = modulus

2.3 Quadratic Congruential Generator Algorithm
This is the formula of Cubic Congruential Generator [1]:

$$X_n = (a\, X_{n-1}{}^2 + b\, X_{n-1} + c) \bmod m \qquad (4)$$

Description of formula :

X_n = Random number to i of the series
X_{n-1} = previous random number
a,b,c = multiplier
d = increment
m = modulus

3. Proposed Method

3.1 One Time Pad Algorithm Modified with LCG
In this chapter, we perform the encryption and decryption process with a combination of the modified OTP and LCG. The modified OTP algorithm is not different from the original OTP algorithm. Differences exist in the encryption and decryption processes that do not involve the first plaintext characters. In generating the keys we use a = 3; b = 17; and m = 101.

3.1.1. Encryption Process:
There are several steps to perform the encryption process :
Step – 1: Retrieve the plaintext to be encrypted
Step – 2: Convert the first letter into ASCII code
Step – 3: Make the number in Step -2 as Z0 (seed) to generate a random number
Step – 4: Leaving the first letter unencrypted
Step – 5: Key = Z
Step – 6: Perform encryption process on next letter
Step – 7: Generate a random number Zn using LCG
Step – 8: Return to step 5 until all letters are encrypted

Here is a table forperforming the encryption process :

Table 1. Encryption Process.

No	Px	ASCII Px	Kn	EncryptionResults	Ciphertext in Char
1	K	75	75	C_1 =75 (first letter unencrypted)	K
2	H	72	42	C_2 = (72 + 42) mod 256 = 114	r
3	O	79	44	123	{
4	I	73	50	123	{
5	R	82	68	150	–

3.1.2 Decryption Process :
There are several steps to perform the decryption process :
Step – 1 : Retrieve the ciphertext to be decrypted

Step – 2 : Convert the first letter into ASCII code
Step – 3 : Make the number in Step -2 as Z_0 to generate a random number
Step – 4 : Leaving the first letter undecrypted
Step – 5 : Key = Z
Step – 6 : Perform decryption process on next letter
Step – 7 : Generate a random number Zn
Step – 8 : Return to step 5 until all letters are decrypted

Here is a table forperforming the decryption process :

Table 2. Decryption Process.

No	Cx	ASCII Cx	Kn	Decryption Process	Plaintext in Char
1	K	75	75	P_1 =75 (first letter undecrypted)	K
2	r	114	42	P_2 = (114 - 42) mod 256 = 72	H
3	{	123	44	79	O
4	{	123	50	73	I
5	–	150	68	82	R

3.2 One Time Pad Algorithm Modified with QCG

In this chapter, we perform the encryption and decryption process with a combination of the modified OTP and QCG. The modified OTP algorithm is not much different from the original OTP algorithm. Differences exist in the encryption and decryption processes that do not involve the first plaintext characters. In generating the keys we use a = 3; b = 17; c = 19, and m = 101.

3.2.1. Encryption Process

There are several steps to perform the encryption process :
Step – 1: Retrieve the plaintext to be encrypted
Step – 2: Convert the first letter into ASCII code
Step – 3: Make the number in Step -2 as Z0(seed) to generate a random number
Step – 4: Leaving the first letter unencrypted
Step – 5: Key = Z
Step – 6: Perform encryption process on next letter
Step – 7: Generate a random number Zn using QCG
Step – 8: Return to step 5 until all letters are encrypted

Here is a table forperforming the encryption process :

Table 3. Encryption process.

No	Px	ASCII Px	Kn	EncryptionResults	Ciphertext in Char
1	K	75	75	C_1 =75 (first letter unencrypted)	K
2	H	72	90	C_2 = (72 + 90) mod 256 = 162	¢
3	O	79	94	173	-
4	I	73	47	120	x
5	R	82	72	154	š

3.2.2.Decryption Process :

There are several steps to perform the decryption process :

Step – 1 : Retrieve the ciphertext to be decrypted
Step – 2 : Convert the first letter into ASCII code
Step – 3 : Make the number in Step -2 as Z_0 to generate a random number
Step – 4 : Leaving the first letter undecrypted
Step – 5 : Key = Z
Step – 6 : Perform decryption process on next letter
Step – 7 : Generate a random number Zn
Step – 8 : Return to step 5 until all letters are decrypted

Here is a table forperforming the decryption process :

Table 4. Decryption process.

No	Cx	ASCII Cx	Kn	Decryption Process	Plaintext in Char
1	K	75	75	P_1 =75 (first letter undecrypted)	K
2	¢	162	90	P_2 = (162 - 90) mod 256 = 72	H
3	-	173	94	79	O
4	x	120	47	73	I
5	š	154	72	82	R

4. Result and Discussion

The testing is done in the text file (.txt) thus get all the characters that have been symbolized on ASCII code. In reality, there are still many file types that need to be tested. Testing is also done on text files with large size and test results are really stable. This is influenced by the use of modulo 255 in the encryption process and decryption process. So all of ASCII characters can be encrypted and decrypted properly.

Table 5. Comparison encryption time process.

No	Length	Time Test	Encryption Time Average		Time Difference
			OTP + LCG	OTP + QCG	
1	100	10	0,000200033	0,000200033	0
2	1000	10	0,00079999	0,000999975	0,000199986
3	10000	10	0,01079998	0,012199974	0,001399994
4	100000	10	0,098200035	0,118799973	0,020599937
5	1000000	10	1,238600016	1,416599989	0,177999973

The process through which Encryption and Decryption are the same, then decryption test is not performed.

5. Conclusions

Based on the above experiments it is concluded that the LCG process is faster than CCG. Future work needs to be tested about how the treatment of documents that have special codes such as the use of font size, italic, bold, underline, and also the use of color. Thing need to be tested on the spreadsheet file and how it affects the formula on the spreadsheet and also against the existing table. In the future are still many more need to be tested with OTP using this dynamic key.

References

[1] B. Schneier, Applied Cryptography Protocol, Algorithms, and Source Code in C, 2nd ed., California: Jhon Wiley & Sons, Inc, 1996, p. 17.
[2] P. P. Kumar, J. M. Kumar, and M. Neelima, "A new method for generating a key for cryptography," International Journal of Engineering Trends and Technology (IJETT), vol. 4, no. 9, pp. 3915-3920, 2013.
[3] C. Matt and U. Maurer, "The One-Time Pad Revisited," in International Symposium on Information Theory, Istanbul, 2013.
[4] E. H. A. Mendrofa, . E. Y. Purba, B. Y. Siahaan and R. W. Sembiring, "Collaborative Encryption Algorithm Between Vigenere Cipher, Rotation of Matrix (ROM), and One," Advances in Science, Technology and Engineering Systems Journal Vol. 2, No. 5, pp. 13-21, 2017.
[5] M. Iqbal, M. A. S. Pane and A. P. U. and Siahaan, "SMS EncryptionUsing One-Time Pad Cipher," IOSR Journal of Computer Engineering (IOSR-JCE), vol. 18, no. 6, pp. 54-58, 2017.
[6] R. Shukla, H. O. Prakash, R. Bhushan, S. Venkataraman and G. Varadan, "Unconditionally Secure and Authenticated One Time Pad Cryptosystem," in International Conference on Machine Intelligence and Research Advancement, India, 2013.
[7] E. B. B. Samah M.H. Alwahbani, "Speech Scrambling Based on Chaotic Maps and," in International Conference on Computing, Electrical and Electronic Engineering (ICCEEE), Sudan, 2013.
[8] B. J. Saha, K. K. Kabi, Arun and C. Pradhan, "Non-Blind Watermarking Technique using Enhanced," in 5th ICCCNT, Hefei, 2014.
[9] N. Khairina and M. K. Harahap, "Analisis Algoritma One Time Pad Dengan Algoritma Cipher Transposisi Sebagai Pengamanan Pesan Teks," SinkrOn (Jurnal & Penelitian Teknik Informatika), pp. 58-62, 2017.
[10] R. Maheshwari, S. Gupta, V. Sharma and V. Chauhan, "VRS algorithm A Novel Approach to Generate Pseudo," in International Advance Computing Conference (IACC), Gurgaon, 2014.
[11] J. Clawdia, N. Khairina and M. K. Harahap, "Implementasi Algoritma Kriptografi One Time Pad (OTP) dengan Dyamic Key Linear Congruential Generator (LCG)," Komik, vol. 1, no. 1, pp. 12-14, 2017.
[12] Z. Mahmood, J. Rana, and A. Khare, "Symmetric Cryptography using Dynamic Key and Linear Congruential Generator (LCG)," International Journal of Computer Applications, vol. 50, no. 19, pp. 7-11, 2012.
[13] M. Mishra and V. Mankar, "Text Encryption Algorithms based on Pseudo-Random Number Generator," International Journal of Computer Applications, vol. 111, no. 2, pp. 1-6, 2015.
[14] M. Sokouti, B. Sokouti, S. Pashazadeh, M. R. F. Derakhshi and S. Haghipour, "Genetic-based random key generator (GRKG): a new method," Neural Comput & Applic, pp. 1667 - 1675, 2013.
[15] J. Eichenauer and Herrmann, "Quadratic congruential pseudorandom numbers:," Journal of Computational and Applied Mathematics,pp. 75-85, 1997.
[16] M. K. Harahap, "Analisa Perbandingan Algoritma Kriptografi Klasik Vigenere Cipher dan One Time Pad," Info Tekjar, vol. 1, no. 1, pp. 61-64, 2016.

IOP Conf. Series: Journal of Physics: Conf. Series **1007** (2018) 012007 doi:10.1088/1742-6596/1007/1/012007

Attribute Weighting Based K-Nearest Neighbor Using Gain Ratio

A A Nababan[1], O S Sitompul[2*], and Tulus[3]

[1]Graduate School of Computer Science

[2]Faculty of Computer Science and Information Technology, Universitas Sumatera Utara, Medan, Indonesia

[3]Department of Mathematics, Universitas Sumatera Utara, Medan, Indonesia

E-mail: *opim@usu.ac.id

Abstract. K- Nearest Neighbor (KNN) is a good classifier, but from several studies, the result performance accuracy of KNN still lower than other methods. One of the causes of the low accuracy produced, because each attribute has the same effect on the classification process, while some less relevant characteristics lead to miss-classification of the class assignment for new data. In this research, we proposedAttribute Weighting Based K-Nearest Neighbor Using Gain Ratio as a parameter to see the correlation between each attribute in the data and the Gain Ratio also will be used as the basis for weighting each attribute of the dataset. The accuracy of results is compared to the accuracy acquired from the original KNN method using 10-fold Cross-Validation with several datasets from the UCI Machine Learning repository and KEEL-Dataset Repository, such as abalone, glass identification,haberman, hayes-roth and water quality status. Based on the result of the test, the proposed method was able to increase the classification accuracy of KNN, where the highest difference of accuracy obtained hayes-roth dataset is worth 12.73%, and the lowest difference of accuracy obtained in the abalone dataset of 0.07%. The average result of the accuracy of all dataset increases the accuracy by 5.33%.

1. Introduction

K-Nearest Neighbor (KNN) is one of the effective, simple and performs well method for classification [1-3], but from several studies, the result performance accuracy of KNN is lower than other methods. One of them is in the study by [4] which compared performance between support vector machine (SVM) and KNN. The result of their research seen that performance of SVM better than KNN, where the value accuracy obtained by SVM of 82.54% while the value obtained from KNN of 79.22%. Another study by [5] compared K-Nearest Neighbor (KNN) and Artificial Neural Network (ANN). The results of their study seen that the performance of ANN (with 5 hidden layers) better than KNN, where the result of accuracy is 90.5%. Research by [6] compared the performance of Naïve Bayes, Decision Tree and K-Nearest Neighbor (KNN). The results show that Naïve Bayes has the best accuracy in classification compared to Decision Tree Tree and K-Nearest Neighbor (KNN) with an average accuracy of 73.7%. whilethe average accuracy of Decision Tree and KNN respectively 58.9% and 56.7%. In the study by [7] compared KNN and Naïve Bayes in diagnosing heart disease. Results obtained by the value of accuracy Naïve Bayes of 79.62% while the average value of accuracy on KNN is only 64.85% when $k = 10$.

From some research above seen that the accuracy of KNN method is still lower than other methods. One of the causes of the low value of accuracy produced, because each attribute has the same effect on the classification process. The solution to this problem is to give weight to each attribute [8]. In this research, we propose attribute weighting based K-Nearest Neighbor using Gain Ratio to increase the accuracy value of K-Nearest Neighbor (KNN) method by giving weight to each attribute, whereby the weights obtained are calculated using the normalization equation min-max, then calculatedthe data similarity using the distance model proposed by [1]. The rest of this paper is structured as follows. Section 2 will summarize previous studies on the theoretical foundation regarding the topic. In Section 3 we will provide the result and discussions and in Section 4 will provide with conclusions

2. Feature Weight K-Nearest Neighbor (FWKNN)

K-Nearest Neighbor (KNN) is a nonparametric learning method and sensitive to distance function due to the inherent sensitivity of irrelevant attributes [1]. So to get FWKNN modeling based on attribute weighting.FWKNN determines the weight of the attribute by identifying the nearest k-neighbor that has the highest similarity with the test data. Therefore the FWKNN requires a model to measure the distance between test data and training sample data.

To measure the distance between data, one of the most popular options is the Euclidean distance model [1]. FWKNN uses a modified distance using the weighted feature for standard Euclidean distance expansion. The equations used in FWKNN are as follows:

$$d^w(X' - X) = \sqrt{\sum_{i=1}^{y} W_y \times (x_i' - x_i)^2} \quad (1)$$

where $d^w(X'-X)$is the Euclidean distance from the test data and training data, y is the number of attributes, the attribute weights,x_i'andx_i the attribute values of test data and training data. The detail of FWKNN algorithm is as follows:

Step 1: Determine the weight of each feature
Step 2: Determine the value parameter k
Step 3: Calculate the distance using equation (1)
Step 4: Sorting results obtained by ascending (sequentially from high to low).
Step 5: Calculate the number of each class based on the k-nearest neighbor.
Step 6: The majority class is used as a class for test data.

FWKNN makes a difference to the characteristics by giving weight to its attributes, meaning that the more important attributes have a greater effect on distance determination [8]. Thus the error in the classification process can be reduced [1].

3. Gain Ratio

Algorithm C4.5 is a decision tree method wherein the selection of attributes based on Gain Ratio. Gain Ratio (GR) is a modification of Information Gain which reduces its bias. Gain Ratio improves the information gain by taking intrinsic information from each attribute [9]. The steps in determining the Gain Ratio are as follows:

Step 1: Calculate the value of Entropy on each attribute, with the equation :

$$\text{Entropy (S)} = \sum_{i=1}^{n} - pi * log_2 pi \quad (2)$$

where :
S = Set of Cases
n = Number of Partitions S
pi = Proportions ofSiS

Step 2: Calculate the value of the information gain in each attribute by the equation :

$$Information\ Gain\ (S, A) = Entropy(S) - \sum_{i=1}^{n} \frac{|S_i|}{|S|} \times Entropy(S_i) \quad (3)$$

where :
S = Overall Dataset

A = Subset Attribute
$| S_i |$ = Subset Size of Dataset
$| S |$ = Subset Size of Dataset owned attribute on A i-partition

Step 3: Calculate the value of Split Information for each attribute with the equation below :

$$SplitInfo_A(D) = -\sum_{j=1}^{v} \frac{|D_j|}{|D|} \times log_2(\frac{D_j}{D}) \quad (4)$$

where :
D = Overall Dataset
A = Subset Attribute
v = Number of Partition Attributes A
$|D_j|$ = Subset Size of Dataset Owned attribute on A j-partition
$|D|$ = Size of Number of Cases in Dataset

Step 4: Calculate the Gain Ratio of each attribute with the equation :

$$Gain\ Ratio\ (A) = \frac{Gain\ \ (A)}{SplitInfo\ \ (A)} \quad (5)$$

The Gain Ratio is found in the C4.5 algorithm, where the gain ratio is used to calculate the effect of attributes on the target of a data [11]. Gain Ratio is the development of information gain, where the gain ratio eliminates the bias value of each attribute.

4. Previous Research

Research work by [12] comparing K-Nearest Neighbor (KNN) with Support Vector Machine (SVM) for water quality status classification. The test results show the highest accuracy value found SVM, which is 92.40% while the average value of accuracy on KNN only 71.28% with k=7.The accuracy of KNN method is still lower than other methods. One of the causes of the low value of accuracy produced, because each attribute has the same effect on the classification process.

Weighting attribute has been proposed by [8]However, the weighting of attributes is less effective, because it gives the same weight for each of characteristics, while in research by [13], his research suggests consideration of KNN method by using different weighting functions.

In addition, attribute weighting is also performed on research work by [1]. in his research suggested the method of Information Gain as a basis for weighting attributes on the KNN algorithm known as Feature Weight K-Nearest Neighbor (FWKNN). This method is considered more effective than original KNN.

Finally, [14] in his research using Gain Ratio as the basic weighting attribute on KNN. From their research, it is seen that KNN by using Gain Ratio is considered more intuitive and easy to understand. The results obtained in the study, weighting attributes by using Gain Ratio able to increase the highest accuracy of 5%

5. The Proposed Method

To describe the proposed method, it will be explained step by step in this chapter. The stages can be seen in figure 1:

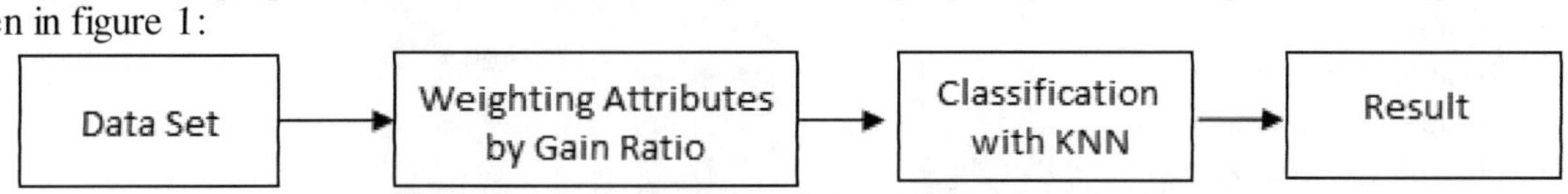

Figure 1. Detail stages of the proposed method.

From figure 1 it can be explained that the proposed method has several stages, among others :

Step 1: Weighting Attributes by Gain Ratio

The weight is calculated using the normalization equation min-max [10], where the lowest weight after normalized is 0.1 and the highest weight after normalized is 1. The equation used is as follows:

$$W_i = \frac{(G_i - Min\ (G))}{Max\ (G) - Min\ (G)} \times (0.9) + 0.1 \quad (6)$$

where :

W_i = Weight of the *i*- Attribute

G_i = TheGain ratio -*i*

$Min(G)$ = The Lowest Value of The Gain Ratio

$Max(G)$ = The Highest Value of The Gain Ratio

Step 2: Classification with KNN

Furthermore, the process of classification with KNN, after the weight of each attribute is obtained it will be classified by KNN. In determining the proximity between data, each attribute will have different effects based on the weight value of the attribute using the equation (1).

This research will use the gain ratio as a parameter to see the correlation between each attribute in the data and the gain ratio also will be used as the basis for weighting each attribute of the data set. The higher the gain ratio of an attribute the correlation to the data class will be greater, so the weight of the attribute is also higher. This research will give analysis result to the performance of KNN method with weighting attribute using gain ratio on the performance of KNN method conducted by [11].

6. Result and Discussion

This research uses data sets obtained from UCI Machine Learning Repository and KEEL-Dataset Repository such as abalone, glass identification,haberman, hayes-roth and one dataset of water quality status derived from research conducted by [11]. The details of the data used can be seen in table 1:

Table 1. Detail of Data.

Data	Attributes	Type	Class	Total Data
Abalone	8	Categorical, Integer, Real	3	4177
Glass Identification	10	Real	6	214
Haberman	3	Real, Integer, Nominal	2	306
Hayes-roth	4	Real, Integer, Nominal	3	160
Water Quality Status	8	Integer, Real	4	120

6.1 Testing Using Data Set

This study used a 10-fold cross validation on datasets, so the total dataset was randomly divided into 10 datasets with the same comparison values, Then do 10 trials, where each trial using a data partition to-kas the testing of data and utilize the remaining more partitions as training data.The average accuracy of each data can be seen in figure 2:

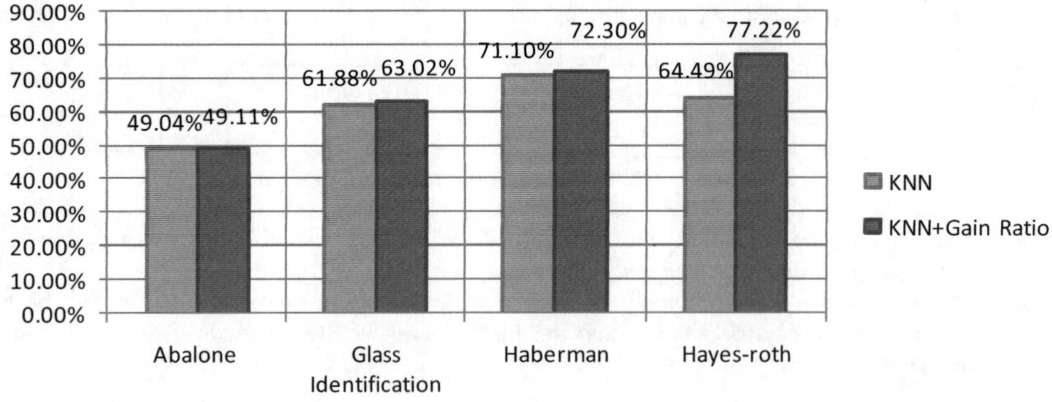

Figure 2. The average result of the accuracy of the dataset.

The figure 2 shows the KNN using Gain Ratio has a higher accuracy value than the original KNN, where the highest difference of accuracy obtained hayes-roth data set is worth 12.73%, and the lowest difference of accuracy obtained in the abalone dataset of 0.07%. The average result of the accuracy ofall dataset increases the accuracy of 3.79%.

6.2 Testing Using Water Quality Status

To know the weighting of attributes using gain ratio can improve the accuracy of KNN, then the next test will be conducted by using the water quality status datasets derived from research by [11]. The result of comparison between KNN and KNN with Gain Ratio can be seen in table 2.

Table 2. Comparison of Accuracy of All *k* KNN against KNN with Gain Ratio.

k	Comparison Of Accuracy (%)		Difference of Accuracy
	KNN	KNN+Gain Ratio	
1	87.5%	94.2%	6.7%
2	87.5%	92.5%	5%
3	85.8%	95%	9.2%
4	85.8%	90%	4.2%
5	85.8%	92.5%	6.7%
6	87.5%	87.5%	0%
7	82.5%	90.8%	8.3%
8	83.3%	88.3%	5%
9	83.3%	89.2%	5.9%
10	83.3%	85.8%	2.5%
Average	85.23%	90.58%	5.35%

Results of testing using KNN and KNN with Gain Ratio classification contained in table 2. The table shows originalKNN with the highest difference of accuracy grade are in grades k = 1, 2, 6 that is 87.5% and the lowest difference of accuracy grade are in grade k = 7 that is 82.5%, while the results KNN with Gain Ratio is obtained the highest difference of accuracy grade are in grades k = 3 that is 95% and the lowest difference of accuracy grade are in grade k = 10 that is 85.8% .The comparison of the accuracy of KNN and KNN with Gain Ratio shows that the average result of the accuracy of all k increases the accuracy by 5.35% against accuracy value of original KNN.

7. Conclusion

Based on the findings and discussion that has been described previously, it can be concluded as follows:

1. The test using 10-fold cross validation proved that Attribute weighting based K-Nearest-Neighbor using Gain Ratio has a higher accuracy value than the original KNN, where the highest difference of accuracy obtained hayes-roth data set is worth 12.73%, and the lowest difference of accuracy obtained in the abalone dataset of 0.07%. The average result of the accuracy of all dataset increase the accuracy by4.09%
2. The test results weighting of attributes using gain ratio on thewater quality dataset can improve the accuracy of KNN with the highest difference of accuracy grade are in grades k = 3 that is 9.2% and the average result of the accuracy of all k increase the accuracy by 5.35% against accuracy value of original KNN.
3. KNN is a good classifier but when we implementthis method over textual data (nominal data) it's all performance parameters are varied according to the size of the dataset. KNN performs poor results as the size of data set increases it is the best fit for a small dataset.

References

[1] Chen, Y., Hao, Y. 2017. A Feature Weighted Support Vector Machine and K-Nearest Neighbor Algorithm for Stock Market Indices Prediction. Expert Systems With Applications (2017), pp. 340-355

[2] Song, Y., Liang, J., Lu, J. & Zhao, X. 2017. An Efficient Instance Selection Algorithm For K Nearest Neighbor Regression. *Neurocomputing*, pp. 26-34.

[3] Suguna, N., Thanushkodi, K. 2010. An Improved k-Nearest Neighbor Classification Using Genetic Algorithm. *ICJSI International Journal of Computer Science Issues,* pp. 18-21.

[4] Raikwal, J.S., Saxena, K. 2012. Performance Evaluation of SVM and K-Nearest Neighbor Algorithm over Medical Data set. *International Journal of Computer Applications*, pp.35-39.

[5] Moosavian, A., Ahmadi, H., Tabatabaeefar, A., Khazaee, M. 2012. Comparison of two classifiers; K-nearest neighbor and artificial neural network, for fault diagnosis on a main engine journal-bearing. *Shock and Vibration*, pp. 263-272.

[6] Ashari, A., Paryudi, I., Tjoa, A.M. 2013. Performance Comparison between Naïve Bayes Decision Tree and k-Nearest Neighbor in Searching Alternative Design in an Ene Simulation Tool. *(IJACSA) International Journal of Advanced Computer Science a Applications,* pp. 33-39.

[7] Meiyar, V.K., Shanmugasundaram, D. 2015. The Comparative Study for Diagnosing Heart Disease Using KNN and Naïve Bayes. *(IJARCSMS) International Journal Advanced Research in Computer Science and Management Studies*,pp. 9-19.

[8] Kuhkan, M. 2016. A Method to Improve the Accuracy of K - Nearest Neighbor Algorithm. *International Journal of Computer Engineering and Information Technology,* pp. 90-95.

[9] Priyadarsini, R.P., Valarmathi, M.L., Sivakumari, S. 2011. Gain Ratio Based Feature Selection Method For Privacy Preservation. *ICTACT Journal on Soft Computing,* pp. 201-205.

[10] Saranya, C & Manikandan, G. 2013. A Study on Normalization Techniques for Privacy Preserving Data Mining. *International Journal of Engineering and Technology (IJET)*, pp. 2701-2704.

[11] Mitchell, T.M. 1997. Machine Learning. McGraw-Hill Science/Engineering/Math; (March 1, 1997). ISBN: 0070428077.

[12] Danades, A., Pratama, D., Anggraini, D., Anggriani, D. 2016. Comparison of Accuracy Level K-Nearest Neighbor Algorithm and Support Vector Machine Algorithm in Classification Water Quality Status.*International Conference on System Engineering and Technology*, pp. 137-141. DOI 10.1109/FIT.2016.7857553.

[13] Demidova, L., Sokoloba, Y. 2017. A Novel SVM-kNN Technique for Data Classification. *2017 6th Mediterranean an Conference on Embedded Computing (MECO),* pp.11-15.

[14] Duneja, A., Puyalnithi, T. 2017. Enhancing Classification Accuracy of K-Nearest Neighbours Algorithm Using Gain Ratio. *International Research Journal of Engineering and Technology (IRJET)*, pp. 1385-1388.

Performance of the Lax-Wendroff finite volume method for solving the gravity wave-model equations

Cecilia Heru Purwitaningsih[1], Sudi Mungkasi[2*]

[1]Department of Mathematics Education, Faculty of Teacher Training and Education, Sanata Dharma University,Mrican, Tromol Pos 29, Yogyakarta 55002, Indonesia
[2]Department of Mathematics, Faculty of Science and Technology, Sanata Dharma University,Mrican, Tromol Pos 29, Yogyakarta 55002, Indonesia

E-mail: *sudi@usd.ac.id

Abstract. The gravity wave-model equations are simplifications of the Saint-Venant equations by neglecting the convective term. This neglectionis realistic as long as the gravity effect is much more significant than the convective effect in the system. In this paper, we present the performance (behaviour) of the standard Lax-Wendroff finite volume method used to solve the gravity wave-model equations. This is the first work in discussing the aforementioned method's performance in solving the gravity wave-model equations. We obtain that the standard Lax-Wendroff method is suitable for solving problems without discontinuity in the solution. When there is a discontinuity, the standard Lax-Wendroff method produces artificial oscillation in the solution.

1. Introduction

One of mathematical models for free-surface flows is the gravity wave-model equations [1]. Gravity wave-model equations are simplifications of the Saint-Venant equations by neglecting the convective terms [1-5]. The Saint-Venant (shallow water) equations consist of two partial differential equations, namely the mass equation and momentum equation [6-8]. The Saint-Venant equationsinvolving horizontal topography are:

$$\frac{\partial}{\partial t}h + \frac{\partial}{\partial x}q = 0 \tag{1}$$

and

$$\frac{\partial}{\partial t}q + \frac{\partial}{\partial x}\left(\frac{q^2}{h}\right) + \frac{g}{2}\frac{\partial}{\partial x}h^2 = 0 \tag{2}$$

where h is fluid depth; q is unit-discharge; gis the gravitational acceleration; t is the time variable and x is the space variable. By neglecting the convective term of the Saint-Venant equations, we obtain the gravity wave-model equations, which can be written as

$$\frac{\partial}{\partial t}h + \frac{\partial}{\partial x}q = 0 \tag{3}$$

and

$$\frac{\partial}{\partial t}q + \frac{g}{2}\frac{\partial}{\partial x}h^2 = 0. \tag{4}$$

IOP Conf. Series: Journal of Physics: Conf. Series **1007** (2018) 012008 doi:10.1088/1742-6596/1007/1/012008

All quantities are assumed to be in SI units.

Real-world problems that have been modeled mathematically needs to be solved. The difficulty in determining the analytical solutions to the mathematical model can be overcome by doing numerical calculations to get the approximate solution of the model. In this paper, we discuss about the numerical solution to the gravity wave-model equations using the Lax-Wendroff finite volume method. On the one hand, variants of the Lax-Wendroff method have been broadly applied in solving numerous problems of hyperbolic conservation laws [9-15]. On the other hand, the gravity wave-model equations were attempted to be solved numerically by several authors, such as Apriani and Mungkasi [16] as well asMartins et al. [17]. However, an accurate and efficient solver for the gravity wave-model equations still needs to be sought. The aim of this paper is to find such solver. Therefore, we investigate the performance of the Lax-Wendroff finite volume method used to solve the gravity wave-model equations.

The rest of this paper is organised as follows. We present the Lax-Wendroff finite volume method for solving the gravity wave-model equations in Section 2. Numerical results are provided in Section 3 together with the discussion about them. We close the paper with concluding remarks in Section 4.

2. Lax-Wendroff finite volume method

The gravity wave-model equations (3) and (4) areconservation laws having the form

$$\bar{U}_t + \bar{f}(\bar{U})_x = \bar{0} \tag{5}$$

where

$$\bar{U}_t \equiv \frac{\partial \bar{U}(x,t)}{\partial t} and \ \ \bar{f}(\bar{U})_x \equiv \frac{\partial \bar{f}(\bar{U})}{\partial x}. \tag{6}$$

Here

$$\bar{U} = \begin{pmatrix} h \\ q \end{pmatrix} \tag{7}$$

And

$$\bar{f}(\bar{U}) = \begin{pmatrix} q \\ \frac{g}{2} h^2 \end{pmatrix}. \tag{8}$$

We can write the gravity wave-model equations as

$$\begin{pmatrix} h \\ q \end{pmatrix}_t + \begin{pmatrix} q \\ \frac{g}{2} h^2 \end{pmatrix}_x = \begin{pmatrix} 0 \\ 0 \end{pmatrix}. \tag{9}$$

In another form, gravity wave-model equations are

$$h_t + f_1^{'} q_x = 0 \tag{10}$$

And

$$q_t + f_2^{'} h_x = 0 \tag{11}$$

where$\mathrm{f}_1^{'} = 1$ and $\mathrm{f}_2^{'} = \mathrm{gh}$.

We can create a discrete form from (5) as follows [7, 8]:

$$\frac{Q_i^{n+1} - Q_i^n}{\Delta t} + \frac{F_{i+\frac{1}{2}}^n - F_{i-\frac{1}{2}}^n}{\Delta x} = 0 \tag{12}$$

And

$$Q_i^{n+1} = Q_i^n - \frac{\Delta t}{\Delta x}\left(F_{i+\frac{1}{2}}^n - F_{i-\frac{1}{2}}^n\right) \tag{13}$$

where $Q_i^n \approx \bar{U}(x_i, t^n)$; $F^n_{i+\frac{1}{2}} \approx \bar{f}\left(\bar{U}\left(x_{i+\frac{1}{2}}, t^n\right)\right)$; Δt is the time step; $\Delta x$is the space step.

The general standard Lax-Wendroff fluxes are

$$F^n_{i+\frac{1}{2}} = \frac{1}{2}\left(f(Q^n_{i+1}) + f(Q^n_i)\right) - \frac{a^n_{i+\frac{1}{2}}\Delta t}{2\Delta x}\left(f(Q^n_{i+1}) - f(Q^n_i)\right) \tag{14}$$

And

$$F^n_{i-\frac{1}{2}} = \frac{1}{2}\left(f(Q^n_i) + f(Q^n_{i-1})\right) - \frac{a^n_{i-\frac{1}{2}}\Delta t}{2\Delta x}\left(f(Q^n_i) - f(Q^n_{i-1})\right) \tag{15}$$

where $a^n_{i+\frac{1}{2}} = f'\left(\frac{Q^n_i + Q^n_{i+1}}{2}\right)$and $a^n_{i-\frac{1}{2}} = f'\left(\frac{Q^n_{i+1} + Q^n_i}{2}\right)$.

The discrete form for the law of massconservation is

$$h^{n+1}_i = h^n_i - \frac{\Delta t}{\Delta x}\left({}^hF^n_{i+\frac{1}{2}} - {}^hF^n_{i-\frac{1}{2}}\right). \tag{16}$$

The Lax-Wendrofffluxes for the law of mass conservation are

$${}^hF^n_{i+\frac{1}{2}} = \frac{1}{2}(q^n_{i+1} + q^n_i) - \frac{{}^ha^n_{i+\frac{1}{2}}\Delta t}{2\Delta x}(q^n_{i+1} - q^n_i) \tag{17}$$

And

$${}^hF^n_{i-\frac{1}{2}} = \frac{1}{2}(q^n_i + q^n_{i-1}) - \frac{{}^ha^n_{i-\frac{1}{2}}\Delta t}{2\Delta x}(q^n_i - q^n_{i+1}). \tag{18}$$

Here, ${}^hF^n_i$ is the flux for the mass conservation. In addition,

$${}^ha^n_{i+\frac{1}{2}} = f_1' = 1 \quad \text{and} \quad {}^ha^n_{i-\frac{1}{2}}\Delta t = f_1' = 1.$$

The discrete form for the law of momentum conservation is

$$q^{n+1}_i = q^n_i - \frac{\Delta t}{\Delta x}\left({}^qF^n_{i+\frac{1}{2}} - {}^qF^n_{i-\frac{1}{2}}\right). \tag{19}$$

The Lax-Wendroff fluxes for the momentum conservationare

$${}^qF^n_{i+\frac{1}{2}} = \frac{1}{2}\left(\left(\frac{g}{2}h^2\right)^n_{i+1} + \left(\frac{g}{2}h^2\right)^n_i\right) - \frac{{}^qa^n_{i+\frac{1}{2}}\Delta t}{2\Delta x}\left(\left(\frac{g}{2}h^2\right)^n_{i+1} - \left(\frac{g}{2}h^2\right)^n_i\right) \tag{20}$$

And

$$^{q}F^{n}_{i-\frac{1}{2}}=\frac{1}{2}\left(\left(\frac{g}{2}h^{2}\right)^{n}_{i}+\left(\frac{g}{2}h^{2}\right)^{n}_{i-1}\right)-\frac{^{q}a^{n}_{i-\frac{1}{2}}\Delta t}{2\Delta x}\left(\left(\frac{g}{2}h^{2}\right)^{n}_{i}-\left(\frac{g}{2}h^{2}\right)^{n}_{i+1}\right). \quad (21)$$

Here, $^{q}F^{n}_{i}$ is the flux for the momentum conservation. In addition,

$$^{q}a^{n}_{i+\frac{1}{2}}=(f_{2}^{'})^{n}_{i+\frac{1}{2}}=(gh)^{n}_{i+\frac{1}{2}}=\frac{(gh)^{n}_{i+1}+(gh)^{n}_{i}}{2}$$

and

$$^{q}a^{n}_{i-\frac{1}{2}}=(f_{2}^{'})^{n}_{i-\frac{1}{2}}=(gh)^{n}_{i-\frac{1}{2}}=\frac{(gh)^{n}_{i}+(gh)^{n}_{i-1}}{2}.$$

By using (14) and (15), for the thegravity wave-model where $Q^{n}_{i}\approx\begin{bmatrix}h^{n}_{i}\\ q^{n}_{i}\end{bmatrix}$, $F^{n}_{i}\approx\begin{bmatrix}q^{n}_{i}\\ \frac{g}{2}(h^{n}_{i})^{2}\end{bmatrix}$, Δt is the time step, Δx is the space step, $a^{n}_{i+\frac{1}{2}}=f'\left(\frac{Q^{n}_{i}+Q^{n}_{i+1}}{2}\right)$ and $a^{n}_{i-\frac{1}{2}}=f'\left(\frac{Q^{n}_{i-1}+Q^{n}_{i}}{2}\right)$ we obtain

$$F^{n}_{i+\frac{1}{2}}=\frac{1}{2}\begin{bmatrix}q^{n}_{i+1}+q^{n}_{i}\\ \frac{g}{2}((h^{n}_{i+1})^{2}+(h^{n}_{i})^{2})\end{bmatrix}-\frac{a^{n}_{i+\frac{1}{2}}\Delta t}{2\Delta x}\begin{bmatrix}q^{n}_{i+1}-q^{n}_{i}\\ \frac{g}{2}((h^{n}_{i+1})^{2}-(h^{n}_{i})^{2})\end{bmatrix} \quad (22)$$

And

$$F^{n}_{i-\frac{1}{2}}=\frac{1}{2}\begin{bmatrix}q^{n}_{i}+q^{n}_{i-1}\\ \frac{g}{2}((h^{n}_{i})^{2}+(h^{n}_{i-1})^{2})\end{bmatrix}-\frac{a^{n}_{i-\frac{1}{2}}\Delta t}{2\Delta x}\begin{bmatrix}q^{n}_{i}-q^{n}_{i-1}\\ \frac{g}{2}((h^{n}_{i})^{2}-(h^{n}_{i-1})^{2})\end{bmatrix} \quad (23)$$

as the Lax-Wendroff numerical fluxes in the vector form including the mass and momentum fluxes.

3. Numerical results and discussion

In this section we provide our numerical results and some discussions.

3.1. Surface wave simulation

Surface wave is simulated by using the initial water depth function $h=2+0.25\cos(x)$ if $-\pi\leq x\leq\pi$ and 1.75 otherwise. We denote $u=u(x,t)$ for the fluid velocity. Note that water discharge is $q=uh$. In this simulations we use the number of cells $N=500$, space step $\Delta x=0.1$, time step $\Delta t=0.001\Delta x$, and gravitational constant $g=9.81$. The simulation is stopped at $t=5$. The initial condition for fluid velocity is $u(x,0)=0$ for all x. We set $L=25$. Boundary conditions at $x=-L$ are

$$h(-L,t)=1.75, \qquad u(-L,t)=0$$

and those at $x=L$ are

$$h(L,t)=1.75, \qquad u(L,t)=0.$$

The initial water surface is shown in Figure 1. After we solve the problem using the Lax-Wendroff method, the water surface at $t=5$ is shown in Figure 2. These results are realistic as initially we have one bump of water in the middle of domain, as there is gravity effect, we then have two waves. One goes to the left and another one propagates to the right direction.

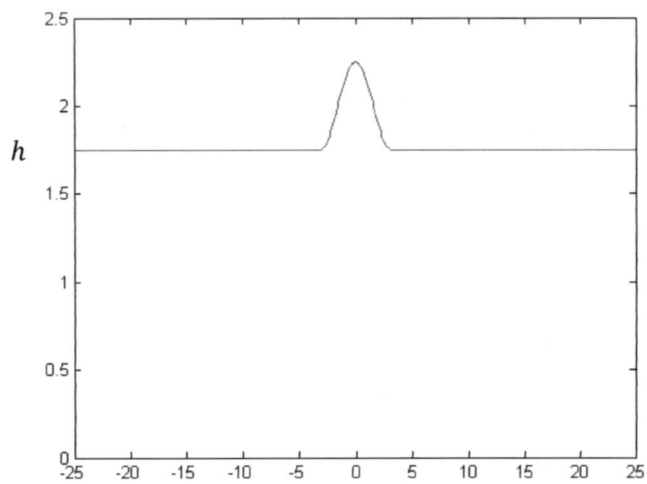

Figure 1. Illustrationof the initial water surface.The horizontal axis is for x variable.

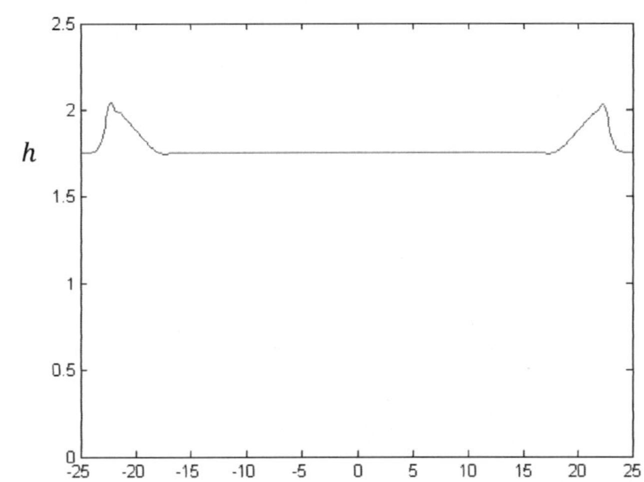

Figure 2. Simulated water surfaceat t = 5. The horizontal axis is for x variable.

3.2. Dam-break simulation

In this simulation we use the number of cells $N = 200$, space step $\Delta x = 0.1$, time step $\Delta t = 0.005\Delta x$, gravitational constant $g = 1$. The simulation is stopped at $t = 1$. We assume the initial for fluid depth is $h_1 = 10$ for all negative values of $x$and $h_0 = 5$ for all positive values of x. The initial condition for fluid velocity is $u(x,0) = 0$, for all x. We set $L = 10$. Boundary conditions at $x = -L$ are

$$h(-L,t) = 10, \qquad u(-L,t) = 0$$

and those at $x = L$ are

$$h(L,t) = 5, \qquad u(L,t) = 0.$$

IOP Conf. Series: Journal of Physics: Conf. Series **1007** (2018) 012008 doi:10.1088/1742-6596/1007/1/012008

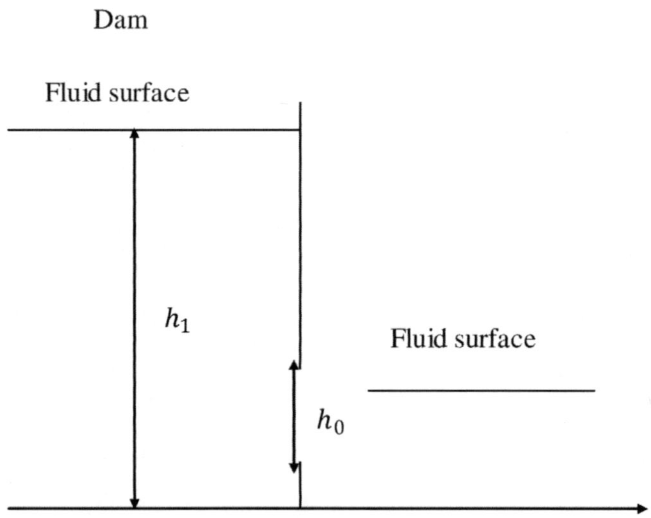

Figure 3. Initial water surface for the dam break problem.

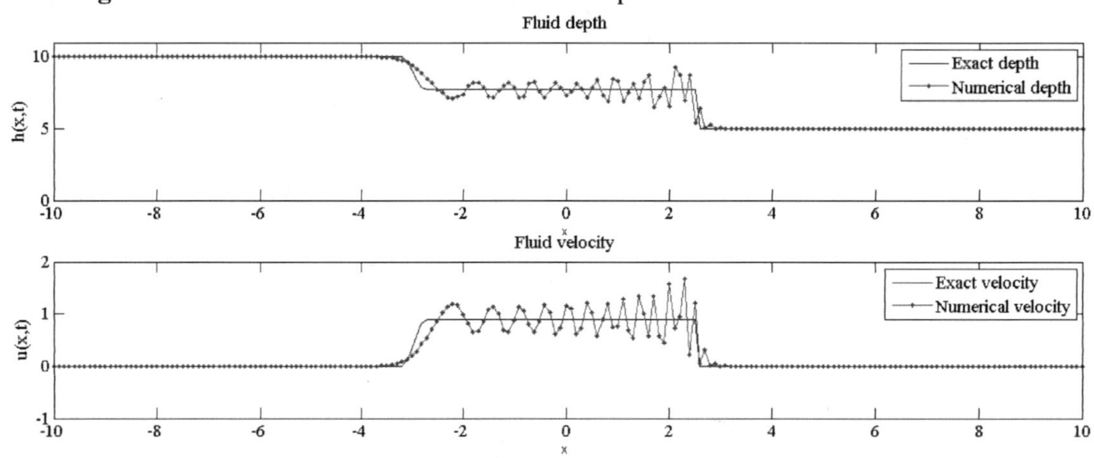

Figure 4. Analytical (exact) and numerical solutions for the dam break problem.

The initial condition of water surface for the dam break problem is given in Figure 3. There is discontinuity in the water depth. At time $t = 0$, we assume that the dam wall is completely removed. The simulation is conducted to observe how the Lax-Wendroff finite volume solution behaves for this discontinuous problem.

This test problem has been solved analytically by Martins et al [1]. We use the analytical (exact) solution to investigate the performance of the Lax-Wendroff finite volume method. We find that in comparison with the analytical solution, for this discontinuous problem, the Lax-Wendroff finite volume method produces artificial oscillation (see Figure 4), no matter how small we take the time step value. This is due to the order of accuracy of the method, which is larger than one. We do not implement any flux or slope limiter in the numerical method.

4. Conclusion

The Lax-Wendroff finite volume method has been used to solve the gravity wave-model equations. The method is a standard method without any flux or slope limiter. We obtain that the method can be solved successfully for a smooth problem, but produces artificial oscillations in solving a discontinuous problem. Future research direction is to seek a way to alleviate artificial oscillations in solving discontinuous problems.

References

[1] Martins R, Leandro J and Djordjević S 2016 Analytical solution of the classical dam-break problemfor the gravity wave-model equations Journal of Hydraulic Engineering**142**06016003
[2] Aronica G, Tucciarelli T and NaselloC 1998 2D multilevel model for flood wave propagation in flood-affected areas Journal of Water Resources Planning and Management**124**210
[3] Bates P D, Horritt M S, Fewtrell T J 2010 A simple inertial formulation of the shallow water equations for efficient two-dimensional flood inundation modelling Journal of Hydrology**387** 33
[4] de Almeida GA M, Bates P, Freer J E and Souvignet M 2012 Improving the stability of a simple formulation of the shallow water equations for 2-D flood modelingWater Resources Research**48** W05528
[5] Seyoum S D, Vojinovic Z, Price R K and Weesakul S 2012 Coupled 1D and noninertia 2D flood inundation model for simulation of urban floodingJournal of Hydraulic Engineering**138** 23
[6] Chung W and Kang Y 2006 Classifying river waves by the Saint Venant equations decoupled in the Laplacian frequency domain Journal of Hydraulic Engineering**132**666
[7] LeVeque R J, 2004 Finite Volume Methods for Hyperbolic Problems (Cambridge: Cambridge University Press)
[8] LeVeque R J, 1992Numerical Methods for Conservation Laws (Berlin:Birkhäuser)
[9] Bürger R, Kenettinkara S K and Zorío D 2017 Approximate Lax–Wendroff discontinuous Galerkin methods for hyperbolic conservation lawsComputers & Mathematics with Applications**74** 1288
[10] Fridrich D, Liska R and Wendroff B 2017 Cell-centered Lagrangian Lax–Wendroff HLL hybrid method for elasto-plastic flowsComputers & Fluids**157** 164
[11] Lee K, Leong Z Q and Nguyen H D 2017 Characterisation of waves via the Lax–Wendroff methodJournal of Ocean Engineering and Marine Energy**3** 247
[12] Li T, Shu C-W and Zhang M 2017 Stability analysis of the inverse Lax–Wendroff boundary treatment for high order central difference schemes for diffusion equationsJournal of Scientific Computing**70** 576
[13] Manshoor B, Khalid A, Sapit A, Zaman I and Nizam A 2017 Numerical solution of Burger's equation based on Lax-Friedrichs and Lax-Wendroff schemesAIP Conference Proceedings**1831** 020025
[14] Shu C-W 2017 Runge-Kutta and Lax-Wendroff discontinuous Galerkin methods for linear conservation lawsAIP Conference Proceedings**1863** 320010
[15] Zorío D, Baeza A and Mulet P 2017 An approximate Lax–Wendroff-type procedure for high order accurate schemes for hyperbolic conservation lawsJournal of Scientific Computing**71** 246
[16] Apriani C M and Mungkasi S 2017 A staggered grid finite difference method for solving the gravity wave-model equationsJournal of Physics: Conference Series**909** 012046
[17] Martins R, Leandro J and Djordjević S 2015 A well balanced Roe scheme for the local inertial equations with an unstructured meshAdvances in Water Resources**83** 351

Acknowledgements

Both authors thank Sanata Dharma University for the financial support to this work.

IOP Conf. Series: Journal of Physics: Conf. Series **1007** (2018) 012009 doi:10.1088/1742-6596/1007/1/012009

Bellman Ford algorithm-in RoutingInformation Protocol (RIP)

Oris Krianto Sulaiman[1*], Amir Mahmud Siregar[2], Khairuddin Nasution[1], Tasliyah Haramaini[1]

[1]Faculty of Engineering, Universitas Islam Sumatera Utara, Indonesia
[2]Faculty of Technology and Computer Science, Universitas Prima Indonesia, Indonesia

E-mail: *oris.ks@ft.uisu.ac.id, amirmahmud@unprimdn.ac.id, khairuddin_nst@uisu.ac.id, tasliyah@ft.uisu.ac.id

Abstract. In a large scale network need a routing that can handle a lot number of users, one of the solutions to cope with large scale network is by using a routing protocol, There are 2 types of routing protocol that is static and dynamic, Static routing is manually route input based on network admin, while dynamic routing is automatically route input formed based on existing network. Dynamic routing is efficient used to network extensively because of the input of route automatic formed, Routing Information Protocol (RIP) is one of dynamic routing that uses the bellman-ford algorithm where this algorithm will search for the best path that traversed the network by leveraging the value of each link, so with the bellman-ford algorithm owned by RIP can optimize existing networks.

1. Introduction

Needs of people by use the technology gradually increasing so that it is reasonably necessary to the development of technology enhanced. Almost all of this technology is now using the internet access as communication media, good internet access and efficient can increase performance of the system that used the technology. For time efficiency and workmanship at the time of application usage on the network then need an optimal network performance. Network with large scale is usually between one and the other are not directly connected, but it went through some other network devices. A good design and architecture is necessary in anticipation of problems that may occur when the network traffic is solid. In heavy of network traffic may need to data packets can reach the destination by using the best line from every kind of path. Bellman-Ford algorithm is one of the best routes search algorithm utilizing the value of each existing link, this algorithm is used in routing protocol RIP, metric is hop i.e. value of each router are in her past. Routing Information Protocol (RIP) is a dynamic routing protocol where the entry route will automatically be formed if there are new routers that are not yet known. Routers connected usingRouting Information Protocol (RIP) will not suffered a looping because one of the links will be used as a backup link. By using the Bellman-Ford algorithm, then the best path that will be going through the package data will be achieved so that it can optimize existing networks

IOP Conf. Series: Journal of Physics: Conf. Series **1007** (2018) 012009 doi:10.1088/1742-6596/1007/1/012009

2. Review of literature

2.1 *Bellman-Ford Algorithm*

Bellman-Ford algorithm is a graph search algorithm for searching shortest route. Bellman-ford algorithm can find negative weights from each edge. In Figure 1 is an example of the bellman-ford algorithm, where there are 5 vertex i.e. A, B, C, D, and E. Each edge has a weight value i.e. directed edge A toward B is -3 and the other edge is one. N vertices Search iterations N – 1, in the example images there are 5 vertices so that there are 4 iteration, But if there's a vegative cyle or value weights are negative then it should be in iteration again called Nth [1][2][3].

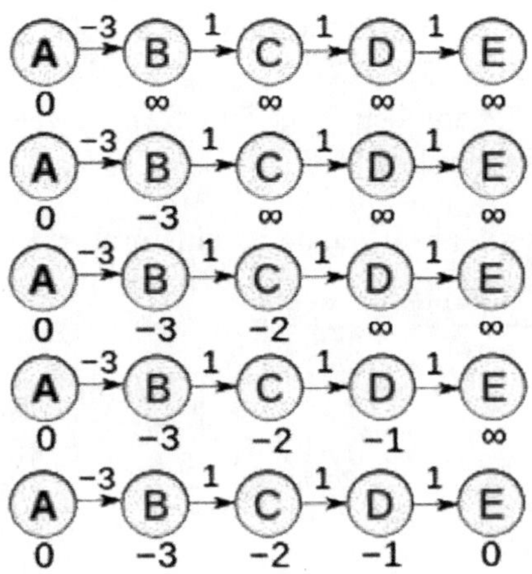

Figure 1. Example of Bellman-Ford algoritm.

As for the source code from the Bellman-ford algorithm is as follow:

```
function BellmanFord(list vertices, list edges, vertex source)
   ::distance[],predecessor[]
// Step 1: initialize graph
for each vertex v in vertices:
       distance[v] := inf
       predecessor[v] := null
   distance[source] := 0
// Step 2: relax edges repeatedly
for i from 1 to size(vertices)-1:
for each edge (u, v) with weight w in edges:
if distance[u] + w < distance[v]:
               distance[v] := distance[u] + w
               predecessor[v] := u
// Step 3: check for negative-weight cycles
for each edge (u, v) with weight w in edges:
if distance[u] + w < distance[v]:
error "Graph contains a negative-weight cycle"
return distance[], predecessor[]
```

2.2 Routing Information Protocol

The process of connecting two or more different network so-called routing. A device that connects two different network is called with a router. Each internet network requires protocol to process data communication, router has a protocol with routing protocol. Routing protocol served to regulate data

IOP Conf. Series: Journal of Physics: Conf. Series **1007** (2018) 012009 doi:10.1088/1742-6596/1007/1/012009

traffic in a network and can search for the best path for each of the data points in the network. Each Routing protocol has an entry route i.e. database storage route that will pass, entry route lies in the routing table that lists ip address all of the objectives that can it traversed.[4]
A Routing Protocol is divided into two, Static routing and Dynamic routing. Static routing is the process of connecting two or more different network with entry routes manually input, While the dynamic route process connects two or more different network where entry routes in the routing table is formed automatically.

There are several types of routing protocol, including:

1. Routing Information Protocol (RIP)
2. Interior Gateway Routing Protocol (IGRP)
3. Enhanced Interior Gateway Routing Protocol (EIGRP)
4. Open Shortest Path Firs (OSPF)
5. Intermediate System-to-Intermediate System (IS-IS)
6. Border Gateway Protocol (BGP)

Figure 2 the following is a division between the routing protocol:

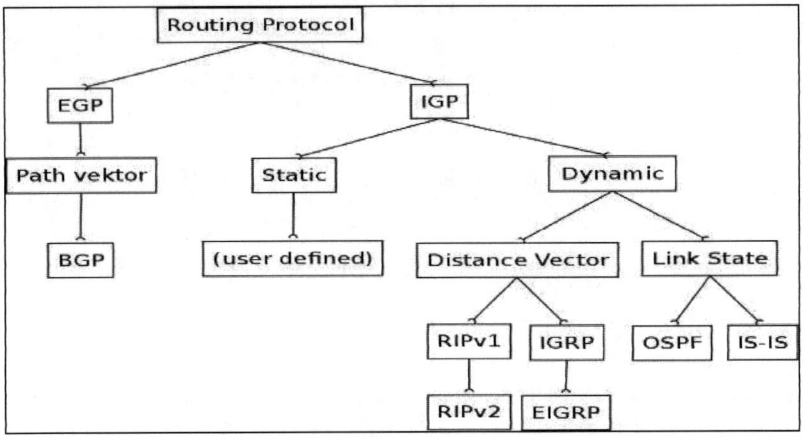

Figure 2. Types Of Routing Protocol.

Routing Information Protocol or collectively RIP is the routing protocol with dynamic types using the bellman-ford algorithm for searching shortest path. Metric of the RIP is using hop count i.e the number of routers that kind. Every router that is passed then the hop count will be calculated 1. In the RIP, the vertex is router, edge is the link, and weight of edge is the number of hop count from each track path. the number of Count hops that can be traversed routing protocol RIP is 15. [5][6]

3. Design of network

This research will use the 2 network topology, the first topology is trials for the bellman-ford algorithm and the second topology is a test for the Routing Information Protocol (RIP). Figure 3 and table 1 is the topology and addressing table for testing the network by using the Routing Information Protocol (RIP). Trials will be carried out by means of simulation, simulation program used is cisco packet tracer.

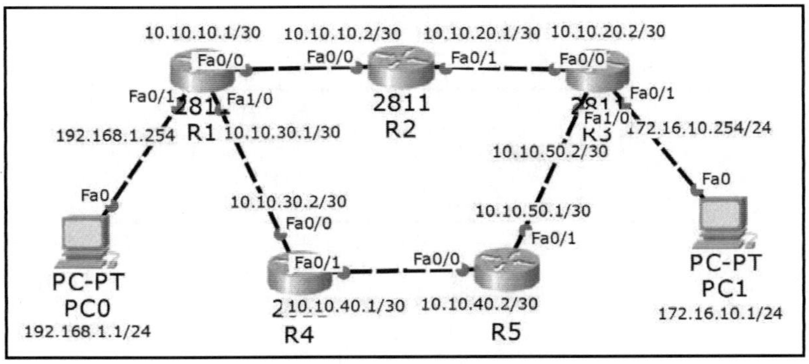

Figure 3. Topology Design.

Table 1. Addressing.

Device	Interface	IP address	Subnet Mask	Default Gateway
R1	Fa 0/0	10.10.10.1	255.255.255.252	N/A
	Fa 0/1	192.168.1.254	255.255.255.0	N/A
	Fa 1/0	10.10.30.1	255.255.255.252	N/A
R2	Fa 0/0	10.10.10.2	255.255.255.252	N/A
	Fa 0/1	10.10.20.1	255.255.255.252	N/A
R3	Fa 0/0	10.10.20.2	255.255.255.252	N/A
	Fa 0/1	172.16.10.254	255.255.255.0	N/A
	Fa 1/0	10.10.50.2	255.255.255.252	N/A
R4	Fa 0/0	10.10.30.2	255.255.255.252	N/A
	Fa 0/1	10.10.40.1	255.255.255.252	N/A
R5	Fa 0/0	10.10.40.2	255.255.255.252	N/A
	Fa 0/1	10.10.50.1	255.255.255.252	N/A
PC0	NIC	192.168.1.1	255.255.255.0	192.168.1.254
PC1	NIC	172.16.10.1	255.255.255.0	172.16.10.254

4. Results and discussion

4.1 *Bellman-Ford Testing*

Discussion of using 3 router as vertex, i.e. R1, R2 and R3, Search route u = 1. Each edge has a value does it weigh each which have been listed in table 1.

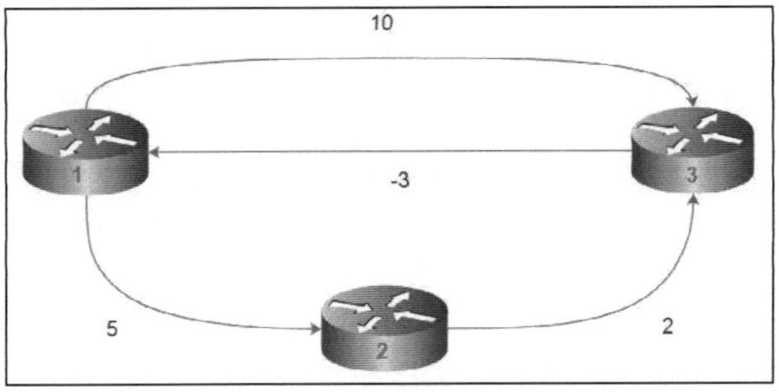

Figure 4. Vertex and weighting value.

Table 2. Weighting value.

Edge	Weight
1 – 2	10
1 – 3	5
2 – 1	- 3
3 – 2	2

Relax (u,v):
If v.d > u.d + w (u,v)
v.d = u.d + w (u,v)
v.p = u

ITERATION I

Table 3. Array d and p.

	1	**2**	**3**
D	0	∞	∞

	1	**2**	**3**
P	-	-	-

Calculate the edge 1 – 2
u.d = 1.d = 0
v.d = 2.d = ∞
w (u,v) = w (1,2) = 10

Relax (u,v):
2.d> 1.d + w (1,2)
∞> 0 + 10
TRUE
v.d = 2.d = 0 + 10 = 10
v.p = 2.p = 0

Table 4. Results of calculation edge 1 – 2.

	1	**2**	**3**
D	0	10	∞

	1	**2**	**3**
P	-	0	-

Calculate the edge1 – 3
u.d = 1.d = 0
v.d = 3.d = ∞
w (u,v) = w (1,3) = 5

Relax (u,v):
3.d> 1.d + w (1,3)
∞> 0 + 5
TRUE
v.d = 3.d = 0 + 5 = 5
v.p = 3.p = 0

Table 5. Results of calculation edge 1 – 3.

	1	2	3
d	0	10	5

	1	2	3
p	-	0	0

Calculate the edge2 – 1
u.d = 2.d = 10
v.d = 1.d = 0
w (u,v) = w (2,1) = -3

Relax (u,v):
1.d> 2.d + w (2,1)
0 > 10 + -3
ELSE
Next Edge

Calculate the edge3 – 2
u.d = 3.d = 5
v.d = 2.d = 10
w (u,v) = w (3,2) = 2

Relax (u,v):
2.d> 3.d + w (3,2)
10 > 5 + 2 = 7
TRUE
v.d =2.d = 5 + 2 = 7
v.p = 2.p = 5

Table 6. Results of calculation edge 3 – 2.

	1	2	3
D	0	7	5

	1	2	3
p	-	5	0

ITERATION II
Calculate the Edge 1 – 2
u.d = 1.d = 0
v.d = 2.d = 7
w (u,v) = w (1,2) = 10

Relax (u,v):
1.d> 2.d + w (2,1)
0 > 17
FALSE

Calculate the edge 1 – 3
u.d = 1.d = 0
v.d = 3.d = 5
w (u,v) = w (1,3) = 5

Relax (u,v):

IOP Conf. Series: Journal of Physics: Conf. Series **1007** (2018) 012009 doi:10.1088/1742-6596/1007/1/012009

3.d> 1.d + w (1,3)
5> 0 + 5
FALSE
Next Edge

Calculate the edge 2 – 1
u.d = 2.d = 7
v.d = 1.d = 0
w (u,v) = w (2,1) = -3

Relax (u,v):
1.d> 2.d + w (2,1)
0 > 7 + -3
FALSE
Next Edge

Calculate the edge 3 – 2
u.d = 3.d = 5
v.d = 2.d = 7
w (u,v) = w (3,2) = 2

Relax (u,v):
2.d> 3.d + w (3,2)
7> 5 + 2 = 7
FALSE

Final result :

Table 7. Result of last iteration.

	1	2	3
d	0	7	5

	1	2	3
p	-	5	0

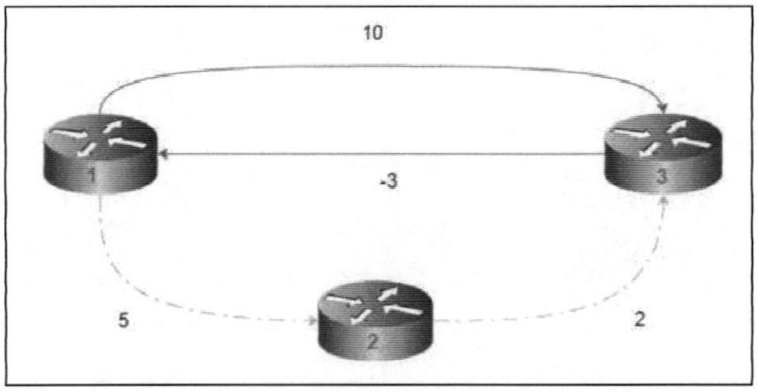

Figure 5. The shortest path is obtained.

Shortest path traveled from vertex 1 to other vertex is weighting value 5 and 2.

4.2 *RIP Testing*
Bellman-ford algorithm used in RIP will be on implementation of topology in Figure 6 below:

IOP Conf. Series: Journal of Physics: Conf. Series **1007** (2018) 012009 doi:10.1088/1742-6596/1007/1/012009

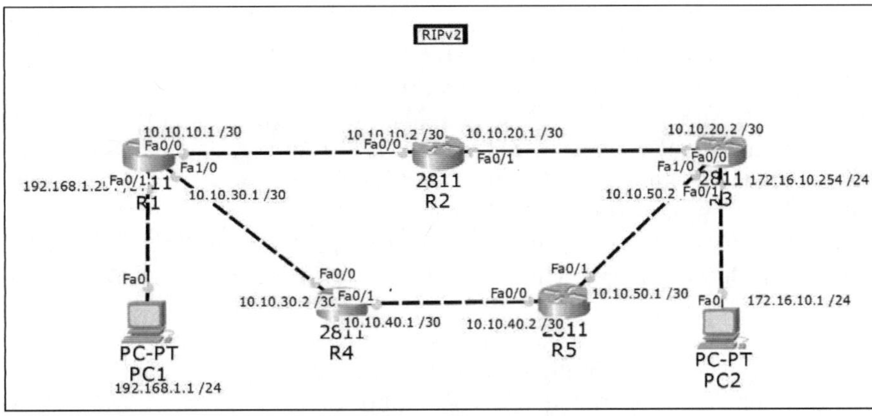

Figure 6. Shortest path.

The image of R1, R2, R3, R4 and R5 as the vertex and each link (edge) have metric i.e. hop count on RIP is the weights value from each edge.

R1#show ip route rip
10.0.0.0/30 is subnetted, 5 subnets
R 10.10.20.0 [120/1] via 10.10.10.2, 00:00:23, FastEthernet0/0
R 10.10.40.0 [120/1] via 10.10.30.2, 00:00:13, FastEthernet1/0
R 10.10.50.0 [120/2] via 10.10.30.2, 00:00:13, FastEthernet1/0
[120/2] via 10.10.10.2, 00:00:23, FastEthernet0/0
172.16.0.0/24 is subnetted, 1 subnets
R 172.16.10.0 [120/2] via 10.10.10.2, 00:00:23, FastEthernet0/0
From R1 look his pass routes to reach across a network is:

- From R1 to network 10.10.20.0 through ip10.10.10.2 (R2) with the number ofhop 1.
- From R1to network 10.10.40.0 throughip10.10.30.2 (R4) with the number ofhop 1.
- From R1 to network 10.10.50.0 through ip 10.10.30.2 (R4) with the number ofhop 2.
- From R1 to network 172.16.10.0 through ip 10.10.30.2 (R4) with the number ofhop 2.

From the trials obtained the optimum results to achieve throughout the network with routing protocol RIP which uses the Bellman-ford algorithm.

5. Conclusions and suggestions

The following is the conclusion from the results of this research are:

1. Bellman-Ford algorithm is used by the routing protocol RIP to search the best path in a network.
2. Weighting value of routing protocol RIP by default is worth 1.
3. The maximum number of vertices in the routing protocol RIP is 15.

The following is a suggestion from the results of this research are:

1. Can be expanded by using the others shortest path search algorithm.
2. Can be developed by using the others routing protocol.

Reference

[1] S. Tayeb and S. Latifi, "An Evaluative Analysis of DUAL, SPF, and Bellman-Ford," Journal of Software Networking, pp. 1-22, 2017

IOP Conf. Series: Journal of Physics: Conf. Series **1007** (2018) 012009 doi:10.1088/1742-6596/1007/1/012009

[2] A. A. P, S. H. Pramono and A. M. Muslim, "Optimasi Jalur Tercepat dengan Menggunakan Modifikasi Algoritma Bellman Ford (Studi Kasus Lintasan antar Kecamatan Kota Malang)," Jurnal EECCIS , vol. 9, no. 2, pp. 168-172, 2015
[3] F. Anggraini and S. Mingparwoto, "Penerapan Metode Algoritma Bellman – Ford Dalam Aplikasi Pencarian Lokasi Perseroan Terbatas Di Pt.Jakarta Industrial Estate Pulogadung (PT.JIEP)," Jurnal Teknologi, vol. 7, no. 1, pp. 28-34, 2015
[4] R. Towidjojo, Mikrotik Kung Fu Kitab 4, Jakarta: Jasakom, 2016
[5] M. N. Shah, "Implementation of Graph Theory in Computer Networking To Find an Efficiect Routing Algorithm," International Journal of Innovative Research in Computer and Communication Engineering, vol. 4, no. 1, pp. 12-20, 2016
[6] A. Verma and N. Bhardwaj, "A Review on Routing Information Protocol (RIP) and Open Shortest Path First (OSPF) Routing Protocol," International Journal of Future Generation Communication and Networking, vol. 9, no. 4, pp. 161-170, 2016

Image Steganography In Securing Sound File Using Arithmetic Coding Algorithm, Triple Data Encryption Standard (3DES) and Modified Least Significant Bit (MLSB)

A B Nasution[1*], S Efendi[1], and S Suwilo[2]

[1]Faculty of Computer Science and Information Technology Universitas Sumatera Utara, Medan, Indonesia
[2]Department of Mathematics, Universitas Sumatera Utara, Medan, Indonesia

E-mail : *buyung.cool@gmail.com

Abstract. The amount of data inserted in the form of audio samples that use 8 bits with LSB algorithm, affect the value of PSNR which resulted in changes in image quality of the insertion (fidelity). So in this research will be inserted audio samples using 5 bits with MLSB algorithm to reduce the number of data insertion where previously the audio sample will be compressed with Arithmetic Coding algorithm to reduce file size. In this research will also be encryption using Triple DES algorithm to better secure audio samples. The result of this research is the value of PSNR more than 50dB so it can be concluded that the image quality is still good because the value of PSNR has exceeded 40dB.

1. Introduction

Delivery of data or messages from one place to another is plagued with confidential issues. Many ways can be done to hide data or messages to be sent. The first by using cryptographic techniques, namely by encoding data/messages by using a particular algorithm. Another technique with steganography techniques is the art of hiding the existence of the message so as not to attract attention to secret messages, then third parties or illegal people can't detect the message [1]. Steganography secures data with an insertion object in the form of audio, images and video and is considered to be a complement from cryptography [2]. In general, digital audio files have a size (size) larger than the text file. So to reduce its size can be done by way of compression. Audio files with WAV format have uncompressed data so that all audio samples are stored all on the hard drive (not coding). Moreover, the concealment of audio files has its own challenges because human hearing is more sensitive than vision [3].

Least Significant Bit algorithm can be used to hide data to cover like image, audio or video. However, LSB requires 8 bits of data to store 1 byte of data [4]. This method is simple method and easy to implement steganography concept [5].The amount of data inserted in the form of audio samples that use 8 bits with LSB algorithm. The impact is a low PSNR value and resulted alteration of image quality from the insertion result. [6]. In Kaul's research in 2013 has used Wavelet Transform and LSB (Least Significant Bit) methods. The results of this study indicate that the value of Peak Signal to Noise Ratio (PSNR) decreases as the number of votes increases so that there is a change in image quality of the insertion result [7]. So in this research will be inserted audio samples using 5 bits with MLSB algorithm to reduce the number of data insertion where previously the audio sample will

IOP Conf. Series: Journal of Physics: Conf. Series **1007** (2018) 012010 doi:10.1088/1742-6596/1007/1/012010

be compressed with Arithmetic Coding algorithm to reduce file size. In this research will also be encryption using Triple DES algorithm to better secure audio samples.

2. Audio Files

Wav file format (Waveform audio format) is one format used to store sound in the frequency range 20 Hz to 20 kHz. File with wave format using pulse code modulation (PCM) method to store analog voice into digital data on the computer. PCM is one way of representing analog data in the digital form where the analog signal data is sampled at each interval of a certain period and then converted into a value on a digital system [8].

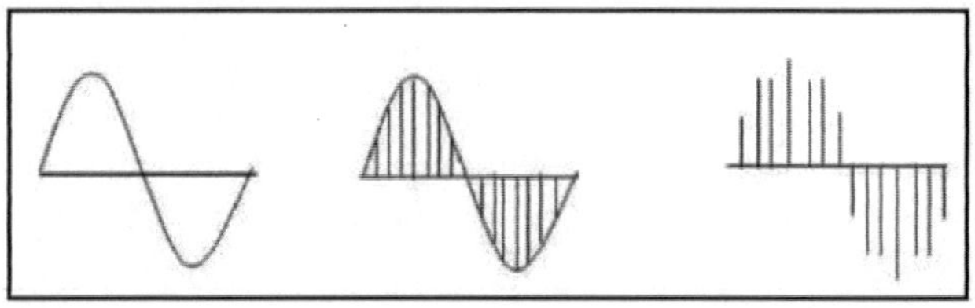

Figure 1. Analog Waves Turned into Digital Waves [9].

3. Arithmetic Coding Algorithm

Compression with Arithmetic Coding algorithm is by replacing one row of input symbols with a floating point number. The output of the encoding Arithmetic Coding is a number smaller than the number 1 and greater or equal to 0 [10]. To generate the output number, each symbol to be encoded is assigned a set of probability values. Once the probability of each character is known, each symbol / character will be given a certain range whose value is between 0 and 1 [11]. For example, suppose that in the audio sample "00 3e 1f 00 9a 00 1f 9a 00 3e 00 1f 00 3e 1f" will be encoded, we will get a probability table like Table 1.

Table 1. Table of Probability Ranges.

Character	Frequency	Probability	Range
00	6	6/15=0,4	0 ≤ 00 <0,4
1f	4	4/15=0,3	0,4 ≤ 1f <0,7
9a	2	2/15=0,1	0,7 ≤ 9a <0,8
3e	3	3/15=0,2	0,8 ≤ 3e <1,0

One thing to note from this table is that each character covers the mentioned range, except for a high number. Thus, the symbol "3e" actually has a range from 0.8 to 0.9999 ... Furthermore, to perform the encoding process, the following algorithm is used.

1. Set low = 0,0
2. Set high = 1.0
3. While do
4. Take the input symbol.
5. CR = high – low.
6. High = low + CR*high_range
7. Low = low + CR*low_range
8. End while
9. Print *low*

Here "low" is the output of the arithmetic coding process.

4. Triple Data Encryption Standard

Triple Data Encryption Standard (3DES) is a development algorithm of DES (Data Encryption Standard) algorithm. Basically, the algorithm used the same, only in 3DES developed by performing encryption with DES algorithm implementation three times [12]. 3DES has three 168-bit keys (three 56-bit keys from DES). In the 3DES algorithm is divided into three stages, each stage is an implementation of the DES algorithm. The first stage, the inputted plaintext is operated with the first external key (K1) and performs the encryption process using the DES algorithm. So as to produce the first pre ciphertext. The second stage, the first pre-ciphertext generated in the first stage, is then operated with a second external key (K2) and performs the encryption or decryption process (depending on the way the encryption is used) using the DES algorithm. So as to produce a second pre ciphertext. The final stage, the second pre ciphertext generated in the second stage, is operated with the third external key (K3) and performs the encryption process using the DES algorithm, resulting in ciphertext (C).

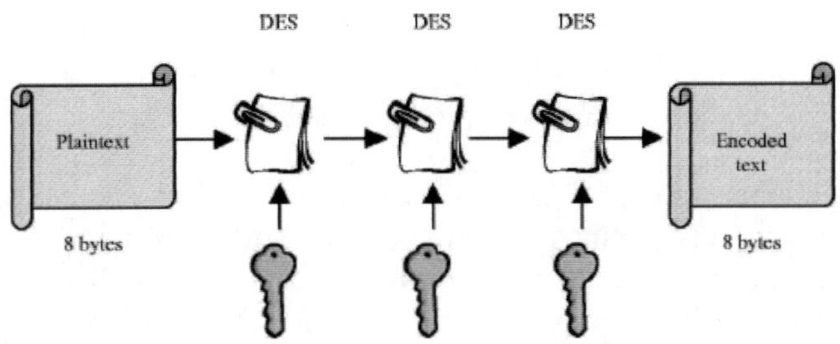

Figure 2. 3DES Algorithm.

5. Modified Least Significant Bit

Modified Least Significant Bit (MLSB) or modification of the LSB algorithm is used to encode an identity into a file. MLSB uses the manipulation of several levels of insert bits before encoding the message. Modify messages with MLSB algorithm where message bits that should have 1 character have an 8 bit ASCII code will be modified to 5 bits. Each message character will be treated with their 8-bit ASCII code, and then the code will be converted to 5-bit code using the MLSB technique. [13] In this algorithm characters and numbers are represented in 5 bits which will then be inserted into the original file by technique LSB [14].

6. Fidelity Measurements

Measurement of steganography fidelity can be calculated by calculating the value of MSE (Mean Squared Error) and PSNR (Peak Signal to Noise Ratio). PSNR is the ratio between the maximum values of the image as measured by the magnitude of noise affecting the image. PSNR is usually measured in decibels. PSNR is used to determine the quality (validation) file processing. To determine PSNR, we must first determine the mean value of the square of the error (MSE - Mean Square Error). The smaller the MSE the better. The larger the PSNR parameter means the more similar it is to the original image. The image with the value of PSNR> 40 dB can be said to have good quality [15].

MSE and PSNR can be calculated by equations (1.1) and (1.2). In equation (2.1), I (i, j) is the image sample value at position (i, j), I 'is the value of the image sample value inserted at positions (i, j), X and Y are image file sizes. In equation (2.2), m is the maximum sample value that may be possessed by audio [16].

$$\text{MSE} = \frac{1}{XY}\sum_x \sum_x [I(x,y) - I'(x,y)]^2$$

$$\text{PSNR} = 10 \log \frac{m^2}{MSE}$$

7. Proposed Method

The proposed method can be seen as in Figure 2.

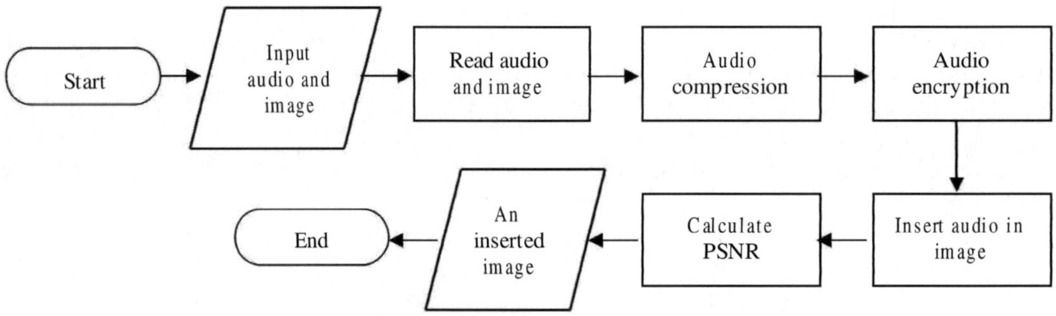

Figure 3. Proposed Method.

In Figure 3 is a general description of the audio insertion step into the image. The first step is to do audio and image where the audio input format. Wav and image format .jpg. Then the program will read the audio and the input image to get the data. The next audio data will be compressed using Arithmetic Coding algorithm. The compressed audio is then encrypted with 3DES to obtain ciphertext. The obtained ciphertext is then inserted into the image file with the MLSB algorithm. The image that has been inserted audio then calculated the value of PSNR to obtain optimization results. Furthermore, the image that has been inserted audio and PSNR value is displayed.

8. Result and Discussion

This research has 3 performance that is performance arithmetic coding algorithm, 3DES and MLSB. The performance result of arithmetic coding algorithm can be seen in table 2.

Table 2. Performance of Arithmetic Coding Algorithm.

File Audio	File Size	Size After Compress	Ratio	Time
Flute.wav	48 KB	47 KB	11.09%	0.033 s
Baby.wav	117 KB	107 KB	8.65%	0.045 s
SMS.wav	246 KB	236 KB	4.41%	0.075 s
Police.wav	356 KB	341 KB	4,29%	0.097 s
Bird.wav	471 KB	436 KB	1.56 %	0,110 s

In table 2 shows that the highest ratio value of 11.09% with time 0.033s and the lowest ratio value of 1.56% with time of 0.110s. So it can be collected that the bigger the audio file size the smaller the ratio. However, the larger the audio file size, the greater the time it takes. The result of the compressed audio file will be the encryption and shown in table 3.

Table 3. Performance of 3DES.

File Audio	File Size	Size After Encryption	Time
Kom_Flute.wav	47 KB	47 KB	1 s
Kom_Baby.wav	107 KB	107 KB	1 s
Kom_SMS.wav	236 KB	236 KB	2 s
Kom_Police.wav	341 KB	341 KB	2 s
Kom_Birds.wav	436 KB	436 KB	2 s

Table 3 shows that the highest time is 2s and the lowest is 1s. In this table also shows that the file size has no effect in the encryption process. Then the encrypted file will be inserted, the insertion result is shown in table 4.

Table 4. Performance of MLSB.

File Audio	File Size	File Image	File size	Size After Inserted	MSE	PSNR (dB)
Kom_Flute.wav.3des	47 KB	Beach.jpg	49 KB	92 KB	4.480	58.382
Kom_Baby.wav.3des	107 KB	Palm.jpg	127 KB	223 KB	1.159	52.475
Kom_SMS.wav.3des	236 KB	Tree.jpg	179 KB	396 KB	6.749	60.162
Kom_Police.wav.3des	341 KB	Pond.jpg	504 KB	845 KB	6.649	60.096
Kom_Birds.wav.3des	436 KB	Mountain.jpg	530 KB	992 KB	3.411	57.198

In Table 4 it shows that the size of the inserted image file rises close to the number of audio file sizes plus the image file size, it indicates that the audio is inserted into the image file. In the table also shows that the PSNR value is more than 50dB with the lowest PSNR value of 52.475 and the highest PSNR value is 60.162 so it can be concluded that the picture quality is very good because the PSNR value is greater than 35dB.

9. Conclusion

The conclusion in this research are as follows.

1. In the process of arithmetic coding algorithm, the larger the audio file size the higher the time. However, inversely proportional to the ratio, the larger the audio file size the lower the ration value. This can be seen in Table 2 where the smallest audio file size has a high ratio of 11.09% with the lowest time of 0.033s. While the largest audio file size has the lowest ratio of 1.56% with the highest time of 0.110s.
2. In the 3DES process, the time required for encryption is 1s to 2s and in this process the size of the audio file does not change in size.
3. In the MLSB process, the PSNR value is more than 50dB with the lowest PSNR value of 52.475 and the highest PSNR value is 60.162 so it can be concluded that the picture quality is very good because the PSNR value is greater than 40dB

Future work, the expected testing process can be done by testing the reliability of the insertion algorithm (robustness) with the addition of noise into the image steganography and extraction of audio data again.

References

[1] Dilawar, A. 2014. Image Steganography: For Hiding Audio Messages Within Grayscale Images Using LSB, DCT And AES Algorithm. *Thesis.* International Islamic University, Sector H-10 : Islamabad

[2] Kumar A. & Pooja, K. 2010. Steganography- A Data Hiding Technique. *International Research of Computer Aplication.* pp. 19.23

[3] Cvejic. N. 2004. Algorithms for Audio Watermarking and Steganography. *Linier Networks and Systems (Book Style).* pp. 123-135

[4] Rachmawati, D., Amalia, A. and Surya, J. 2016. Combination of Huffman Coding Compression Algorithm and Least Significant Bit for Image Hiding. *International Conference on Computing and Applied Informatics.* pp. 1-4

[5] Kekre, H.B., Athawale, A. & Halarnkar, P.N. 2008. Increased Capacity of Information Hiding in LSB'S Method for Text and Image. *International Journal of Computer, Electrical, Automation, Control and Information Engineering*. pp. 1497-1500

IOP Conf. Series: Journal of Physics: Conf. Series **1007** (2018) 012010 doi:10.1088/1742-6596/1007/1/012010

[6] Por, L.Y., Beh, D., Ang, T.F. & Ong, S.Y. 2013. An Enhanced Mechanism for Image Steganography Using Sequential Colour Cycle Algorithm. *The International Arab Journal of Information Technologi.* pp. 51-60

[7] Kaul, N. and Bajaj, N. 2013. Audio in Image Steganography based on Wavelet Transform. *International Journal of Computer Application.* pp. 7-10

[8] Husain, M.P. & Rafat, K.F. 2016. Enhanced Audio LSB Steganography for Secure Communication. *International Journal of Advanced Computer Science and Applications*. pp. 340-347

[9] Kumar, S., Barnali, B. & Banik, G. 2012. LSB Modification and Phase Encoding Technique of Audio Steganography Revisited. *International Journal of Advanced Research in Computer and Communication Engineering.* pp . 1-4

[10] Habib, A. & Chowdhury, D. 2015. An Efficient Compression Technique Using Arithmetic Coding. *Journal of Scientific Research & Reports 4 (1).* pp. 60-67

[11] Sentilkumar, M. & Mathivanan, V. 2016. Analysis of Data Compression Techniques using Huffman Coding and Arithmetic Coding. *International Journal of Advanced Research in Computer Science and Software Engineering.* pp. 930-936

[12] Kumar, P. & Rajaanadan, N.S. 2016. Data Encryption and Decryption Using By Triple DES Performance Efficiency Analysis of Crypto System. *International Journal of Innovative Research in Computer and Communication* Engineering. pp. 4030-4040

[13] Shreya, M.S. & Khumar, S. 2013. Separable Reversible Data Hiding In Encrypted Image Using Modified Least Significant Bit and Virtual Embedding. *International Journal of Science and Research.* pp. 3099-3106

[14] Zaher, M.A. 2011. Modified Least Significant Bit (MLSB*). Computer and Information Science.* pp. 60-67

[15] Laskar, S.A. & Hamachandran, K. 2012. High Capacity Data Hiding Using LSB Steganography and Encryption. *International Journal of Database Management System.* pp. 57-68

[16] Prasetiyo, B. 2013. Kombinasi Steganografi Bit Matching dan Kriptografi DES untuk Pengamanan Data. *Thesis*. Universitas Diponegoro : Semarang

IOP Conf. Series: Journal of Physics: Conf. Series **1007** (2018) 012011 doi:10.1088/1742-6596/1007/1/012011

Variational estimate method for solving autonomous ordinary differential equations

Sudi Mungkasi

Department of Mathematics, Faculty of Science and Technology, Sanata Dharma University, Mrican, Tromol Pos 29, Yogyakarta 55002, Indonesia

E-mail: sudi@usd.ac.id

Abstract. In this paper, we propose a method for solving first-order autonomous ordinary differential equation problems using a variational estimate formulation. The variational estimate is constructed with a Lagrange multiplier which is chosen optimally, so that the formulation leads to an accurate solution to the problem. The variational estimate is an integral form, which can be computed using a computer software. As the variational estimate is an explicit formula, the solution is easy to compute. This is a great advantage of the variational estimate formulation.

1. Introduction

A number of methods are available in the literature for solving ordinary differential equations, either autonomous or non-autonomous [1-5]. Some analytical methods are available in the work of Haberman [1]. Variational iteration methods involving Lagrange multiplier were discussed by He and Wu [2], Inokuti et al. [3], Mungkasi et al. [4-5], and still many others. In this paper, we shall solve first-order autonomous ordinary differential equations of the form

$$x'(t) + f(x) = 0 \tag{1}$$

with the initial condition

$$x(0) = \alpha. \tag{2}$$

We limit our discussion to one-dimensional problems with variable x is dependent on variable t. The discussion could be extended to higher dimensions, but it is out of the scope of this paper. Some real-world problems are governed by equations (1) with the initial condition (2), such as investment problems and the dynamics of population of a species [1]. Therefore, finding a solving method for this problem that is reliable is our interest. In this paper, we propose a variational estimate method for solving the problem (1)-(2). The method can be considered as an alternative solver to the usual numerical methods, such as those discussed by Mungkasi and Christian [6]. We note that the term of variational estimate itself has also been used in physics [7-12], engineering [13-15], and statistics [16]. The rest of this paper is organised as follows. We provide a general formulation for the variational estimate method in Section 2. Computational results and discussion are given in Section 3. We conclude the paper in Section 4 with some remarks.

2. Variational estimate method

In this section, we follow the work of Inokuti et al. [3] to derive a general formulation of the variational estimate method. Let us consider problem (1)-(2). Suppose that we want to compute $x(T)$, that is the solution to the problem at point $t = T$. We assume that t is the free variable and x is the dependent variable with respect to t. The variational estimate is given by

$$x(T)_{\text{est}} = x(T) - \int_0^T \lambda(t)[x'(t) + f(x)]\, dt. \tag{3}$$

With this variational estimate, the differential equation (1) is treated as a condition on $x(t)$ that must be satisfied for all $t \in [0, T]$. To get the solution estimate $x(T)$, we can set $x(t)$ to be the initial condition. We note that choosing $x(t) = x(0)$ to be set into the variational estimate formula may not always lead to the most accurate results. However, it shall still produce a quite accurate solution to the problem.

3. Computational results and discussion

In this section, we write our computational results and some discussions.
We consider $f(x) = x^2$ and $\alpha = 1$, then we have the following ordinary differential equation

$$x'(t) + x(t)^2 = 0 \tag{4}$$

with the initial condition

$$x(0) = 1. \tag{5}$$

This problem is the same as the one considered by He and Wu [2] as well as Inokuti et al. [3]. He and Wu [2] solved the problem using the variational iteration method, but their standard variational iteration method has a demerit as they come to unnecessary calculation for the solution. Inokuti et al. [3] provided the solution only at point $t = 1$. We solve problem (4)-(5) using the variational estimate method for the domain of $t \in [0,1]$. We follow the step by step of Inokuti et al. [3]. We estimate the solution at $t = T$ using the variational estimate

$$x(T)_{\text{est}} = x(T) - \int_0^T \lambda(t)[x'(t) + x(t)^2]\, dt. \tag{6}$$

We take a trial function $x(t) = x_0(t) + \delta x(t)$, where $x(0) = 1$ such that $\delta x(0) = 0$. With these data, we obtain

$$x(T)_{\text{est}} = x_0(T) + \delta x(T) - \int_0^T \lambda(t)[x_0'(t) + \delta x'(t)]\, dt \\ - \int_0^T \lambda(t)\left[x_0(t)^2 + 2x_0(t)\, \delta x(t) + \delta x(t)^2\right] dt. \tag{7}$$

Calculating the optimal Lagrange multiplier, we obtain

$$\lambda(t) = \exp\left(2\int_1^x x(\tau)\, d\tau\right). \tag{8}$$

Therefore, the variational estimate formula for the problem is

IOP Conf. Series: Journal of Physics: Conf. Series **1007** (2018) 012011 doi:10.1088/1742-6596/1007/1/012011

$$x(T)_{est} = x(T) - \int_0^T \exp\left(2\int_1^x x(\tau)\, d\tau\right)[x'(t) + x(t)^2]\, dt\,. \tag{9}$$

This simplifies to

$$x(T)_{est} = \frac{1}{2}(1 + e^{-2T}) \tag{10}$$

which is the estimate solution x at point $t = T$.

Table 1. Variational estimate solution and the exact solution.

t	Estimate solution	Exact solution	Absolute error
0.0	1.0000	1.0000	0.0000
0.1	0.9094	0.9091	0.0003
0.2	0.8352	0.8333	0.0018
0.3	0.7744	0.7692	0.0052
0.4	0.7247	0.7143	0.0104
0.5	0.6839	0.6667	0.0173
0.6	0.6506	0.6250	0.0256
0.7	0.6233	0.5882	0.0351
0.8	0.6009	0.5556	0.0454
0.9	0.5826	0.5263	0.0563
1.0	0.5677	0.5000	0.0677

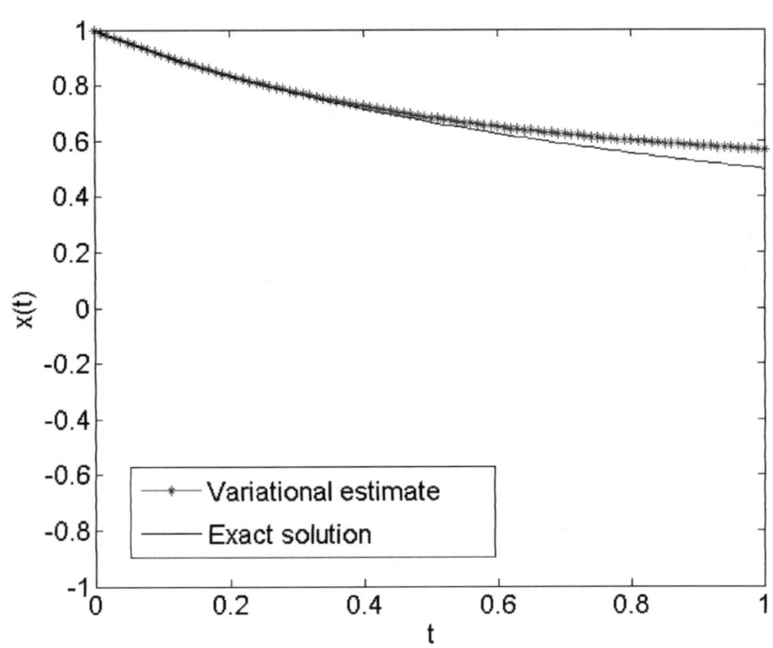

Figure 1. Computational results of variational estimate solution and the exact solution.

IOP Conf. Series: Journal of Physics: Conf. Series **1007** (2018) 012011 doi:10.1088/1742-6596/1007/1/012011

In order to investigate the accuracy of our results, we recall that the exact solution to problem (4)-(5) is given by

$$x(t) = \frac{1}{1+t}. \tag{11}$$

Table 1 summarises some computational results with respect to the exact solution. Figure 1 shows the graphic comparison between the variational estimate solution and the exact solution. Even though the absolute error gets larger as the free variable increases, the error is still quite small. We obtain that the variational estimate solution is accurate and is reliable. Therefore, we believe that it will be successful to be used to solve other first-order autonomous ordinary differential equation problems.

4. Conclusion

We have proposed a variational estimate method for solving first-order autonomous ordinary differential equation problems. The variational estimate solution is quite accurate. It is a promising method for ordinary differential equations with the initial condition is known. Our research focuses on one-dimensional autonomous problems. Future research direction could seek for a variational estimate method for solving non-autonomous and/or higher dimensional problems.

References

[1] Haberman R 1977 *Mathematical Models: Mechanical Vibrations, Population Dynamics, and Traffic Flow* (Englewood Cliffs: Prentice-Hall)
[2] He J-H and Wu X-H 2007 Variational iteration method: New development and applications *Computers and Mathematics with Applications* **54** 881
[3] Inokuti M, Sekine H, and Mura T 1978 General use of the Lagrange multiplier in nonlinear mathematical physics, in: *S. Nemat-Nasser (Ed.), Variational Method in the Mechanics of Solids* (New York: Pergamon Press) pp 156-162
[4] Setianingrum P S and Mungkasi S 2017 Variational iteration method used to solve the one-dimensional acoustic equations *Journal of Physics Conference Series* **856** 012010
[5] Yuliyanto B D and Mungkasi S 2017 Variational iteration method for solving the population dynamics model of two species *Journal of Physics: Conference Series* **795** 012044
[6] Mungkasi S and Christian A 2017 Runge-Kutta and rational block methods for solving initial value problems *Journal of Physics: Conference Series* **795** 012040
[7] Corboz P 2016 Improved energy extrapolation with infinite projected entangled-pair states applied to the two-dimensional Hubbard model *Physical Review B* **93** 045116
[8] Ebrahimi F and Barati M R 2016 Magnetic field effects on buckling behavior of smart size-dependent graded nanoscale beams *The European Physical Journal Plus* **131** 238
[9] Neupert T, Santos L, Ryu S, Chamon C and Mudry C 2012 Noncommutative geometry for three-dimensional topological insulators *Physical Review B* **86** 035125
[10] Sachdev S and La Placa R 2013 Bond order in two-dimensional metals with antiferromagnetic exchange interactions *Physical Review Letters* **111** 027202
[11] Sandri M and Fabrizio M 2013 Nonequilibrium dynamics in the antiferromagnetic Hubbard model *Physical Review B* **88** 165113
[12] Vaikuntanathan S and Geissler P L 2014 Putting water on a lattice: the importance of long wavelength density fluctuations in theories of hydrophobic and interfacial phenomena *Physical Review Letters* **112** 020603
[13] deBotton G and Shmuel G 2010 A new variational estimate for the effective response of hyperelastic composites *Journal of the Mechanics and Physics of Solids* **58** 466
[14] Lakhno V D 2012 Translation-invariant bipolarons and the problem of high-temperature superconductivity *Solid State Communications* **152** 621

[15] Rudykh S and deBotton G 2012 Instabilities of hyperelastic fiber composites: micromechanical versus numerical analyses *Journal of Elasticity* **106** 123
[16] Bickel P, Choi D, Chang X and Zhang H 2013 Asymptotic normality of maximum likelihood and its variational approximation for stochastic blockmodels *The Annals of Statistics* **41** 1922

Acknowledgements
The author thanks Sanata Dharma University for the financial support to this research.

Ontology Design of Influential People Identification Using Centrality

Rolly Maulana Awangga, Muhammad Yusril, Helmi Setyawan

Applied Bachelor Program of Informatics Engineering, Politeknik Pos Indonesia, Indonesia

E-mail: awangga@poltekpos.ac.id, yusrilhelmi@poltekpos.ac.id

Abstract. Identifying influential people as a node in a graph theory commonly calculated by social network analysis. The social network data has the user as node and edge as relation forming a friend relation graph. This research is conducting different meaning of every nodes relation in the social network. Ontology was perfect match science to describe the social network data as conceptual and domain. Ontology gives essential relationship in a social network more than a current graph. Ontology proposed as a standard for knowledge representation for the semantic web by World Wide Web Consortium. The formal data representation use Resource Description Framework (RDF) and Web Ontology Language (OWL) which is strategic for Open Knowledge-Based website data. Ontology used in the semantic description for a relationship in the social network, it is open to developing semantic based relationship ontology by adding and modifying various and different relationship to have influential people as a conclusion. This research proposes a model using OWL and RDF for influential people identification in the social network. The study use degree centrality, betweenness centrality, and closeness centrality measurement for data validation. As a conclusion, influential people identification in Facebook can use proposed Ontology model in the Group, Photos, Photo Tag, Friends, Events and Works data.

1. Introduction

The semantic web standard proposed by World Wide Web Consortium (W3C) as the next-generation web. Ontology chose as a knowledge-based semantic description for base formalisation. Ontology used to communicate between website on the Internet. Ontologies give more power on conceptual with domain and relation to describe the problems. Many research on software engineering use ontology as a modelling language for formalisation design. Especially multiple sites for communicating between designs in remote team member [1] to avoid miss-configuration of software and mediating mechanism for accessing heterogeneous data and information sources [2]. If machine learning and software framework can solve healthy problem[3][4]. Data science can solve warehouse[5] and knowledge management problem[6]. Ontology is a solution for software engineering knowledge representation [1]. There was a classification of Ontology for Software Development Life Cycle phase [7]. Ontology is presenting more metadata schemas, offering a controlled vocabulary more than Unified Modelling Language (UML) [8].

Ontology can improve Social Network Analysis(SNA) measurement [9]and software engineering knowledge representation [10]. In the graph theory and network analysis, centrality used to determine the relative importance of a node within the graph [11]. Research topics in different aspects [12] using centrality analysis to get influential people. Beside for political issues [13] and terrorism information [14] on the social network [15], an online market business with higher values for three measurement degrees(degree, betweenness, closeness) obtain users who can target as productive customer for selling a product [16].

All the user-related graph on a social network mapped to semantic description model on ontology. Ontologies are a useful tool for modelling, sharing and reusing knowledge [10]. Ontologies are closely related to modern object-oriented software engineering with richer information model compared to UML and as a shared conceptualisation, naturally easy to develop in future work [8]. Centrality[17] use as graph theory with social network analysis[18] [19]. A person can be influential observed by how many friends list that he has, his ability to connect two networks of friends, or a person who knew a lot of others influential person. Degree, betweenness, and closeness centrality as popular centrality concept adopted in measuring graph. A graph consists a node and an edge. Node is defined person; edge defines a relation between nodes.

The objective of this research is modelling influential people from social network using OWL and RDF. Semantic description will use to modelling the concept of influential people. The relationships model used from a popular social network Facebook, which have ability providing user data in Application Programming Interface(API). The data gathered from Facebook using API[20]. Facebook allows third-party apps to take a user data from Facebook account who was interested in the application.

2. Methods

In the social network analysis, the data designed as structural characteristics of a social organisation. The Analysis need more information about the relation between nodes and edge[21]. Social network semantic make a computer understand the social network. In real social status, every person has a different relationship with different level and function. The Semantic describe more than attributes data and relation like characteristics of nodes and connections. Ontology is a philosophical concept to describe things. It is a modelling tool in artificial intelligence, which can describe conceptual models of information system in semantic and knowledge level [22] as a formal specification of a shared conceptualisation [23]. The definition relations in ontology illustrated as actor and relation attributes. The actor could be people, organisation or company and others [9]. Relation attributes describe various connection formed in social activities interact.

A Graph concept use for studying social phenomena. Actors and relationships are forming social networks data, the detailed explanation explains by the application of Wasserman and Faust (1994)[11] or Scott (1991)[24]. The most important thing in the analysis of social networks is vertices in the graph (Bavelas, 1984; Sabidusi, 1966; Freeman, 1979)[25]. The design was interpreting the interests into the graph. The formulation of centrality [10] obtained of the most influential people in a community. Centrality calculation divide into several calculation formulas.

Degree centrality is the most fundamental measure of centrality computation. Degree centrality has a formula essentially measures the amount of edge connection as a vertex of an object (equation 1).

IOP Conf. Series: Journal of Physics: Conf. Series **1007** (2018) 012012 doi:10.1088/1742-6596/1007/1/012012

$$C_D(v) = deg(v). \tag{1}$$

A nodes that could be a broker node or interface between two different clusters in a graph. This node is a bridge between clusters. There will be no flow of information between two different clusters without this bridge. Betweenness Centrality seeks each vertex connects directly to the vertex then measure it. $dG(s, t)$ is the notation of the distance between vertex s to vertex t that the value of $dG(s, s) = 0$ for all $s \in V$, and $dG(s, t) = dG(t, s)$ for $s, t \in V$. There are exist variations of the notation that indicates the importance of a vertex in a graph. If $\sigma_{st} = \sigma_{st}$ a notation of the number of shortest paths from $s \in V$ to $t \in V$, where the agreement $\sigma_{ss} = 1$ then $\sigma_{st}(v)$ indicates the number of the shortest path from s to t where there is $v \in V$ between them. Based on the standard measurement of centrality, Betweenness Centrality explains in equation 2.

$$C_B(v) = \sum_{B \neq v \neq t \in V} \frac{\sigma_{st}(v)}{\sigma_{st}} \tag{2}$$

s and t is a sum of any possibility of vertices in the graph. $C_B(v)$ is Betweenness Centrality on vertices. $\sigma_{st}(v)$ is all the number of shortest paths through v from s to t. σ_{st} is the number of shortest paths from s to t.

Calculation of Closeness Centrality measure the fastest and individuals who can monitor the network. There is a node which is directly connected with the highest degree centrality value and connected with the highest points on the Betweenness Centrality.Closeness Centrality is the average distance (shortest path) between vertices connection. Closeness Centrality formula describes in Equation 3. CC (v) is Closeness centrality for vertex v. dG(v, t) is the distance between the vertex v to t.

$$C_C(v) = \frac{1}{\sum_{t \in V} d_G(v, t)} \tag{3}$$

3. Experiment and Result

The ontology defines actor and relation attributes into a property. Actor in this research identified as Social Network user account. One actor has many relation attributes. Facebook data have several relationship attributes. There is: HasGroup, HasPhotos, HasSchool, HasWork, HasFriend, HasAttentEvent, HasTagged, HasLocation, HasHometown. Facebook user is an actor object as influential people. Two superclass is in a relationship as properties bridging actor with another object as classes. Classes is an object related to the user(figure 1).

Three Facebook classes create the domain on the social network (figure 2). The user is an account who use a social network. Area describes the city or town where the user comes from and lives in recently. Activities describe user activity in Facebook. Under activity class, there is a group where the user joins in. Photo tags where the user was in the photo with someone else. Events which are user come to an invitation. The school who is user getting a school there.

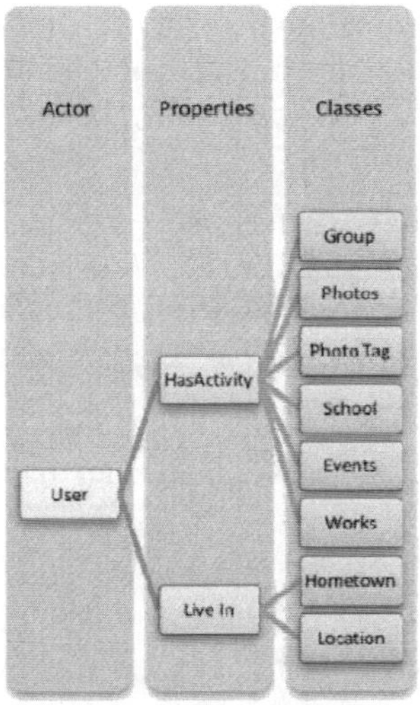

Figure 1. Facebook data design into Properties and Classes.

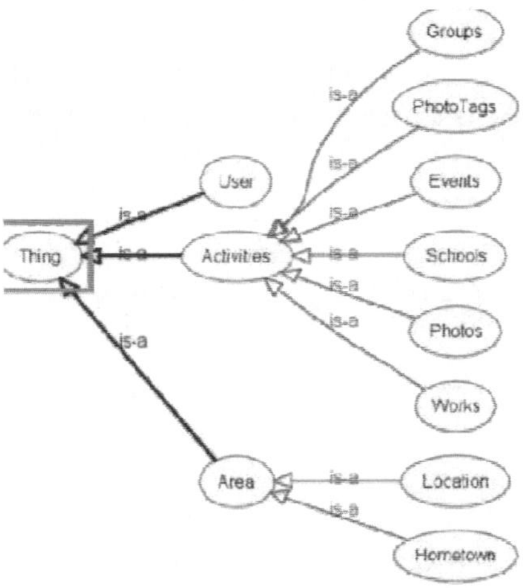

Figure 2. Secial Network Ontology.

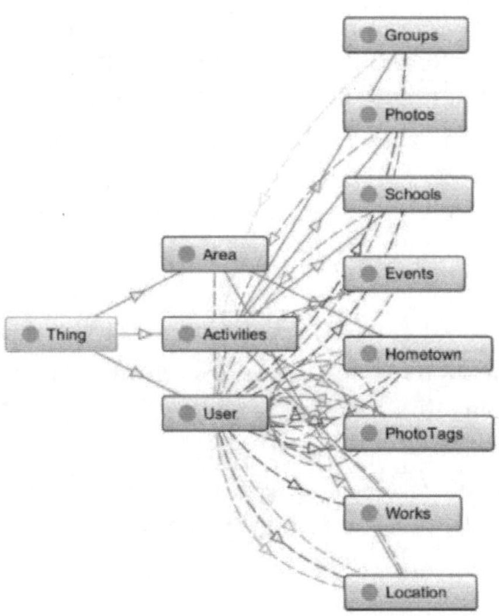

Figure 3. Ontology design.

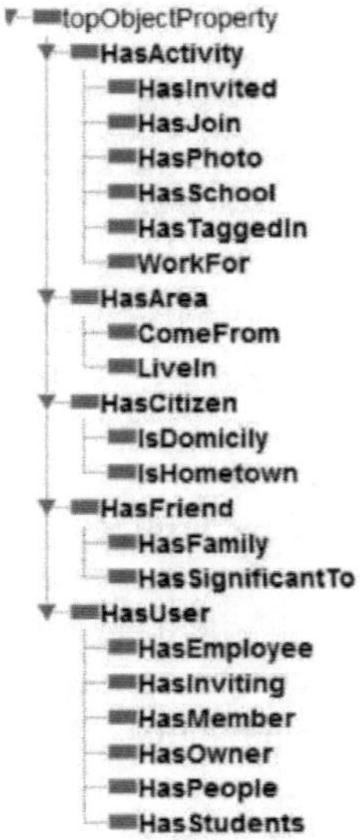

Figure 4. Property of Social Network.

Photos is a list of user photos. Work is the occupation of the user. We can look clearly at Onto-graph figure 3. Onto-graph is visualisation for domain and relation ontology, the connection between domain has many properties like in figure 4.

Super-properties in the social network ontology describing in five properties. HasActivity is a relation between users to the Activities superclasses, with HasUser as inverse properties. HasArea is a functional relation between user and Area Classes, with HasCitizen as inverse function. Also, HasFriend is a symmetric relation between users. There is a restriction in ontology, especially for the user. Property of come from has maximally related to one object in HasSignificantTo and LiveIn.

Restriction rules added in user class. The user has relation to hometown maximal is one relation, same as location and significantTo. The semantic description of ontology for influential people identification generated from this ontology. The research use HasFriend and HasActivity or HasUser as properties and user and activities as classes. The Ontology is explaining in figure 5.

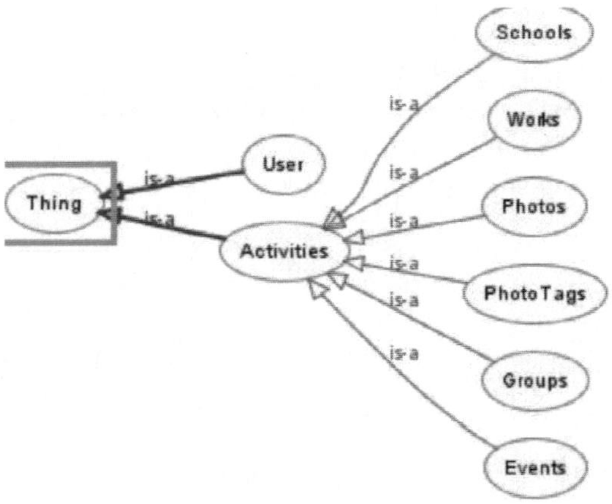

Figure 5. Influential People Identification Ontology.

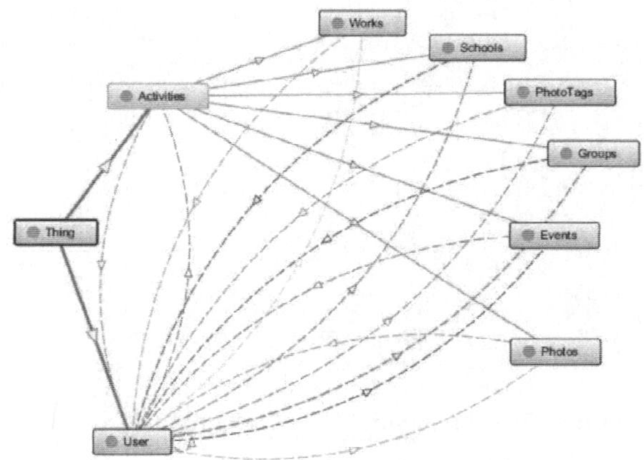

Figure 6. OntoGraf of Influential People Identification.

Activities object class is superclasses with six classes under it. School, Works, Photos, PhotoTags, Groups, and Event is activities class related to the user with HasUser and HasActivity super properties. HasUser is an asymmetric relation between Activities class to User class. HasUser is super properties for HasEmployee, HasInviting, HasMember, HasOwner, HasPeople, HasStudents classes. Inverse object properties of HasUser are HasActivity. HasActivity are super properties for HasInvited, HasJoin, HasPhoto, HasSchool, HasTaggedIn, WorkFor. User class has symmetric irreflexive properties called HasFriend.

In figure 6 the blue box is a class, a solid arrow showing subclasses. Dashed Arrow is showing properties of each class well connected each other. This ontology used by Social Network Analysis calculation. Measuring influential people using Centrality on this ontology to get a value of importance user as nodes in social activities and users graph.

Centrality calculation was implementing into Facebook user data. The data gathered by Facebook Application which have 120283 friend connection. The highest frequency of appearance of account in Degree Centrality data measurement, sorted by frequency of occurrence. The Result is account A get 3, account IMB get 3, account JRJS get 2 and account AW get 1.

For Each measurement of betweenness centrality in Social Network Data. The top three nodes selected by frequency sorting. The result is: account IMB get 2, account AL get 2, account TIH get 2, account AN get 2, account AM get 1.

The result from every closeness centrality measurement. The data sorting by frequency. Top three selected from population data. Account AL gets 3; account MI get 2, account ER get 2, account ES get 2, account KYT get 1.

4. Conclusion

The highest Centrality calculation using Equation 1, 2, and 3 have results for influential people identification using ontology model in Figure 5 and 6. Ontology model in Figure 5 and 6 in the Group, Photos, Photo Tag, Friends, Events, and Works data shows the name of a person, this identified as influential people. The centrality calculation from Location, Hometown data shows the name of City so cannot identify influential people. Education Shows the name of an institution so cannot identify influential people. As a result, influential people identification in Facebook can use ontology on Figure 6 in the Group, Photos, Photo Tag, Friends, Events and Works data.

Influential people identification can be more varied with different social network data. Centrality measurement for influential people identification can expand with other calculation method using fuzzy logic or another intelligence system. The model can combine with another technique besides centrality for evaluation purpose on influential people identification. Another aspect of individual influential people measurement can use statistical methods.

References

[1] Wongthongtham P, Kasisopha N, Chang E and Dillon T 2008 *Convergence and Hybrid Information Technology, 2008. ICCIT'08. Third International Conference on* vol 2 (IEEE) pp 668–675

[2] Dillon T S, Chang E and Wongthongtham P 2008 *Software Engineering, 2008. ASWEC 2008. 19th Australian Conference on* (IEEE) pp 13–23

[3] Yulita I N, Purwani S, Rosadi R and Awangga R M 2017 *Advanced Informatics, Concepts, Theory, and Applications (ICAICTA), 2017 International Conference on* (IEEE) pp 1–5

[4] Awangga R M, Fathonah N S and Hasanudin T I 2017 *2017 2nd International conferences on Information Technology, Information Systems and Electrical Engineering (ICITISEE)* (IEEE) pp 255–260 ISBN 978-1-5386-0658-2 URL http://ieeexplore.ieee.org/document/8285506/

[5] Pane S F, Awangga R M and Azhari B R 2018 *TELKOMNIKA (Telecommunication Computing Electronics and Control)* **16**

[6] Awangga R M, Pane S F, Tunnisa K and Suwardi I S 2018 *TELKOMNIKA (Telecommunication Computing Electronics and Control)* **16**

[7] Zhao Y, Dong J and Peng T 2009 *IEEE Transactions on Services Computing* **2** 303–317

[8] Siricharoen W V 2008 *Computer Science and its Applications, 2008. CSA'08. International Symposium on* (IEEE) pp 320–325

[9] Li T, Yang H, He J and Ai Y 2010 *Knowledge Acquisition and Modeling (KAM), 2010 3rd International Symposium on* (IEEE) pp 258–261

[10] Hilera J R and Fernandez-Sanz L 2010 *Software Engineering Advances (ICSEA), 2010 Fifth International Conference on* (IEEE) pp 380–383

[11] Wasserman S and Faust K 1994 *Social network analysis: Methods and applications* vol 8 (Cambridge university press)

[12] Jamali M and Abolhassani H 2006 *Web Intelligence, 2006. WI 2006. IEEE/WIC/ACM International Conference on* (IEEE) pp 66–72

[13] Ya-ting L and Jing-min C 2011 *Consumer Electronics, Communications and Networks (CECNet), 2011 International Conference on* (IEEE) pp 5441–5444

[14] Li B x, Li M j, Sun D y, Li J and Li W j 2012 *Intelligence and Security Informatics (ISI), 2012 IEEE International Conference on* (IEEE) pp 193–193

[15] Wennerberg P O 2006

[16] Kumar P and Zhang K 2007 *e-Business Engineering, 2007. ICEBE 2007. IEEE International Conference on* (IEEE) pp 363–367

[17] Warmbrodt J, Sheng H and Hall R 2008 *Hawaii International Conference on System Sciences, Proceedings of the 41st Annual* (IEEE) pp 291–291
[18] Martin T, Zhang X and Newman M 2014 *Physical review E* **90** 052808
[19] Scott J 2017 *Social network analysis* (Sage)
[20] Catanese S A, De Meo P, Ferrara E, Fiumara G and Provetti A 2011 *Proceedings of the international conference on web intelligence, mining and semantics* (ACM) p 52
[21] Chen L, Wei S and Qingpu Z 2010 *E-Business and E-Government (ICEE), 2010 International Conference on* (IEEE) pp 1936–1939
[22] Dorr C 2010
[23] Borst W N 1997 *Construction of engineering ontologies for knowledge sharing and reuse* (Universiteit Twente)
[24] Scott J 1991 *Annual review of sociology* **17** 181–203
[25] Zalzala L 2013 *ICMLG2013-Proceedings of the International Conference on Management, Leadership and Governance: ICMLG 2013* (Academic Conferences Limited) p 348

Acknowledgement
Thanks to Politeknik Pos Indonesia to support the publication.

An application of viola jones method for face recognition for absence process efficiency

Rudolfo Rizki Damanik*, **Delima Sitanggang, Hendra Pasaribu, Hendrik Siagian, Frisman Gulo**

Faculty of Technology and Computer Science, Universitas Prima Indonesia, Indonesia

E-mail: *echos.damanik@gmail.com

Abstract. Absence was a list of documents that the company used to record the attendance time of each employee. The most common problem in a fingerprint machine is the identification of a slow sensor or a sensor not recognizing a finger. The employees late to work because they get difficulties at fingerprint system, they need about 3 – 5 minutes to absence when the condition of finger is wet or not fit. To overcome this problem, this research tried to utilize facial recognition for attendance process. The method used for facial recognition was Viola Jones. Through the processing phase of the RGB face image was converted into a histogram equalization face image for the next stage of recognition. The result of this research was the absence process could be done less than 1 second with a maximum slope of ± 700 and a distance of 20-200 cm. After implement facial recognition the process of absence is more efficient, just take less 1 minute to absence.

1. Introduction

Absence was a way to know the number of attendance to a person. Absenteeism could be done in various ways. First could be used by using the absence sheet or a tool to detect fingerprint [1]. The development of a more modern absentee system with respect to the honesty side needs to be held as a substitute for manual absence by using paper at the same time can become a new standard [2]. The use of a fingerprint presence system will reduce the problems posed by the use of manual attendance systems. With a biometric fingerprint system, frequent fraud rates such as data manipulation and preset care could be reduced [3].

Actually with the application of fingerprint system for attendance, could minimize the level of fraud that could be done in attendance system. However, it should be underlined that the fingerprint system still cannot be applied perfectly if there is someone who has physical limitations. In view of the existing weaknesses in the fingerprint system that cannot be used if a person has physical limitations, a system that can be applied to everyone without exception, and reduces the level of fraud as big as possible [4]. A fingerprint system can be took long time for process absence when the sensor is dirty or position of finger is not fit. In the field of computer vision research, facial recognition processing takes the initial stages before the facial recognition process stage itself. The process is the process of detection of human face (face detection) [5].

Based on the description above, the author was interested in designing attendance applications by utilizing face recognition. Face detection is one of the most complex and challenging problem in the field of computer vision, due to the large variations caused by the changes in facial appearance, lighting, and facial expression [6].Face recognition plays a vital role in many applications, which are crucial and integral part of life, and hence a high identification rate is desired [7].The method used for face recognition was Viola Jones. Using Viola Jones method, fast processing speed and high accuracy of face detection were important factors in face recognition. In this algorithm, already define the distance between two eye when any image will be coming then processing the algorithm and matching the eyes distance and pupil distance then eyes will be detected [8].The approach was used to construct a face detection system which is approximately 15 faster than any previous approach [9].Viola-Jones algorithm with some specific threshold value gives the result with fast detection rate and high accuracy, the average detection rate is 97.41% [10].Viola Jones method was a fast and accurate method in image processing [11 – 15]. Face Recognition was considered to reduce the shortcomings of the fingerprint system and help users to perform faster attendance.

2. Method

Viola Jones classified the image of the value of simple features and uses three types of features, which were square features, three-square features, and a four-square feature. The value of these features was the difference between black and white regions. In each sub-window image, the total number of Feature Haar was very large, much larger if compared to the number of pixels. To ensure that classification could be done quickly, the learning process should eliminate the majority of features available, and focus on a small set of necessary features. AdaBoost aimed to form face templates [16].

The facial object was searched using Viola Jones which refers to the plot as shown in Figure 1, where the grayscale image would be scanned per sub-window to look for positive features with AdaBoost and Cascade Classifier. If a face was detected, a rectangular image would be drawn on the face. The object's detection group casts an image based on the value of a simple feature. The basic operation of a feature was much faster than pixel processing. Some Haar Features represent the rectangular region of the image and add up all the pixels in the area [16].

A classification method that was used multiple levels of selection. At each level performs the selection using the AdaBoost algorithm that has been trained by using the Haar Feature. The selection was useful for separating between sub-windows containing positive objects (images that are detected to have the desired object) with negative objects (the detected images do not have the desired object) [4]. The Viola-Jones method combined four general keys: Haar Like Feature, Integral Image, Adaboost learning and Cascade classifier. Haar Like Feature was the difference of the number of pixels from the area inside the rectangle. An example of Haar Like Feature was presented in Figure 1.

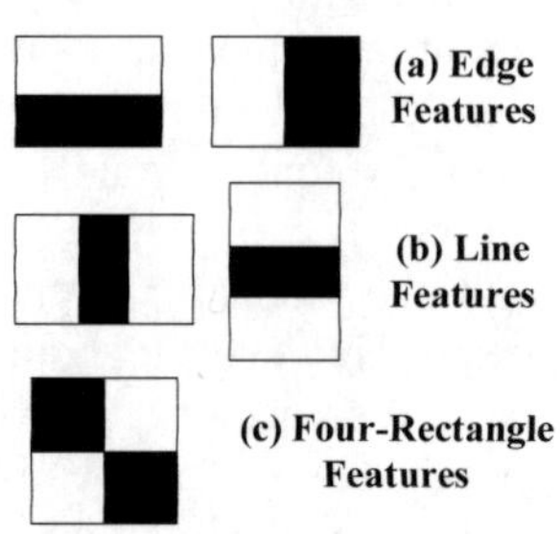

Figure 1. Haar Like Feature.

Haar Like Feature values were derived from the difference in the number of dark area pixel values by the number of bright area pixels:

$$F(Haar) = \sum F_{white} - \sum F_{black} \quad (1)$$

F(Haar) was the total feature value, $\sum F_{white}$ was the feature value on the brightest area and $\sum F_{black}$ was the feature value on the dark area. Haar features are composed of either two or three rectangles. Face candidates are scanned and searched for Haar features of the current stage. The weight and size of each feature and the features themselves are generated using a machine learning algorithm from AdaBoost [17]. Integral image was a technique for calculating the feature value quickly by changing the value of each pixel into a new image representation, as shown in Figure 2.

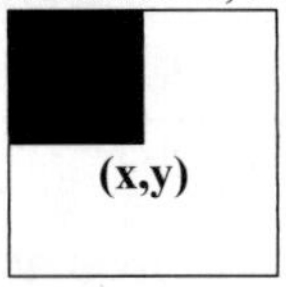

Figure 2. Integral image (x,y).

Based on Figure 3, integral image in *(x,y)* (*ii*(*xy*)) could be find used the formula:

$$ii(x, y) = \sum x' \leq x, y' \leq y^{i(x',y')} \quad (2)$$

where *ii* (*x, y*) was the integral image at location *x, y* and *i* (*x ', y'*) was the pixel value at the original image location. The calculation of the value of a feature could be done quickly by computing the integral image value at four points as shown in Figure 3. If the integral value of image of point 1 was A, point 2 was *A* + *B*, point 3 was *A* + *C*, and at the point 4 was *A* + *B* + *C* + *D*, then the number of pixels in region D could be known by 4 + 1 (2 + 3).

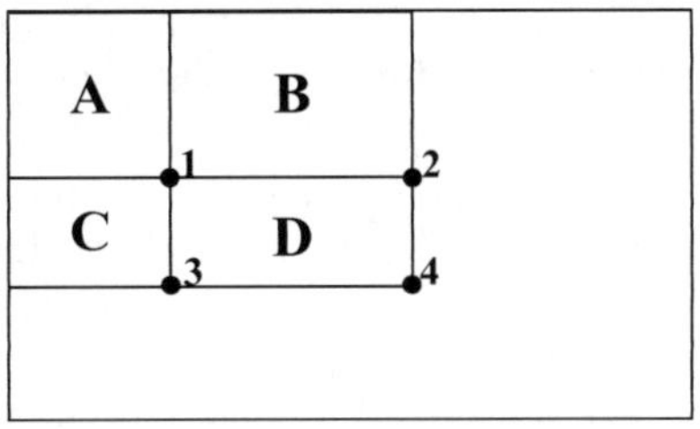

Figure 3. The Figure Score Count.

The Adaboost learning algorithm, used to improve classification performance with simple learning to combine many weak classifier into one powerful classifier. Weak classifier was a correct answer with a less accurate truth level. A weak classifier was stated below [4] :

$$h_j(x) = \begin{cases} 1, & if\ p_j f_j < p_j \theta_j\ (x) \\ 0, & other \end{cases} \quad (3)$$

where $h_j(x)$ was a weak classification, p_j is parity to *j*, θ_j was threshold to *j* and *x* is sub image dimension such as *24 x 24*. The steps to get a strong classifier were expressed in an algorithm as follows, giving figure as example $(x_1 y_1), \dots (x_n y_n) y_i = 0$ the positive example and $y_i = 1$ for negative example. Weight initation $y_{i,1} = \frac{1}{2m'} \frac{1}{2l'}$; *m* and l was the number of negative and positive, and for $t = 1, \dots, T$ then weight normal so $w_{t,i} \leftarrow \frac{w_{t,i}}{\sum_{j=1}^{n} w_{t,j}}$ for every figure, *j* classifier exercises h_j, for every individual figure. Fault $\epsilon_j = \sum_i w_i \left| h_j(x_i) - y_i \right|$ Classifier Choose *C C* h_t with smallest error where $e_i = 0$ for x_i was the true classification, and $e_i = 1$ for others weight update: $w_{t+1,i} = w_{t,i} \beta_t^{1-e_i}$ where $\beta_t = \frac{\epsilon_t}{1-\epsilon_t}$. There got to be a strong Classifier $h(x) = \begin{cases} 1, & \sum_{t=1}^{T} \alpha_t h_t(x) \geq \frac{1}{2} \sum_{t=1}^{T} \alpha_t \\ 0, & other \end{cases}$ where $\alpha_t = \log \frac{1}{\beta_t}$

The Cascade classifier was a method to combine complex classifier in a multilevel structure that could increase the speed of object detection by focusing on only possible imagery areas. The cascade classifier structure was presented in Figure 4.

IOP Conf. Series: Journal of Physics: Conf. Series **1007** (2018) 012013 doi:10.1088/1742-6596/1007/1/012013

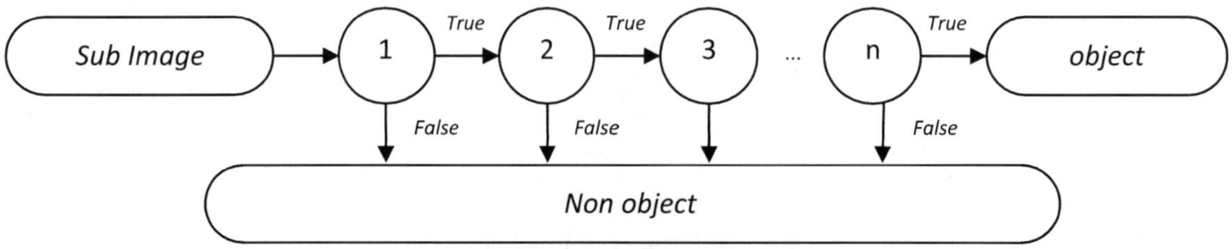

Figure 4. Cascade Classifier.

The workings of face detection was using Viola Jones, in Figure 5 below was a scheme of how face detection works using the Viola-Jones algorithm from the beginning of the detection process to the results of the detection process. The first process was to read the sample of the face image of an image or read the sample of the face image that was facing the camera. From the image that had been captured it would be done Haar feature reading by processing the image into boxes to get the difference of threshold value of the dark areas and bright areas of the image. If the difference between the dark and the bright areas above the threshold or threshold value, then it could be said that the feature exists.

Further to determine whether or not there were hundreds of Haar features on an image and on a different scale efficiently used Integral Image. In general, the integration adds small units simultaneously. In this case the small units were pixel values. The integral value for each pixel was the sum of all the pixels from top to bottom. Starting from the top left to the bottom right, the entire image could be summed with multiple integer operations per pixel. Then to select the specific Haar feature to be used and to set the threshold value was used a machine learning method called AdaBoost. AdaBoost combines many weak classifiers to create a powerful classifier. By combining some AdaBoost classifiers as filter circuits that were efficient enough to classify the image area. Each filter was a separate AdaBoost classifier that consists of a weak classifier or a Haar filter. During the filter process, if any of the filters fail to pass an image area, then the area was immediately classed as non-face. However, when the filter passes an image area and passes through all filter processes in the filter circuit, the image area was classified as the face. The next stage was cascade classifier. The order of filters on the cascade was determined by the weight given AdaBoost. The largest weighted filter was placed in the first process, aiming to erase the non-face image area as quickly as possible. The last stage was showed the object of the sample image that has been detected face or not face

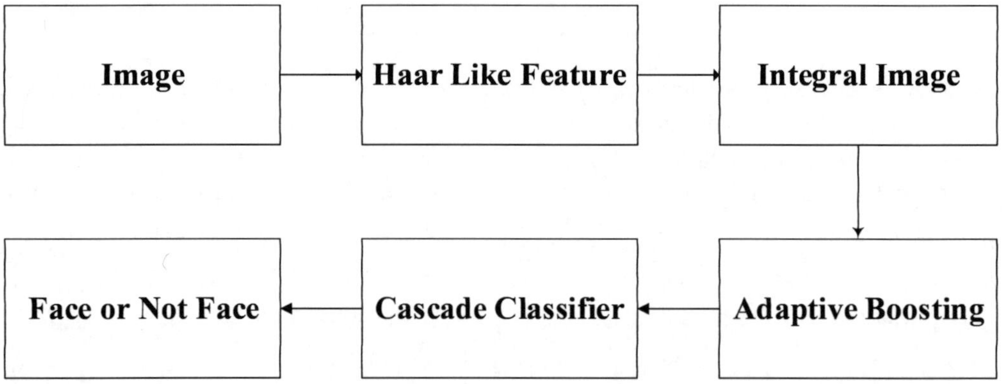

Figure 5. Cascade Scheme Classifier Face Detection Process with Viola-Jones Method.

IOP Conf. Series: Journal of Physics: Conf. Series **1007** (2018) 012013 doi:10.1088/1742-6596/1007/1/012013

3. Design and development

The absence process needed a web camera or webcam as an input data media, and then the system would be processing the input data with Viola Jones method. In this method, the picture of face should be passing some steps to become valid or invalid like figure 6 below.

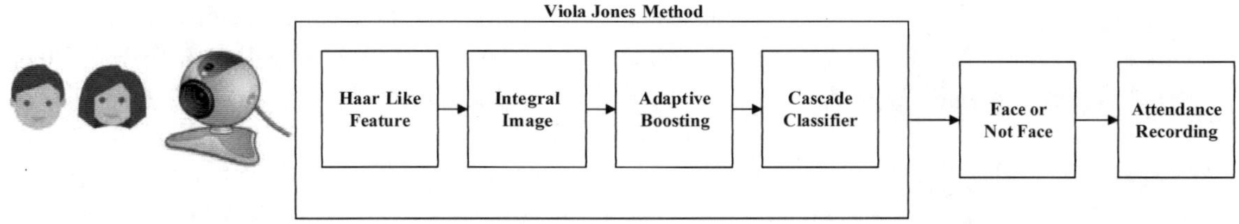

Figure 6. Block Diagram of Viola-Jones Method.

This process had been tried in Computer and Technology Faculty expecially for staffs and lecturers. There were 15 correspondents that become participants in this try an error where there were 10 lecturers and 5 staffs. The test was done by asking 3 staffs to use glasses and did the absence, the result was taken that this system could known the face eventhough the saving data in database didnot use the glasses. The rest was 12 people was tested by reduce light intention when the detection process, moving the distance and face detection corner. By the testing of this process, the system still detect the face and match with database without any trouble. For light intention, if the condition of light was dim or dark so the system would not detect the face process. The time that needed to do the process of detecting face from 20 to 50 second and of cource the presention process would be faster than using fingerprint.

4. Results and Discussions

A better recognition rate was obtained by maximizing the number of images used as the face matching reference and governing the conditions at which the facial image retrieval image and test face image had the same conditions. Also at the time of face recognition was done using a webcam should not be too many interference behind the user. But the accuracy of the attendance system through Viola Jones method would be worse if the image of the face in training has much in common. Compared with other studies [6] where the results obtained tolerance angle for face recognition is 0 - 5 degrees, whereas in this research angle tolerance for face recognition reaches 70 degrees with improvement ratio 1:14. For the face distance from the camera in the study [6] only 20-60 cm, while in this study reached 20 - 200 cm. Applications built on this research are also able to recognize the obstructed face of objects such as eyeglasses and on dim lighting.

Table 1.The Result with criteria.

Figure	Example Genres	Result	Conclusion
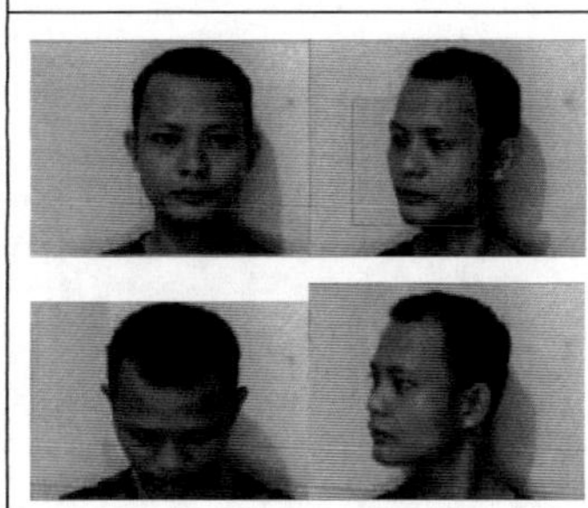	Slant	Detected Lose	The front-facing position of the webcam with accuracy is 100% with detection time of less than 1 second, while the maximum slope of $\pm$ 70 °.

IOP Conf. Series: Journal of Physics: Conf. Series **1007** (2018) 012013 doi:10.1088/1742-6596/1007/1/012013

	Light	Detected	In dark or bright lights are still able to detect the presence of objects.
	Distance	Detected	The closest distance ± 20cm, at a distance of ± 200cm could still be detected.
 	Stunted	Detected Lose	Able to detect facial objects even if obstructed as long as the contour was the same as the facial contours.

The following interface was a display of attendance application with face recognition system, which users only need to face the fore face of the webcam for absenteeism and absenteeism home. First of all the users came and performs attendance by facing the fore face of the webcam, if the face of the user is declared to match the image of face training contained in the database, then the absence successfully done. The same thing was also done when the user perform absenteeism home. Display attendance system face recognition could be seen in Figure 7.

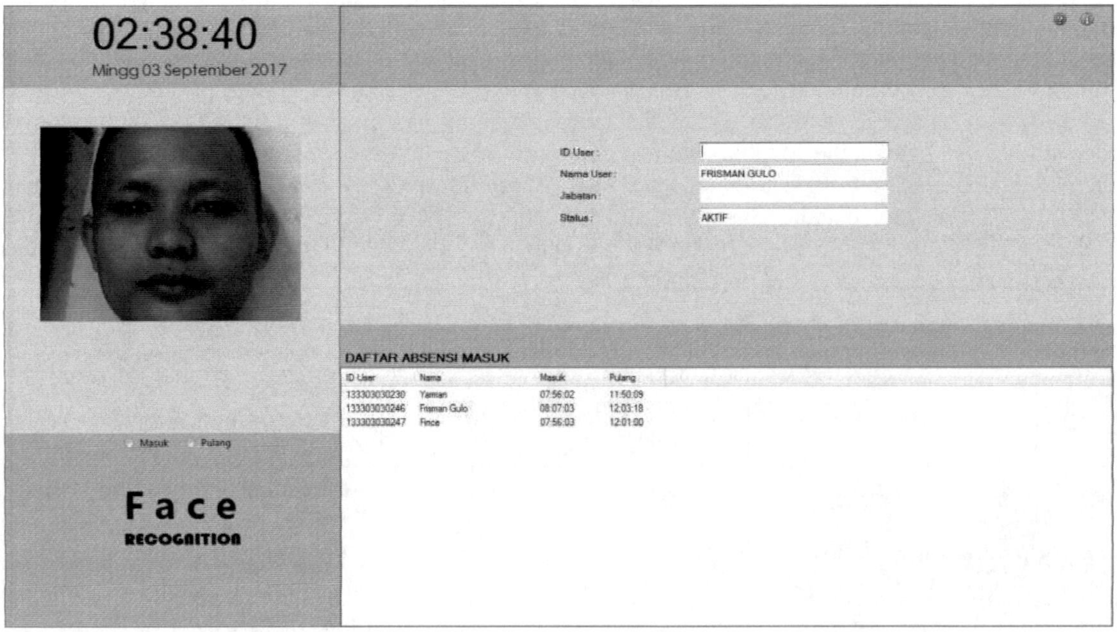

Figure 7. Face Recognition Process.

In this attendance application had provided the features of training data used as a reference matching face, as in figure 8 below:

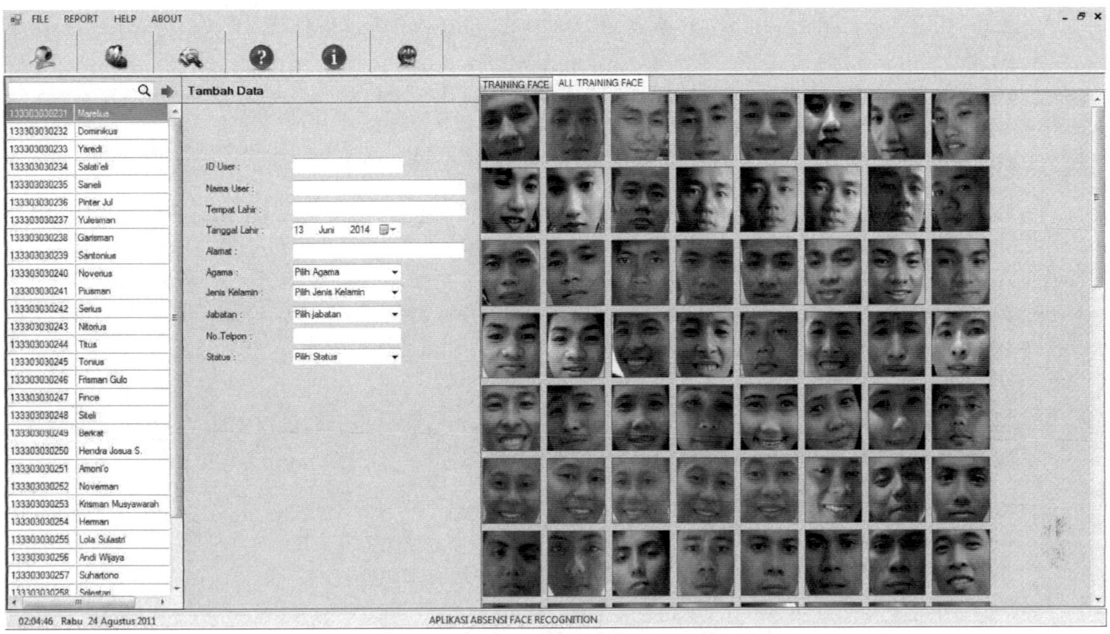

Figure 8. Data Training Process.

5. Conclusion

Based on the results of tests performed, author could draw some conclusions were: with the absence application using face recognition, it helped users who have physical limitations such as not having a finger or wet fingers. The absence of face system was successful implemented and moved the process of fingerprint, the test was done to 20 correspondence and the result was shown that 20 faces was successful detection with time speed from 20 to 50 seconds.

Acknowlegdement

This research was supported by the Faculty Technology and Computer Science Universitas Prima Indonesia

References

[1] A. T.Kurniawati andA. Ruli Dwi Rama, 2012, "Aplikasi pengenalan wajah menggunakan metodeeigenface dengan bahasa pemrograman java", *Seminar Nasional Sains dan Teknologi Terapan III, 2015.*

[2] N. Haris Masruriand A. Sunyoto, "Aplikasi cambot menggunakan metode viola jones sebagai sistem penjejak objek", *JURNAL DASI **Vol.** 13, No. 3.*

[3] C. Fifti AnasSari and L. Yulianto, 2013, "Perancangan sistem informasi absensi menggunakan finger print di badan perencanaan pembangunan daerah dan penanaman modal kabupaten pacitan", *Seminar Riset Unggulan Nasional Informatika dan Komputer FTI UNSA, No. 1.*

[4] D. Ari Prasetya and I. Nurviyanto, 2012, "Deteksi wajah metode *viola jones* pada *opencv* menggunakan pemrograman *python", Simposium Nasional RAPI XI FT UMS.*

[5] Yi-Qing Wang, 2012, "An Analysis of the Viola-Jones Face Detection Algorithm", *Image Processing Online.*

[6] M. K Dabhi and B. K Pancholi, 2016, "Face Detection System Based on Viola-Jones Algorithm", *International Journal of Science and Research(IJSR)* ***Vol. 5****, Issue 4, April 2016.*

[7] A. Mohsen Abdul Hossen, R. Abd Alsaheb Ogla, M. Mahmood Ali, 2017, "Face Detection by Using OpenCV's Viola Jones Algorithm based on coding eyes", *Iraqi Journal of Science,* ***Vol. 58,*** *No. 2A.*

[8] N. T. Deshpande and S. Ravishankar, 2016, "Face Detection and Recognition using Viola-Jones algorithm and fusion of LDA and ANN", *IOSR Journal of Computer Engineering (IOSR-JCE)* ***Vol. 18****, Issue 6.*

[9] A. Gupta and R. Tiwari, 2015, "Face Detection Using Modified Viola Jones Algorithm", *International Journal of Recent Research in Mathematics Computer Science and Information Technology* ***Vol. 1,*** *Issue 2. October 2014 – March 2015.*

[10] S. Tikoo and N. Malik, 2016, "Detection of Face using Viola Jones and Recognition using Back Propagation Neural Network", *International Journal of Computer Science and Mobile Computing* ***Vol. 5****, Issue 5. May 2016.*

[11] A. Sahitya, M. TN, V. N, 2016, "A Survey on Face Recognition Technology – Viola Jones Algorithm", *International Journal of Computer Applications.*

[12] I. Khan, H. Abdullah and M. Shamian Bin Zainal, 2014, "Efficient Eyes and Mouth Detection Algorithm Using Combination of Viola Jones and Skin Color Pixel Detection", *International Journal of Engineering and Applied Sciences* ***Vol. 3****, No. 4, June 2014.*

[13] M. Kumar Mathur, P. Bhati, 2017, "Face Object Detection in still image using Viola-Jones algorithm through mathlab tools", *International Journal of Innovative Research in Computer and Communication Engineering* ***Vol. 5****, Issue 2. February 2017.*

[14] M. Chaudhari, S. Sondur, G. Vanjare, 2015, "A review on Face Detection and study of Viola Jones method", *International Journal of Computer Trends and Technology* ***Vol. 25****, No. 1, Juli 2015.*

[15] R. Boda and M. Jasmine Pemeena Priyadarsini, 2016, "Face Detection And Tracking Using Klt And Viola Jones", *ARPN Journal of Engineering and Applied Sciences Vol. 11, No. 23, December 2016.*

[16] A. Hendro Triatmoko, S. Hadi Pramono, H. S. Dachlan, 2014, "Penggunaan metode *viola-jones* dan algoritma *eigen eyes* dalam sistem kehadiran pegawai", *Jurnal EECCIS* ***Vol. 8****, No. 1, Juni 2014.*

[17] R. N. Dascoudhary and R. Tripathy, 2014, "Real time face detection and tracking using haar classifier on soc", *Proceedings of SARC-IRF International Conference, April 2014.*

IOP Conf. Series: Journal of Physics: Conf. Series **1007** (2018) 012014 doi:10.1088/1742-6596/1007/1/012014

Combination of advanced encryption standard 256 bits with md5 to secure documents on android smartphone

Hendra Pasaribu, Delima Sitanggang, Rudolfo Rizki Damanik, Alex Chandra Rudianto Sitompul

Faculty of Technology and Computer Science, Universitas Prima Indonesia, Indonesia

E-mail: *kevinbadboyz@gmail.com

Abstract. File transfer by using a smartphone has some security issues like data theft by irresponsible parties. To improve the quality of data security systems on smartphones, in this research the integration of AES 256 bit algorithm by using MD5 hashing is proposed. The use of MD5 aims to increase the key strength of the encryption and decryption process of document files. The test results show that the proposed method can increase the key strength of the encryption and decryption process in the document file. Encryption and decryption time by using AES and MD5 combination is faster than using AES only on * .txt file type and reverse results for *.docx, *.xlsx, *.pptx and *.pdf file files.

1. Introduction

Smartphone is a communication tool that used by human every day. As the increasing use of smartphones, tablets, and computers, information security becomes a very important issue. Information theft by strangers aims to read and change information when the transmission of information occurs. Cryptography is a technique for protecting information from unauthorized persons. Cryptography has always been an important task, where the main purpose of cryptography is data security [1]. Cryptography uses fairly complex mathematical calculations in the process of encryption and decryption. Cryptography can be used to secure document files. Cryptography is widely used on applications such as chatting, communication system, mobile devices, and various other embedded applications. To secure communication in chatting, it can be overcome by using cryptography technique end to end encryption [2]. For embedded applications, it can be applied on employee monitoring system. This monitoring system can track information from an employee. This application can provide data protection from the tappers who can steal employee information without their knowledge by using AES cryptography techniques [3]. Cryptographic usage is also available on mobile devices such as encryption and decryption applications using AES algorithms on Android phones [2-4].

There are several cryptographic techniques such as DES, 3DES, Blowfish, RSA and others. DES has a short key length. 3DES uses three keys and key sizes longer than DES, but the process is slower than other cryptographic algorithms. AES is not only suitable for securing data but also faster in terms of encryption and decryption processes. AES can be implemented on various platforms from smartphones [4]. A message digest is a cryptographic hash function that contains a series of digits created with a one-way hashing formula. Message digest is designed to protect the integrity of a piece of data or media to detect changes in the message section [5]. Based on the above description, AES cryptography techniques can be used to secure document files from attacks or bugs. Applications designed by researchers also use a combination of AES cryptography techniques and hashing MD5. Addition of MD5 hashing is used when the key is inputted by the user then the key will go through MD5 hashing process. This applies during encryption and decryption processes. The purpose of adding MD5 hashing is to strengthen cryptographic techniques of an attack and the application is implemented on Android-based smartphones.

2. Method

Cryptography is used on digital communications system. It is used to secure data. Based on the keys used in encryption and decryption, cryptographic keys use two key types: symmetric keys and asymmetric keys. A symmetric key is a key used for both encryption and decryption. While asymmetric keys are keys used for encryption and decryption are not the same, where the sender uses public key and receiver using private key [6]. AES uses symmetric keys in encryption and decryption [7]. AES is a symmetrical block cipher with 128 bit bit size [8]. The 128 bit AES key length is 128 bits and uses 10 rounds. The 192 bit AES key length is 192 bits and uses 12 rounds. The key length of AES 256 bits is 256 bits and uses 14 rounds [9,10,11-12]. Each round on AES 128, 192 and 256 bits in each encryption and decryption process has identical rotation, except the last round. The AES encryption and decryption process consists of four types of byte transformations:

- Sub Byte Transformation: using an S-box table to substitute every byte of a message block. This non linear layer is useful for resistance to differential and linear cryptanalysis attacks.
- Shift Row Transform: shifts every byte of the message block. The byte shift is done from the leftmost byte to the most right byte. This transformation is applied on the second, third and fourth lines, while the first line does not shift. The second row will experience a one-time shift, the third row undergoes two shifts and the fourth row undergoes three shifts.
- Mix Column Transform: substitution that utilizes GF arithmetic (28).
- Add Round Key (ARK): a round key added to a state with bitwise XOR operations.

In cryptography there is a function used to authenticate the security and integrity of a message called a hash function. The hash function is a function that receives a string input of any length and converts it into a fixed length length string. There are two hash functions: Message Digest (MD5) and Secure Hash Algorithm 1 (SHA1). MD5 generates message digest 128 bit and SHA1 produces 160 bit message digest [5,9]. The MD5 algorithm uses a mathematical function to generate a message digest with different input strings. MD5 algorithm has the advantage that MD5 will generate a different message digest if one of the bits has changed the input string. Another advantage is that the message digest size is smaller than the document size generated by MD5 [13]. The following are the steps of making MD5 hashing [14-15]: i) The addition of the wedge bits. The purpose of adding bit bits is for a congruent message length of 448 modulo 512. Messages converted to binary form are added by bit "1" followed by bit "0" to length equal to 448 modulo 512; ii) Increase the value of the original message length. The message that has been given the bits of the wedge then added again with 64 bits that state the length of the original message. Once added with 64 bits, the message length now becomes 512 bits; iii) Initialize the MD buffer. MD requires four buffers each of which is 32 bits in length. The buffer is named A, B, C, D. The total length of the buffer is 128 bits. Each buffer is initialized with values in hexadecimal notation in the form of A: 01234567, B: 89ABCDEF, C: FEDCBA98, D: 76543210; iv) The functions of fF, fG, fH, fI each contain 16 basic operations of the input. The function fF, fG, fH, fI is a function to manipulate inputs A, B, C, D with 32 bit size. Each function is:

$$f_F = \mathrm{F(B, C, D)} = (B \wedge C) \vee (\sim B \wedge D) \quad (1)$$

$$f_G = \mathrm{G(B, C, D)} = (B \wedge D) \vee (C \wedge \sim D) \quad (2)$$

$$f_H = \mathrm{H(B, C, D)} = B \oplus C \oplus D \quad (3)$$

$$f_I = \mathrm{I(B, C, D)} = C \oplus (B \wedge \sim D) \quad (4)$$

v) The output of the message digest is the result of connecting bits in A, B, C, D.

Procedures of encryption and decryption of files on AES 256 bit and MD5 are:

IOP Conf. Series: Journal of Physics: Conf. Series **1007** (2018) 012014 doi:10.1088/1742-6596/1007/1/012014

- File encryption algorithm with AES 256 bit and MD5: i) The user selects the plaintext file to be encrypted; ii) User entering key; iii) The application will hashing the key with MD5; iv) AES 256 bit encryption process; v) The application will generate a ciphertext file.
- File decryption algorithm with AES 256 bit and MD5: i) User selects the ciphertext file to be decrypted; ii) User entering key; iii) The application will hashing the key with MD5; iv) AES 256 bit decryption process; v) The application will generate a plaintext file.

3. Results and Discussions

AES 256 bit application that allows users to encrypt and decrypt documents on Android smartphones like txt files, doc, docx, xls, xlsx, ppt, pptx, pdf. This application uses the bottom navigation menu, so users simply click on the menu located at the bottom of the application. This app is built using the Android Studio editor. For AES encryption and decryption process using AES class developed by Popa Tiberiu 2011 using Java programming language. The following packages and classes are used in this program:

import com.example.user.enkripsideskripsifile.aes.AES; // the class that used for encryption and decryption

The implementation of AES encryption and decryption class show in Figure 1.

```
while(fileSumber.getFilePointer() < fileSumber.length()){
    baca=fileSumber.read(plaintext);

    for(i=baca;i<MAX_BUFFER;i++){
        plaintext[i]=92; //Character padding that less from 16
    }

    for(i=0;i<MAX_BUFFER;i++){
        if(i==15){
            ciphertext=AES.encrypt(plaintext, kunci.getBytes());
            for(j=0;j<MAX_BUFFER;j++){
                temp[j]=ciphertext[j];
            }
        }
    }
    //Save ciphertext file
    fileTujuan.write(temp);
}
```

```
while(fileSumber.getFilePointer() < fileSumber.length()){
    baca=fileSumber.read(ciphertext);

    for(i=0;i<MAX_BUFFER;i++){
        if(i==15){
            plaintext=AES.decrypt(ciphertext, kunci.getBytes());
            for(j=0;j<16;j++){
                temp[j]=plaintext[j];
            }
        }
    }
    //Save plaintext file
    fileTujuan.write(temp);
```

Figure 1. Implementation of encryption and decryption AES 256 class.

Here are the packages and classes used on MD5:

```
import java.math.BigInteger;
import java.security.MessageDigest;
import java.security.NoSuchAlgorithmException;

public class MD5 {
    public static String getMD5(String input){
        try{
            MessageDigest md= MessageDigest.getInstance("MD5");
            byte[] messageDigest=md.digest(input.getBytes());
            BigInteger number=new BigInteger(1, messageDigest);
            String hashText=number.toString(16);
            while (hashText.length()<32){
                hashText="0"+hashText;
            }
            return hashText;

        }catch (NoSuchAlgorithmException nsae){
            throw new RuntimeException(nsae);
        }
    }
}
```

Figure 2. The MD5 class.

Implementation of MD5 class shown in Figure 3 and Figure 4

IOP Conf. Series: Journal of Physics: Conf. Series **1007** (2018) 012014 doi:10.1088/1742-6596/1007/1/012014

```
String kunci_awal=ekunci_enkripsi.getText().toString();
while (kunci_awal.length()<32){
    kunci_awal="]"+kunci_awal;
}
String kunci=MD5.getMD5(kunci_awal);
```

Figure 3. MD5 implementation in AES encryption.

```
String kunci_awal=ekunci_dekripsi.getText().toString();
while (kunci_awal.length()<32){
    kunci_awal="]"+kunci_awal;
}
String kunci=MD5.getMD5(kunci_awal);
```

Figure 4. MD5 implementation in AES decryption.

Encryption Menu: As it is shown in the screen shot in Figure 5, the user interface is created and designed by researchers using the .xml layout in Android Studio and name it layout_enkripsi.xml. This menu is used for the encryption process, the user simply select the plaintext file and enter a key or password to encrypt the document. The time required for the decryption process is also displayed at this stage.

Decryption Menu: As it is shown in the screen shot in Figure 6, the user interface is created and designed by researchers using the .xml layout in Android Studio and name it layout_dekripsi.xml. This menu is used for the decryption process, the user simply select the ciphertext file and enter a key or password to decrypt the document. The time required for the decryption process is also displayed at this stage.

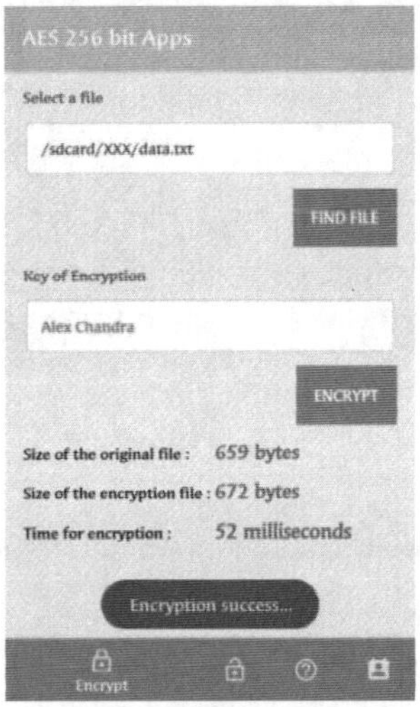

Figure 5. Encrypt the file.

Figure 6. Decrypt the file.

In this section the researcher conducted an encryption experiment on the type of * .txt file with the size of 659 bytes shown in Figure 7. The ciphertext file will generate a 672-byte * .ct file extension shown in Figure 8. In the experiment's experiment, the ciphertext file the decrypted looks the same as the original file, but the ciphertext file is 672 bytes. All the encryption and decryption process files are shown in Figures 7, 8, 9 which are opened by using the Notepad ++ applsication.

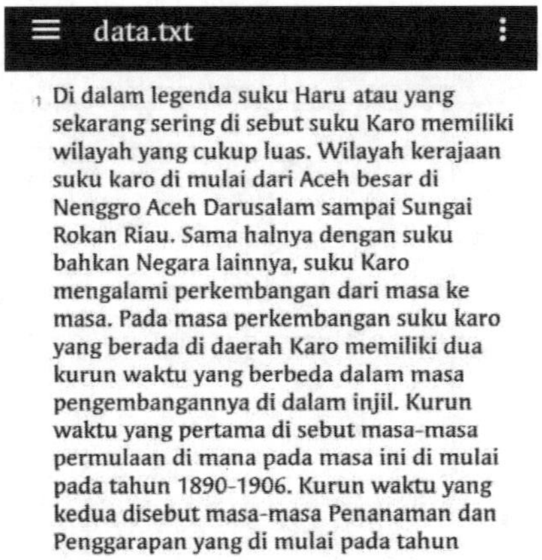
data.txt

Di dalam legenda suku Haru atau yang sekarang sering di sebut suku Karo memiliki wilayah yang cukup luas. Wilayah kerajaan suku karo di mulai dari Aceh besar di Nenggro Aceh Darusalam sampai Sungai Rokan Riau. Sama halnya dengan suku bahkan Negara lainnya, suku Karo mengalami perkembangan dari masa ke masa. Pada masa perkembangan suku karo yang berada di daerah Karo memiliki dua kurun waktu yang berbeda dalam masa pengembangannya di dalam injil. Kurun waktu yang pertama di sebut masa-masa permulaan di mana pada masa ini di mulai pada tahun 1890-1906. Kurun waktu yang kedua disebut masa-masa Penanaman dan Penggarapan yang di mulai pada tahun

Figure 7. Input original file *.txt before encryption.

Figure 8. Encrypted File *.txt.

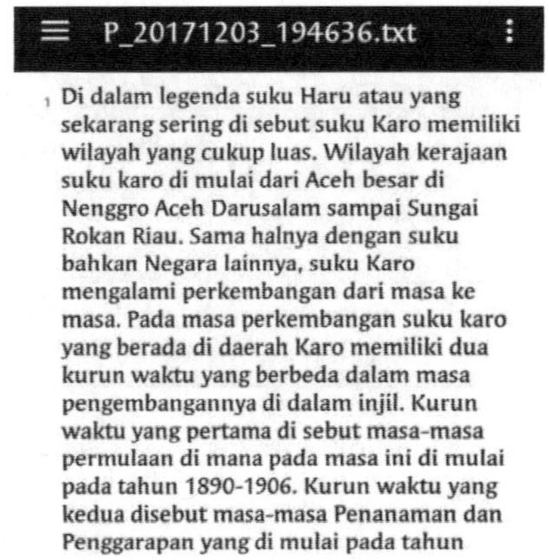
P_20171203_194636.txt

Di dalam legenda suku Haru atau yang sekarang sering di sebut suku Karo memiliki wilayah yang cukup luas. Wilayah kerajaan suku karo di mulai dari Aceh besar di Nenggro Aceh Darusalam sampai Sungai Rokan Riau. Sama halnya dengan suku bahkan Negara lainnya, suku Karo mengalami perkembangan dari masa ke masa. Pada masa perkembangan suku karo yang berada di daerah Karo memiliki dua kurun waktu yang berbeda dalam masa pengembangannya di dalam injil. Kurun waktu yang pertama di sebut masa-masa permulaan di mana pada masa ini di mulai pada tahun 1890-1906. Kurun waktu yang kedua disebut masa-masa Penanaman dan Penggarapan yang di mulai pada tahun

Figure 9. Decrypted File *.txt after decryption.

In this section the researcher also permorfmed encryption experiments on the * .docx file type with the size of 864465 bytes shown in Figure 10. The ciphertext file will produce a file extension * .ct of size 864480 bytes shown in Figure 11. In the experiment, the decrypted ciphertext looks the same as the original file, but the ciphertext file is 864480 bytes. All the encryption and decryption process files are shown in Figure 10, 12 which are opened using the Office WPS application and Figure 11 opens with the Notepad ++ application.

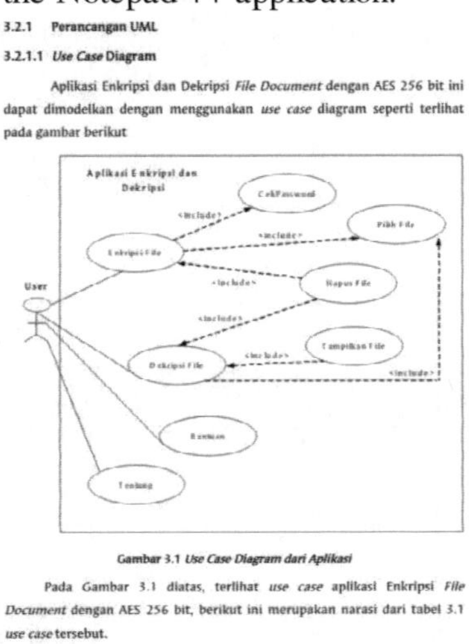

Figure 10. Input original file *.docx before encryption.

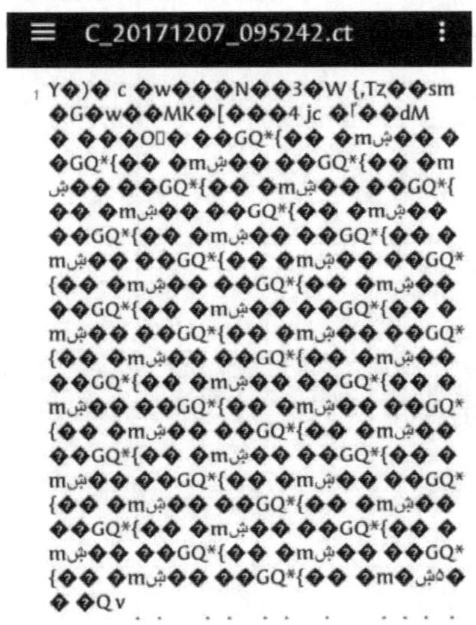

Figure 11. Encrypted File *.docx.

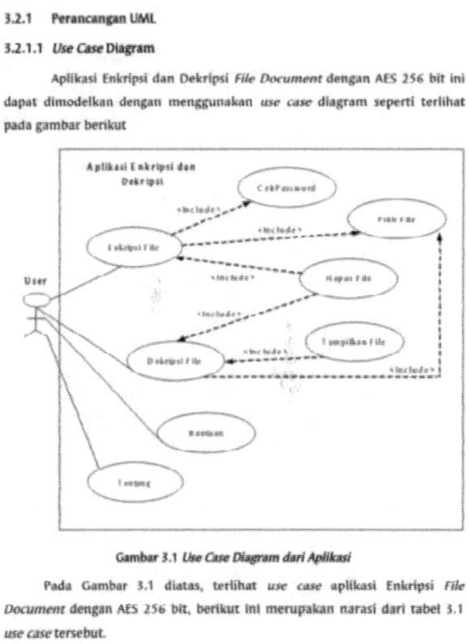

Figure 12. Decrypted File *.docx after decryption.

Encryption and decryption applications have been installed and tested on smartphone devices based on the Android operating system. The smartphone specs used during the experiment were Asus Zenfone 6, Intel Atom 2 GHz CPU Model, 2 GB RAM and Android 5.0 Lollipop. The results of encryption and decryption of files using AES 256 bits are shown in Table 1.

Table 1. The results of encryption and decryption of files using AES 256 bit.

No	File Name	The size of the original	File Encryption	Duration of Encryption	File Size Decryption	Duration of Decryption

		file (byte)	Size (byte)	(ms)	(byte)	(ms)
1	data.txt	659	672	107	672	43
2	Data Teknisi.xlsx	12,877	12,880	533	12,880	969
3	readme.txt	15,637	15,648	789	15,648	1,329
4	Chapter I.docx	23,556	23,568	1039	23,568	2,040
5	Slide.pptx	71,847	71,856	2,960	71,856	7,876
6	Student data.xlsx	97,585	97,600	3,915	97,600	12,330
7	SQL Tutorial.pdf	506,113	506,128	21,529	506,128	152,161
8	Chapter III.docx	864,465	864,480	40,383	864,480	390,521
9	Writing guidelines.pdf	867,691	867,696	36,382	867,696	386,393
10	PPS TI.pptx	1,632,198	1,632,208	72,274	1,632,208	1,217,620

Table 2. The results of encryption and decryption of files using AES 256 bit combination with MD5

No	File Name	The size of the original file (byte)	File Encryption Size (byte)	Duration of Encryption (ms)	File Size Decryption (byte)	Duration of Decryption (ms)
1	data.txt	659	672	52	672	40
2	Data Teknisi.xlsx	12,877	12,880	618	12,880	1,068
3	readme.txt	15,637	15,648	751	15,648	1,138
4	Chapter I.docx	23,556	23,568	1,089	23,568	2,014
5	Slide.pptx	71,847	71,856	2,817	71,856	8,067
6	Student data.xlsx	97,585	97,600	4,815	97,600	12,166
7	SQL Tutorial.pdf	506,113	506,128	23,022	506,128	395,923
8	Chapter III.docx	864,465	864,480	37,435	864,480	386,988
9	Writing guidelines.pdf	867,691	867,696	37,570	867,696	1,602,469
10	PPS TI.pptx	1,632,198	1,632,208	75,136	1,632,208	200,981

Table 1. shows that the decryption process takes considerably longer than the encryption process. The larger the file size, the encryption and decryption process will take a long time and this applies to all document file types as well. The process of encrypting and decrypting a text file has a fairly short duration of time. Encryption and decryption of *. xlsx files has a fairly fast time duration. While encryption and decryption of document files such as *.docx, *.pptx, *.pdf where in the file there are text and images that have a long duration of time. Table 2 shows that the results are similar to Table 1. In some experiments conducted by researchers with the same document file, the results of encryption and decryption in document files such as data.txt and readme.txt files show faster results using a

combination of AES 256 bits with MD5 than just using AES only. While the document file containing graphic data such as tables, images, lines in the file *.docx, *.xlsx, *.pptx, *.pdf then encryption using only AES 256 bit method faster than using AES 256 combination bit and MD5.

4. Conclusion

In this paper, it shows that the successful implementation of file document encryption as well as decryption on Android smartphones. Experience from experimental results was conducted by researchers, the encryption process is faster than decryption. It shows that the combination of AES 256 bit encryption and decryption algorithms and MD5 runs faster on Android smartphones. This app provides better mobile security than unauthorized access. This application ensures the security of end to end data transfer without experiencing corrupt data, but for file size, this is larger than the original file. The time required for encryption and decryption for document files using a combination of AES and MD5 is faster than just using AES. While the files contain graphic data such as tables, lines, images of encryption time and decryption faster by using AES alone than using a combination of AES and MD5.

References

[1] Nishika and R. Kumar Yadav,"Cryptography on Android Message Applications – A Review", International Journal on Computer Science and Engineering (IJCSE) ISSN : 0975-3397, vol. 5 no. 05, May 2013

[2] A.H Ali and A. Makki Sagheer, 2017, "Design and Implementation of Secure Chatting Application with End to End Encryption", Journal of Engineering and Applied Sciences 12 (1) ISSN : 1816-949X, 156-160

[3] A. Pandey and U. Kumar Lilhore, 2017, "An Improved AES Cryptosystem Based Genetic Method on S-Box, With, 256 Key Sizes and 14-Rounds", International Journal of Advanced Engineering Research and Science (IJAERS), ISSN: 2349-6495(P) | 2456-1908(O) vol.4 no. 3 March 2017, pp. 166-171.

[4] S. Tayde and S. Siledar, 2015, "File Encryption, Decryption Using AES Algorithm in Android Phone", International Journal of Advanced Research in Computer Science and Software Engineering, vol. 5 no. 05 May 2015, pp. 550-554.

[5] S.Mishra, Shikha Mishra, N.Kumar, 2013, "Hashing Algorithm: MD5", IJSRD - International Journal for Scientific Research & Development ISSN (online): 2321-0613 vol. 1, no.9, 2013, pp.1931-1933.

[6] R. Verma and Shikha Agrawal, 2013, "Data Security For Any Organization By Using Public Key Infrastructure Components And MD5, RSA Algorithms", IJRET: International Journal of Research in Engineering and Technology ISSN: 2319-1163 , vol.2 Issue: 05, May-2013

[7] A. Pandey and U. Kumar Lilhore, 2017, "An Improved AES Cryptosystem Based Genetic Method on S-Box, With, 256 Key Sizes and 14-Rounds", International Journal of Advanced Engineering Research and Science (IJAERS), ISSN: 2349-6495(P) | 2456-1908(O) vol.4 no. 3 March 2017, pp. 166-171.

[8] Preeti Singh et al, 2014, "Symmetric Key Cryptography: Current Trends", International Journal of Computer Science and Mobile Computing ISSN 2320–088X, vol.3 Issue.12, December-2014, pp. 410-415.

[9] N.K. Jharbade† and R. Shrivastava, 2012, "Network based Security model using Symmetric Key Cryptography (AES 256– Rijndael Algorithm) with Public Key Exchange Protocol (Diffie-Hellman Key Exchange Protocol), IJCSNS International Journal of Computer Science and Network Security, vol.12 No.8, August 2012, pp. 69-73.

[10] M.Gavhane, S.More, G.Ghodke, Patil S.S., Nimbalkar P.P, 2015, "Spy Android App", International Journal of Emerging Technologies and Engineering (IJETE) ISSN 2348 – 8050 vol.2 no.2, February 2015, pp.33-37.

IOP Conf. Series: Journal of Physics: Conf. Series **1007** (2018) 012014 doi:10.1088/1742-6596/1007/1/012014

[11] S. El Adib and N. Raissouni, 2012, "AES Encryption Algorithm Hardware Implementation Architecture: Resource and Execution Time Optimization", International Journal of Information & Network Security (IJINS) ISSN: 2089-3299 , vol.1, no.2, June 2012, pp. 110-118
[12] S.R.Paddhan, T.R. Wagh, S.N. Shelke, 2014, "Secured Wireless Communication Using AES-256", Global Journal of Advanced Engineering Technologies, Special Issue (CTCNSF-2014) ISSN (Online): 2277-6370 & ISSN (Print): 2394-092
[13] S. Ranjan Lenka and B. Nayak, 2014, "Enhancing Data Security in Cloud Computing Using RSA Encryption and MD5 Algorithm', International Journal of Computer Science Trends and Technology (IJCST), ISSN : 2347-8578 vol. 2 no. 3 June 2014, pp.60-64.
[14] P. Walia and V.Thapar, 2014, "Implementation of New Modified MD5-512 bit Algorithm for Cryptography", International Journal of Innovative Research in Advanced Engineering (IJIRAE) ISSN: 2349-2163, vol.1 issue:6, July 2014
[15] A. Kumar Kasgar, M.Kumar Dhariwal, N. Tantubay, H.Malviya, 2013,"A Review Paper of Message Digest 5 (MD5)", International Journal of Modern Engineering & Management Research ISSN: 2320-9984, vol.1, Issue 4, December 2013

Acknowledgements
Researchers give many thanks to the Faculty of Technology and Computer Science, Universitas Prima Indonesia. Moreover, special thanks to the Abdi Dharma (Dean of Faculty of Technology and Computer Science) and Arjon Turnip for supporting us in this research.

The diagnose of oil palm disease using Naive Bayes Method based on Expert System Technology

Marlince Nababan*, **Yonata Laia, Delima Sitanggang, Oloan Sihombing, Evta Indra, Saut Siregar, Windania Purba, Roy Mancur**

Faculty of Technology and Computer Science, Universitas Prima Indonesia, Indonesia

E-mail: *marlince@unprimdn.ac.id

Abstract. Expert system is dealt with system that used computer-based human intelligence to overcome particular problem which is commonly conducted by an expert. Frequent problem faced by the farmers of oil palm is the difficulty in defining the type of plant disease. As a result, the delay treatment of plant disease brings out the declining of farm products. An application system is needed to deal with the obstacles and diagnosing the type of oil palm plant disease. The researcher designed an intelligence-based application with input-output plan which is able to diagnose the type of oil palm plant disease by applying naive bayes method. Based on the research result by conducting bayes method with recognized symptom, diagnose of oil palm plant disease could be accomplished. The data of symptoms found are leaves turned yellow 0.4, dead leaves 0.4, black and brown color among the veins of leaves 0.5, young and old fruit with whole space 0.4, and decay of bunches is 0.3. The roots are tender in the amount of 0.5, and damage on sheath is 0.3. Through the chosen symptoms as mentioned above, the value of bayes is 80% with the type of disease is rotten bunch.

1. Introduction

Palm oil contributes almost half of vegetable oil in entire world, which most of them are from mesocarp of oil palm of Africa, ElaeisguineensisJacq. Unfortunately, oil palm plantation in Southeast Asia, especially Malaysia and Indonesia are threatened by Basal Stem Root (BSR). BSR causes the extinct of 50% of oil palm product [1-2]. Technology that can adopt the process and way of thinking of human which is called artificial intelligence. In the prior research, expert system was applied as developing device in agriculture which was aimed to increase the growth of farm efficiently, including with decision making. When the expert system was applied, the result of research showed that the value of accuracy level is 84.15%. In other words, expert system could be used to as an effective training device in agriculture sector. Further, the research also shown that there was an increase of work performance when expert system was applied [3]. The most common obstacle invades oil palm is plant disease and epidemic pest or infection. Infected oil palm plant is usually indicated by various symptoms such as spot or colorful scratch on the leaves, stem, and seed of the plant.

Nowadays, in agriculture sector, diagnosing the type of plant disease is based on the expertise of human intelligence that tended to use conventional method, in the sense that the device is usually tested in laboratory. Therefore, the need for new approach to classify the type of plant disease is the main priority. This research is purposed to focus on innovative method based on technical processing of image. It is done to classify the shortage of nutrition which usually invades the leaves of oil palm by

IOP Conf. Series: Journal of Physics: Conf. Series **1007** (2018) 012015 doi:10.1088/1742-6596/1007/1/012015

analyzing the surface. The feature vector will be achieved as the input for fuzzy classifier. Overall, fuzzy method that proposed is beneficial for oil palm industry in gratifying industrial demand [4-7]. This research offers a description of how to apply WSN in the field and how the learning model machine is designed to predict the plant pest or disease by conducting Algorithm of Naive Bayes [8-12]. Earlier research was purposed to identify and classifying parts of plant disease pattern that consists of color and oil palm texture. Therefore, comparative study that related to segmentation of oil palm leaf figure taken from oil palm to be evaluated and validated by using Global Otsu Method by displaying that threshold of Global Otsu is the best segmentation based on tested image [13].

Prior research of expert system was mostly used in agriculture sector, especially oil palm. Some methods have been applied are fuzzy and Global Otsu Method. Fuzzy was used to classify the type of plant disease [14]. While Global Otsu Method was conducted to analyze the discriminator, in the sense that distinguishing between two or more groups that arose naturally. Discriminator will maximize the variables to divide the object of both foreground and background. The application of Global Otsu Method differentiate or classifying the pattern of oil palm plat disease [15]. In the earlier research, naive bayes was used to forecast the plant disease. Furthermore, naive bayes method was applied in the research to diagnose the plant disease of oil palm, coffee, onion, and assisting doctor in diagnosing any particular plant diseases [16-20]. Naive bayes classifier works very well comparing with another classifier model. It is indicated by Xhemali, Hinde and Stone in their prominent journal entitled "Naive bayes vs. Decision Trees vs. Neural Networks in the Classification of Training Web Pages" that elaborated that "Naive Bayes Classifier possesses much better level of accuracy than another classifier model [21].

2. Methodology

Data obtained was all about plants diseases that are commonly suffered by oil palm, symptoms, and ways of treating them. Disease symptom with probability level (P) is indicated drying leaves (0.4), leaves turned yellow (0.4), dead leaves (0.5) black and brown color among the veins of leaves (0.5), root turned tender (0.4), leaves are wilted (0.3), damage on the frond (0.3), torn leaves and totally torn (0.7), dead plant (0.5), hand of flower or closed spear of flower (0.4), tip of leaves turned into other color (0.6), decay on the stem around 2 meters above the ground turned into grey (0.5), decay of bunch (0.7), young or old leaves look holey (0.6), delayed formation of flower (0.3), tip of plant decay and turned brown (0.3), tip turned bended and curved (0.5), spot and fleck on the leaves (0.4), leaves fallen (0.3), tiny yarn (mycelium) turned into shiny white (0.5), pericarp faded and decay (0.4), fruit color turned and turned into brownish (0.3), damage in the plant in the age of 3 to 10 years (0.4), and decay on the main stem (0.4).

Expert system [22] is a computerized system that accumulates human decision making capacity which works for all things as human expert. Expert system arises from the very earlier of working process in overcoming the problems, especially when specific domain of intelligence is required. The knowledge of expert must be obtained from specialists or another relevant expert, such as texts, journal, and sources of data. Moreover, expert system receives facts from users and replies it back with expertise. The users will interact with the system through face to face users, then building by using menus, natural language or any other interactions. All the rules collected from the domain of expert are decoded in the form of intelligence basis. Then, inferential machine will conclude all the things from the intelligence basis and facts given by users. The expert system may or may not possess learning components. Series of expert advisory systems [23-24] can be developed in agriculture, and has been conducted in India.

Naive bayes is a statistical classification that can be used to predict the membership probability of particular class. Naive bayes is grounded from Bayestheorem that has classification ability as decision tree and neural network have. Naive bayes empirically proven has an accurate and high speed when it is applied in the database with huge number of data [8], with the formula (1), where X: data with unknown class, H: hypothesis data X comprises a specific class, P(H|X), P(H): probability of hypothesis H (prior probability), P(X|H): probability of X based on the condition of hypothesis H, P(X): probability of X.

$$P(C \mid X) = \frac{P(X \mid C)P(C)}{P(X)} \quad (1)$$

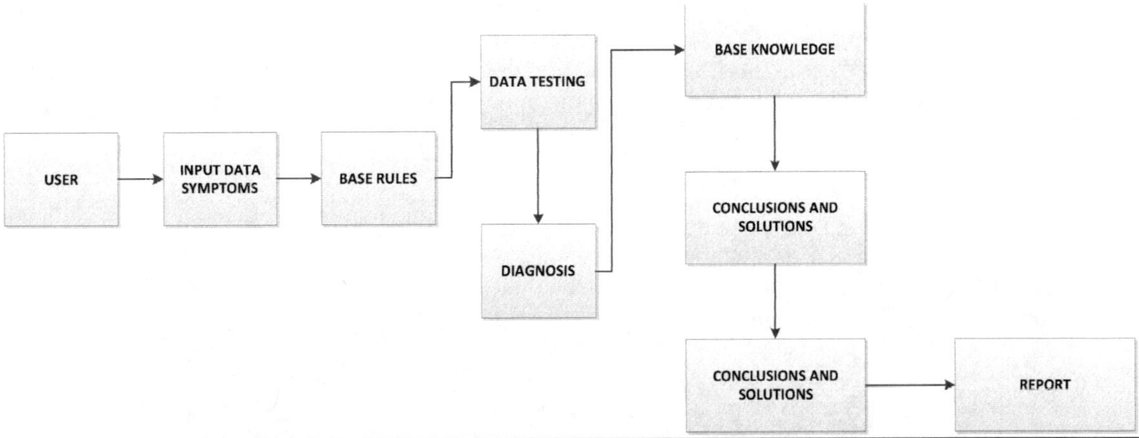

Figure 1. Diagram of Block system with bayes method.

3. Result and Analysis

The figures below elaborate the process within the application of expert system which has been designed by the researcher. Login button functions as security device when the data is needed to be edited or completed by the administrator. In other words, all the process should be through this phase. Further, if the login phase is correct, the main page can be entered which provides some menu such as Home, Types of plant disease, Symptoms, Training, and Password. Another important button is Logout menu which has been filled with some links connected to particular pages.

If diagnosing is going to be done, first, Symptoms button must be clicked then the display would be as figure number 2 below. Then, we have to click the symptom which is to be diagnosed. Below is the example of yellow leaves, dead leaves, brown and black among the bones, young or old fruit looks with all spaces, decay in bunches, the roots are tender, and damage on sheath is 0.3. Through the chosen symptoms as mentioned above, the value of bayes is 80% with the type of disease is rotten bunch.

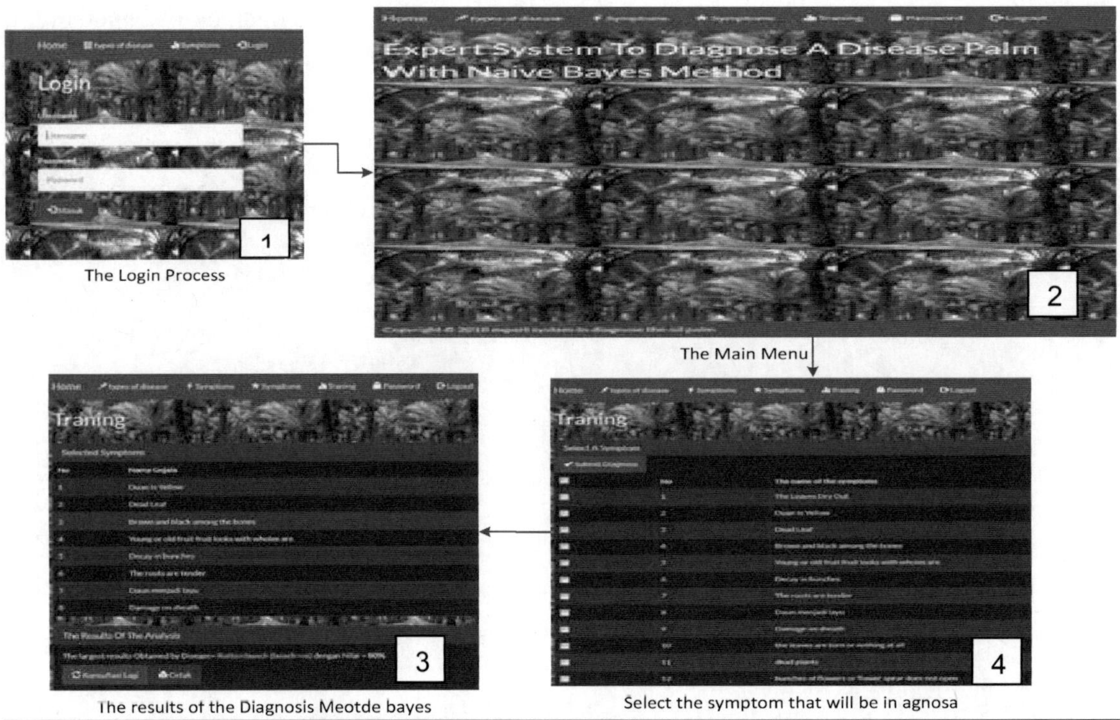

Figure 2. Process of diagnose of naive bayes.

4. Conclusion

Based on the research result by conducting bayes method with recognized symptom, diagnose of oil palm plant disease could be accomplished. The data of symptoms found are like yellow leaves are 0.4, dead leaves are 0.4, brown and black among the bones are 0.5, young and old fruit with whole space are 0.4, and decay in bunches is 0.3. The roots are tender in the amount of 0.5, and damage on sheath is 0.3. Through the chosen symptoms as mentioned above, the value of bayes is 80% with the type of disease is rotten bunch.

Refrences

[1] Pilotti, 2005, C.A. Stem rotsof oil palm caused by Ganodermaboninense: Pathogen biology and epidemiology. Mycopathologia 129–137

[2] Singh, R ; Ong-Abdullah, M; Low, E.T; Manaf, M.A; Rosli, R; Nookiah, R; Ooi,L.C.; Ooi, S.E.; Chan, K.L.; Halim, 2013, Oil palm genome sequence reveals divergence of interfertile species in old and new worlds. Nature 335–339

[3] Santoso, H.; Gunawan, T.; Jatmiko, R.; Darmosarkoro, W.; Minasny, B. Mapping and identifying

[4] Yu-Te Chang and Bor-Sen Chen, 2010, A Fuzzy Approach for Robust Reference-TrackingControl Design of Nonlinear Distributed Parameter Time-Delayed Systems and Its Application, IEEE transactions on fuzzy systems, vol. 18, no. 6

[5] Qiu, G. Feng, and J. Yang, 2009, A new design of delay-dependent robustH∞ filtering for discrete-time T–S fuzzy systems with time-varying delay, IEEE Trans. Fuzzy Syst, vol. 17, no. 5, pp. 1044–1058

[6] Muhammad AsrafHairuddin, NooritawatiMdTahir, Shah Rizam Shah Baki, 2011, Overview of image processing approach for nutrient deficiencies detection in ElaeisGuineensis, System Engineering and Technology (ICSET), IEEE International Conference on, 978-1-4577-1255-5

[7] Kusumadewi, Sri. 2003. Artificial Intelligence (TeknikdanAplikasinya). GrahaIlmu.Yogyakarta
[8] Rish, I, 2006, An empirical study of the Naive Bayes Classifier, International Joint Conference on Artificial Intelegence, California, 41-46
[9] Silachan K and Tantasanawong P, 2011, Evaluation Information Extraction for Health Text Categories Using C4.5 and Naive Bayes, Proceedings of ICIME Conferences, Toronto, Ryerson University, 403-411
[10] Prof. M.S. Prasad Babu, J. Anitha and K. Hari Krishna, 2010, A Web Based Sweet Orange Crop Expert System using Rule Based System and Artificial Bee Colony Optimization Algorithm, International Journal of Engineering Science and Technology ,volume. 2
[11] Prof. M.S. Prasad Babu, N.ThirupathiRao, 2010, Implementation of Parallel Optimized ABC Algorithm with SMA Technique for Garlic Expert Advisory System, International Journal of Computer Science, Engineering and Technology (IJCSET), Volume 1, Issue 3
[12] Dr. B D C N Prasad, P E S N Krishna Prasad and Y Sagar, 2011, A Comparative Study of Machine. Learning Algorithms as Expert Systems in Medical Diagnosis (Asthma)– CCSIT, Part I, CCIS 131, pp 570– 576
[13] MansinghGunjan, Reichgelt Han and Kweku-MuataOsei Bryson, 2007, *CPEST: An expert system for the management of pests and diseases in the Jamaican coffee industry*, Expert Systems with Applications, Vol. 32(1):184–192
[14] Kumar Ashish a, AnishDasari a, SubhagataChattopadhyay b, NirmalBaranHui a, 2018, Genetic-neuro-fuzzy system for grading depression, Applied Computing and Informatics 14, 98–105
[15] Muhammad AsrafHairuddin, NooritawatiMdTahir, Shah Rizam Shah Baki, 2011, *Overview of image processing approach for nutrient deficiencies detection in ElaeisGuineensis,* System Engineering and Technology (ICSET), IEEE International Conference on
[16] MansinghGunjan, Reichgelt Han and Kweku-MuataOsei Bryson, 2007, *CPEST: An expert system for the management of pests and diseases in the Jamaican coffee industry*, Expert Systems with Applications, Vol. 32(1):184–192
[17] Dr. B D C N Prasad, P E S N Krishna Prasad and Y Sagar, 2011, A Comparative Study of Machine. Learning Algorithms as Expert Systems in Medical Diagnosis (Asthma)–CCSIT, Part I, CCIS131, pp 570– 576
[18] Muhammad AsrafHairuddin, NooritawatiMdTahir, Shah Rizam Shah Baki, 2013, *ElaeisGuineensis leaf image segmentation: A comparative study and analysis*, Shah Alam Malaysia, System Engineering and Technology (ICSET), IEEE 3rd International Conference on, 978-1-4799-1030-4.
[19] Prof. M.S.Prasad Babu, N.V. RamanaMurty, S.V.N.L.Narayana, A Web Based Tomato Crop Expert Information System Based on Artificial Intelligence and Machine learning algorithms, International Journal of Computer Science and Information Technologies, Vol. 1, ISSN: 0975 -9646, pp6-15.
[20] Daniela Xhmeli, ChristhoperJ.Hinde& Roger G.Stone, 2009, Naive bayesvs Decision Tree vs Neural Network in the classification of training web pages, IJSI Intenational Journal of Computer Science Issus, Vol.4, No.1.
[21] Mulyanto, Edy, T. Sutojo, Vincent Suhartono, 2011, KecerdasanBuatan. Andi.Yogyakarta.
[22] Putra, Darma, 2010. Pengolahan Citra Digital, Yogyakarta:Andi.
[23] Rafea,A. and K. Shaalan, 1996, *Using Expert Systems as a Training Tool in the Agriculture Sector in Egypt*, Expert Systems with Applications, Elsevier Science Ltd, Vol. 11: 343-349.
[24] HemantkumarWani, NilimaAshtankar, 2017, *An appropriate model predicting pest/diseases of crops using machine learning algorithms*, Coimbatore, India, Advanced Computing and Communication Systems (ICACCS), 17137092

IOP Conf. Series: Journal of Physics: Conf. Series **1007** (2018) 012016 doi:10.1088/1742-6596/1007/1/012016

File Cryptography with AES and RSA for Mobile Based on Android

Yonata laia[1]*, Marlince Nababan[1], Oloan Sihombing[1], Siti Aisyah[1], Delima Sitanggang[1], Saut Parsaoran[1], Niskarto Zendato[2]

[1]Faculty of Technology and Computers Science, Universitas Prima Indonesia, Indonesia
[2]Faculty STMIK Neumen Kristen Indonesia, Indonesia

E-mail :*yonatan_li@yahoo.co.id

Abstrak. The users of mobile based on android were increasing currently even now mobile was almost the same computer one of which could be used to be done by every users mobile was save the individual important data.Saving the data on mobile was very risk because become hackers' target. That was the reason of researchers want to add cryptography which the combination between Advance Encryption System (AES) dan Ron Rivest, Adi Shamir dan Len Adleman (RSA). The result of the second method above could do cryptography data on mobile. With different encryption time where the file size; 25.44 KB, encryption time 4 second, 200 KB, 5 second, 600 KB 7 second, 2.29 MB, 10 second. Where decryption 25.44 KB, encryption 2 second, 200 KB, 1.5 second, 600 KB 2.5 second, 2.29 MB, 2.7 second.

1. Introduction

Mobile which was using application of android will be needed to satisfy social expectations which was connecting and social needs now days the reason of increasing Smartphone usage. One of the leader of Smartphone was Google Android system operation. It can be possible that Android will be installed on many phones in the near future [1-3].Android has every system smartphone operation, security was one of attention mobile users nowdays, smartphone mobile very often virus attact etc [4]. Every person was using Smartphone like PC to save private our individual and data PC to save and individualdata or secret.

Smartphone become the main point of hackers attack particularly android operation system of Smartphone. The previous researchers have done the comparison RSA, AES and DES and the result of this research showed that AES and DES needed little time and little memory, while RSA needed the long time and the using big memory, but in the giving strong key than DES and AES method [5]. The previous researchershave done in connecting algorithm of DES and RSA and showing stronger than using one method only [6-12]. Based on increase in mobile application in many platforms, security was the most necessary in issue, monitoring, security, reliability and accurate [13]. The previous researchers have ever done research to compare AES and RSA to encrypt the data in web [14-15].

In this research, there would be make an application that did the security data in mobile that android with RSA and AES method, the application could be download in play store so it was easier to Smartphone user to download on the individual mobile and developing the previous EKG.

2. Method

The RSA algorithm is the most popular and proven asymmetric key cryptographic algorithm; it can contrary to password attack to present. RSA comes from the founder name, the first algorithm encryption and digital signature. The security is in the two prime number count math from two big of the number. In RSA, two algorithmnumberusedtomaking two public and private key. It was hard to know the original message from the signal key. It was safe from brute force attack too [16-19].

3. Algorithm Advanced Encryption

AES standard was block cipher where the length of this block was 128 bit. AES enable to three the different key long: 128, 192 or 256 bit. Encryption consists of 10 rounds processing to 128 key bit, 12 round for 192 bit and 14 round for 256 bit. Every round processing consists of one single – byte substitution based on steps, wisdom-line, step permutation, wisdom column mix stepand beside of the general round.

The four steps were running in different encryption and decryption. It was not the same like DES, algorithm different decryption based on substantial from algorithm encryption. Eventhough from all steps were the same step that used in encryption, the steps that done was different with before [20-21]. All the planning of algorithm was public (not a secret) the long of key flexible; 128, 192 and 256 bit, the block measure that encryption was 128 bit, algorithm could be implicated in software or whadware [22].

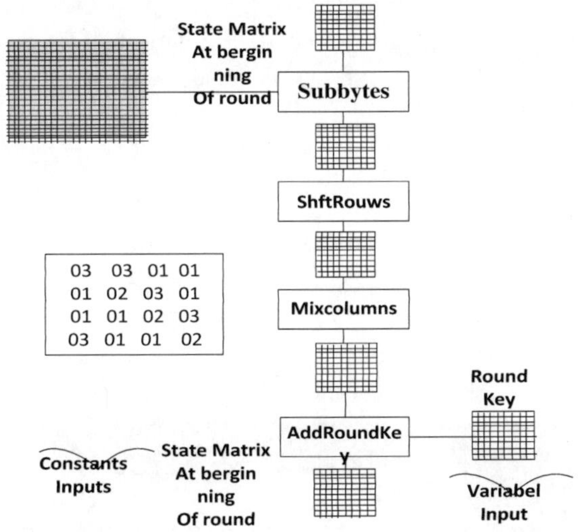

Figure 1. Input For Single AES Around [23].

4. Design and development

Figure 2 was explain how the process of giving key toward data with RSA and AES method. The first step done was determining the file that key giving. After the data was determine so the next was the process of key RSA and AES method.

IOP Conf. Series: Journal of Physics: Conf. Series **1007** (2018) 012016 doi:10.1088/1742-6596/1007/1/012016

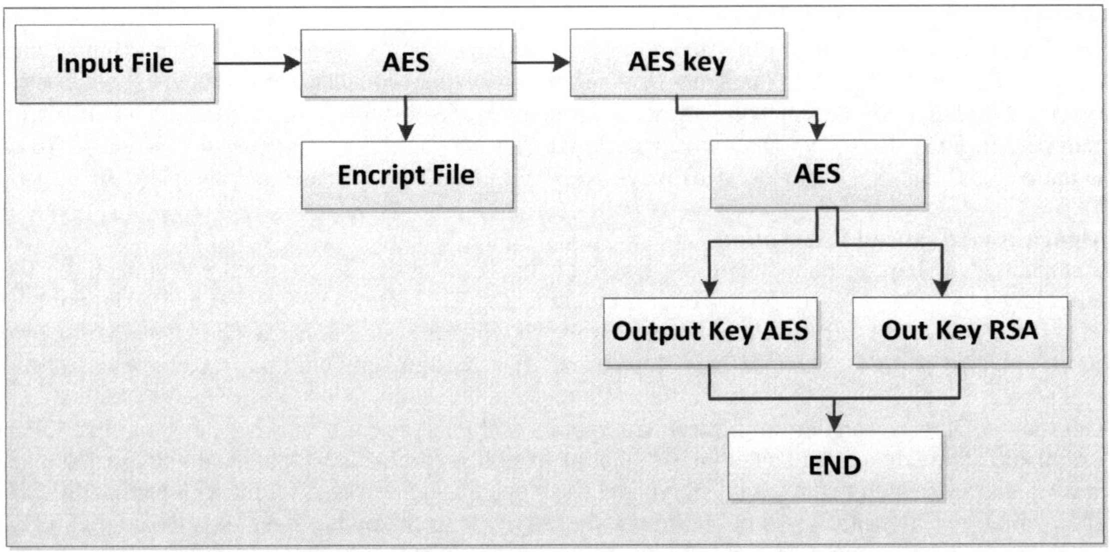

Figure 2. The Combined Block Diagram Method Of RSA And AES.

Figure 3 was explain how the process of giving key toward data with RSA and AES method. The first step done was determining the file that key giving. After the data was determine so the next was the process of key RSA and AES method.

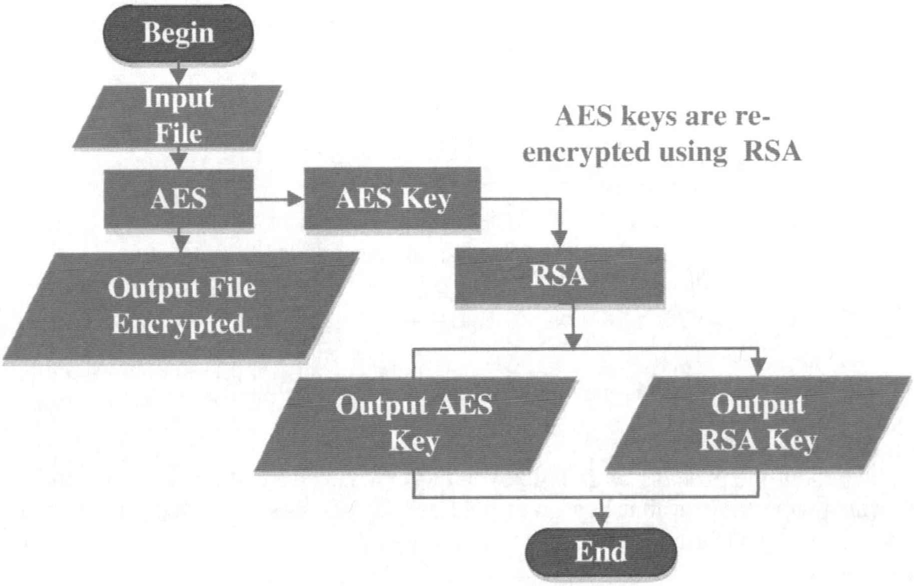

Figure 3. Flowchart Enkripsi.

The user insert the file in file system will be decrypt, RSA private, and AES. The system would be decrypted by AES using RSA private, AES that has decrypted would be use to decrypt file, the file that has been decrypted would be export.

IOP Conf. Series: Journal of Physics: Conf. Series **1007** (2018) 012016 doi:10.1088/1742-6596/1007/1/012016

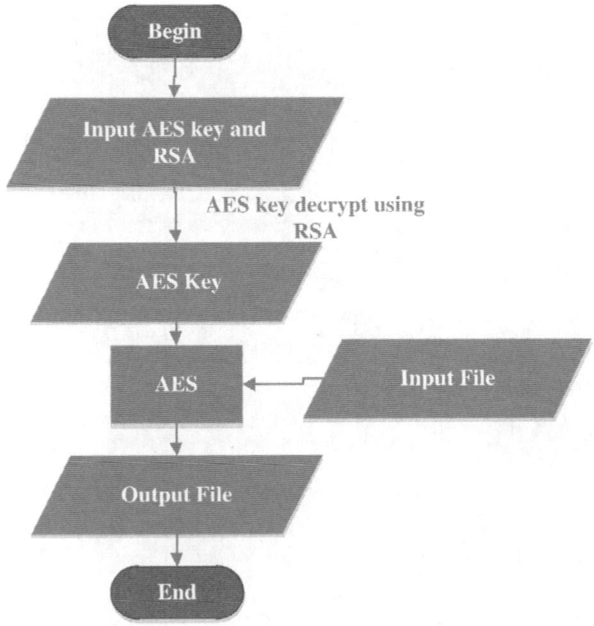

Figure 4. Flowchart Decripsion.

5. Discussion

Figure 5 was explained how the process of making RSA and AES in file that has been determined. The first process was download of application in play store, second step installation process, third step has run and could choose the file which encryption, fourth step choosing the encryption file, fifth step showing the file performance that has been determined.

Based on result of this research could be seen how the process of cryptography in android system.Therefore, found the file first would be encrypting, after it assign the number in increasing key based on algorithm which applied, whereas the key would be used in opening file that has been encrypting become decrypting. Below could be seen in figure 6 the file that have been encryption figure 7 was file that has been encryption in safety file which was used android system.

IOP Conf. Series: Journal of Physics: Conf. Series **1007** (2018) 012016 doi:10.1088/1742-6596/1007/1/012016

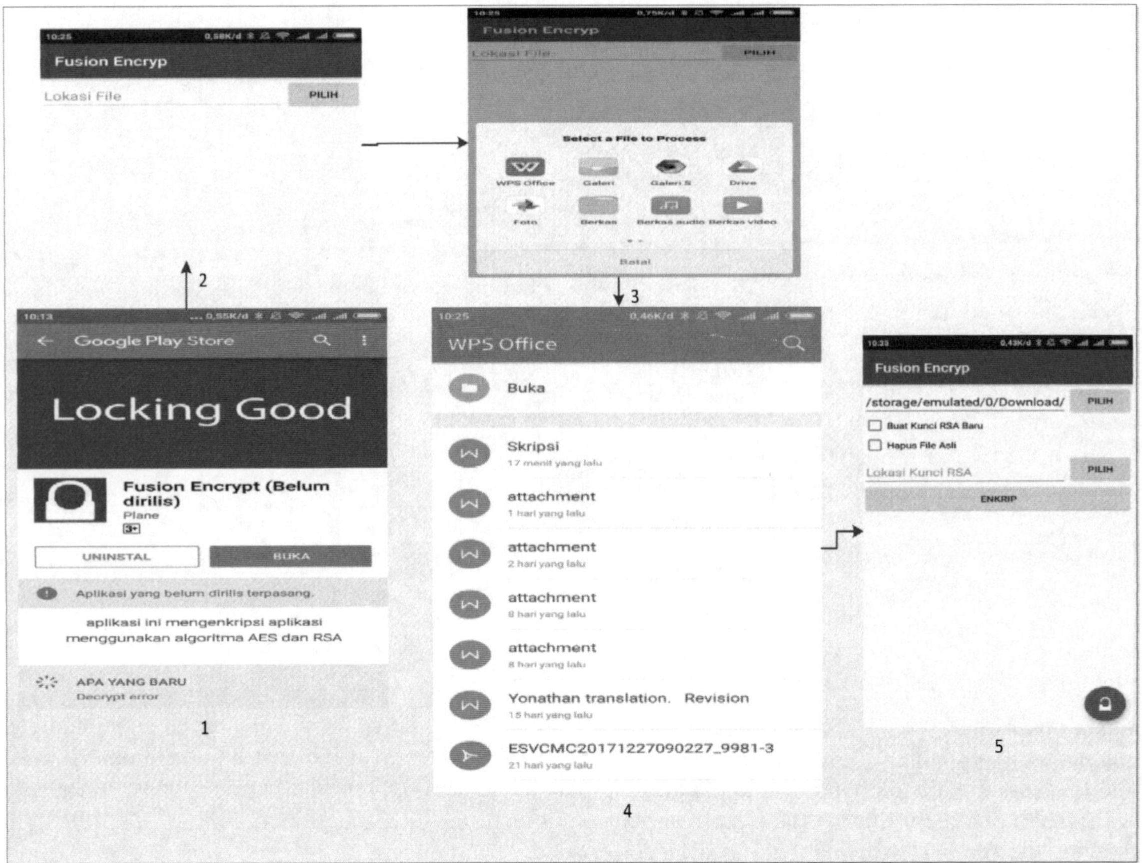

Figure 5. The process of Enkrip the File RSA and AES.

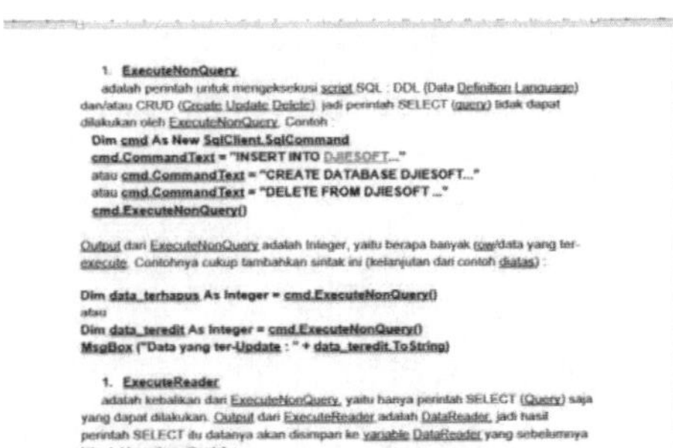

1. ExecuteNonQuery
adalah perintah untuk mengeksekusi script SQL : DDL (Data Definition Language) dan/atau CRUD (Create Update Delete) jadi perintah SELECT (query) tidak dapat dilakukan oleh ExecuteNonQuery. Contoh :

```
Dim cmd As New SqlClient.SqlCommand
cmd.CommandText = "INSERT INTO DJIESOFT..."
atau cmd.CommandText = "CREATE DATABASE DJIESOFT..."
atau cmd.CommandText = "DELETE FROM DJIESOFT ..."
cmd.ExecuteNonQuery()
```

Output dari ExecuteNonQuery adalah Integer, yaitu berapa banyak row/data yang ter-execute. Contohnya cukup tambahkan sintak ini (kelanjutan dari contoh diatas) :

```
Dim data_terhapus As Integer = cmd.ExecuteNonQuery()
atau
Dim data_teredit As Integer = cmd.ExecuteNonQuery()
MsgBox ("Data yang ter-Update : " + data_teredit.ToString)
```

1. ExecuteReader
adalah kebalikan dari ExecuteNonQuery, yaitu hanya perintah SELECT (Query) saja yang dapat dilakukan. Output dari ExecuteReader adalah DataReader, jadi hasil perintah SELECT itu datanya akan disimpan ke variable DataReader yang sebelumnya

Figure 6. Before Encryption.

Figure 7. After Encryption.

Based on the analysis process of encryption and decryption in some files which has been test from files looking from different files, so could be seen the different in encryption and decryption process which in android system from time that has gotten. Below could be seen in figure 8 encryption process and figure 9 decryption which was test in some different files that has using in cryptography with decryption and encryption process that has different file size 25.44 KB 200 KB, 600 KB, and 2.29 MB

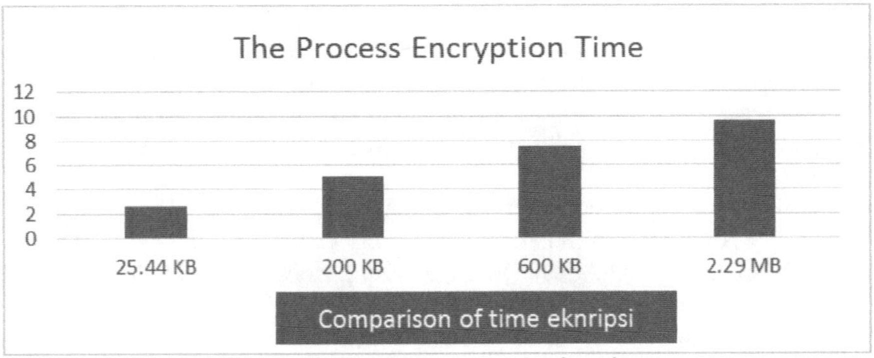

Figure 8. The process encryption time.

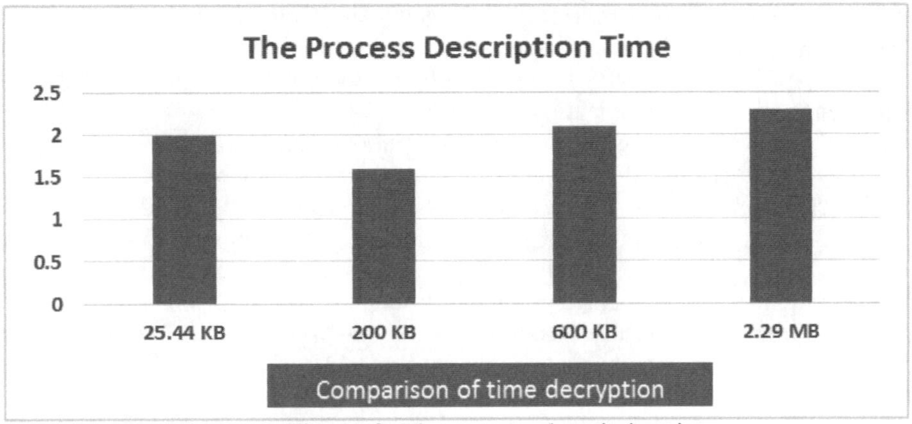

Figure 9. The process description time.

6. Result

Based on result of cryptography file has been done with AES and RSA could give cryptography in kinds of files in mobile. There were size of file data that have done in test. It found the different encryption. Some of file size 25.44 KB, encryption time 4 second, 200 KB, 5 second, 600 KB 7 second, 2.29 MB, 10 second. Where decryption 25.44 KB, encryption 2 second, 200 KB, 1.5 second, 600 KB 2.5 second, 2.29 MB, 2.7 second.

Refrences

[1] Muneer Ahmad Dar and Javed ParvezEvaluating Smartphone Application Security: A Case Study on Android, International Journal of Advance Research, IJOAR .org ISSN 2320-9194, Volme 1, ISSN 2320-9194

[2] Tiwari Mohini, Srivastava Ashish Kumar and Gupta Nitesh Review on Android and Smartphone Security, Research Journal of Computer and Information Technology Sciences, ISSN 2320 – 6527, Vol. 1(6), 12-19, November (2013)

[3] Zhiyi F, Ying W, Zelin D and Fan Y. (2013) The Research on NSSC Key Service Mechanism in Mobile Network, 2nd International Conference on Science and Social Research (ICSSR 2013)

[4] David Lisonek and Martin Drahansky 2008, SMS Encryption for Mobile Communication, IEEE International Conference on Security Technology, 198-201

[5] Mary Agoyi and Devrim Seral 2010, SMS Security: An Asymmetric Encyrption Approach, IEEE International Conference on Wireless and Mobile Communication, 448-452

[6] Bin Liu, and Bevan M. Baas (2013) "Parallel AES Encryption Engines for Many-Core Processor Arrays " Journal IEEE Transactio ns on Computers, 2(3), 536-547

[7] Das Debasis, MisraRajiv. "Programmable Cellular Automata Based Efficient Parallel AES Encryption Algorithm". International Journal of Network Security & Its Applications (IJNSA), Vol.3, No.6, November 2011, pp. 204

[8] Rohan Rayrikar,Sanket Upadhyay and Priyanka Pimpale,"SMS Encryption using AES Algorithm on Android",International Journal of Computer Applications Vol. 50–No.19, July 2012

[9] Jawahar Thakur and Nagesh Kumar 2011, DES, AES and Blowfish: Symmetric Key Cryptography Algorithms Simulation Based Performance Analysis, International Journal of Emerging Technology and Advanced Engineering, (ISSN 2250-2459, Volume 1, Issue 2)

[10] Kalpana Parsi, Singaraju Sudha. "Data Security in Cloud Computing using RSA Algorithm". International Journal of Research in Computer and Communication technology, IJRCCT, ISSN 2278-5841, Vol 1, Issue 4, September 2012. pp. 145

[11] Kalpana Parsi, Singaraju Sudha. "Data Security in Cloud Computing using RSA Algorithm". International Journal of Research in Computer and Communication technology, IJRCCT, ISSN 2278-5841, Vol 1, Issue 4, September 2012

[12] Marko Hassinen 2008, SafeSMS- End-to-End Encrption for SMS Messages, IEEE International Conference on Telecommunications, 359-365

[13] Padmapriya, Dr.A, Subhasri, P. "Cloud Computing: Security Challenges & Encryption Practices". International Journal ofAdvanced Research in Computer Science and Software Engineering, ISSN: 2277 128X, Volume 3, Issue 3, March 2013, pp. 257

[14] Padmapriya, Dr.A, Subhasri, P. "Cloud Computing: Security Challenges & Encryption Practices". International Journal ofAdvanced Research in Computer Science and Software Engineering, ISSN: 2277 128X, Volume 3, Issue 3, March 2013, pp. 257

[15] Prashanti.G, Deepthi.S & Sandhya Rani.K. "A Novel Approach for Data Encryption Standard Algorithm". International Journal of Engineering and Advanced Technology (IJEAT) ISSN: 2249 – 8958, Volume - 2, Issue - 5, June 2013, pp. 264

[16]. Singh Narjeet, Raj Gaurav. "Security On Bccp Through Aes Encryption Technique". International Journal Of Engineering Science & Advanced Technology Volume-2, Issue-4, 813 – 819. pp. 817

[17] Shraddha Soni, Himani Agrawal, Dr. (Mrs.) Monisha Sharma. (2012). *Analysis and Comparisonbetween AES and DES Cryptographic Algorithm Shraddha*. International Journalof Engineering and Innovative Technology (IJEIT) Volume 2, Issue 6. December 2012

[18] WilliamStallings. (2011).*CryptographyAnd Network Security:Principles And Practice.*Fifth Edition. Pearson Education,Inc., publishing as Prentice Hall

[19] J. Yenuguvanilanka and O. Elkeelany, "Performance Evaluation of Hardware Models of Advanced Encryption Standard (AES) Algorithm, ," Southeastcon, 2008 IEEE, pp.222-225,2008

[20] Nagesh kumar, jawahar thakur, arvind kalia on "performance analysis of symmetric key cryptography algorithms:des , aes and blowfish " in an international journal of engineering sciences issn: 2229-6913 issue sept 2011, vol. 4 ,pp.28-37

[21] Kalpana Parsi, Singaraju Sudha. "Data Security in Cloud Computing using RSA Algorithm". International Journal of Research in Computer and Communication technology, IJRCCT, ISSN 2278-5841, Vol 1, Issue 4, September 2012. pp. 145

[22] Rajdeep Bhanot , Rahul Hans. (2015). *A Review andComparative Analysis of Various Encryption Algorithms*. International Journal ofSecurity and Its Applications Vol. 9, No. 4.1Mei 2015. pp. 289-306

IOP Conf. Series: Journal of Physics: Conf. Series **1007** (2018) 012017 doi:10.1088/1742-6596/1007/1/012017

Mining association rule based on the diseases population for recommendation of medicine need

M Harahap*, A M Husein, S Aisyah, F R Lubis, B A Wijaya

Faculty of Technology and Computer Science, Universitas Prima Indonesia, Indonesia

E-mails: *mawaddah@unprimdn.ac.id, amirmahmud@unprimdn.ac.id

Abstract. Selection of medicines that is inappropriate will lead to an empty result at medicines, this has an impact on medical services and economic value in hospital. The importance of an appropriate medicine selection process requires an automated way to select need based on the development of the patient's illness. In this study, we analyzed patient prescriptions to identify the relationship between the disease and the medicine used by the physician in treating the patient's illness. The analytical framework includes: (1) patient prescription data collection, (2) applying k-means clustering to classify the top 10 diseases, (3) applying Apriori algorithm to find association rules based on support, confidence and lift value. The results of the tests of patient prescription datasets in 2015-2016, the application of the k-means algorithm for the clustering of 10 dominant diseases significantly affects the value of trust and support of all association rules on the Apriori algorithm making it more consistent with finding association rules of disease and related medicine. The value of support, confidence and the lift value of disease and related medicine can be used as recommendations for appropriate medicine selection. Based on the conditions of disease progressions of the hospital, there is so more optimal medicine procurement.

1. Introduction

The hospital is one of the health healing and recovery facilities for the patient [1]. Most hospitals in the implementation of health care activities have implemented Hospital Management Information System (HMIS), so in health services to patients have been recorded in the database, ranging from registration of the process of payment of health costs. However, the implementation of the process of medicine management in the hospital pharmacy installation is still not optimal because there is still often a vacancy at medicine stock. Medication is a good need for sick people with 50-60% of the overall budget of the hospital which is used at medication and medical equipment. Management of non-optimal medicine handling will adversely affect the hospital both medically and economically [2].

Selection of medicine need is the first phase at the cycle planning of procurement medicine [3] in [2]. The process of medicine selection is based on the patient's disease population. The patient's disease population may change which based on the patient's diagnostic volumes of data stored in the database, so it is need to require an automated way to select medicine requirement based on disease progression. Data mining is one of useful technique for extract and defining patterns of datasets in databases into information [4]. Application of data mining methods of the health field is proposed by many researchers, such as predicting heart disease [5], [6], health insurance [7], disease identification

[6], hypertension [8], insurance cheating, low cost patient medical solutions, related disease detection, treatment and other treatment methods [9].

Apriori is one of the Mining Association methods of Data Mining to find all related adjustment items in a database transaction that fill minimum of rules and limits or another limit [10]. Apriori algorithm that is proposed by many researchers like finding the associaton rules of Chinese traditional medicine [11], Chinese herbal medicine [12], diagnosing patients diseases with hypertensive symptoms [8], disease identification [6], detection of heart disease factors for men and women [5], identification of symptoms with traditional Korean therapy [13]. Apriori algorithm can reduce the number of candidates that must be calculated by butchering method, it has good performance [14], with the required scanning process of each iteration will increase the high time computation.

Some researchers use comparative methods to optimize time computation of apriori algorithms, one of them is the application of the k-means clustering method and it proves to be very accurate [15], [16], [17]. K-Means Clustering is one of the hard partition classification techniques, it is efficient in grouping large and fast datasets in [18] calculations, but limited at numerical data [19].

At this research, we analyze the patient prescriptions based on a doctor's diagnosis in the hospital database to identify the relationship between the disease and the medicine used by the physician in treating the patient's illness. The application of the k-means clustering method was used to find the 10 more dominant diseases in health dataset in year 2015 and 2016, then we use an apriori algorithm to find useful relationships and information between disease and related medicine based on support, confidence and lift values. Our paper is structured as follows: section 2 of the related research, the proposed method is described in Section 3. Section 4 is results and conclusion is in Section 5.

2. Related research

Research [15], applied a combination of k-means clustering algorithms with a priori in consumer data to find association rules, from the results of the study of consumer data clustering with k-means algorithm showed a significantly better and consistent influence on a priori based on value support. So it provides useful information service providers to offer the right products / ads to the right consumer. [16] Applied on clustering algorithm is to improve the performance of a priori algorithms to find solutions by generating different items each site on a cloud-based network. [17], proposed by a novel method of clustering using k-means and a priori with the aim of allocating unique id for the object of the cluster. Each objects of the group has a certain position which may vary depending on the circumstances, the id is allocated then applied to k-means clustering method along with the a priori algorithm. [20] Applying k-means clustering is to analyze goals treatment in breast cancer based on user behavior, datasets using UCI with 569 data and 32 attributes. [11] proposed modification of the Association Rules Mining method to study the structural character of the Traditional Chinese medicine (TCM) pairs with a dataset source of 625 medicine data onto 347 medicines and 5 types of cold, hot, warm, cool and normal properties. The application of a priori algorithms is used to find out some specific medicine or properties more commonly used in medicine pairs by comparing a priori by proposed method, from the test results based on statistical tests, optimal proposed method finding association rules on medicine than previous methods.

[21] proposed the mining of health data to find the pattern of illness that occurs to patients by seeking symptom relation of disorders in the medical database. Yan Yan's research, Wang Chunyan, Li Min developed a multi-model based on a priori algorithm at the hospital to extract data from the database to produce useful information in medical decision making, [20] proposed a priori algorithm to find the characteristics of headache on traditional medicine, making it easier about doctors' decision on recipes for various kinds for headache sufferers.

IOP Conf. Series: Journal of Physics: Conf. Series **1007** (2018) 012017 doi:10.1088/1742-6596/1007/1/012017

3. Methodology

Apriori Algorithm is the most famous algorithm for finding patterns of a database that has a frequency or support above a certain threshold called the minimum support term. A priori algorithm consists of several stages of iteration, each iterations will generate a calculated frequency pattern by scanning the database to obtain support of each items, items that have support above the minimum support are selected into high frequency patterns of length one or often called 1-itemset . K-itemset is a term of a set consisting of k items. In the second iteration process will produce 2-itemset which each set has two items [22]. The rules of association are the implications of the form X → Y, where X is the antecedent and Y is the consequence of the rule. Thus X∩Y = Φ. The support of the item set is defined as the ratio of the number of transactions containing items set to the total number of transactions. Trust of association rule X → Y is the probability that Y transaction contains an association rules mining X algorithm.

Support ot the association rule X→Y :

$$Support\,(X,Y) = P(X,Y) = \frac{\{medicine\ who\ bought\ X\ and\ Y\}}{\{medicine\}} \quad (1)$$

Confidence of association rule X→Y:

$$Confidence\,(X \to Y) \equiv P(Y|X) = \frac{P(X,Y)}{P(X)} = \frac{\{medicine\ who\ bought\ X\ and\ Y\}}{\{medicine\}} \quad (2)$$

Lift, also known as interest of association rule $X \to Y$:

$$Lift\,(X \to Y) = \frac{P(X,Y)}{P(X)\ P(Y)} = \frac{P\,(X|Y)}{P\,(XY)} \quad (3)$$

K-Means algorithm is one of the most popular clustering algorithms used cause it has a simple algorithm, easy to implement and efficient in its complexity [23]. The grouping of k-means is based on proximity to each other according to the Euclidean distance. It takes k as an input parameter and partition a set of n objects (1) from k cluster. The average value of the object (2) is taken as the resemblance (3) to the parameter to form the cluster. Cluster mean or center is formed by random selection of object k. Comparing most similarities (4) of other objects is assigned to the cluster. For each data vector the algorithm calculates the distance between the data vector and each clan centroid using the equation [24]. The steps in the K-means algorithm are as follows:

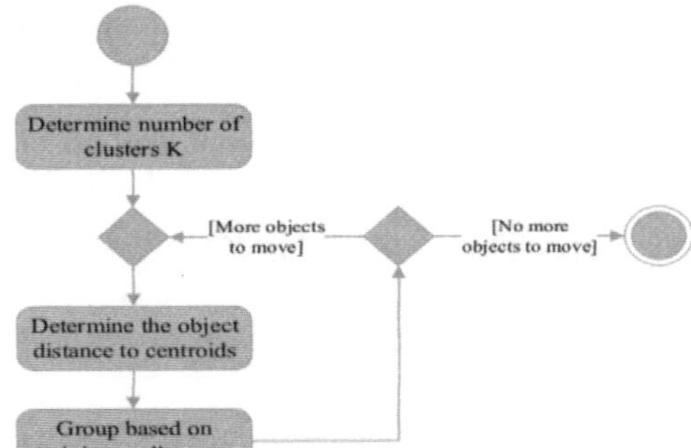

Figure 1. Data and control flow of *K*-Means algorithm.

In this study we used patient prescription datasets in 2015 and 2016 from two hospitals, we apply these datasets source to the MySQL database to facilitate the process of cleaning and transformed data. After the process of cleaning the noise data, we concluded 651.378 prescriptions with 12.015 patient data and 1.945 medicine type data for 2015, when the 2016 patient prescription dataset amounted to 956.152 prescriptions, 18.416 patients and 1.835 medicine type, like at the first table.

Table 1. Datasets source.

Year	Prescription	Patient	Medicine Type
2015	651.378	12.015	1.945
2016	956.152	18.416	1.835

The main objective of this study was to classify the 10 more dominant disease populations based on the patient's disease progression using the k-means algorithm on the patient prescription dataset. From this clustering, we apply an a priori algorithm to establish the relationship between disease and related medicine based on the value of support, confidence and lift. This knowledge can be a recommendation of appropriate medicine selection of the procurement in medicine to be more optimal to avoid the occurrence vacancy at stock of medicine in pharmacy hospital.

4. Results and Discussion

Tests were conducted to find the association for disease with medicine in patient prescription dataset, on dataset 2015 and 2016 used, we made as material Analysis and identification process of patients disease pattern. The initial step of the patient's prescription dataset will be grouped with the k-means algorithm of the dataset by 2015 consisting of 651, 378 prescriptions, 12,015 patients and 1,945 medicines, when the 2016 patient prescription dataset consists of 965,152 prescriptions, 18,416 patients and 1,835 medicaments. In grouping the disease, we use 3 (three) variables, those are age, gender and disease. Gender (man and women), age (infant, toddler, children, adult and elderly) the disease variables use ICD10 with 21,591 kinds of diseases. In table 1 is the result of cluster and number of instances, table 2 results cluster of attributing to patient prescription dataset 2015 consisting of 10 clusters.

Table 2. Cluster dataset 2015.

	# of instance	ratio
Cluster0	816	15%
Cluster1	753	14%
Cluster2	490	9%
Cluster3	487	9%
Cluster4	437	8%
Cluster5	369	7%
Cluster6	351	6%
Cluster7	306	6%
Cluster8	1,193	22%
Cluster9	207	4%

Table 3. Result attribute dataset 2015.

Age	Gender	Disease Code
Elderly	Women	L10
Adult	Women	A15
Adult	Women	K30
Adult	Women	E11
Adult	Women	O000
Elderly	Women	M15
Elderly	Man	I69.3
Elderly	Man	A01
Elderly	Women	H26
Elderly	Women	M53

In table 1 is the result of grouping of dataset year 2015 with three attributes that is age, gender and disease code ICD10. The result of grouping of three attributes is seen in table 1. In cluster has the highest value that is 22% with Elderly age attribute, gender women and disease code H26.9 compared with 10 other diseases. The results of the clustering of 10 diseases in 2016 dataset are six similar diseases of the 2015 dataset and four new diseases, but disease code H26.9 has the highest 40% compared with 9 other diseases. This provides information that cataract disease is the highest disease suffered by patients in 2015 and 2016. Results of grouping dataset in 2016 are as in table 3 and table 4.

Table 4. Cluster dataset 2016.

	# of instance	Ratio
Cluster0	1,250	40%
Cluster1	377	12%
Cluster2	298	10%
Cluster3	126	4%
Cluster4	92	3%
Cluster5	85	3%
Cluster6	387	12%
Cluster7	222	7%
Cluster8	208	7%
Cluster9	78	2%

Table 5. Result cluster attribute dataset 2016.

Age	Gender	Disease Code
Elderly	Women	H269
Adult	Women	A151
Adult	Women	H612
Adult	Man	H661
Elderly	Man	A012
Adult	Women	H608
Adult	Women	EL1
Elderly	Women	I10
Adult	Women	J30.3
Adult	Women	O000

Based on the results of the grouping of 10 of the highest diseases in the dataset of 2016, we used association rules to find the relationship between medicine-related illnesses by forming binary matrices in which columns were medicine and rows represented 10 of the highest diseases and each cells had 0 and 1. We analyzed without considering the dosage and the way of the medicine used due to varying doses. Table 5 shows the support, trust, adoption of association rules between the top ten diseases (antecedent) and the related medicine(consequent) which has a minimum limit value of Support 20% and confidence 65%. Trust and lift values can be used to assess the rules of association. Medicine that has high confidence and lifting values has a relationship of diseases such as illness with ICD code H26.9 have a trust value and support for the medicine Ciprofloxacin 500 Mg Tablet means Ciprofloxacin 500 Mg medicine the tablet is most commonly used for unspecified Cataract disease, but the value of lift to this medicine is relatively low. This implies that the Ciprofloxacin 500 Mg Tablet medicine is often used in other diseases, the Polidemisin Eye Drop medicine has a high lift value, it means that the Polidemisin Eye Drop medicine is a special remedy for unspecified Cataract disease.

Table 6. Association rules between diseases and related medicine (Min Sup: 20%, Min Conf : 65%).

Antecedent (disease Code)	Disease name	Consequent (Medicine)	Support	Confidence	Lift
H26.9	UNSPECIFIED CATARACT	CIPROFLOXACIN 500 MG TABLET	3.5	48.2	1.6
		CENDO VITROLENTA MINIDOSE	4.7	31.9	3.7
		CENDO CATARLENT MINIDOSE	2.6	76.5	2.6
		POLIDEMISIN EYE DROP	2.5	70.4	4.2
		VIGAMOX TETES MATA	4.1	48.2	2.8
		NEVANAC 0,1%	2.7	86.2	2.6
		CIPROFLOXACIN 500 MG TABLET	2.2	71.3	2.5
		CRAVIT 5ML	2.7	67.6	3.4
		FLAMAR TETES MATA	2.3	80.5	3.2
A15.1	TUBERCULOSIS OF LUNG	ALPRAZOLAM 0.5 MG TABLET	2.6	59.5	2.8
		BECOM C KAPLET	2.4	85.9	2.3
		NATRIUM DIKLOFENAC 50 MG TABLET	2.4	63.8	1.1
		LEVOFLOXACIN FC 500 MG TABLET	2.6	47.3	2.2
		MELOXICAM 15 MG TABLET	3.2	27.2	2.2

IOP Conf. Series: Journal of Physics: Conf. Series **1007** (2018) 012017 doi:10.1088/1742-6596/1007/1/012017

NOVALGIN INJEKSI	2.9	75.6	2.1
OMEPRAZOLE TABLET	2.9	24.4	1.5
OPICEF 500 MG KAPSUL	2.8	41.9	2.4
PARACETAMOL TABLET	2.1	57.8	2.2
RANITIDINE 25 MG INJEKSI	2.9	57.1	2.2
TEGADERM PAD 9X25 CM	2.3	44.4	1.5
TRAMIFEN TABLET	2.7	68.9	2.1
GENTAMYCIN 0.1% SALEP	2.7	69.3	2.3
METRONIDAZOLE 500 MG TABLET	2.3	89.9	0.3
CEFADROXIL 500 MG KAPSUL	2.6	52.1	2.1
SARUNG TANGAN NON STERIL (M)	2.2	47.1	2.1
MASKER EARLOOP (KARET)	2.8	48.2	1.1
SPUIT 3 CC	3.1	54.3	1.3
SPUIT 5 CC	2.5	34.1	1.3
SPUIT 10 CC	2.2	54.1	2.7
CEFTRIAXONE 1 GR INJEKSI	2.5	36.5	1.2
ALCOHOL SWAB	3.2	55.3	1.4
ANSEL GAMEX NO 7,5	2.4	85.1	1.4
CIPROFLOXACIN 500 MG TABLET	2.5	55.4	1.1
HI-BONE CAPSUL	3.4	40.3	2.9
VERBAND GULUNG SWALOW 15 CM	2.6	42.2	2.7
ASAM MEFENAMAT 500 MG KAPLET	2.5	60.2	1.7
LOMATUELL	2.9	33.8	1.3
TEGADERM PAD 9X10 CM	2.6	59.7	1.2
SARUNG TANGAN NON STERIL (S)	2.9	38.8	2.2
CEBEX KAPSUL	3.7	52.7	2.2
NEURODEX TABLET	0.6	37.4	2.1
GENTAMYCIN INJEKSI	1.4	84.3	2.4
VITAMIN B KOMPLEX	2.2	75.9	1.5
CO AMOXICLAV TABLET	1.1	36.6	1.6
VITAMIN C TABLET	3.9	88.3	2.2
CEFADROXIL 125MG/60 ML SIRUP	1.8	71.1	1.4
CALCIUM LACTAT TABLET	0.6	63.2	2.2
METHYLPREDNISOLONE 4 MG TABLET	1.9	50.4	1.5
MELOXICAM 7,5 MG TABLET	1.3	61.6	2.1
LEVOFLOXACIN FC 500 MG TABLET	2.4	25.7	2.6
CEFIXIME 100 MG KAPSUL	3.1	31.9	2
KETOROLAC 3% INJEKSI	2.4	24.4	2.3
ONDANSETRON 4 MG INJEKSI	2.7	67.3	2.2
ASAM MEFENAMAT 500 MG KAPLET	1.1	22.9	2.9
RANITIDINE 150 MG TABLET	3.2	89.5	2.1
VITAMIN C TABLET	1.9	78.9	1.3
ETABION (NOVABION) TABLET	3.1	80.6	1.7
CTM TABLET	1.6	42.6	1.1
AMOXICILLIN 500 MG TABLET	0.2	34.7	1.3
CIPROFLOXACIN 500 MG TABLET	2.9	42.5	1.9
METRONIDAZOLE 500 MG TABLET	2.2	34.4	2.2
CEFADROXIL 500 MG KAPSUL	1.1	67.8	2.1
CALCIFAR PLUS(KALSIUM LAKTAT) KAPLET	1.6	60.9	2.1
GENTAMYCIN 0.1% SALEP	1.5	38.6	1.1
PARACETAMOL TABLET(PACETIK TABLET)	1.5	81.8	2.8
MECOBALAMIN 500 MG KAPSUL	1.3	60.4	1.4
STERIL WATER 25 ML OTSU	3.1	34.6	1.4
NACL 500 ML WIDA	3.2	22.4	2.7
RINGER LACTAT 500 ML WIDA	3.8	31.4	1.2
RANITIDINE 25 MG INJEKSI	0.4	78.4	1.7

		SPUIT 3 CC	2.7	35.9	1.5
H61.2	IMPACTED CERUMEN	CETIRIZINE 10 MG TABLET	2.1	28.9	0.9
		METHYLPREDNISOLONE 4 MG TABLET	1.9	57.3	0.9
		H2O2 50%	1.9	27.9	0.6
		BOTOL 30 CC	0.5	64.9	1.9
		LEVOFLOXACIN FC 500 MG TABLET	2.2	85.2	1.6
		PIPET TETES KACA	1.5	33.5	1.1
		CEFADROXIL 500 MG KAPSUL	1.2	30.2	1.4
		CIPROFLOXACIN 500 MG TABLET	1.4	71.8	0.7
		LANSOPRAZOLE 30 MG TABLET	2.8	55.1	0.6
		METRONIDAZOLE 500 MG TABLET	0.4	82.4	1.3
		OMEPRAZOLE TABLET	1.2	24.1	0.7
		NATRIUM DIKLOFENAC 50 MG TABLET	1.9	80.6	0.8
		AVAMYS NASAL SPRAY	1.6	45.1	1.6
		CTM TABLET	1.8	24.3	1.5
		VITAMIN B KOMPLEX	2.7	89.5	0.7
		ILIADIN SPRAY 0.05% DEWASA	2.3	68.9	1.2
		SARUNG TANGAN NON STERIL (M)	2.4	52.8	1.5
		PARACETAMOL TABLET(PACETIK TABLET)	2.7	70.9	1.5
		CEFIXIME 200 MG KAPSUL	3.3	41.1	0.6
		MASKER EARLOOP (KARET)	0.4	67.3	1.1
		MECOBALAMIN 500 MG KAPSUL	2.4	81.3	0.2
		ILIADIN SPRAY 0.05% DEWASA	0.5	64.2	0.5
		AZITROMYCIN 500 MG TABLET	0.1	88.7	0.5
		CEFIXIME DRY SYRUP	0.7	24.7	0.3
H66.1	CHRONIC TUBOTYMPANIC SUPPURATIVE OTITIS MEDIA	CETIRIZINE 10 MG TABLET	4.4	26.2	3.8
		LEVOFLOXACIN FC 500 MG TABLET	3.9	73.4	3.4
		METHYLPREDNISOLONE 4 MG TABLET	4.8	85.4	1.9
		METRONIDAZOLE 500 MG TABLET	4.9	59.6	0.1
		CIPROFLOXACIN 500 MG TABLET	4.5	71.3	3.2
		H2O2 50%	4.8	24.9	2.8
		CTM TABLET	3.4	73.6	1.9
		BOTOL 30 CC	4.4	72.8	0.5
		PIPET TETES KACA	4.5	37.4	3.8
		DEXAMETHASON TABLET (DIOMETA-FLACOID)	4.3	34.5	0.5
		CLINDAMYCIN 300 MG KAPSUL	.3.0	59.7	1.6
A01.2	PARATYPHOID FEVER B	CONCOR 2.5 MG TABLET	2.1	61.5	1.7
		FUROSEMIDE 40 MG TABLET	0.8	67.3	1.9
		APTOR TABLET	2.0	27.2	1.3
		SPIRONOLACTONE 25 MG TABLET	1.7	31.2	0.7
		MICARDIS 80 MG TABLET	2.2	40.3	3.9
		NITROCAF RETARD FORTE	2.2	48.4	2.2
		CLOPIDOGREL 75 MG TABLET	0.8	82.3	0.2
		MICARDIS 40 MG TABLET	2.4	66.2	1.2
		SIMVASTATIN 20 MG TABLET	0.7	27.1	3.5
		ISOSORBIDE DINITRAT 5 MG TABLET	1.6	57.1	1.0
		ADALAT OROS 30 MG TABLET	1.6	72.9	3.5
		ALPRAZOLAM 0.5 MG TABLET	2.0	42.7	1.6
		RAMIPRIL 5 MG TABLET	0.4	21.3	0.4
		APTOR TABLET	1.9	59.3	2.5
H60.8	OTHER OTITIS EXTERNA	CETIRIZINE 10 MG TABLET	1.3	87.6	3.6
		LEVOFLOXACIN FC 500 MG TABLET	1.7	89.2	3.8
		METHYLPREDNISOLONE 4 MG TABLET	1.5	29.9	3.5
		H2O2 50%	0.7	47.5	3.6
		BOTOL 30 CC	0.1	78.3	3.3
		SARUNG TANGAN NON STERIL (M)	0.2	86.7	3.4

IOP Conf. Series: Journal of Physics: Conf. Series **1007** (2018) 012017 doi:10.1088/1742-6596/1007/1/012017

		PIPET TETES KACA	0.9	35.2	3.2
		MASKER EARLOOP (KARET)	1.6	49.4	3.6
		OMEPRAZOLE TABLET	1.6	75.9	3.8
E11	TYPE 2 DIABETES MELLITUS	METFORMIN 500 MG TABLET	1.8	88.9	0.8
		ALCOHOL SWAB	2.0	30.1	0.7
		GLIMEPIRIDE 2 MG TABLET	1.9	25.3	1.7
		DIAMICRON MR 60 MG TABLET	0.1	49.3	1.0
		MICARDIS80 MG TABLET	1.9	78.4	0.7
		OMEPRAZOLE TABLET	0.3	43.6	1.1
		GABAPENTIN 300 MG KAPSUL	1.1	73.9	1.4
		GLIQUIDONE TABLET	2.0	81.7	0.2
		ANTASIDA DOEN 60 ML SYRUP	1.7	41.4	1.4
		AMLODIPINE BESYLATE 10 MG TABLET	0.7	36.9	1.1
		MECOBALAMIN 500 MG KAPSUL	1.8	63.6	1.7
		APTOR TABLET	1.5	37.9	0.1
		ACARBOSE 50 MG TABLET	1.0	54.8	1.5
		MELOXICAM 15 MG TABLET	0.2	21.1	0.5
		GLUCOSAMINE TABLET 500 MG	0.2	42.5	1.6
		ALPRAZOLAM 0.5 MG TABLET	1.9	58.7	0.7
		SIMVASTATIN 20 MG TABLET	1.6	80.5	0.7
		SIMVASTATIN 10 MG KAPLET	1.6	80.8	0.7
		MICARDIS40 MG TABLET	0.3	47.4	0.6
		LANSOPRAZOLE 30 MG TABLET	2.1	54.1	0.3
		AMITRIPTYLINE TABLET	0.6	75.9	0.8
		CANDESARTAN 8 MG TABLET	0.6	64.9	1.0
		NOVORAPID	1.5	68.1	1.1
		ADALAT OROS 30 MG TABLET	1.5	43.1	1.7
		CURCUMA FCT TABLET	1.0	59.9	0.8
		CILOSTAZOL 100MG TABLET	0.1	58.7	2.0
		CLOPIDOGREL 75 MG TABLET	0.1	71.8	1.8
		BETAHISTIN MESILAT 6 MG TABLET	1.4	55.2	0.1
		CETIRIZINE 10 MG TABLET	0.4	65.1	1.4
		GABAPENTIN 300 MG KAPSUL	1.7	69.1	1.3
		ETABION (NOVABION) TABLET	0.4	43.2	1.7
		SIMVASTATIN 10 MG KAPLET	1.4	53.3	0.6
		GLIMEPIRIDE 2 MG TABLET	1.3	50.5	1.9
		RETAPHYL SR 300 MG TABLET	0.8	32.1	1.2
		FLUNARIZINE TAB(SEREMIG)	1.4	76.1	2.1
		DIAMICRON MR 60 MG TABLET	1.2	30.8	0.8
		EPERISONE HCL 50 MG TABLET	1.7	61.6	0.2
		CONCOR 2.5 MG TABLET	1.7	30.7	0.5
		VALESCO 80 MG KAPSUL	1.3	79.3	1.5
		AMLODIPINE 5 MG TABLET	0.1	48.7	0.6
		LEVOFLOXACIN FC 500 MG TABLET	0.4	73.2	1.4
		CANDESARTAN 16 MG TABLET	0.2	34.2	0.8
		NOVOMIX	0.9	65.2	2.3
		KETOPROFEN 100 MG TABLET	1.1	66.7	1.1
		CAMIDRYL EXP 60 ML SYRUP	0.7	59.3	2.4
I10	ESSENTIAL (PRIMARY) HYPERTENSION	MICARDIS80 MG TABLET	0.2	78.2	4.7
		AMLODIPINE BESYLATE 10 MG TABLET	0.8	65.3	5.4
		OMEPRAZOLE TABLET	0.9	67.1	4.8
		ADALAT OROS 30 MG TABLET	0.2	48.5	4.9
		SIMVASTATIN 20 MG TABLET	0.7	81.5	5.8
		CONCOR 2.5 MG TABLET	1.5	58.7	4.5
		ANTASIDA DOEN 60 ML SYRUP	0.3	67.2	5.1
		GLUCOSAMINE TABLET 500 MG	1.2	24.9	1.3

		MELOXICAM 15 MG TABLET	1.2	30.8	4.3
		MICARDIS 40 MG TABLET	3.9	57.1	4.6
		ALPRAZOLAM 0.5 MG TABLET	1.2	35.1	4.5
		LANSOPRAZOLE 30 MG TABLET	1.9	42.5	4.8
		APTOR TABLET	2.1	22.7	5.5
		CANDESARTAN 8 MG TABLET	3.0	49.9	4.6
		CANDESARTAN 16 MG TABLET	3.8	86.2	5.2
		SIMVASTATIN 10 MG KAPLET	3.1	54.7	4.8
		EPERISONE HCL 50 MG TABLET	3.3	38.8	4.4
		GABAPENTIN 300 MG KAPSUL	3.2	36.9	4.1
		AMLODIPINE 5 MG TABLET	2.3	31.4	4.1
		SIMVASTATIN 10 MG KAPLET	3.5	21.1	5.3
		VALESCO 80 MG KAPSUL	1.6	51.1	4.2
		MECOBALAMIN 500 MG KAPSUL	2.8	50.9	4.9
		BETAHISTIN MESILAT 6 MG TABLET	1.4	84.5	5.9
		ULSAFATE SUSPENSI 100 ML	1.9	36.2	4.2
		BISOPROLOL FUMARATE 5 MG TABLET	2.5	45.4	4.9
		METHYLPREDNISOLONE 4 MG TABLET	3.1	44.4	4.6
		ALCOHOL SWAB	3.1	44.7	5.3
		VALESCO 160 MG KAPSUL	2.2	36.9	4.3
		FLUNARIZINE TAB(SEREMIG)	3.8	60.5	5.1
J30.3	OTHER ALLERGIC RHINITIS	CETIRIZINE 10 MG TABLET	1.8	73.1	1.4
		METHYLPREDNISOLONE 4 MG TABLET	3.1	22.2	2.1
		AVAMYS NASAL SPRAY	3.9	77.4	1.2
		CTM TABLET	3.2	55.8	2.1
		SARUNG TANGAN NON STERIL (M)	2.3	29.7	1.3
		OMEPRAZOLE TABLET	3.3	60.1	1.6
		ILIADIN SPRAY 0.05% DEWASA	2.2	46.1	1.1
		MASKER EARLOOP (KARET)	2.5	23.6	1.6
		CEFADROXIL 500 MG KAPSUL	4.7	50.7	1.9
		LANSOPRAZOLE 30 MG TABLET	3.5	82.1	1.1
		ILIADIN SPRAY 0.05% DEWASA	4.3	63.3	2.4
		PARACETAMOL TABLET(PACETIK TABLET)	3.9	27.7	2.7
		TREMENZA TABLET	3.6	83.5	1.1
		NASACORT AQ	2.2	89.3	1.8
O00.0	ABDOMINAL PREGNANCY	PROMAVIT KAPSUL	2.3	79.8	2.9
		ETABION (NOVABION) TABLET	4.5	37.4	3.9
		ASAM MEFENAMAT 500 MG KAPLET	4.4	31.8	3.8
		ABBOCATH NO 18 TERUMO	4.5	51.3	2.6
		INFUSION SET ADULT (EASY FUSION VENTED)	4.6	82.2	3.8
		3-WAY BD CONNECTA PLUS 3 WHITE	4.2	67.5	2.8
		STERIL WATER 25 ML OTSU	4.9	22.6	3.3
		CEFADROXIL 500 MG KAPSUL	4.7	79.3	2.5
		CEFTRIAXONE 1 GR INJEKSI	4.6	22.9	4,4
		SPUIT 10 CC	4.4	80.4	3.5
		GILLETTE GOAL II	2.5	69.1	2.7
		LOMATUELL	3.3	78.1	3.8
		PEMBALUT WANITA BIO PANCA	3.4	81.8	2.8
		TRAMADOL 100MG/2ML INJEKSI	3.1	70.2	2.8
		SPUIT 3 CC	4.4	22.6	2.7
		NEOLUS DISP NEEDLE NO.26 TERUMO	2.4	60.1	3.4
		METRONIDAZOLE 500 MG TABLET	4.9	39.8	3.4
		SARUNG TANGAN NON STERIL (M)	4.8	38.9	3.8
		CALCIFAR PLUS(KALSIUM LAKTAT) KAPLET	4.9	24.8	3.2
		RINGER LACTAT 500 ML WIDA	3.4	63.6	2.1
		PROFENID 100 MG SUP	4.5	44.4	2.3

IOP Conf. Series: Journal of Physics: Conf. Series **1007** (2018) 012017 doi:10.1088/1742-6596/1007/1/012017

5. Conclusion

The application of a priori algorithms in this study aims to extract useful information about the patient prescription database sourced from two different hospitals. We use Association Rules to find the relationship between disease and related medicine based on the grouping disease using k-means algorithms. From the results of the tests, the k-means algorithm accurately classifies 10 dominant diseases in patient prescription datasets in 2015 and 2016, thus significantly affect the a priori algorithm, it is more consistently to find association rules between disease and related medicine. The value of support, confidence and lift between medicine related diseases can be useful as a recommendation of appropriate medicine selection based on the condition of disease progression of the patient, so the procurement of medicine in the hospital is more optimal.

References

[1] Dep Kes RI, 2004. *Keputusan Menteri Kesehatan Republik Indonesia Tentang Standar Pelayanan Farmasi Di Rumah Sakit dan Apotek*, Jakarta

[2] Satibi, (2014), Manajemen Obat di Rumah Sakit, Fakultas Farmasi Universitas Gadjah Mada Yogyakarta

[3] Quick, J.P., Rankin, J.R., Laing, R.O., O'Cornor, R.W. (2012). "Managing Drug Supply, the selection, procurement, distribution and use of pharmaceutical, third edition". Kumarin Press, Conecticus, USA

[4] Sanjeev Rao, Priyanka Gupta. 2012. "Implementing Improved Algorithm Over APRIORI Data Mining Association Rule Algorithm". IJCST Vol. 3, Issue 1. 489-493

[5] Jesmin Nahar, Tasadduq Imam, Kevin S. Tickle, Yi-Ping Phoebe Chen, 2013, "Association rule mining to detect factors which contribute to heart disease in males and females, Expert Systems with Applications", 1086–1093

[6] M. Ilayaraja, T. Meyyappan, 2013, "Mining Medical Data to Identify Frequent Diseases using Apriori Algorithm". Proceedings of the 2013 International Conference on Pattern Recognition, Informatics and Mobile Engineering, February 21-22, 194-199

[7] R.J. Kuo*, C.W. Shih, 2007, "Association rule mining through the ant colony system for National Health Insurance Research Database in Taiwan, Computers and Mathematics with Applications", 1303–1318

[8] A Mi Shin, RN, et al. 2017, "Diagnostic Analysis of Patients with Essential Hypertension Using Association Rule Mining", The Korean Society of Medical Informatics, 77-81. doi:10.4258/hir.2010.16.2.77

[9] H. C. Koh and G. Tan, "Data Mining Application in Healthcare", Journal of Healthcare Information Management, vol. 19, no. 2, (2005)

[10] Kotsiantis S, Kanellopoulos D (2006) "Association rules mining: A recent overview". GESTS International Transactions on Computer Science and Engineering 1: 71-82

[11] Shang E*, Duan J, Fan X, Tang Y and Ye L, 2017, "Association Rules Mining and Statistic Test Over Multiple Datasets on TCM Drug Pairs", International Journal of Biomedical Data Mining. Volume 6, Issue 1, 1000126. doi: 10.4172/2090-4924.1000126

[12] Pei-Shan Chien, Yu-Fang Tseng, Yao-Chin Hsu, Yu-Kai Lai and Shih-Feng Weng, 2013, BMC Complementary and Alternative Medicine, 2-7

[13] Yang DH, Kang JH, Park YB, Park YJ, Oh HS, et al. (2013) "Association Rule Mining and Network Analysis in Oriental Medicine". PLoS ONE 8(3)

[14] Badri Patel, Vijay K Chaudhari, Rajneesh K Karan, YK Rana. 2011. "Optimization of Association Rule Mining Apriori Algorithm Using ACO". International Journal of Soft Computing and Engineering (IJSCE) Volume-1, Issue-1. 24-26. ISSN: 2231-2307

[15] Nergis Yılmaz and Gülfem Işıklar Alptekin, 2013, The Effect of Clustering in the Apriori Data Mining Algorithm: A Case Study Proceedings of the World Congress on Engineering 2013 Vol III, ISSN: 2078-0958 (Print); ISSN: 2078-0966 (Online)

[16] Nisha Rani, Yamini Chouhan, 2015, "Improving the Performance of Apriori Algorithm by Combining with Clustering Techniques", International Journal of Emerging Technologies in Engineering Research (IJETER), Volume 3, Issue 2, 13-15
[17] Syed Zishan Ali, Nikhil Tiwari and Sushmita Sen, 2016, "A Novel Method for Clustering using k-means and Apriori Algorithm", International Conference on Advances in Electrical, Electronics, Information, Communication and Bio-Informatics (AEEICB16), 978-1-4673-9745-2
[18] Ghosh, S., Dubey, K. S. (2013). "Comparative Analysis of K-Means and Fuzzy C Means Algorithms". ((IJACSA) International Journal of Advanced Computer Science and Applications, Vol. 4, No. 4, 2013
[19] Bai, L., Liang, J., & Dang, C. (2011). "An initialization method to simultaneously find initial cluster centers and the number of clusters for clustering categorical data". Knowledge-Based Systems, 24(6), 785–795
[20] Miao Wang, Lei Zhang, Zihao Zhang, Chen Xu, Guobin Chen, Huiliang Shang, 2014. The "Application Characteristics of Traditional Chinese Medical Science Treatment on Headache Based on Data-Mining Apriori Algorithm". IEEE International Conference on Bioinformatics and Biomedicine. 153-157
[21] K. Zuhtuogullari. N. Allahverdi. 2011. An Improved Itemset Generation Approach for Mining Medical Databases. IEEE. 39-43. 978-1-61284-922-5
[22] Goswami D.N, Chaturvedi Anshu, Raghuvanshi C.S, 2010. "An Algorithm for Frequent Pattern Mining Based On Apriori",(IJCSE) International Journal on Computer Science and Engineering Vol. 02, No. 04, 942-947
[23] Aggarwal, N., & Aggarwal, K. (2012). "Comparative Analysis of kmeans and Enhanced K-means clustering algorithm for data mining", International Journal of Scientific & Eninnering Research
[24] Saurabh Shah & Manmohan Singh "Comparison of A Time Efficient Modified K-Mean Algorithm with K-Mean and K-Medoids algorithm", International Conference on Communication Systems and Network Technologies, 2012

Acknowledgments

This research was supported by Universitas Prima Indonesia and one of private hospital in Medan city.

IOP Conf. Series: Journal of Physics: Conf. Series **1007** (2018) 012018 doi:10.1088/1742-6596/1007/1/012018

Performance analysis of AES-Blowfish hybrid algorithm for security of patient medical record data

Amir Mahmud H[1]*, Bayu Angga W[1], Tommy[2], Andi Marwan E[2], and Rosyidah Siregar[2]

[1]Departement of Computer Science, Faculty of Technology and Computer Science, Universitas Prima Indonesia, Indonesia
[2]Department of Computer Science, Faculty of Technic and Computer Science, Universitas Harapan, Indonesia

E-Mail: *amirmahmud@unprimdn.ac.id

Abstract. A file security is one method to protect data confidentiality, integrity and information security. Cryptography is one of techniques used to secure and guarantee data confidentiality by doing conversion to the plaintext (original message) to cipher text (hidden message) with two important processes, they are encrypt and decrypt. Some researchers proposed a hybrid method to improve data security. In this research we proposed hybrid method of AES-blowfish (BF) to secure the patient's medical report data into the form PDF file that sources from database. Generation method of private and public key uses two ways of approach, those are RSA method f RSA and ECC. We will analyze impact of these two ways of approach for hybrid method at AES-blowfish based on time and Throughput. Based on testing results, BF method is faster than AES and AES -BF hybrid, however AES-BF hybrid is better for throughput compared with AES and BF is higher.

1. Introduction

The development of internet technology nowadays in digital era becomes the point of attention, especially issues which are about data confidentiality, integrity and information security [1]. The internet utilization in the current era can facilitate human, one of them is easy to deliver data, so needs of data security will be vital and significant [2]. Cryptography is one of techniques that used for secure and ensure data confidentiality [3]. Technically, cryptography is conversion of plaintext (original message) to ciphertext (hidden message) with two important processes to secure message, those are encryption and decryption [4]. Cryptographic techniques are divided into two categories: Symmetric and Asymmetric Key Cryptography. Symmetric key Cryptography referred to as a Private Key Cryptography using key encryption and decryption at the same data. Symmetric techniques like DES, IDEA, Blowfish, RC4, RC2, RC5, Triple DES, and AES. Blowfish algorithm is faster than other algorithms, but AES algorithm is more secure and superior than DES [5]. While Asymmetric Key Cryptography is known as Public Key Cryptography which uses two different keys for encryption and decryption [6], like RSA, Rabin Cryptosystem, ElGamal Cryptosystem and Elliptic Curve Cryptosystem (ECC). ECC is faster than RSA. Some researchers have applied hybrid method to overcome weaknesses of cryptographic algorithms, such as the Fast method of Chaos-based, AES,

RSA and ElGamal[7]. AES-RSA [8], RSA-AES [9], [10], RSA-AES and RSA [11], AES-DES-RSA [12], AES-Blowfish [5] AES-DES [13].

Some researscher have applied hybrid method to resolve the weakness of cryptography algorithm, like research [12], did a performance comparison on algorithms AES, DES, RSA with a hybrid method of AES-DES-RSA with four parameters, namely avalanche effect encryption time, CPU usage and Throughput. [5] developed a hybrid method of AES-Blowfish algorithm with the use of Elliptic Curve Cryptography (ECC) for the management and key exchange, testing was done in the format of text files, image files, audio files and video files. [10] proposed a hybrid method of AES-RSA data security in order to support cloud computing, [9] implemented RSA-AES hybrid methods for security of data exchange between two clients using web services as an intermediary with encrypt content blocks that can only be decrypted by the client side, and not on the web server.

RSA and AES hybrid methods propose [11] to increase data security. AES Method for encryption and decryption data process based on key generation method using RSA-128 bytes. [14] proposed a method of hybrid RSA-AES for encryption at the IP address stored in the DSN by comparing results with RSA algorithms hybrid RSA-AES. [13] proposed a hybrid method of AES-DES for the security of data transmission based on 128-bit key with testing on file format text, image, audio and video.

In this study, we aimed to analyze the performance of a public and private key generation RSA and ECC method to secure the patient's medical report on data sourced from a database. Both of them approach key generation method, we analyze the influence of time and Throughput at the encryption process and decryption algorithm on hybrid AES-Blowfish. This paper is served as follows: part 2 proposed model framework, the result will be served in part 4 and part 5 is conclusion.

AES algorithm was developed by Joan Daemen and Vincent Rijment in 2001 through a contest organized by the National Institute of standards and technology (NIST). AES relies on patterns, basic principles related to substitution permutations system. AES has four methods: shift the Mix of columns, rows, substitute bytes, and add an all-around sub-key [12].

AES encryption process using button on a special called the round keys. This key is applied with other operations such as Sub Shift Bytes, Rows etc in plaintext (original data) to get the cipher text (encrypted data). For the process of decryption, the inverse process is used at the stage of transformation. The process starts from the InvSubBytes, InvShiftRows, and ends in InvMixColumns. Therefore, the S-box to encryption and decryption is different. In the process of decryption, the S-box used is reverse S-box substitution and operations permutations in block size 128 bits, and the key sizes from 128, 192, or 256 bits [12].

Blowfish is one of encryption method that is included in the Symmetric Cryptosystem. Symmetric key cipher block algorithm was designed in 1993 by Bruce Schneider to replace the DES (Data Encryption Standard). Blowfish is a block cipher, the encryption process and the decryption, Blowfish divides a message into blocks of equal size in length, i.e. 64 bits. This algorithm consists of two parts, namely the expansion of the key (the key expansion) and data encryption. Key expansion is stage to expanding the key length is 32 bits to 448 bits into several subkey arrays (subkey) with a total 4168 bytes. Data encryption occurs in as many as 16 rounds of feistily network where each round consists of a permutation lock and substitution of data. All operations are addition and XOR on 32-bit variables. Other operations are merely additional of four tables search (lookup table) for each round. Blowfish uses a large subkey where each keys must be computed before encryption or decryption of data begins.

RSA algorithm was invented by Ron Rivest, Adi Shamir and Leonard Aldeman and it is public key encryption. RSA method is widely used to secure data transmission with public key for encryption and decryption based on two prime numbers. The second prime number is to keep secret and only the authorized user can decode the message, the following steps [5].(a) Establishment of Key: Step 1 : take the two prime numbers tell a and b, Step 2 : Calculate n such that n = a * b, Step 3 : Calculate total n, (n) = (a-1) * (b-1), Step 4 : choose value e s.t 1 < e < (n) and e and (n) must relative prime,

IOP Conf. Series: Journal of Physics: Conf. Series **1007** (2018) 012018 doi:10.1088/1742-6596/1007/1/012018

Step 5 : Count d in such a way, de = 1 (mod (n)). (b) Message Encryption, C = me (mod n), where m is plaintext. (c) Message Decryption, The plaintext calculated from text password by M = cd (mod n)

Elliptic Curve Cryptography (ECC) is a system for public-key cryptography that makes the use of elliptic curve equation. The algorithm was designed and proposed by Neal Koblitz and Victor s. Miller. ECC method is considered more superior than other methods, because it uses a key that is much smaller, but it can provide the same level of security with an asymmetric algorithm using larger keys. With the smaller size of the key and the same high level of security, the implementation of ECC becomes more efficient.

SHA-256 is one of cryptographic hash functions designed by the NSA (U.S National Security Agency). SHA stands for Secure Hash Algorithm. Cryptographic hash function is a mathematical operation is performed on a digital data; by comparing the computed "hash" (the output of the execution algorithm) to the hash value that is known and expected, one can determine the integrity of data. SHA-256 hash computed by the words of 32 bits [9].

2. Proposed model framework

In the research, we propose hybrid cryptography method of AES and Blowfish, both of these symmetric methods will increase the time process of encryption and decryption on the patient's medical record data taken from databases with the format portable document file (PDF). Figure 1 and figure 2 are an illustration process Framework hybrid cryptography and decryption which we are proposing. Document activation uses SHA 256 while the key generation of private process and public key uses two ways of approaching, namely the method of RSA and ECC.

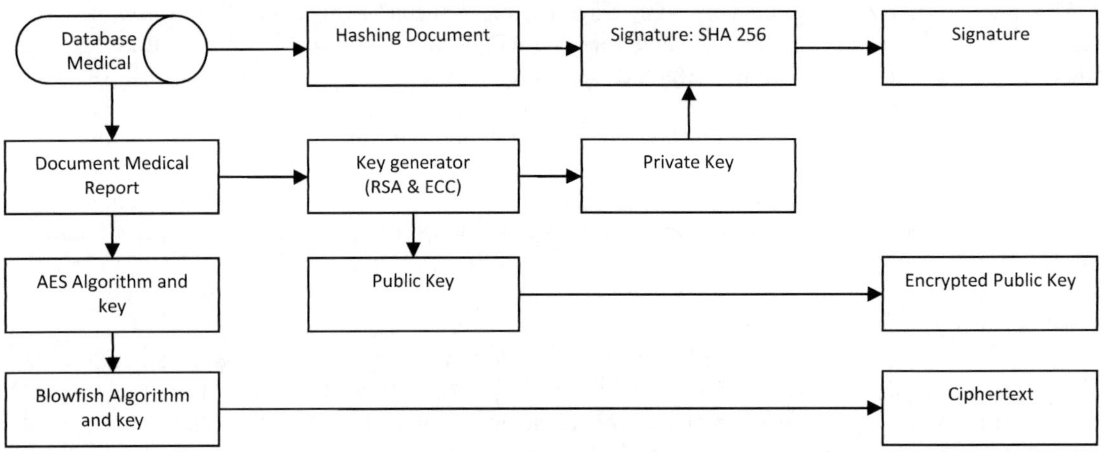

Figure 1. The proposed hybrid encryption architecture.

In figure 1 is a proposed hybrid encryption process, medical record data will be encrypted using AES and Blowfish, follows the process used: Generate data of medical record into a format portable document file (PDF); Apply a hashing document with the method of SHA 256 resulting signature; Generate a public and a private key using two ways of approaching, namely: RSA and ECC; Encryption of patient's medical record file with AES and Blowfish uses key; Sends files (1) encrypted file (2) encrypted key and (3) the signature to the recipient.

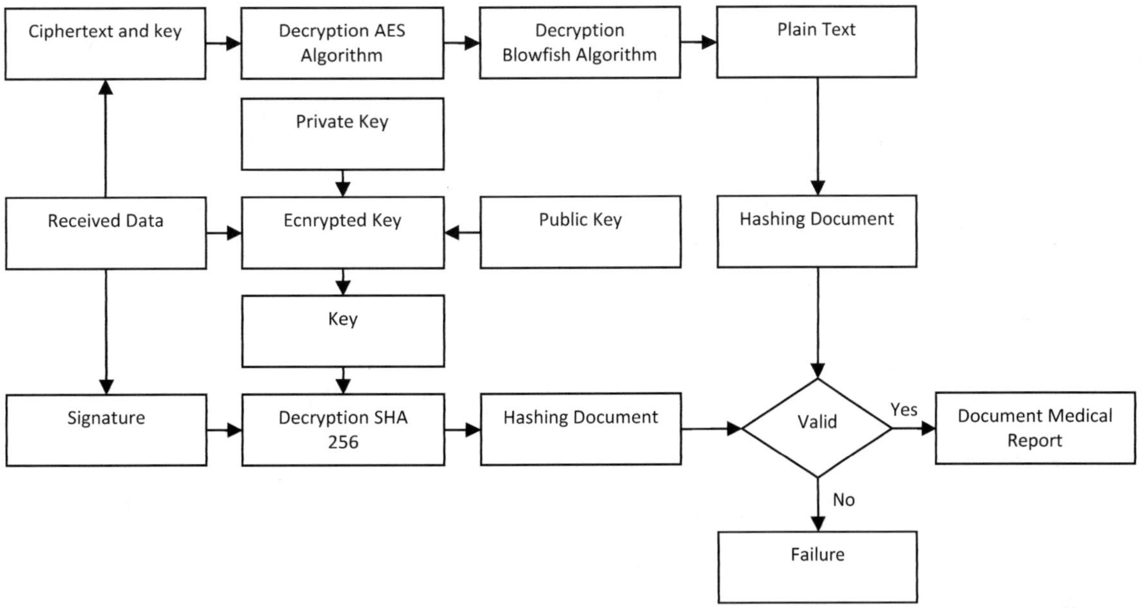

Figure 2. The proposed hybrid decryption architecture.

In figure 2 describes the hybrid decryption process proposed, patient record data which has been already encrypted will be decrypted again so that it can be read by the recipient, the following decryption process: The recipient will receive a three-block file password: (1) block key, (2) signature block and (3) the encrypted file; Implement private key on block to open the public key, apply the public keys to open the file signature; Enter AES key and Blowfish key to unlock the encryption file so that it results Hashing document file; Then it is compared with the result of hashing signature by hashing the document, if it works then your medical record file will perform normally. If not, then the files will appear in encrypted.

The main objective in this study is to analyze the effect of key generation RSA and ECC against AES-Blowfish hybrid methods to secure the patient's medical record file based on speed time and Throughput. In General, throughput is the rate of production or rate at which something can be processed. In the context of Encryption throughput is the rate at which data are increasingly encrypted. It is better to have a high encryption throughput. As a high encryption level (throughput) and safe, it is used to transmit data securely over the network. We have taken the approach to calculate the throughput using the formula: (Total number of Bits that are accessed)/(Total CPU time).

3. Result and Discussion

An experiment is done with the platform specification of intel core i5 2.5 GHz CPU, 8 Gb of RAM and using an operating system WIN 10 64-bit from the client side, whereas the specification of the database server uses Intel Xeon Processor E3-1200 v5 Series; Core i3, 64 GB DDR4 Memory with 7 operating system Centos Linux and applications are created using C # 2010. In testing, before medical record data, process of encryption is done, medical record data will be processed from a database by using query then the results will be displayed in the form of a report with the PDF format, the PDF report files in encrypted with AES and Blowfish which 256 SHA hashing functions applied for integrity. Key generation method of private and public keys uses RSA and ECC method, both of these methods will be analyzed its effects on the results of the encryption based on speed and Throughput. In table 1 is a result of data encryption processes on medical record of the patient, from the results of testing method of Blowfish (BF) is faster than AES, this result is appropriate literature, whereas AES-BF hybrid methods proposed to take higher than AES methods.

The effect application of the public and the private key on a hybrid AES-BF is proposed to give different results from the two ways of approaching which are used. On the ECC method is faster than RSA method, the results of the comparison of the two methods are shown in figure 1 using the method of RSA and ECC method using figure 2. Decryption process results in the method proposed can be seen in table 3 and figure 3 and figure 4 is a graphics display comparison results of both ways of approaching duplicate medical record file into PDF format.

Table 1. Hybrid encryption runtime approach.

File Name	File Size (kb)	Encryption Runtime (t)					
		RSA Generate Key			ECC Generate Key		
		BF	AES	AES-BF	BF	AES	AES-BF
Medical Report 1	372.67	124.22	165.89	244.22	111.22	152.89	231.22
Medical Report 2	367.33	122.44	164.11	242.44	109.44	151.11	229.44
Medical Report 3	424.00	141.33	183.00	261.33	128.33	170.00	248.33
Medical Report 4	214.67	71.56	113.22	191.56	58.56	100.22	178.56
Medical Report ..	296.67	98.89	140.56	218.89	85.89	127.56	205.89

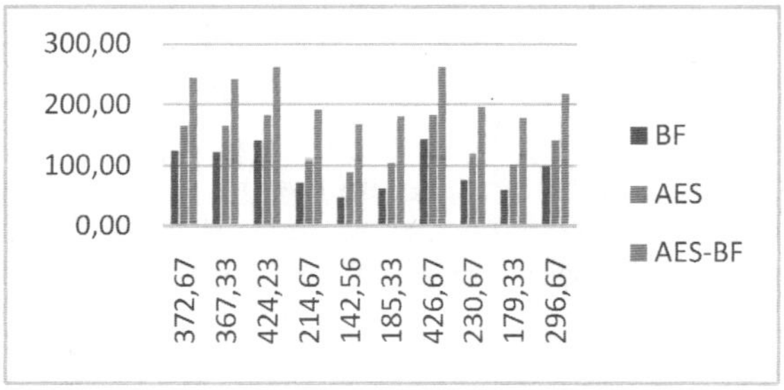

Figure 3. Graph of RSA encryption key.

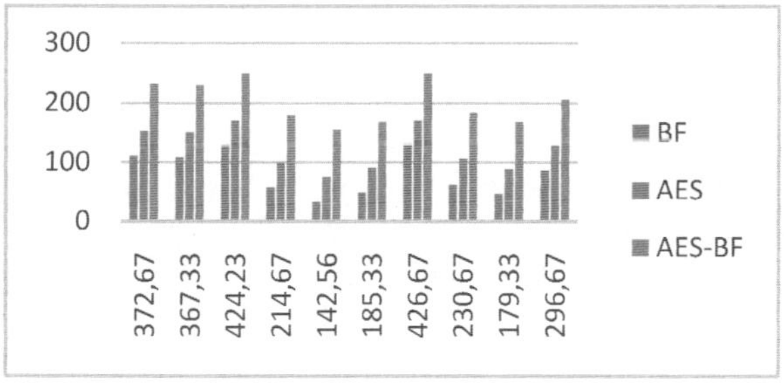

Figure 4. Graph of encryption ECC key.

Table 2. Hybrid decryption runtime approach.

File Name	File Size (kb)	Decryption Runtime (t)					
		RSA Generate Key			ECC Generate Key		
		BF	AES	AES-BF	BF	AES	AES-BF
Medical Report 1	372.67	135.89	177.55	255.8867	119.55	157.89	232.89
Medical Report 2	367.33	134.11	175.77	254.1067	117.77	156.11	231.11
Medical Report 3	424.23	153.00	194.66	272.9967	136.66	175.00	250.00
Medical Report 4	214.67	83.23	124.89	203.2277	66.89	105.23	180.23
Medical Report ..	296.67	110.56	152.23	230.5587	94.22	132.56	207.56

In table 2 above, is the result of the decryption process using the hybrid method of AES-BF the time where the process is higher than data encryption, but the decryption key using the private key RSA and public method takes more than a method of ECC.

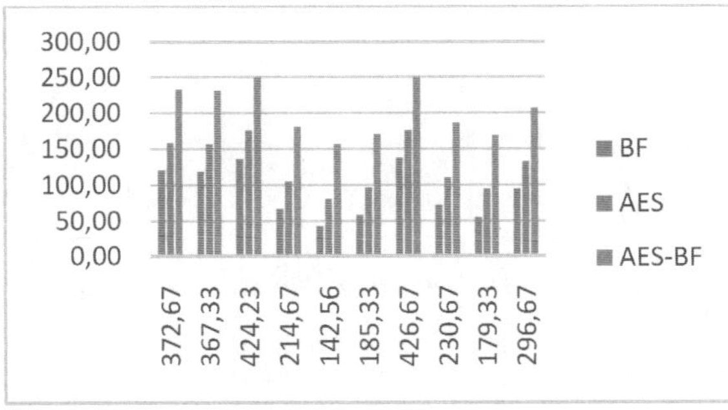

Figure 5. Graph of decryption RSA Key.

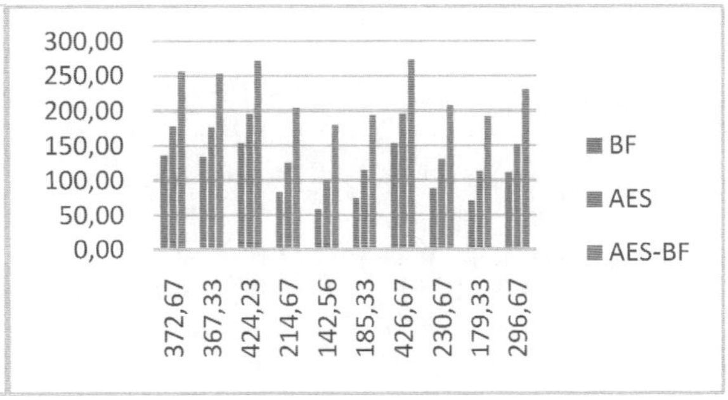

Figure 6. Graph of decryption ECC Key.

For the results of the performance in terms of Throughput shown in table 3 and table 4, Throughput Performance Analysis for the encryption process and duplicate is measured in Bits per Second. The process of encryption and decryption of data files of patient is on medical record, BF method is faster than AES, AES-BF but a combination is higher than AES. Results for Throughput, method of AES-is more efficiently than BF and AES, whereas AES is better than BF. The results of the comparison in the form of a graph are like in figure 5 and 6.

4. Result and Discussion

An experiment done with the platform specification intel core i5 2.5 GHz CPU, 8 Gb of RAM and is using an operating system WIN 10 64-bit from the client side, whereas the specification of the database server using Intel Xeon Processor E3-1200 v5 Series; Core i3, 64 GB DDR4 Memory with 7 operating system Centos Linux and applications are created using C # 2010.

In testing, before medical record data encryption, medical record data will be processed from a database by using query then the results will be displayed in the form of a report with the PDF format, the PDF report files in encrypted with AES and Blowfish which 256 SHA hashing functions applied for integrity. Key generation method of private and public keys using RSA and ECC method, both of these methods will be analyzed its effects on the results of the encryption based on speed and Throughput. In table 1 is a result of data encryption processes medical record of the patient, from the results of testing method of Blowfish (BF) faster than AES, this result appropriate literature, whereas AES-BF hybrid methods proposed takes higher than AES methods.

The effect application of the public key and the private key on a hybrid AES-BF proposed gives different results from the two approaches are used. On the ECC method faster than RSA method, the results of the comparison of the two methods are shown in figure 1 using the method of RSA and ECC method using figure 2. Decryption process results in the method proposed can be seen in table 3 and figure 3 and figure 4 is a graphics display comparison results of both approaches duplicate medical record file into PDF format.

Table 3. Hybrid encryption runtime approach.

File Name	File Size (kb)	Encryption Runtime (t)					
		RSA Generate Key			ECC Generate Key		
		BF	AES	AES-BF	BF	AES	AES-BF
Medical Report 1	372.67	124.22	165.89	244.22	111.22	152.89	231.22
Medical Report 2	367.33	122.44	164.11	242.44	109.44	151.11	229.44
Medical Report 3	424.00	141.33	183.00	261.33	128.33	170.00	248.33
Medical Report 4	214.67	71.56	113.22	191.56	58.56	100.22	178.56
Medical Report ..	296.67	98.89	140.56	218.89	85.89	127.56	205.89

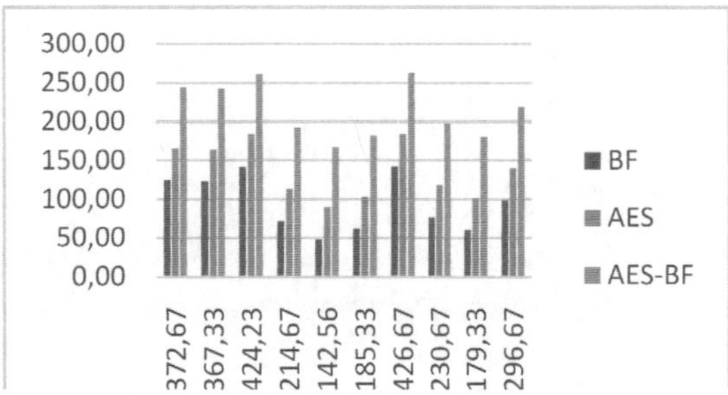

Figure 7. Graph of RSA encryption Key.

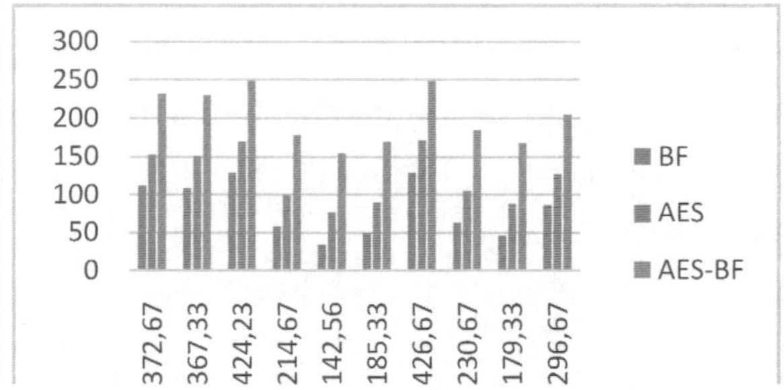

Figure 8. Graph of encryption ECC Key.

Table 4. Hybrid decryption runtime approach.

File Name	File Size (kb)	Decryption Runtime (t)					
		RSA Generate Key			ECC Generate Key		
		BF	AES	AES-BF	BF	AES	AES-BF
Medical Report 1	372.67	135.89	177.55	255.8867	119.55	157.89	232.89
Medical Report 2	367.33	134.11	175.77	254.1067	117.77	156.11	231.11
Medical Report 3	424.23	153.00	194.66	272.9967	136.66	175.00	250.00
Medical Report 4	214.67	83.23	124.89	203.2277	66.89	105.23	180.23
Medical Report ..	296.67	110.56	152.23	230.5587	94.22	132.56	207.56

In table 4 above, is the result of the decryption process using the hybrid method of AES-BF the time where the process is higher than data encryption, but the decryption key using the private key RSA and public method It takes more than a method of ECC.

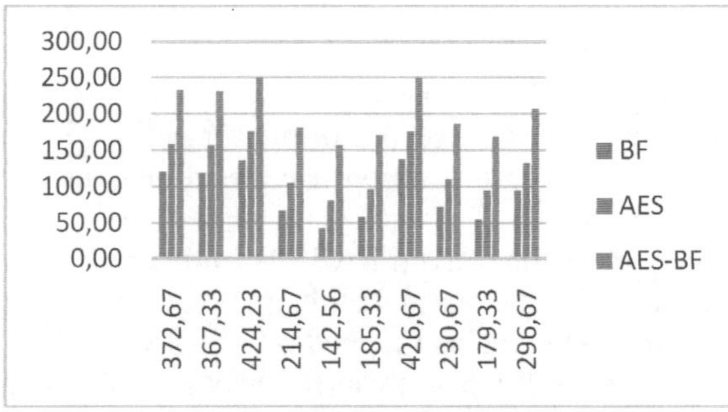

Figure 9. Graph of decryption RSA Key.

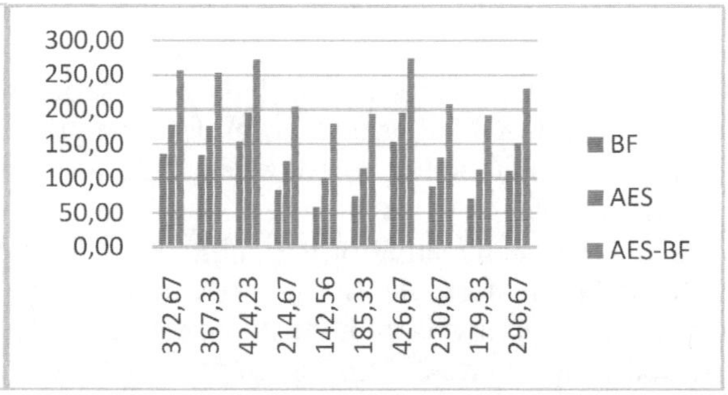

Figure 10. Graph of decryption ECC Key.

For the results of the performance in terms of Throughput shown in table 3 and table 4, Throughput Performance Analysis for the encryption process and duplicate is measured in Bits per Second. The process of encryption and decryption of data files of patient medical record, BF method faster than AES, AES-BF but a combination higher than AES. Results for Throughput, method of AES-more efficiently than BF and AES, whereas AES better than BF. The results of the comparison in the form of a graph like figure 5 and figure 6.

Table 5. Hybrid encryption throughput approach.

File Name	File Size (kb)	Encryption Throughput					
		RSA Generate Key			ECC Generate Key		
		BF	AES	AES-BF	BF	AES	AES-BF
Medical Report 1	372.67	403378	242027	145216	355922	213553	128132
Medical Report 2	367.33	397598	238559	143135	350822	210493	126296
Medical Report 3	424.23	459187	275512	165307	405165	243099	145859
Medical Report 4	214.67	232359	139415	83649	205022	123013	73808
Medical Report ..	296.67	321116	192669	115602	283337	170002	102001

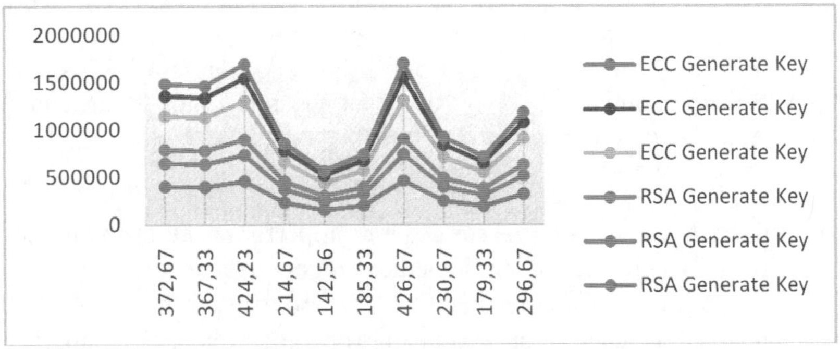

Figure 11. Graph of hybrid encryption throughput approach.

In table 3 is the result of Hybrid encrypting Throughput of medical record data of the patient, the establishment of a public key and a private key using the RSA method produces a higher computing time compared methods of ECC. Throughput results of Hybrid method AES -BF proposed better than BF and AES method, whereas the application of the private key and public key RSA method is still higher than the method of ECC it is not much different from the results of the encryption process.

Table 6. Hybrid decryption throughput approach.

File Name	File Size (kb)	Decryption Throughput					
		RSA Generate Key			ECC Generate Key		
		BF	AES	AES-BF	BF	AES	AES-BF
Medical Report 1	372.67	3025335	2337759	2337837	2269001	1753319	1753394
Medical Report 2	367.33	2981985	2304261	2304339	2236489	1728196	1728271
Medical Report 3	424.23	3443899	2661195	2661273	2582924	1995896	1995971
Medical Report 4	214.67	1742691	1346625	1346703	1307018	1009969	1010044
Medical Report ..	296.67	2408367	1861011	1861089	1806275	1395758	1395833

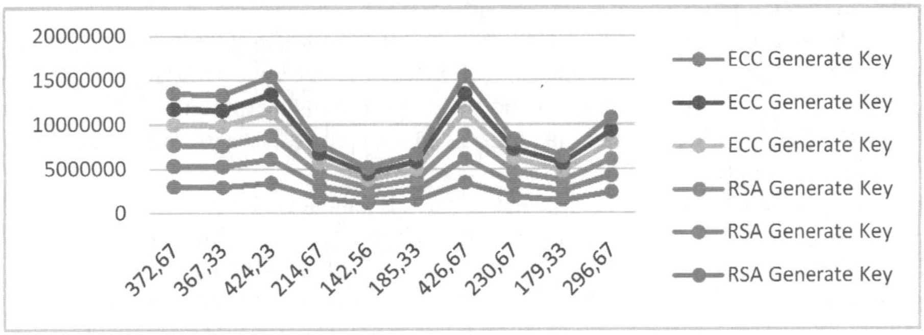

Figure 12. Hybrid decryption throughput approach.

IOP Conf. Series: Journal of Physics: Conf. Series **1007** (2018) 012018 doi:10.1088/1742-6596/1007/1/012018

Based on testing performed, RSA method takes computing with average time process of encryption 94.62 seconds and decryption 106.29 seconds at BF method. AES method for encryption process is 136.34 second and 147.96 seconds for decryption while the hybrid method for encryption is 214.62 second and for decryption is 226.29 second, while ECC method with encryption time is 81.62 second and decryption is 89.96 second at BF method. Encryption of AES method is 123.36 second and Decryption is 147.96 second. AES-BF hybrid takes higher than BF and AES in which time of encryption process is 201.62 second and 203.29 to decryption process.

The value of average Throughput on testing is as shown in table 3 Encryption Throughput and table 5 Decryption Throughput, AES- BF Hybrid methods is better than BF and AES with average throughput time BF encryption is 307416 bps and decryption is 2305618 bps. The average throughput of AES encryption is 188449 bps and decryption is 1781614 bps and AES-BF hybrid encryption is 110670 bps and decryption is 1781692 bps on RSA method, While the average ECC method of throughput BF method for encryption is 271249 bps and decryption is 17281692 bps. The average throughput of the AES encryption is 162749 bps and decryption is 1336210 bps and AES- BF hybrid encryption is 97650 bps and decryption is 1336285 bps.

5. Conclusion

Our primary goal in this study is to analyze the generation performance of public and private key at RSA and ECC method to secure the patient's medical report data in the form of a PDF file that sourced from a database. From the testing results, ECC method is faster than RSA method as shown in tables. Table 1 is comparison of encryption processing time and table 2 is comparison of decryption processing time on BF, AES and AES-BF hybrid method. Thus it can be drawn as the conclusion of ECC method which gives highest strength per bit of any cryptosystem with the smaller size of the key by generating a faster calculation, power consumption and low memory, it also provides a methodology to get protocol implementation and high speed key authentication, efficient and measurable.

References

[1] Rani R, Sharma G, 2017, "Review Paper on Data Hiding In 3D Barcode Image Using Steganography", International Journal of Advanced Research in Computer Science, Volume 8, No. 5, May-June 2017, 2271—2276, ISSN No. 0976-5697

[2] Ghodke M, Mali N. 2016, FPGA Based network security using cryptography. International Research Journal of Engineering and Technology. 3(3):469–71

[3] Bhuvaneshwari M, Tenmozhi S. 2016, "A VLSI architecture for security based stenographic processor with AES algorithm". International Journal of Electrical and Computer Engineering; 1–6

[4] Sukhraliya V, Chaudhary S, Solanki S. 2013, "Encryption and decryption using ASCII values with substitution array approach". International Journal of Advanced Research in Computer and Communication Engineering. 2(8):3094–7

[5] A. P Shaikh, V. kaul, 2014, "Enhanced Security Algorithm using Hybrid Encryption and ECC", IOSR Journal of Computer Engineering (IOSR-JCE), Volume 16, Issue 3; PP 80-85

[6] Garg SK. 2014, "Modifed encryption and decryption using symmetric keys at two stages: Algorithm SKG 1.2". International Journal of Advanced Research in Computer Science and Electronics Engineering. 4(6):778–80

[7] M Enriquez, D W Garcia and E Arboleda, 2017, "Enhanced Hybrid Algorithm of Secure and Fast Chaos-based, AES, RSA and ElGamal Cryptosystems", Indian Journal of Science and Technology, Vol 10(27), DOI: 10.17485/ijst/2017/v10i27/105001. 1-4

[8] Ch.Vijayalakshmi, L.Lavanya, Ch.Navya, 2016, "A Hybrid Encryption Algorithm Based On AES and RSA, International Journal of Innovative Research in Computer and Communication Engineering", Vol. 4, Issue 1, 909-917

[9] K G Kadam and V Khairnar, 2015, "Hybrid RSA-AES Encryption for Web Services, International Journal of Technical Research and Applications", Special Issue 31(September, 2015), PP. 51-56
[10] B Kumar, J Boaddh and L Mahawar, 2016, "A hybrid security approach based on AES and RSA for cloud data, International Journal of Advanced Technology and Engineering Exploration", Vol 3(17), 43-49
[11] Palanisamy, V. and Jeneba Mary, A, 2011, "Hybrid Cryptography By The Implementation Of RSA and AES, International Journal of Current Research". Vol. 33, Issue, 4, pp.241-244.
[12] S.N.Ghosh, et al. 2015, "Performance Analysis of AES, DES, RSA And AES-DES-RSA Hybrid Algorithm for Data Security", International Journal of Innovative and Emerging Research in Engineering Volume 2, Issue 5, 83-88
[13] J R Patel, R S. Bansode and V Kaul, 2012, "ybrid Security Algorithms for Data Transmission using AES-DES", International Journal of Applied Information Systems (IJAIS), Volume 2– No.2, 15-21
[14] R Behl, G Sehgal, M Kumar, P Gupta, S Garg, 2015, "Experimental comparison between Hybrid RSA-AES and RSA algorithms in IP security", IJMTER, 588-594
[15] Por, L. Y., Beh, D., Ang, T. F., Ong, S. Y., 2013, "An Enhanced Mechanism for Image Steganography Using Sequential Colour Cycle Algorithm", The International Arab Journal of Information Technology, Vol. 10, 51-60

Acknowledgments
This research was supported by Universitas Prima Indonesia, Universitas Harapan Medan, Indonesia and one of private hospital in Medan city.

IOP Conf. Series: Journal of Physics: Conf. Series **1007** (2018) 012019 doi:10.1088/1742-6596/1007/1/012019

Extraction of ECG signal with adaptive filter for hearth abnormalities detection

Mardi Turnip[1*], Rijois. I. E. Saragih[2], Abdi Dharma[1], Dwi Esti Kusumandari[3], Arjon Turnip[3], Delima Sitanggang[1], Siti Aisyah[1]

[1]Faculty of Technology and Computer Science, Universitas Prima Indonesia, Indonesia
[2]Universitas Methodist Indonesia, Indonesia
[3]Technical Implementation Unit for Instrumentation Development Indonesian Institute of Sciences, Indonesia

E-mails: [*]marditurnip@unprimdn.ac.id, 0105128001@methodist.ac.id

Abstract. This paper demonstrates an adaptive filter method for extraction ofelectrocardiogram (ECG) feature in hearth abnormalities detection. In particular, electrocardiogram (ECG) is a recording of the heart's electrical activity by capturing a tracingof cardiac electrical impulse as it moves from the atrium to the ventricles. The applied algorithm is to evaluate and analyze ECG signals for abnormalities detection based on P, Q, R and S peaks. In the first phase, the real-time ECG data is acquired and pre-processed. In the second phase, the procured ECG signal is subjected to feature extraction process. The extracted features detect abnormal peaks present in the waveform. Thus the normal and abnormal ECG signal could be differentiated based on the features extracted.

1. Introduction

Everyone must have known that the heart is the most important organ in the body. Even since the fetus is in the womb, the heartbeat is something eagerly awaited because the heartbeat is a sign of one's life. Therefore, to keep the heart in good health is one of the steps that we should do. But in fact, the variety of physical activity and daily consumption of food and beverages have made us neglect of body health especially heart organ. As a result, most people do not even know if they have a serious illness that can cause complications until the sudden cardiac arrest. Electrocardiogram (ECG) is a graph made by an electrocardiograph, which records the electrical activity of the heart within a certain time [1]. Its name consists of a number of different parts: Electro, as it relates to electronics, cardio, indicates heart, and gram means to write. Analysis of a number of ECG waves and normal vectors of depolarization and repolarization yields important diagnostic information of the heart activity. A typical ECG signal is used to tracks a normal heartbeat (or heart cycle) which consist of P wave, QRS complex and T wave. A small U wave is normally seen at 50-75% in the ECG signal. The baseline voltage of the ECG is known as the isoelectric line. Typically, the isoelectric line is measured as a tracking portion following the T wave and precedes the next P wave.

The distance of hospitals from patients, the limitations of cardiovascular doctors and its equipment, make the number of patients with heart disease especially sudden attacks increased. Since the death rate of patients with heart disease is relatively high, the development and application of the telemonitoring of ECG signal are needed [2, 3]. Therefore, an online ECG signal monitoring system for patients with heart disease by identifying the RR signals. With the development of a remote patient monitoring system, it is expected that each patient's condition can be monitored in online and real-time so that the physical condition data and medical history of the patient can be monitored at any time by a team of hospital doctors without regularly visit the hospital. Furthermore, both patients and families can be informed of alarms regarding the patient's condition, especially during an emergency.Several wireless ECG technologies which could significantly improve the mobility and comfort of patients for monitoring and diagnosis has been developed [4–9]. However, some issues related to current wireless ECG technology still need to be improved such as the speed, accuracy, its application, and the best reliability [10]. Heart function can be measured by the wireless ECG system used by the patient wherever located and the recording data is sent to the remote monitoring station. When a patient moves, this communication needs to be maintained with a wireless link and the quality of data with less noise needs to be maintained with powerful software.

Developing wireless ECG technology is required to provide online and real-time remote cardiography. In the conventionally designed wireless applications, it tends only for data transmission processes such that the ECG monitoring applications require only moderate data rate. In this paper, an adaptive wireless technology based on ECG filters to improve data quality, speed data rate, and the best reliability are being developed. The recorded ECG signal (generated by the stimulation of nerve impulses to the heart) from normal heartbeat consists of P waves, QRS complexes, and T waves. The ECG base voltage is known as the isoelectric line measured as part of track following the T wave and predates the next P wave. The purpose of this study was to develop an extraction algorithm that is a distance of the RR between QRS complex using adaptive filter to support the designed wireless ECG system.

2. Methods

To ensure the quality of life and up-to-date information about the physical and physiological health of the users, monitoring and recording of their physiological status are essential [11 - 13]. Therefore, the ECG technology-based wireless is proposed. The scheme of a wireless ECG technology is given in Figure 1. The developed system consists of electrode sensors (to measure the ECG signal), a wireless ECG recorded module (to preprocessed and transmit the ECG signals via Bluetooth), and an android mobile device. Then, an ECG monitoring algorithm includes the classifier developed in the android mobile device will continuously monitor the subject heart activities. When the abnormal heart rate is detected by a classifier, the classified ECG signals will be sent to the cloud or directly to the physician or family. Based on this mechanism, the cardiac state of the subject can be monitored anywhere in the globe as long as the subject under internet network coverage. The objective of this technology development is to build universal interfaces for medical instrument that are easy to use, cheap, easy to repair, best reliability, and good accuracy.

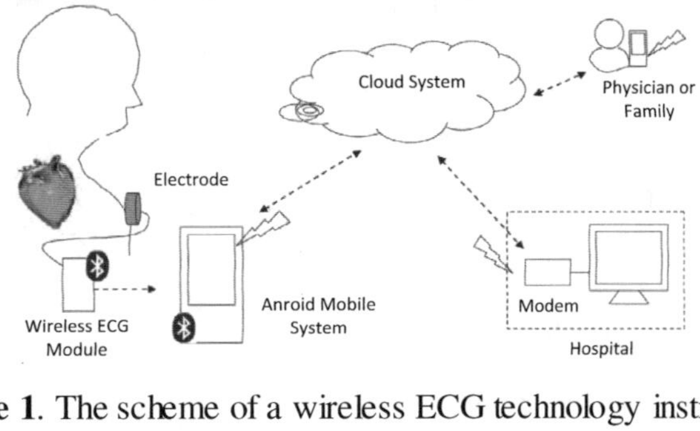

Figure 1. The scheme of a wireless ECG technology instrument.

To test the reliability and the quality of the developed system, the data is used from the previous experiment with experiment setup shown in Figure 2 [14-16].

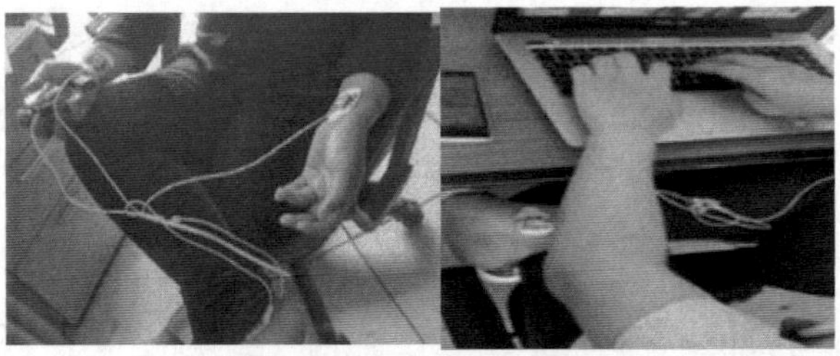

Figure 2. Experiment setup and ECG recorded signals with relaxing and typing conditions.

3. Signal Processing

Signal processing concerns the analysis, synthesis, and modification of signals, which are broadly defined as functions conveying, information about the behavior of some phenomenon such as ECG signals [17]. In experimental sciences, noise is referred to any random fluctuations of data that hinders perception of the desired signal. The process of elimination or reduction of an undesired signal which considered as noise is needed. An adaptive filter is used reduce noise in the recorded ECG signals while preserving the desired signal characteristics.The adaptive filters described in Figure 3 generally consist of a digital filter to represent the dynamic system and an adaptive algorithm to adjust the coefficient of the dynamic filter. (i.e., the desired signal(k) and the reference signal x(k) are simultaneously applied to the design adaptive filter). Commonly, digital filters use fixed coefficients, but in this design filter, the coefficients are adjusted based on an information from desired, reference, and output signals. The obtained signal d(k) is measured using electrode sensors which consist of both the actuals s(k) and also undesired noise signals n(k) (assumed to be independent). If the contaminated noise signal n(k) is known, the actual signal s(k) is obtained by subtracting noise signal n(k) from contaminated signal d(k). Since the undesired or noise signal is difficult to measure, then the calculated using a filter and measurable noise source is used. The estimated desired signal is obtained by subtracting the estimated noise signal from the measured signal d(k).

FIR filter is one of the digital filters that commonly used in an adaptive system. The coefficient of the dynamic filter at time k is arranged by vector w(k) with length of K. A filter with K tap can be considered by K sample of reference input x(k) at time k as a vector as follows [11][18].

$$W(k) = [w_0(k)\, w_1(k) \ldots\ldots w_{M-1}(k)]^T \quad (1)$$

$$x(k)=[x(k)\, x(k-1) \ldots\ldots x(k-M+1)] \quad (2)$$

$$y(k) = \sum_{i=0}^{M-1} w_i(k)x(k-i) = w^T(k)x(k) \quad (3)$$

The Least Mean Square (LMS) is an approximation of the gradient vector with instantaneous value. The error signal e(k) is minimized in the mean-square sense to change the weighting filter. The conventional LMS algorithm can simply replace the cost function [19].

$$\xi(k) = E[e^2(k)] \quad (4)$$

The error estimation

$$e(k) = d(k) - y(k) = d(k) - w(k)x(k) \quad (5)$$

with the coefficient updating equation is

$$w(k+1) = w(k) + \mu e(k)\, x(k) \quad (6)$$

where w(k) and w(k+1) represent the weight and the next weight vectors, respectively, x(k) and e(k) are the input and error signals and μ is an appropriate step size of the convergence factor to be chosen to determine the filter speed [19-21]. The larger of the step sizes could make the coefficients of the filter become wildly fluctuate and eventually become unstable.

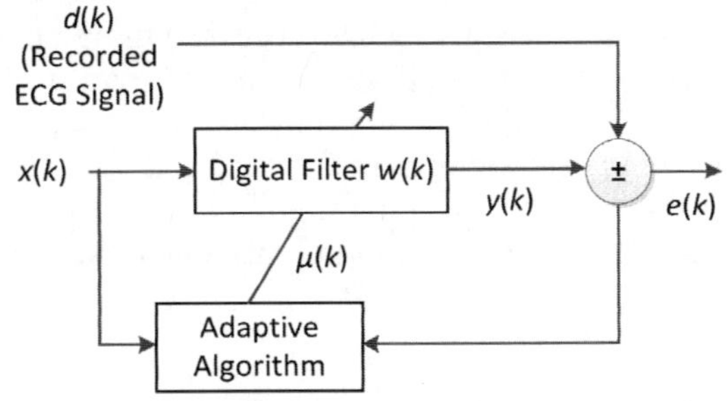

Figure 3. An adaptive filter design for the ECG signal.

To avoid the gradient noise amplification condition especially when μ is large, then the Normalized Least Mean Square (NLMS) algorithm is applied. The convergence factor of the NLMS algorithm in the time-varying form is calculated as

$$\mu(k) = \frac{\alpha}{c + \|x(k)\|^2} \tag{7}$$

where α is the adaptation value (0<α<2) used to improve the convergence rate of the developed algorithm and c is a constant which usually less than 1. Then, the weights vector of the design filter will be updated in the form of

$$w(k+1) = w(k) + \frac{\alpha}{c + \|x(k)\|^2} e(k)u(k) \tag{8}$$

When $x(k)$ is large, the noise amplification problem is diminished by normalization of the term $\|x(k)\|^2$.

4. Results and Discussions

The results of data processing on advanced ECG recording allows doctors and users to easily understand and observe the performance and condition of the heart as well as to make a real-time diagnosis. For the purpose, the signal disturbance in the form of power lines, muscle artifacts, electrode motion artifacts, etc. Should be eliminated as good as possible. The noise signal is mainly due to patient breathing, movement, poor electrode conditions and improper location electrode preparation. All these disturbances will mask the original signal and tend to be smaller amplitude. If this problem can not be solved properly, then it can result in a false diagnosis. In this paper, adaptive filters are implemented.

The algorithm is evaluated on the recorded data from ten healthy volunteers subject aged around 25 ± 3 years old and it functions normally. Figure 4 and 5 show the recorded using developed system (top) and filtered signals using the adaptive filter with the normalized mean square algorithm (bottom). In normal ECG, the time interval between R-R is 0.6-1s, in case of the fast heartbeat the time interval is less 0.6s which is known as sinus tachycardia; in case of a slow heartbeat, the time interval is more than 1 sec that is known as sinus bradycardia. By referring Figure 4 and 5 we can say that the R-R interval for normal case is between 0.6-1s. All signals from all subject and different distance of wireless ECG recording signals are in the standard form of ECG signals. Compared with the data measured by medical instruments in nursing centers and hospitals, our physiological status monitoring system has reached the application level. By improving the signal processing ability for filtering, the quality of filtered signal from a farther distance.

IOP Conf. Series: Journal of Physics: Conf. Series **1007** (2018) 012019 doi:10.1088/1742-6596/1007/1/012019

Initial stage on detecting subject's heart condition is developing an algorithm that can be used to detect the QRS position as in Figure 6. The QRS positioning information on each recording signal will make it easier for doctors or family to know their heart conditions. To enhance the accuracy of peak detection related to the QRS position, the signal is preprocessed using an adaptive filter. The R and S points including RR interval were well detected as shown in Figure 7. The obtained results clearly different from the recorded data of relax than typing subjects. If in the previous results with a general filter algorithm the difference is highly significant but with the adaptive filter, the peak detection accuracy is only slightly different. Then the proposed filter is sufficiently able to overcome the dynamic interference on the signal recording. That difference can be seen in Figure 8 and 9.

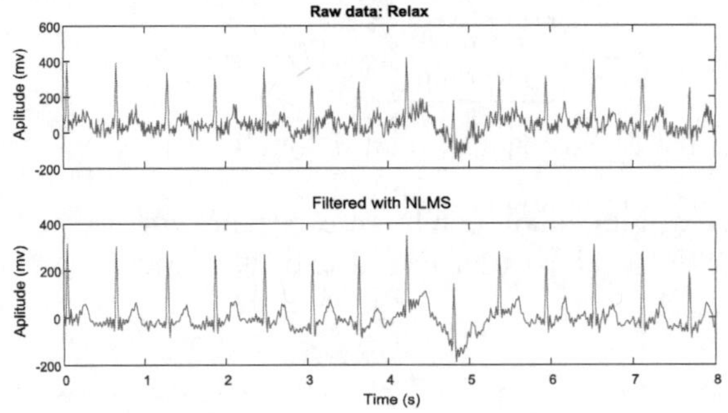

Figure 4. Filtered ECG signals from relaxing subject.

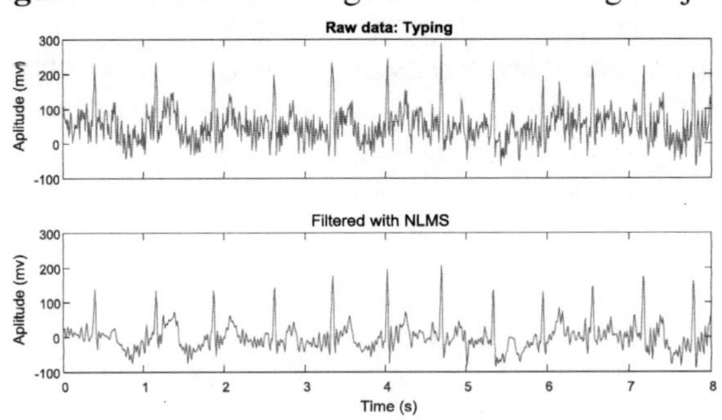

Figure 5. Filtered ECG signals from typing subject.

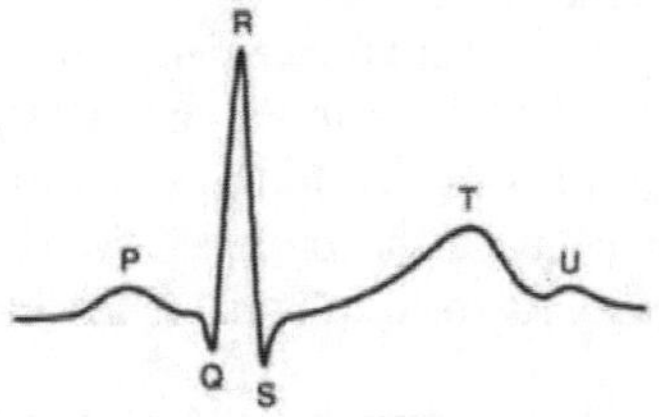

Figure 6. The standard form of ECG Signals [12].

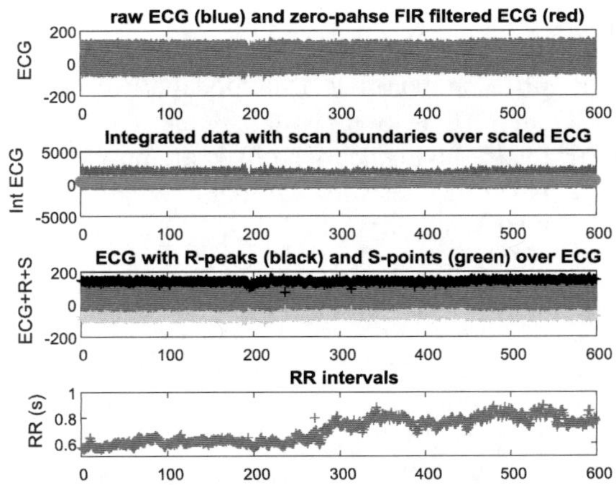

Figure 7. Detected R & S point and RR Interval.

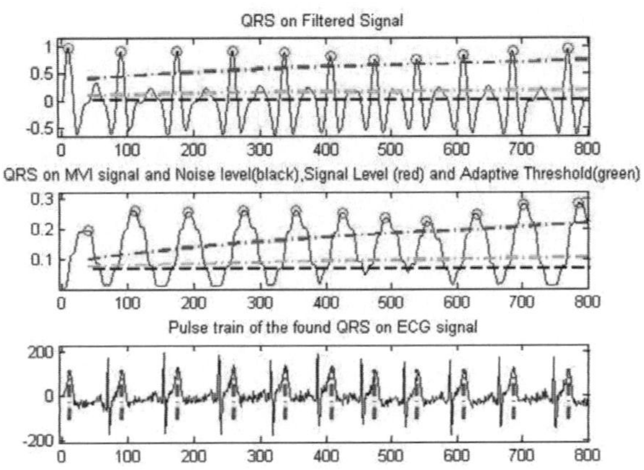

Figure 8. Detected QRS for relax condition.

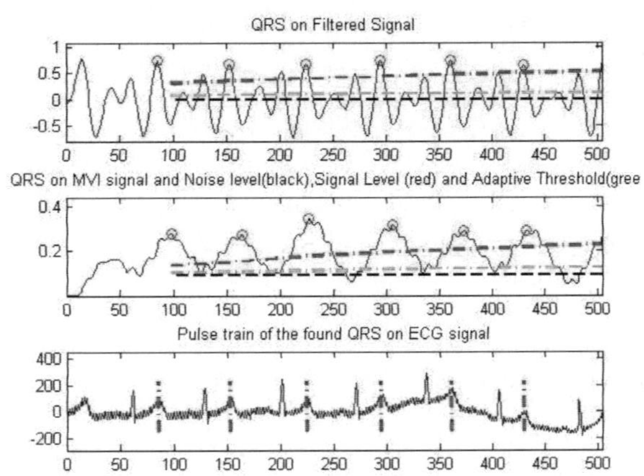

Figure 9. Detected QRS for typing condition.

IOP Conf. Series: Journal of Physics: Conf. Series **1007** (2018) 012019 doi:10.1088/1742-6596/1007/1/012019

5. Conclusions

In this paper, the developed (i.e., wireless ECG system that consists of mobile physiological examination device and wireless base station) and its application in the experiment is evaluated. One of the crucial steps in the ECG analysis is to accurately detect the different form of P, Q, R and S which represent the entire heart cycle. An application of the adaptive filter with the developed system, the higher quality of the ECG signals is achieved. An improvement of QRS peak detection accuracy supported by adaptive filter about 86% is achieved.

References

[1] Rivera-Ruiz M,Cajavilca C and Varon J 1927*Texas Heart Institute journal / from the Texas Heart Institute of St. Luke's Episcopal Hospital, Texas Children's Hospital.* **35** 174-8
[2] Zhao Z and Cui L 2005 Proceeding of the 2005 *IEEE Engineering in Medicine and Biology 27th Annual Conference* (Shanghai, China) p 2145-2148
[3] Blasco S, Uceda J D, Elena M 2008 *International Conference on Complex, Intelligent and Software Intensive Systems.* p 643-648
[4] Varshney U 2009 Pervasive *Healthcare Computing-EMR/EHR, Wireless and Health Monitoring* (New York: Springer)
[5] Scalvini S, Martinelli G, Baratti D, Domenighini D, Benigno M, Paletta L, Zanelli E and Giordano A 2005 *J. Telemedicine Telecare.* **11** p 18–20
[6] Alesanco A and Garcıa J 2010 *IEEE Trans. Inf. Technol. Biomed.* **14** p 1144–1152
[7] Sneha S and Varshney U 2007 *Int. J. Electron. Healthcare.* **3** p 32–50
[8] Cypher D, Chevrollier N, Montavont N and Golmie N *2006 IEEE Commun. Mag.* **44** p 56–63
[9] Kang K, Park K J, Song J J, Yoon C H and Sha L 2011 *IEEE Trans. Inf. Technol. Biomed.* **15** p 260–267
[10] Baker S D 2002 *Health Manag. Technol.* **23** p 100-103
[11] Olguin D, Bouchereau F and Martinez S 2005 *conference on information sciences and systems, the John Hopkins University*
[12] Johnson P and Andrews D C 1996 *J. Telmed. Telecare.* **2** p 107–113
[13] Rodriguez M J, Arredondo M T, Del Pozo F, Gomez E J, Martinez A and Dopico A 1995 *J. Telmed. Telecare.* **1** p 86–94
[14] TurnipA, KusumandariDE, SuhendraMA and HermidaIDP 2016 *International Journal of Information Technology & Computer Science.* **24** p 55-61
[15] TurnipA, RizgyawanMI and KusumandariDE 2016 *International Journal of Pharma Medicine and Biological Sciences.* **5** p 189-193
[16] RizgyawanM I, AmriMF, Pratama P R S and Turnip A 2016 *International Conference on Information Technology, Computer, and Electrical Engineering* (Semarang, Indonesia)
[17] Prandoni P andVetterli M 2018 *Signal Processing for Communications, EPFL Press.*
[18] Varady P, Benyo Z andBenyo B 2002 *IEEE Trans. Inf. Technol. Biomed.* 6 p 95–98
[19] Gacek A andPedrycz W 2012 *ECG Signal Processing, Classification and Interpretation, Springer*
[20] Jiashu Z, Heng-Ming T 2007 *IEEE Transactions.* p 2489-2492
[21] Shah J H, Joshi J M Digital signal processing *University science press laxmi publication company Limited New Delhi*

Acknowledgments

This research was supported by Universitas Prima Indonesia, the thematic program through the Bandung Technical Management Unit for Instrumentation Development, Indonesian Institute of Science and the Insinas Ristek funded by Minister of Research, Technology, and Higher Education, Indonesia.

IOP Conf. Series: Journal of Physics: Conf. Series **1007** (2018) 012020 doi:10.1088/1742-6596/1007/1/012020

Performance Of The IEEE 802.15.4 Protocol As The Marker Of Augmented Reality In Museum

Adi Kurniawan Saputro[1], Surya Sumpeno[1,2] and Mochamad Hariadi[1,2]

[1]Department of Electrical Engineering, Institut Teknologi Sepuluh Nopember, Surabaya, Indonesia
[2]Department of Computer Engineering, Institut Teknologi Sepuluh Nopember, Surabaya, Indonesia

E-mail: adikurniawansaputro@gmail.com, surya@ee.its.ac.id, mochar@ee.its.ac.id

Abstract. Museum is a place to keep the historic objects and historical education center to introduce the nation's culture. Utilizing technology in a museum to become a smart city is a challenge. Internet of thing (IOT) is a technological advance in Information and communication (ICT) that can be applied in the museum. The current ICT development is not only a transmission medium, but Augmented Reality technology is also being developed. Currently, Augmented Reality technology creates virtual objects into the real world using markers or images. In this study, researcher used signals to make virtual objects appear in the real world using the IEEE 802.14.5 protocol replacing the Augmented Reality marker. RSSI and triangulation are used as a substitute microlocation for AR objects. The result is the performance of Wireless Sensor Network could be used for data transmission in the museum. LOS research at a distance of 15 meters with 1000 ms delay found 1.4% error rate and NLOS with 2.3% error rate. So it can be concluded that utilization technology (IOT) using signal wireless sensor network as a replace for marker augmented reality can be used in museum.

1. Introduction

Museum is a place to keep the historic objects and historical education center to introduce the nation's culture. Internet of things is the development of internet technology for human needs in various ways. Utilization of technology in the museum to the city of smart is a challenge. Technology that may be realized is AR. Augmented Reality (AR) combines real and virtual objects in real environments, runs interactively in real time, and there is integration between objects in three dimensions [2]. Combining real and virtual objects is something that might happen with appropriate technology, interactivity is not impossible through good input devices and integration with effective tracking.

The development of AR research is very rapid. The AR object can appear without having to use an image as the point where the virtual object appears. The AR connected to the sensor and the internet can display virtual objects. The virtual object may appear in a location using the Global Positioning System (GPS). The GPS work system is by using a satellite as the signal sender to the GPS receiver and the signal is used by the receiver to determine the location. GPS has been popularly used by Google Maps to determine the user's location. Unfortunately GPS can not be used properly in the room so it can not be implemented in the museum.

Previous research has found that position tracking indoors or known as microlocation has been successful. Some examples of the use of this system are indoor guides, ISAM European ADAMANT project [3], and Wireless Sensor Network application indoors that can replace GPS [4]. There are three wireless positioning algorithms that use ID, triangulation and signal strength [5]. Based on the above descriptions researchers are interested in making research using signals to make virtual objects appear in the real world by using the IEEE 802.14.5 protocol instead of the Augmented Reality marker.

2. Augmented Reality Markers

There are several ways in reading AR. Commonly used AR with images or known as marker AR. Markeles AR is rarely encountered because to make AR appear using GPS coordinates as ID or marker [6]. In this research, the author used a marker by utilizing ID and RSSI on WSN. Marker AR reading AR with markers is a way to bring 3D objects to the real world using images. AR implementation in the smartphone there are several processes: Tresholding, Contour detection, Rectangle Fitting, Undistortion, Pattern checking and Pose estimation [7].

Markeles is a method of reading AR without pictures or drawings as its markers but by utilizing the real environment itself as its marker [6] [8],. The markeles concept is not much different from the marker. Markeles reads the mark using GPS coordinates or captured directly from the surrounding environment. So without using paper, 3D objects can appear in accordance with the coordinates or positions we want. This is easier and does not require much time in AR readings. Wireless Sensor Network (WSN) is a wireless network infrastructure composed of a number of sensor nodes scattered in an area. Each sensor node has radio transceivers and antennas, microcontrollers, electronic interfacing circuits and energy sources [9]. WSN technology can be used to monitor, control, and communicate.

3. Designing The Positioning System

At this stage, augmented reality system will be built on smart museum based on wireless sensor network to recognize objects in the museum by using the IEEE 802.14.5 protocol in recognizing markers to show the object AR. The method in this research is by designing a microlocation system to determine the AR object of marker replacement on augmented reality based on wireless sensor network by using RSSI and Triangulation which will be applied in the museum. In addition, testing will be performed to determine the performance of Wireless Sensor Network Figure1.

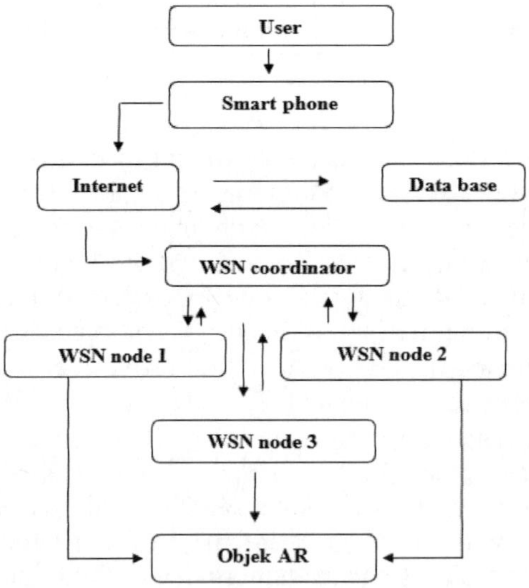

Figure 1. System Design WSN Node And Object WSN.

IOP Conf. Series: Journal of Physics: Conf. Series **1007** (2018) 012020 doi:10.1088/1742-6596/1007/1/012020

Stages of the WSN system starts from the smartphone user connected to the internet in the museum. Then the user will get information from objects that have been placed wireless sensor network. Wireless sensor networks start sending data after sleep mode, then wireless sensor network nodes send a packet of data with hexa letters, where each protocol has a different package structure. Furthermore, the package structure will be sent to the coordinator. Coordinator will be provided with 1 arduino to buffer to get the ID from wireless sensor network. The next step is the data that has been obtained will be continued on the database using ethernet shield. Once the data entered into the database, the data will be taken to be displayed into virtual objects in the Augmented Reality application. The position of AR will use RSSI and triangulation. Where this position is used as the appearance of the AR object.

3.1. Triangulation and RSSI

By measuring the distance using the signal intersection, we get the midpoint of the signal. This method is also used to calibrate the node position with the coordinator. In determining the position with strong signal there are still many obstacles. Therefore we adopt a more sophisticated propagation model [10]. Described in this study with an explanation of our position, shown in 1.

$$RSSI = -(10n\log 10\, d + A) \tag{1}$$

To measure RF parameters (Radio Frequency) with A is defined as absolute energy with dBm units. Where n is the constant for signal transmission. d is the distance from each transmitting node to the coordinator or receiver. The value of RSSI can be measured using several equations. In WSN there is an RSSI value using the XCTU application. To calculate d can use (1). In order to know the third boundary of WSN with the coordinates of the reference nodes A, B, C and determine the distance between the nodes to the WSN coordinator Suppose the WSN coordinates are (x, y), in equation (2) can be set:

$$\begin{aligned} (x-x1)2+(y-y1)2&=d1^2 \\ (x-x2)2+(y-y2)2&=d2^2 \\ (x-x3)2+(y-y3)2&=d3^2 \end{aligned} \tag{2}$$

With the above equation we can determine the coordinates of the user's WSN and predict the AR object to appear.

4. Designing The Positioning System

From several experiments we got some positions where AR objects appear. With a signal of 15 dBm, the distance was estimated at 5.6 m. This distance corresponded to the system when we did the experiment and was supported from some of the data we had with the three WSN nodes. ID had been set 16 for brahma gods, ID 18 for god ganesha and ID 19 for gods of siwa. The data can be viewed in Table 1.

Table 1. Position of AR Object with signal and WSN ID.

ID	x	Y	z	dBm	Distance	AR Object Position
16	3	1	2	15	5.6m	Right
18	1	2	1	13	4.4m	Right
19	2	1	3	17	7m	Left

From the analysis and experiments we did for many times, the position of the object more often appeared in the position of the object in Table 1 with the position of x, y and z in accordance with the condition of data collection in the museum. We used three WSNs for this experiment. With three WSN we estimated where the AR object appeared. The appearance of the AR object can be seen in Figure 2. Furthermore, from this experiment conducted a third performance analysis WSN.

Figure 2. AR Object 3D Position in Museum.

4.1 *Performance of Wireless Sensor Network*

By tested and observed the performance maximum on the wireless sensor network on LOS and NLOS conditions, the researcher will get the results to get the maximum distance on the data transmission. The purpose of distance measurement testing is to determine the quality of data delivery process in Xbee S1 for the system to work optimally and can be used indoors or outdoors museum's Mpu Tantular.

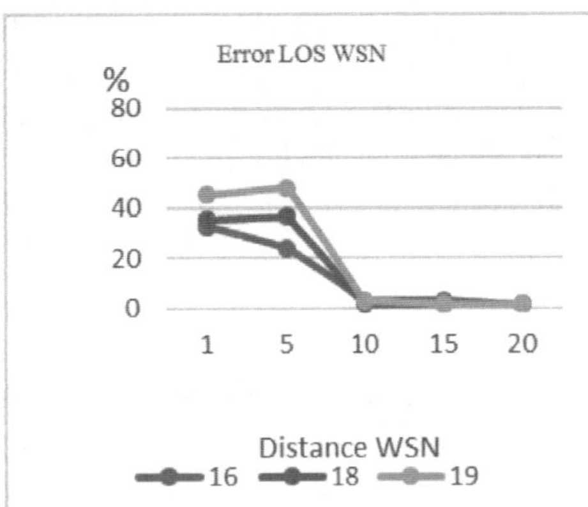

Figure 3. Graph Error in LOS condition.

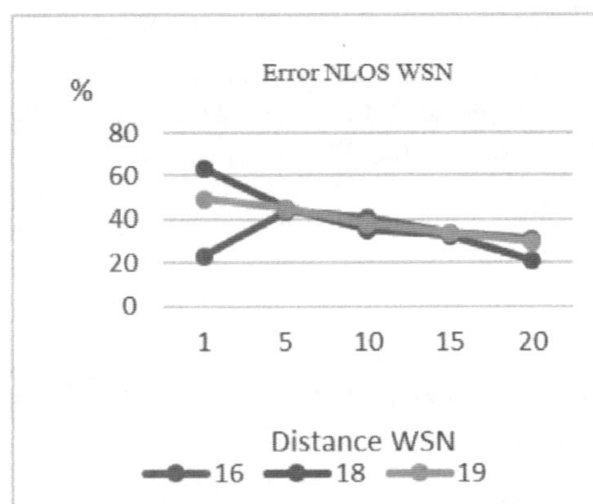

Figure 4. Error Chart under NLOS Conditionsfik.

This test was used to determine the level of system success made to get the error value for the evaluation in the processed of sending data. In LOS testing, the node of the coordinator was further away, the data received by the coordinator got better with 1000ms delay. It can be seen on graphic Figure 3 and Figure 4. However, still need to be tested and delay analysis to know the delay of data because this affect the appearance of AR object. Next will be analyzed the delay to error on WSN performance.

4.2 *Analysis of delay to error*

The delay test was performed to observe the data transmission time from the node to the coordinator. In this test some delay experiments were conducted to observe the effect of AR objects. This time the test was done to start sending data packets with the time from 50ms to 1000ms.

Table 2. Delay to WSN coordinator.

No	ID WSN	*Delay Error (Node - Coordinator)*			
		50 ms	100 ms	500 ms	1000 ms

1	16	54,92%	45,07%	39,43%	1,40%
2	18	66,19%	47,88%	36,61%	1,40%
3	19	59,15%	46,47%	38,02%	1,40%

Table 3. Delay12 against database.

No	ID WSN	*Delay Error (Node - Database)*			
		50 ms	100 ms	500 ms	1000 ms
1	16	57,43%	47,88%	42,25%	2,81%
2	18	69,01%	67,60%	39,43%	2,81%
3	19	60,56%	47,88%	39,43%	1,40%

From experiments and system testing using 3 nodes, the best delay for the appearance of the object was visible at 1000ms. From the test conducted as many as 71 data taken repeatedly.

4.3 Loss analysis

In this test was done Data loss analysis when sent data packets. From the data obtained in Table 4 and Table 5, loss occurs very much at 500ms. The greater the delay, the smaller the data loss and reduced the rate of data transmission failed. The caution was the data accumulation in the coordinator and data base. And too soon the data sent by xbee to the coordinator. When the process of sorting ID occurs data accumulation resulting in loss. The effect of loss will be seen when the data transmission process takes place and result in the AR object does not appear.

Table 4. Loss on The Data Base.

No	ID WSN	*LOSS (Node - Data Base)*			
		50 ms	100 ms	500 ms	1000 ms
1	ID16	4,22%	2,81%	1,40%	0%
2	ID18	2,81%	2,81%	1,40%	0%
3	ID19	4,22%	2,81%	1,40%	0%

Table 5. Loss on Data Coordinator.

No	ID WSN	*LOSS (Node - Coordinator)*			
		50 ms	100 ms	500 ms	1000 ms
1	ID16	4,22%	2,81%	1,40%	0%
2	ID18	2,81%	2,81%	1,40%	0%
3	ID19	4,22%	2,81%	1,40%	0%

5. Conclusion

The use of wireless sensor network in Augmented Reality as ID and position using triangulation method and RSSI run well and can be used in place of AR marker and AR markeles using GPS position. However, there are some advantages and disadvantages with this method. The advantages of this system no longer require images in presenting AR and can be used to determine the position of the user. Besides maximum distance that can be applied to the Augmented Reality Real time Monitoring system is as far as 15 meters with the condition of LOS (Line of Sight) found 1.4% error rate and NLOS (Non Line of Sight) with 2.3% error rate. Because at this distance maximum data can be sent well. In addition it can reduce the loss rate with set delay standarts 1000ms. So that the data sent is not lost and accumulated by new data.

References

[1] I. E. Sutherland, "A head-mounted three dimensional display," Proceedings of the AFIPS '68 (Fall, part I), pp. 757–764, 1968
[2] R. Azuma, R. Behringer, S. Feiner, S. Julier, and B. Macintyre, "Recent Advances in Augmented Reality," IEEE Computer Graphics and Applications, vol. 2011, no.December, pp. 1–27, 2001
[3] Y. Wang and A. Ma, "An Agent-based Passenger Support System over Heterogeneous Wireless Infrastructures in an Airport Environment," IASTED International Conference on Networks and Communication Systems (NCS 2006), pp. 240–244, 2006
[4] F. Darakeh, "An Accurate distributed rage free localization algorithm for WSN l," pp. 2014–2019, 2014
[5] Y. Wang, X. Yang, Y. Zhao, Y. Liu, and L. Cuthbert, "Bluetooth positioning using RSSI and triangulation methods," 2013 IEEE 10th Consumer Communications and Networking Conference, CCNC 2013, pp. 837–842, 2013
[6] Grasset, "Image-Driven View Management for Augmented Reality Browsers," pp. 177–186, 2012
[7] J. Kim and H. Jun, "Implementation of image processing and augmented reality programs for smart mobile device," Proceedings of the 6th International Forum on Strategic Technology, IFOST 2011, vol. 2, pp. 1070–1073, 2011
[8] N. Navab, T. Blum, L. Wang, A. Okur, and T. Wendler, "First deployments of augmented reality in operating rooms," Computer, vol. 45, no. 7, pp. 48–55, 2012
[9] M. Patel and A. Shah, "Energy Efficient Target Tracking in Wireless," International Journal of Distributed and Parallel Systems (IJDPS), vol. 3, no. 2, pp. 542–549, 2012
[10] J. Albowicz, A. Chen, and L. Zhang, "Recursive position estimation in sensor networks," Network Protocols Ninth International Conference on ICNP 2001.pp. 35–41, 2001

Smart home design for electronic devices monitoring based wireless gateway network using cisco packet tracer

Oloan Sihombing[1*], Niskarto Zendrato[2], Yonata Laia[1], Marlince Nababan[1], Delima Sitanggang[1], Windania Purba[1], Diarmansyah Batubara[1], Siti Aisyah[1], Evta Indra[1], Saut Siregar[1]

[1]Faculty of Technology and Computer Science, Universitas Prima Indonesia, Indonesia
[2]Faculty STMIK Kristen Neuman Indonesia, Indonesia

Email: *oloansihombing007@gmail.com

Abstract. In the era of technological development today, the technology has become the need for the life of today's society. One is needed to create a smart home in turning on and off electronic devices via smartphone. So far in turning off and turning the home electronic device is done by pressing the switch or remote button, so in control of electronic device control less effective. The home smart design is done by simulation concept by testing system, network configuration, and wireless home gateway computer network equipment required by a smart home network on cisco packet tracer using Internet Thing (IoT) control. In testing the IoT home network wireless network gateway system, multiple electronic devices can be controlled and monitored via smartphone based on predefined configuration conditions. With the Smart Home can potentially increase energy efficiency, decrease energy usage costs, control electronics and change the role of residents.

1. Introduction

In the era of technological development today, without realizing the technology has become a need that is often used for the life of today's society. Where almost all areas have been able to feel from the development of such technology, as in the field of computer network electronics. In this case, the need for home technology is increasingly required to make a smart home, where during this time in the control of electronic devices at home is done through the switch on or off switch.

Previously some smart home development has been done in monitoring or controlling electronic devices that are in the household but can only monitor or control on one electronic device only and the transmission media is done still using SMS gateway and Bluetooth [1-2], so can not maximize in creating or building a smart home system is good in monitoring or controlling electronic devices that are in the household.

Smart Home is an application of a combination of technology and services devoted to the home environment with specific functions aimed at improving the safety, efficiency, and comfort of its inhabitants. In the smart home system usually consists of monitoring tools, control devices and automatic there are several devices that can be accessed using a computer or smartphone connected to the Internet network [4].

IoT is a structure in which objects, people are provided with an exclusive identity and the ability to move data through a network without requiring two-way inter-human to human source ie the destination or human interaction to the computer. IoT is a very promising development to optimize life based on intelligent sensors and smart appliances that work together over the internet [5].

The emergence of IoT is a major development in the world of the Internet that it is expected that billions of physical things or things will be equipped with various types of sensors connected to the internet through the network as well as technology support such as embedded sensors and actualization, Radio Frequency Identification (RFID), wireless sensor networks, real time and Web services , IoT is often also referred to as a physical Cyber 'system or network of networks with a large number of electronic devices or electronic objects and sensors connected to the internet [6-7].

Smartphones are handheld phones with computer-like capabilities, which are working phones using all software, operating systems that provide standard and fundamental relationships for application fans. Based on the development of smartphones today already has the presentation of advanced features and also has the ability to support the development of computer networking devices that make it a necessity in humans [8].

For that purpose, efforts are made to increase the needs of households to facilitate the use and control of all electronic devices by using remote or remote control, where devices that can be controlled using smart home technology such as lighting devices, access door, fan and various electronic devices others that can be activated or not activated using a smartphone through the home gateway computer network device as a transmission medium or liaison to some electronic devices that are in the household, other than that smart home has several benefits such as providing comfort, safety and security at the residence [9-10].

2. Methodology

In this study apply literature and experimental study methods, which in the literature study obtained from print media and Internet media in the form of journals, e-books and material books. While in experimental done using software and hardware that is designed its performance and done testing to make the application, whether it is as expected. In the configuration and testing of network systems by designing the concept of smart home IoT model which then test each subsystem model, in the end after subsystem testing, testing the overall system model to see the success. level model of smart home systems designed on cisco package trackers. It uses the IoT home gateway as the media path that connects multiple wireless devices wirelessly and provides automatic addressing to devices connected to the home gateway, where all devices connect to the smartphone as an interface medium for controlling and monitoring electronic devices [11].

3. Configuring the Simulation Concepts In Cisco Packet Tracer

In building smart home network can be implemented through the simulation concept designed on cisco packet tracer. Some of the tools needed to build a smart home network are home gateways that function as transmission media paths and provide automatic addressing to multiple devices connected via wireless networks, smartphones that serve as interfaces in controlling and monitoring electronic devices, and some devices to be controlled based on conditions set on smartphone devices [12-14]. In figure 1 configure the smartphone device as an interface in monitoring and controlling electronic devices, then in figure 2 as the intermediate home gateway login on the smartphone device via the web browser, then in figure 3 as the interface list of IoT server electronic devices in the home menu located on the smartphone, then in figure 4 is a condition setting on electronic devices that are set in the menu conditions that are on the smartphone.

IOP Conf. Series: Journal of Physics: Conf. Series **1007** (2018) 012021 doi:10.1088/1742-6596/1007/1/012021

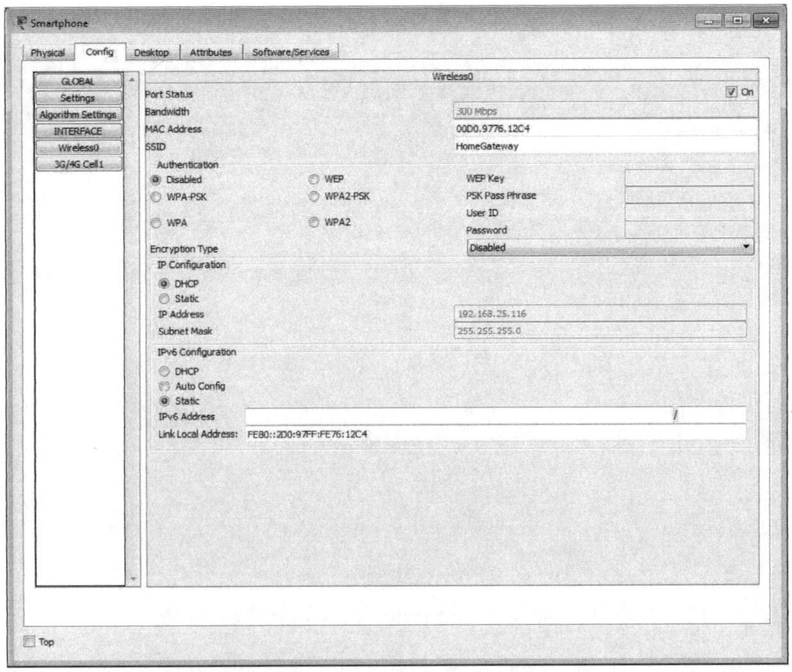

Figure 1. IoT server wireless settings.

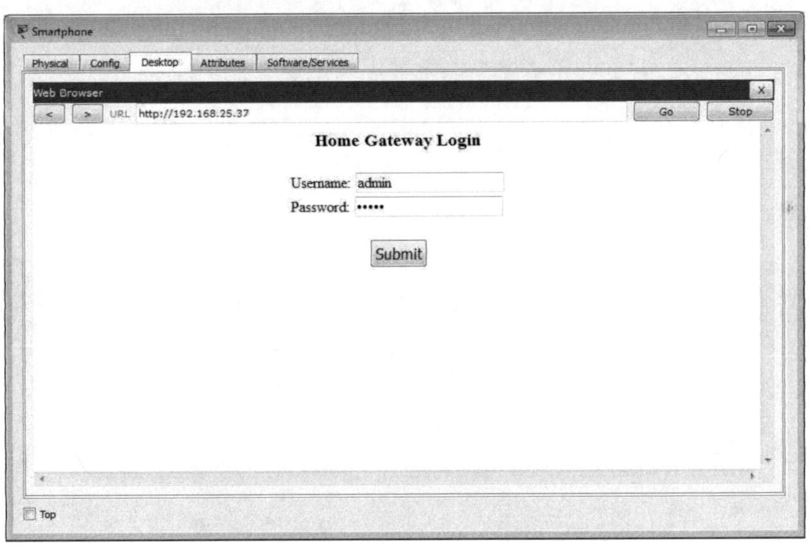

Figure 2. Home gateway smartphone login.

IOP Conf. Series: Journal of Physics: Conf. Series **1007** (2018) 012021 doi:10.1088/1742-6596/1007/1/012021

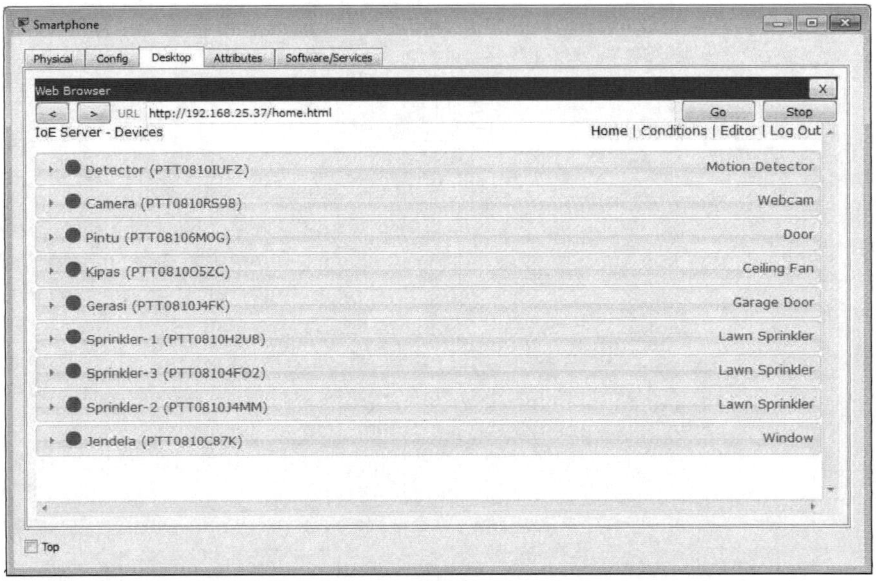

Figure 3. IoT server home device smartphone interface.

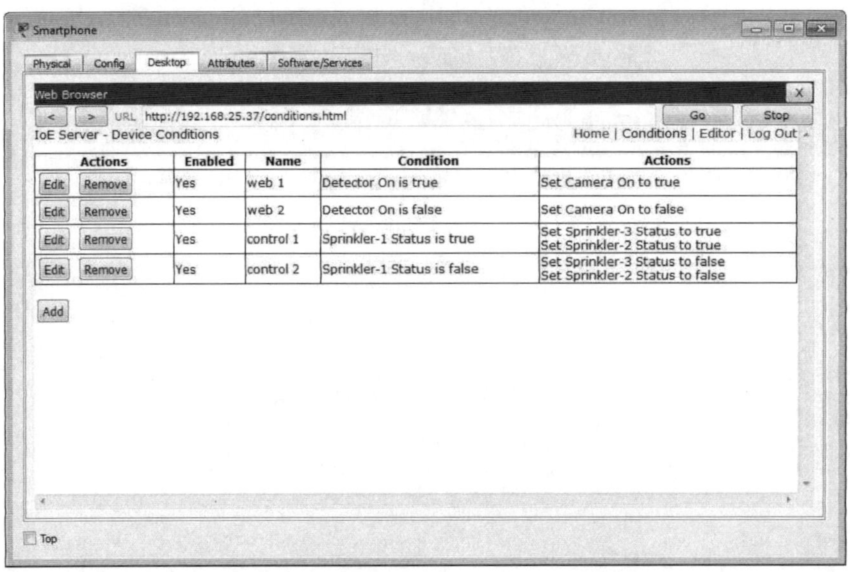

Figure 4. IoT server device conditions smartphone interface.

4. Smart Home Simulation Concept Results In Cisco Packet Tracer

Based on the configuration of the smartphone simulation concept that has been set up in the IoT home gateway network on the Cisco packet tracer, some electronic devices can be controlled and monitored according to system testing, configuration and condition settings on electronic devices specified in in the smartphone system, making it easier in designing and implementing in building the smart home network. In figure 5, we can explain the process of controlling electronic devices through a smartphone-controlled home gateway. In-camera webcam device connected to the device motion detector, where webcam camera will light up when the device motion detector captures a move. So do some conditions on lawn sprinkler devices, when lawn sprinkler 1 is turned on and off by smartphone, then lawn sprinkler 2 and lawn sprinkler 3 will react. And on the window, fan, garage door, and door devices are also controlled by the smartphone through the home gateway transmission media.

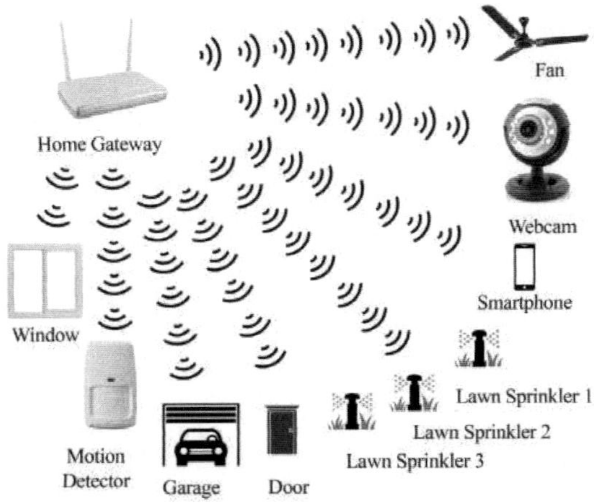

Figure 5. Concept results of smart home simulation.

5. Conclusion

In building a smartphone network we can simulate it using a Cisco package tracker and learn or know how to configure the network and connect multiple electronic devices to connect to each other over the wireless network, in addition, to set up some electronic devices based on their condition. set on smartphones With this simulation, design and implementation planning can be done in building smart home network using IoT home gate and there is a possibility that this simulation can be applied in real world based on current technology development, thus making it a necessity for community life which potency can improve energy efficiency , reducing energy use costs, controlling electronic devices and changing the role of occupants.

References

[1] A. ElShafee and K. A. Hamed 2012 *Design and Implementation of a WiFi Based Home Automation System*, World Academy of Science, Engineering and Technology, pp. 2177-2180.

[2] R.Pivare, M.Tazil 2011 *Bluetooth Based Home Automation System Using Cell Phone*. IEEE 15th International Symposium on Consumer Electronics Singapore, pp.192-195.

[3] Sun, C. (2012) Application of RFID Technology for Logistics on Internet of Things.

[4] Kosmatos, E.A., Tselikas, N.D. and Boucouvalas, A.C 2011 *Integrating RFIDs and Smart Objects into a Unified Internet of Things Architecture*. Advances in Internet of Things: Scientific Research, **1**, 5-12.

[5] Moeinfar, D., Shamsi, H. and Nafar, F 2012 *Design and Implementation of a Low-Power Active RFID for Container Tracking @ 2.4 GHz Frequency*: Scientific Research.

[6] Want, R 2006 *An Introduction to RFID Technology*. IEEE Pervasive Computing, **5**, 25-33.

[7] R. Piyare, 2013 Internet of Things, *Ubiquitous Home Control and Monitoring System using Android based Smart Phone*, International Journal of Internet of Things, **2**, no. 1, pp. 5-11.

[8] A. Rajabzadeh, A. R. Manashty, and Z. F. Jahromi 2010 *A Mobile Application for Smart House Remote Control System*, World Academy of Science, Engineering and Technology, **62**, pp.80-86.

[9] R. Shahriyar, E. Hoque, S. Sohan, I. Naim, M. M. Akbar, and M. K. Khan 2008 *Remote controlling of home appliances using mobile telephony*, International Journal of Smart Home, **2**, pp. 37-54.

IOP Conf. Series: Journal of Physics: Conf. Series **1007** (2018) 012021 doi:10.1088/1742-6596/1007/1/012021

[10] Aggarwal, R. and Lal Das, M 2012 *RFID Security in the Context of Internet of Things*. First International Conference on Security of Internet of Things, Kerala, pp.51-56.
[11] B. Park, 2012. *Mobile IP-Based Architecture for Smart Homes*, International Journal of Smart Home, **6**, pp.29-36.
[12] Gigli, M. and Koo, S 2011 *Internet of Things, Services and Applications Categorization*. Advances in Internet of Things, **1**, 27-31.
[13] R. T. Fielding and R. N. Taylor, 2002 *Principled design of the modern web architecture*, Software Engineering, Proceedings of the 2000 International Conference, pp. 407-416.
[14] U. Sharma and S. R. N. Reddy 2012 *Design of Home/Office Automation Using Wireless Sensor Network*, International Journal of Computer Applications, **43**, pp. 53-60.

A Simple Compression Scheme Based on ASCII Value Differencing

Tommy[1], Rosyidah Siregar[1], Imran Lubis[1], Andi Marwan E[1], Amir Mahmud H[2], and Mawaddah Harahap[2]

[1]Department of Computer Science, Faculty of Technic and Computer Science, Universitas Harapan, Medan, 20217, Indonesia
[2]Departement of Computer Science, Faculty of Technology and Computer Science, Universitas Prima Indonesia, Medan, 20111, Indonesia

E-mail: tommy@stth-medan.ac.id[1], rosyidahsiregar0705@gmail.com[1], imran@stth-medan.ac.id[1], andimarwanelhanafi@gmail.com[1], amirmahmud@unprimdn.ac.id[2], mawaddah@unprimdn.ac.id[2]

Abstract. ASCII characters have a different code representation where each character has a different numeric value between the characters to each other. The characters is usually used as a text message communication has the representation of a numeric code to each other or have a small difference. The value of the difference can be used as a substitution of the characters so it will generate a new message with a size that is a little more. This paper discusses the utilization value of the difference of characters ASCII in a message to a much simpler substitution by using a dynamic-sized window in order to obtain the difference from ASCII value contained on the window as the basis in determining the bit substitution on the file compression results.

1. Introduction

Compression is generally divided into two categories, namely compression lossless and lossy compression which a lossless compression is the reversible while lossy compression is irreversible. Compression of digital data generally use lossless compression due to data that has been compressed must be returned into the original data so that it can be used. There are a few exceptions in the digital multimedia files like digital images, sound, video compressed files which can still be used so that the process is not fully needed decompression which is different to other digital files such as text, documents, etc.

Most of the methods of compression using statistic frequency of occurrence characters and bytes. Methods such as Huffman Codes process symbols in ASCII encodes the digital file into a row of new bit simpler [1]. Other methods such as LZW are also using the same approach, but have slightly different operating [2]. ASCII code is the numeric value that represents the characters or command on a computer device. The character used in a text message is the characters of the alphabet and punctuation characters that has a numeric value of adjacent and have difference or difference which is quite small. The small difference that it can be used as a substitution of the sequence of characters of sources resulting in a row of bit or symbol of the new, more simple. Approach to simplification of the bits of the original symbol based on the proximity of his bit values can be found on other compression

IOP Conf. Series: Journal of Physics: Conf. Series **1007** (2018) 012022 doi:10.1088/1742-6596/1007/1/012022

methods such as Half-Byte Compression. The basic concept of the Half-Byte Compression was eliminated four-bit MSB on a sequence of characters that have four MSB are identical so that on the next symbol using only four bits LSB so as to allow the use of the number of bits that are much more a little though there are several conditions that must be fulfilled before the compression can be done.

The though foundation of the development of a compression scheme which was done on this paper is the approach of the difference value in byte of the symbols on the Delta Encoding [3-4]. Delta Encoding using the difference value or delta as a substitution of symbols or character input that can be reconstructed again using a reference value.

$$f_{new} = f_{old} + f_{\delta}$$

f_{new} is a new file the results of the reconstruction obtained from the sum of the f_{old} and f_{δ}. The Compression operation on delta encoding is counting f_{δ} in such a way smaller so that it can be used in reconstructing f_{new} using f_{old} as the reference value. Delta Encoding is very effective in reducing the consumption of bandwidth on the network that has been implemented in the compression on the web service and HTTP in several studies [6-7]. Utilizing the basic concept of the difference from the value of a byte from an input symbol, on paper was developed the concept of a simple compression on the difference between ASCII value that has a smaller range than with using the concept of the difference between the value of the bytes.

2. Proposed Work: ASCII Value Differencing Based Compression

Differentiation method utilizing differential value of the ASCII characters in the ASCII table. Deferential value low enough can be used as a substitution of characters in a text message. On the application of simple method is not using the table lookup or reference tables used for encoding and decoding process. The process of encoding and decoding full use of the differential value of each character in the message. However, there was a possibility of table lookup implementation to improve the effectiveness of the methods on the various types of digital files.

2.1 Basic Concept

Encoding is done by calculating the numeric value of the lowest and highest characters in the message text using the ASCII table. The next differential will be calculated from the lowest and highest value. The differential value will be divided two to find the middle value of the differential value. The central value will be the value of the reference point or reference point. Encoding process is then carried out by doing the substitution value deference obtained from each character's response to the reference point.

The following is encoding stages of differential ASCII:

1. Looking for a character with ASCII code value of the lowest and highest.
2. Calculate the differential value of the lowest and highest character:

   $d = MaxC - MinC$ (1)

   Where:

   D = Differential value

   MinC : The lowest value of ASCII code of the character message.

   MaxC : The highest value of ASCII code of the character message.
3. Calculate the mid value of differential value by using a rounding over

   $Mid = d/2$ (2)
4. Looking for the value of reference point:

   $p = MinC + Mid$ (3)
5. Calculate the differential value each character against the value of reference point.
6. Do the encoding character added with the value of reference point as an initial byte.

Decoding is done by calculating the origin of ASCII character value using the differential between differential characters are compressed against the value of reference point.The following stages of the process of decoding:

1. Initial the Reference Point.
2. Calculate difference between each bit character encoded with the value of reference point.
3. Conversion differential value to ASCII characters

2.2 Proposed Scheme

On differentiation the basic ASCII, use the value of a reference or references from the midpoint between the lowest value of the character with highest value of character, so that each character will substituting with a bit of differentiation globally so that the bigger value of the differentiation so compression results have bad compression ratios even uncompressed altogether.

1. Take some characters that found in the area of the window.
2. Calculate the differential value of the lowest and highest character of the window
   $d = MaxC - MinC$ (4)
   Where:
   d = Differential value
   MinC : The lowest value of ASCII code of the character message.
   MaxC : The highest value of ASCII code of the character message.
3. If d <= 15 so the size of window can be zoom as needed during d < 15 where the expectation is bit encoded result no more than four bits.
4. Do the encoding against ASCII code of the character based on the value of differential against the value of reference point.
   "123" = {49, 50, 51}
   Min = 49
   Max = 51
   dot P = 49 + ((51-49)/2) = 50
   Differential:
   {49, 50, 51} = {-1, 0, 1}, encode = {11, 00, 01}
   "tommy" = {116, 111, 109, 109, 121}
   Min = 109
   Max = 121
   Point P = 109 + ((121-109)/2) = 115
   Differential:
   {116, 111, 109, 109, 121} = {1, -4, -6, -6, 6}, encode = {0001, 1100, 1110, 1110, 0110}
5. Putting together bits of header compression in order to result in decompression back:
   Header:
   The First Bit to bit into – n = window information
   Bit n+1 to end = bit encoding result
   With the following lineup:

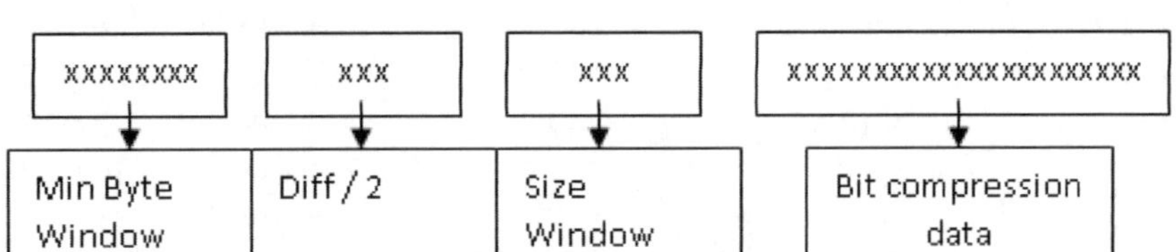

Figure 1. Compression Scheme.

IOP Conf. Series: Journal of Physics: Conf. Series **1007** (2018) 012022 doi:10.1088/1742-6596/1007/1/012022

Example: From the above compression obtained two window, that window 1 {123} and window 2 {tommy}. Window 1 consists of three characters with the value min 49 and differential divided 2 = 1. So a bit of window 1 is:
00110001 001 011 11 00 10

Window 2 consists of 5 characters with the value min 109 and differential 12 divided 2 = 6, so that bit of window 2 is: 01101101 110 101 0001 1100 1110 1110 0110

6. So, the compression results are obtained as follows:
0011000100101111001001101101110101000111001110111100110 = 54 bit

The number of initial bits 64-bit ratios obtained = 15.62 %

3. Performance

An experiment done with the platform specification intel core i5 2.5 GHz CPU, 8 Gb of RAM and is using an operating system WIN 10 64-bit by using a different file type. In Figure 2 is the display of application testing on the file extension .bmp, while Figure 3 with type a file. doc, the results of the test methods proposed for compression of files is shown in table 1.

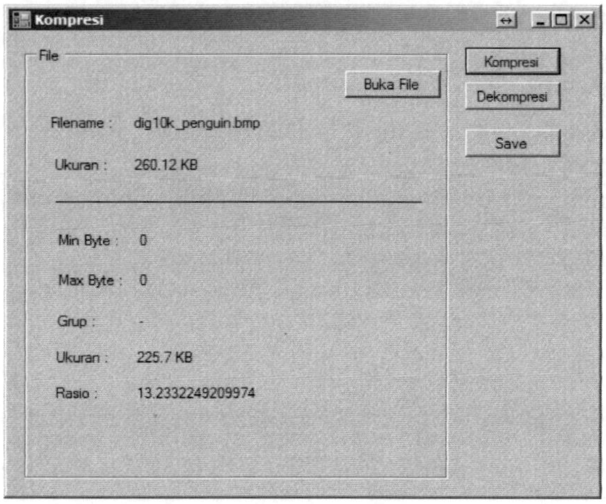

Figure 2. Type of Compresses Image.

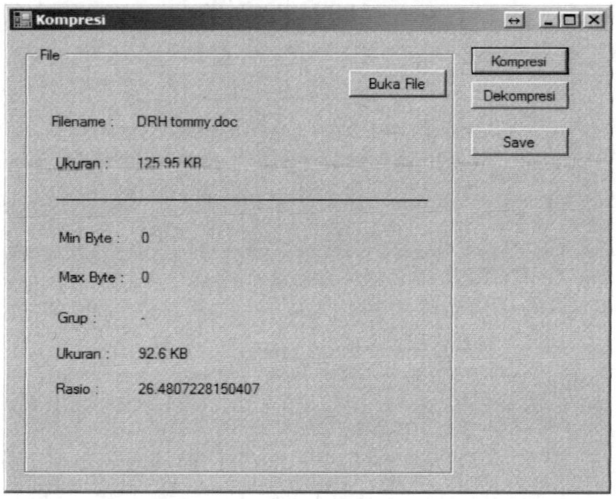

Figure 3. Type of Compresses File.

Table 1. Performance Results.

File Name	Extension	Original Size (kb)	Compressed Size (kb)	Comp. Ratio (%)
File 1	.bmp	260.12	225.7	13.23%
File 2	.html	45.56	45.04	1.14%
File 3	.doc	125.95	92.6	26.48%
File 4	.doc	780.67	534.78	31.50%
File 5	.png	435.56	334.56	23.19%
File 6	.jpg	231.59	211.45	8.70%
File 7	.doc	867.23	672.34	22.47%
File 8	.html	49.82	48.79	2.07%
File 9	.bmp	987.34	879.34	10.94%
File …	.xls	234.21	225.45	3.74%

Testing has been done gives varying compression ratio results were in a file with a uniform byte locally can give quite a good compression ratio. As for the file or files with difference bytes large enough will give you a small ratio or even give a considerable compression results given the existence of additional information for byte – byte which is not compressed and header information for the purposes of decompression. In table 1 is the partial data for test results of compression with a different file type, from the results of test the proposed method produce good compression rate of 14%, but for file with extension .doc type has higher compression lower than in the file extension .html, in Figure 4 and Figure 5 is the display of the results of compression in the form of graphs.

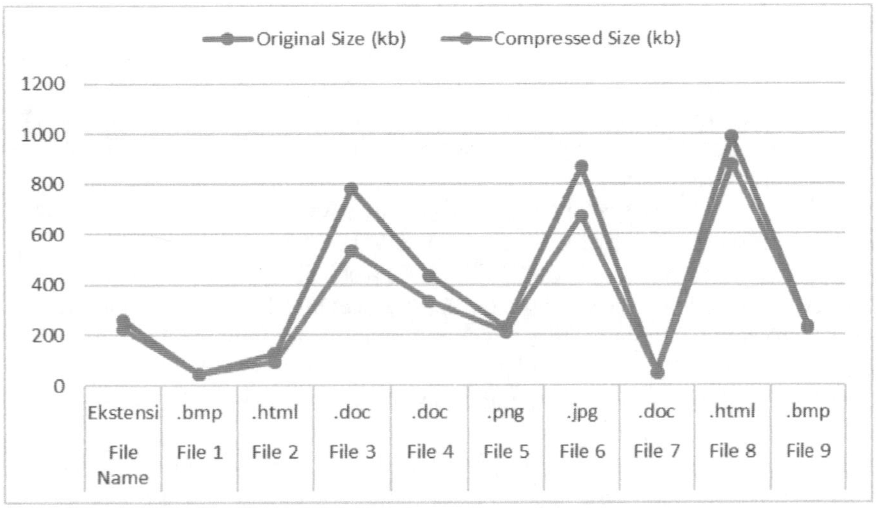

Figure 4. Graph of the Compression Proposed result.

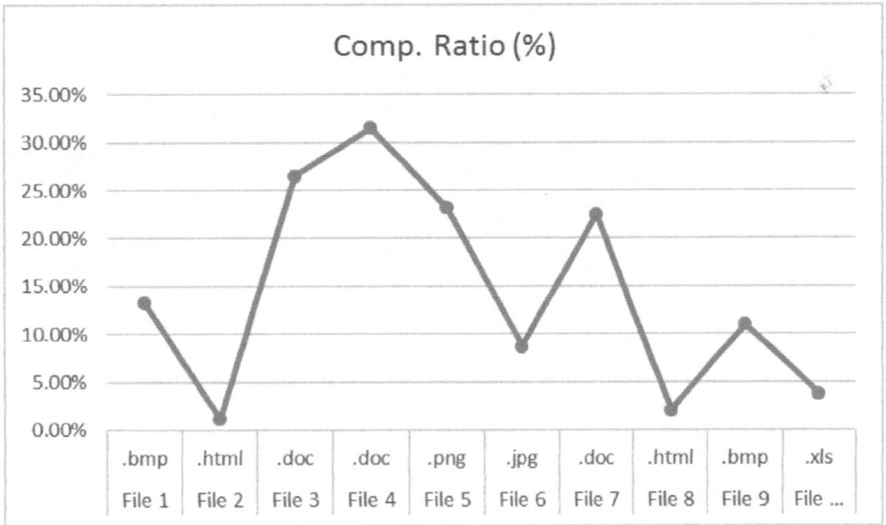

Figure 5. Ratio compression results in a different file type.

Based on the test results statistics, it can be seen that the development differential ASCII can be applied not only on files with text characters but also can be applied to digital files in general although the necessary existence of a development further away from the side of efficiency header information that many spend space compression results.

4. Conclusion

In this paper has discussed a simple compression scheme utilizing difference value from ASCII symbols to produce a simple bit value. Compression ratio of the scheme proposed in this paper is good enough, but relies on a variant of the ASCII symbols of files or messages. There is a weakness in the delta encoding that requires a constant variant symbol or stable so it can be retrieved the optimal compression ratio. The application of window size parameters and limit values for the differential can be adjusted so as to give the different compression results, so more research is needed to be able to see the influence of the use of window size and limit differential value are different.

References

[1] Canovas, R., & Moffat, A. (2013). Practical Compression for Multi-Alignment Genomic Files. Proceedings or the Thirty-Sixth Australasian Computer Science Conference.

[2] Chakraborty, D., Ghosh, D., & Ganguly, P. (2015). A Dictionary based Efficient Text Compression Technique using Replacement Strategy. International Journal of Computer Applications (0975 – 8887).

[3] Sharma, M. (2010). Compression Using Huffman Coding. IJCSNS International Journal of Computer Science and Network Security, VOL.10 No.5.

[4] Schlosberg, A., Lam, B. Y., Yeo, G. S., & Clifton, R. J. (2013). DEFLATE Compression Algorithm Corrects for Overestimation of Phylogenetic Diversity by Grantham Approach to Single-Nucleotide Polymorphism Classification. International Journal of Molecular Sciences, 15, 8491-8508.

[5] Zu, Y., & Hua, B. (2015). Parallelizing the Deflate Compression Algorithm on GPU. Journal of Computational Information Systems 11:17.

[6] Sulochana, V., & Verma. (2012). Design and Implementation of LZW Data Compression Algorithm. International Journal of Information Sciences and Techniques (IJIST) Vol.2, No.4.

[7] Sidhu, A. S., & Garg, E. M. (2014). An Advanced Text Encryption & Compression System Based on ASCII Values & Arithmetic Encoding to Improve Data Security. International Journal of Computer Science and Mobile Computing, Vol. 3, Issue 10.

[8] Singh, U., & Garg, U. (2013). An ASCII Value Based Text Data Encryption System. International Journal of Scientific and Research Publications, Vol. 3, Issue 11.

[9] Klein, S. T., & Shapira, D. (2007). Compressed Delta Encoding for LZSS Encoded Files. Data Compression Conference (DCC '07). ISSN : 1068-0314.

[10] Mogul, J. C., Douglis, F., Feldmann, A., & Krishnamurthy, B. (1997). Potential benefits of delta encoding and data compression for HTTP (Corrected version). WRL Research Report 97/4a. Western Research Laboratory.

[11] Klein, S. T., Serebro, T. C., & Shapira, D. (2006). Modeling Delta Encoding of Compressed Files. Data Compression Conference.

[12] Suel, T., & Memon, N. (2002). Algorithms for Delta Compression and Remote File Synchronization. Academic Press.

[13] Psounis, K. (2002). Class-based Delta-encoding: A Scalable Scheme for Caching Dynamic Web Content. Proceedings of the 22 nd International Conference on Distributed Computing Systems Workshops (ICDCSW'02).

[14] Mogul, J., Krishnamurthy, B., Douglis, F., Feldmann, A., Goland, Y., van Hoff, A., et al. (2002, January). Delta encoding in HTTP. Network Working Group.

[15] Tommy. (2016). Pemanfaatan Diferensiasi ASCII Pada Implementasi Kompresi Teks. Jurnal Logika, Vol. IX, No. 18, ISSN : 2087-3018.

Acknowledgments

This research was supported by Universitas Prima Indonesia and Universitas Harapan Medan, Indonesia.

Developing business strategies using SWOT analysis in a color crackers industry

I Budiman*, U P P Tarigan, A Mardhatillah, A C Sembiring, W Teddy

Industrial Engineering Department, Faculty of Technology and Computer Sciences, Universitas Prima Indonesia, Indonesia

E-mails: *irwanb01@gmail.com, unipratama@gmail.com, ellalapink@gmail.com, anitakembaren22@gmail.com, teddy.09.wt@gmail.com

Abstract. This research was conducted in color crackers industry in Indonesia. Color crackers are snacks colored in red and white, found as additional in many Indonesian traditional foods. The used traditional business strategies are not appropriate for the market condition in the industrial 4.0 era. The aim of this study is to develop business strategies in this market condition. This research was conducted in several steps using SWOT Analysis, weighting assessment for SWOT questions, Internal Factor Analysis Summary, External Factor Analysis Summary, and Creating SWOT Matrix. Data were collected through interviews and questionnaires with internal and external stakeholders. The result of this research is the business positioned in the first quadrant. This gives the explanation that the traditional strategies used already inappropriate with the current condition. They need to use mixed SO (Strength – Opportunity) strategies or called as the aggressive strategy to win the market competition. The developed SO strategies are: creating distribution network with the customers and utilizing renewable technology.

1. Introduction

From a worldwide perspective, it has been recognized that small and medium enterprises (SMEs) play a vital role in economic development, especially in Indonesia, where they have been the primary source of job/employment creation and output growth [1]. The SME also play a vital role in the rural income because they scattered widely and very labor-intensive in the rural area [2]. Tambunan identifies several key issues for improving SME competitiveness in Indonesia, they are human resource, working capital, management and technological skills. These key factors are important to improve SME's business performance [3].

One of the management skills required is developing the appropriate business strategy [4,5]. In developing appropriate business strategies, management has to know the market condition well so that strategies developed can provide benefits as well as preparing the company to anticipate business risks that will be faced [5,6,7]. Implementing a marketing strategy should begin with a SWOT analysis of strengths, weaknesses, opportunities, and threats [5,6]. Many researchers and businessman usually develop business strategies using environment in industrial 3.0 era such as research undertaken by Tulus, Setiawan, and Ariska [2,3,6]. Nowadays, business is running in an open network system or usually called the industrial 4.0 era. The most fundamental difference between the traditional and industrial 4.0 era is the production basis. In the traditional strategies, a number of production was

IOP Conf. Series: Journal of Physics: Conf. Series **1007** (2018) 012023 doi:10.1088/1742-6596/1007/1/012023

based on production planning-generated and usually, the product is ready-made. Meanwhile, in the industrial 4.0 era, a number of production is based on customer and the product is customized. This era has a different condition and giving significant impact to the growth of a company.

This research was conducted in color crackers business in Indonesia. Color crackers are snacks colored in red and white, found as additional in many Indonesian traditional foods. Traditional business strategies used is not appropriate for the market condition in this industrial 4.0 era. Rapidly market changes compared to traditional market conditions cause business need to be more flexible in defining the strategies.The aim of this study is to develop business strategies in this kind of market situation.

2. Method and equipment

This research was done in one of the biggest crackers in Asahan using several steps of SWOT Analysis, Give weighting assessment for SWOT questions, Internal Factor Analysis Summary (IFAS), External Factor Analysis Summary (EFAS), and Creating SWOT Matrix [8, 9, 10, 11, 13, 14]. Data were collected through interview and questionnaire with several stakeholders.

Research was created using several steps as below: [11,12,15]

- Identify Internal Factors
  Internal Factors Analysis was created together with internal stakeholders in the company such as production, marketing, labor union, research, and development, and also the top management of the company. Qualitative data was collected through Focussed Group Discussion (FGD) for Strengths and Weaknesses.
- Identify External Factors
  External Factors Analysis was created together with external stakeholders such as business observer, distributor, customer, and also marketing and top management of the company. Qualitative data was collected through Focussed Group Discussion (FGD) for Opportunities and Threats.
- Internal Factors Data Collection
  IFAS is Internal Factor Analysis Summary by weighting and rating strengths and weaknesses so that we can determine the total score of strengths and weaknesses. IFAS was created through FGD with the internal stakeholders. Steps of determining IFAS are:
  - Creating questionnaire using Strengths and Weaknesses. Respondent can choose 1 (disagree) to 4 (very agree). Collected data was treated as the value of the variable.
  - Interview with internal stakeholders and marketing to set weighting for each variable
- External Factors Data Collection
  EFAS is External Factor Analysis Summary by weighting and rating opportunities and threats so that we can determine the total score of Opportunities and Threats. EFAS is calculated with the same steps as IFAS.
- Creating IFAS and EFAS Analysis and plotted to SWOT Matrix Diagram
  Calculate the total score for each IFAS and EFAS and plot into SWOT Matrix Diagram. (1) When the total score created from IFAS and EFAS plotted in the first quadrant, then the company need to use aggressive strategy; (2) when it plotted in the second quadrant, then the company need to use conservative strategy; (3) when it plotted in the third quadrant, then the company need to use defensive and; (4) when it plotted in the fourth quadrant, the company need to use competitive strategy.
- Develop Chosen Strategy Mix within SWOT
  Chosen strategy, then developed into several business strategies, that need to be developed by the company into programs and activities or usually in Key Performance Indicator.

The Block Diagram for each process in this research can be simply described in Figure 1.

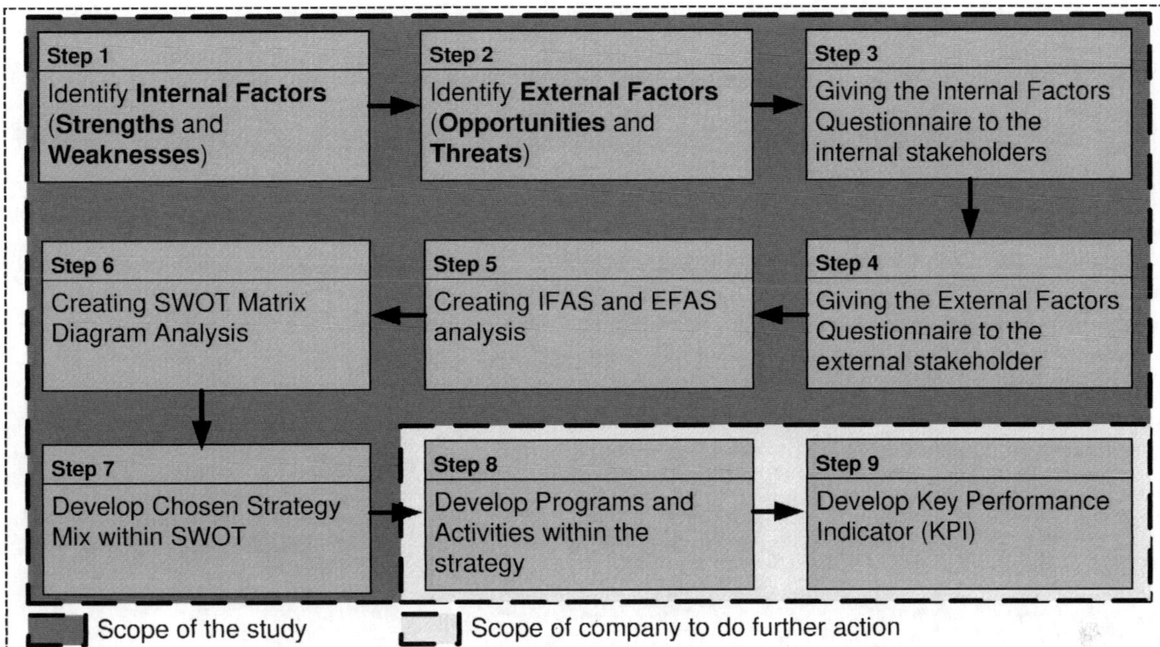

Figure 1. Block Diagram for Developing Business Strategies.

3. Results and Discussions

3.1. SWOT Analysis

SWOT qualitative data was collected through Focussed Group Discussion with internal and external stakeholders. Based on the collection, condition for each SWOT is as follow;

- Strengths determined are: strategic location, product quality, lower price, huge capacity, payment method, delivery service and fast complain procedure;
- Weaknesses are natural drying process, limited drying places, undisciplined labor, high production cost, less promotion, and huge capital needed;
- Opportunities are smooth customer access, lower competitor's product quality, new technology, growth of product demand, good promotion skill, loyal potential customer, easy bank loans;
- Threats are new competitors, customer losses, price competition, changes in customer taste, and nonperforming loan.

3.2. IFAS (Internal Factor Analysis Summary)

Strengths and Weaknesses then gave value and weight based on the results of questionnaire and interviews with internal departments. The result of IFAS is recapitulated per department as can be seen in Table 1.

IOP Conf. Series: Journal of Physics: Conf. Series **1007** (2018) 012023 doi:10.1088/1742-6596/1007/1/012023

Table 1. Internal factor analysis summary.

Dept.	No	Factor	Attribute	Value	Weight	Value x Weight
Production	1	product quality	S	4	0.08	0.32
	2	huge capacity	S	4	0.06	0.24
	3	The natural drying process	W	2	0.08	0.16
	4	limited drying places	W	3	0.08	0.24
	5	high production cost	W	2	0.08	0.16
Finance	1	huge capital needed	W	3	0.08	0.24
Marketing	1	strategic location	S	3	0.08	0.24
	2	lower price	S	4	0.08	0.32
	3	payment method	S	4	0.08	0.32
	4	delivery service	S	4	0.08	0.32
	5	fast complain procedure	S	3	0.06	0.18
	6	less promotion	W	3	0.08	0.24
Human Resource	1	undisciplined labor	W	2	0.08	0.16
		Total			1.00	
S refers to Strength, W refers to Weakness			S			1,94
			W			1.20

As per shown in above table, we can see that the greater impact to a factor is, the higher weighting be. The rating was given based on comparison to main competitors. The higher this business position, the higher total value be. As can be seen in the above table, natural drying process, high production cost, and undisciplined labor is the lower position of this business (serious weaknesses) compared to the main competitor. The difference between strength and weakness is 1.94-1.20=0.74.

3.3. EFAS (External Factor Analysis Summary)

Opportunities and Threats also proceed with the same steps as in IFAS. The result of EFAS is recapitulated per department as can be seen in Table 2.

As per shown in above table, we can see that new competitors, price competition, and changes in the customer taste is the lower position of this business (serious threats) compared to the main competitor. The difference between opportunity and threat is 1.94-0.91=1.03.

Table 2. External factor analysis summary.

Dept.	No	Factor	Attribute	Value	Weight	Value x Weight
Production	1	lower competitor's product quality	O	4	0.08	0.32
	2	new technology	O	4	0.07	0.28
Finance	1	easy bank loans	O	3	0.07	0.21
Sales	1	smooth customer access	O	3	0.07	0.21
	2	the growth of product demand	O	4	0.10	0.40
	3	good promotion skill	O	3	0.08	0.24
	4	loyal potential customer	O	4	0.07	0.28
	5	new competitors	T	1	0.10	0.10
	6	customer losses	T	3	0.10	0.30
	7	price competition	T	1	0.09	0.09
	8	changes in customer taste	T	2	0.09	0.18
	9	nonperforming loan	T	3	0.08	0.24
		Total			1.00	
O refers to Opportunities, T refers to Threats			O			1.94
			T			0.91

3.4. SWOT Matrix Diagram Analysis

Based on calculated data from IFAS, Strength is 1.94, whether Weakness is 1,20 and the differences between them are 0.74. Based on the EFAS, Opportunity is 1.94, whether Threat is 0.91 and the differences between them are 1.03. These data were then plotted to Diagram in Figure 2.

IOP Conf. Series: Journal of Physics: Conf. Series **1007** (2018) 012023 doi:10.1088/1742-6596/1007/1/012023

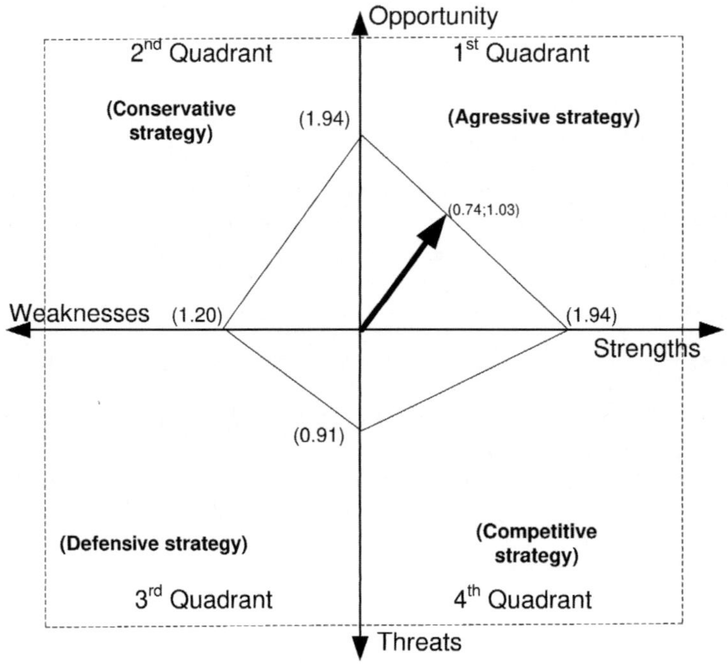

Figure 2. SWOT Matrix Diagram.

As per shown in Figure 2, the chosen strategy is in 1st quadrant (aggressive / SO strategy) and the strategy needed to be executed as follow:

- Creating distribution network with the customer, which has sub-strategies such as: improve payment method, delivery service, complain procedure, and increase marketing area.
- Utilizing renewable technology, which has sub-strategies such as: using new technology (especially for network information system based) to reduce ineffective and inefficiency, and increase production quantity.

The implementation of this strategy, especially the programs and activities developed will make the business survive in the industrial era 4.0 by reducing production cost, easy to access, easy payment method, and efficient delivery service. This will create a strong position for this business in this era.

4. Conclusions

The result of this study is the business positioned in the first quadrant. This gives the explanation that the traditional strategies used already inappropriate with the current condition. They need to use SO (Strength – Opportunity) strategy or called the aggressive strategy to win the market competition in the industrial 4.0 era. The developed SO strategies are: creating distribution network with the customers and utilizing renewable technology.

References

[1] Tulus T, Liu Xiang F 2007 *ERIA Related Joint Research of SME Project, IDE-JETRO: SME Development in Indonesia and China* pp. 1-29.

[2] Tulus T 2006 *Development of Small and Medium-Scale Industry Clusters in Indonesia, Kadin Indonesia-JETRO*.

[3] Setyawan Agus A, Isa M, Wajdi Farid M, Syamsudin, Nugroho Permono S 2015 *J. Competitiveness,* An Assessment of SME Competitiveness in Indonesia **7** 60-74

[4] Iin Sri Karina S 2016 *Strategi Bersaing dalam Pemasaran Melalui Analisis SWOT pada Toko Kaban Di Perumnas Simalingkar Medan* (Medan: University of Sumatera Utara)

[5] Agarwal R, Grassi W, Pahl J 2012 *J. Business Strategy,* Meta-SWOT: introducing a new strategic planning tool **33(2)** 12-21

[6] Ariska I 2016 *Analisis Swot Dalam Upaya Meningkatkan Keunggulan Bersaing Usaha Kecil Menengah* (Medan: University of Sumatera Utara)

[7] Rao Y S, 2017 *J. Library Management, C5 model for the consortium management: SWOT analysis* **38(4/5)** 248–262

[8] Talib M S A 2014 *J. Islamic Marketing,* Halal logistics in Malaysia: a SWOT analysis **5(3)** 322–343.

[9] Piercy N, & Giles W 1989 *J. Marketing Intelligence & Planning,* Making SWOT Analysis Work **7(5/6)** 5–7.

[10] Sukaria S 2011 *Metode Penelitian, Research Methodology*, 1st Ed., (Indonesia: USU Press)

[11] Kamla Ali A B 2014 *J. information and knowledge management systems: VINE,* SWOT of social networking sites for group work in government organizations: An exploratory Delphi study from IT managers' perspective, **44(1)** 121-139.

[12] Vaia T, Christodoulou A 2014 *British Food J.,* SWOT analysis of the truffles market in Greece **116(12)** 1976-1997.

[13] Gajanand G, Rajesh P M 2016 *J. Quality in Maint. Eng.*, A SWOT Analysis of reliability centered maintenance framework, **22(2)** 130-145

[14] Charikleia K, Alkaterini P, Phaedra D, Vangelis M, John P 2016 *Int. J. Energy Sector Man.,* Investigating EU-Turkey Renewable Cooperation Opportunities: a SWOT analysis, **10(3)** 337-362

[15] Saurabh P 2017 *J. Strategic Direction,* Improving the Learning and developmental potential of SWOT Analysis: Introducing the LISA Framework, **33(3)** 12-14

Acknowledgments
In this research, the authors would like to thank profusely to the industry which has been willing to be a place of research for the writing of this paper and highly support researcher for a better result.

Defect Analysis Of Quality Palm Kernel Meal Using Statistical Quality Control In Kernels Factory

M T Sembiring*, N J Marbun

Industrial Engineering Department, University of Sumatera Utara, Almamater St USU Campus Medan 20155, Indonesia

E-mail: *meilita_tryana@yahoo.co.id, ovibanjarnahor@gmail.com

Abstract. The production quality has an important impact retain the totality of characteristics of a product or service to pay attention to its capabilities to meet the needs that have been established. Quality criteria Palm Kernel Meal (PKM) set Factory kernel is as follows: oil content: max 8.50%, water content: max 12,00% and impurity content: max. 4.00% While the average quality of the oil content of 8.94%, the water content of 5.51%, and 8.45% impurity content. To identify the defective product quality PKM produced, then used a method of analysis using Statistical Quality Control (SQC). PKM Plant Quality Kernel shows the oil content was 0.44% excess of a predetermined maximum value, and 4.50% impurity content. With excessive PKM content of oil and dirt cause disability content of production for oil, amounted to 854.6078 kg PKM and 8643.193 kg impurity content of PKM. Analysis of the results of cause and effect diagram and SQC, the factors that lead to poor quality of PKM is Ampere second press oil expeller and hours second press oil expeller.

1. Introduction

Kernel factory is a manufacturing company engaged in the manufacturing Palm Kernel Oil (PKO) and Palm Kernel Meal (PKM) as by-products. PKM many production processes produce a defective product, where the quality of the resulting PKM is considered less good because it does not meet the quality standards set by the company. Quality criteria PKM set Factory kernel is as follows: oil content: max 8.50%, water content: max 12,00% and impurity content: max. 4.00% While the average quality of the oil content of 8.94%, the water content of 5.51%, and 8.45% impurity content.

Quality is a systematic approach that the products or services in accordance with the specified requirements [1]. It is also an operating system as the corrective measures to produce economic goods and services and meet customer needs.

Quality consists of the development, design, production, marketing, and service of products and services [2]. Each of these activities is done well would be up to the consumer with excellent product quality and at a low cost. It needs to continue to do that with the continuous development of quality. Statistical Quality Control (SQC) is a branch or part of the quality control in which the control process is to collect, analyze and interpret the data used in quality control activities [3, 4, 5, 6].

The problem-solving process and quality improvements using quality control tools through the problem-solving process so that it becomes more rapidly and systematically. Quality control tool can be used by professionals to facilitate the process of quality improvement [7,8]. In this study addressed

IOP Conf. Series: Journal of Physics: Conf. Series **1007** (2018) 012024 doi:10.1088/1742-6596/1007/1/012024

the use of Statistical Quality Control Plant Kernel. Variables used in this research is the quality of PKM. This study aims to identify product defects that occur.

2. Methodology/Experimental

Type of research is applied research because this research was done to solve real problems that occur in the plant that is the problem of quality. This research will design improvements to reduce variations in the characteristics are there [9].

2.1. Data Collection Methods

Data collection methods used in this research is through direct observation and interviews with the company. Data processing method used in this study are the implementation of quality control tool to know the data PKM are still within the limits of control and identification of factors that cause deviation PKM quality.

2.2. Data Processing Methods

Methods of data processing techniques in this study with a tool quality control use the statistical quality control to determine the quality of data that is still within the limits PKM control. The results of quality control tools are as follows: Stratification, CheckSheet, Histogram, Scatter Diagrams, Map control, and analysis Cause and effect diagram.Stratification is the process of grouping data defects that occur on the production floor. The criteria selected for the stratification of defects PKM production. Check Sheet or Sheet Inspection is a means of collecting and analyzing data to facilitate the data collection process so that it can present the data to be more communicative and can be converted into information. The examination can be determined by the number of PKM product defective and not defective in units of kilograms. Scatter Diagrams are used to determine the distribution of data each PKM production quality. Map this control is used to see whether the data is the percentage of oil content and impurity content in the production of PKM homogeneous or not. Analysis of cause and effect diagram is used to determine the cause of the high percentage of oil content and degree of dirt on the production of PKM in detail.

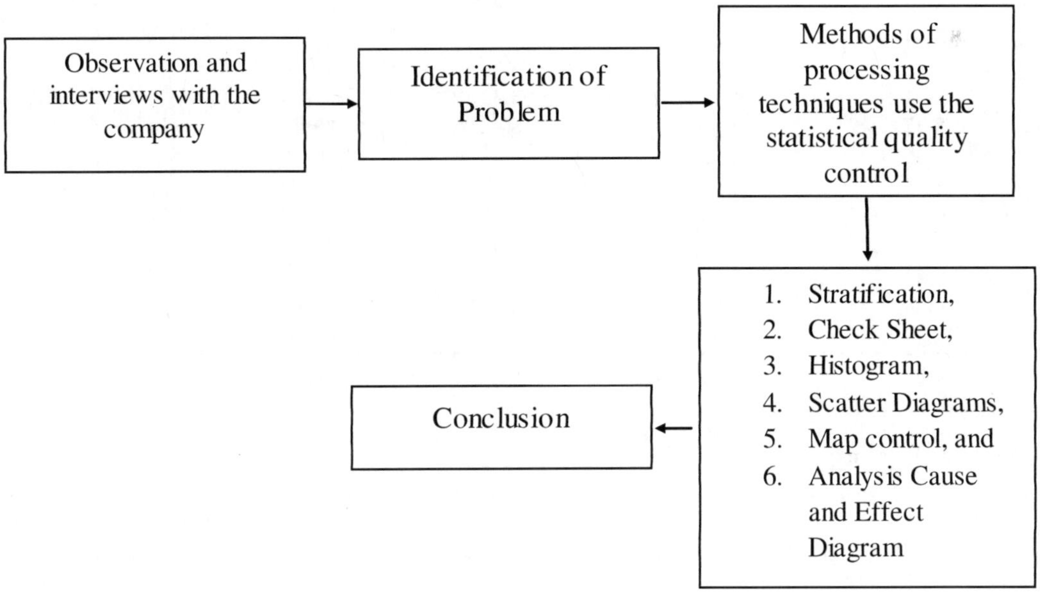

Figure 1. Block Diagram methods of data processing.

3. Results and Discussion

3.1. Statistical Quality Control (SQC)

The Statistical Quality Control (SQC) is a control process that is used to collect, analyze and interpret the data used in quality control activities [10,11]. The control process consists of seven tools of quality control, namely:

1. Stratification

Stratification is the process of grouping data defects that occur on the production floor. In addition, stratification also keeps records of the number of defects caused by each type of defects. To determine the number of production defects PKM, do multiplication excess oil and dirt content percentage of the daily production amount PKM. PKM disability stratification can be seen in Table 1.

Table 1. Production of defects stratification PKM.

No	Quality of PKM Kernels Factory			Number of PKM Production	Number of PKM Defects
	Oil Content (maks 8,5%)	Water Content (maks 12,00%)	Impurities Content (maks 4,00%)		
1	9.54	5.65	8.49	218963	12108.65
2	9.20	5.55	8.48	228779	11850.75
3	9.30	5.55	8.44	217252	11384.00
4	9.36	5.50	8.47	226180	12055.39
5	9.52	5.47	8.46	212608	11650.92
6	9.55	5.53	8.48	219401	12132.88
7	8.48	5.55	8.44	212207	9379.549
8	8.49	5.52	8.46	166555	7411.698
9	8.49	5.47	8.47	178450	7958.87
10	8.50	5.54	8.44	201274	8936.566
11	8.49	5.51	8.50	218258	9799.784
12	8.43	5.48	8.50	228308	10114.04
13	8.44	5.49	8.42	222911	9718.92
14	8.43	5.50	8.44	200022	8740.961
15	8.46	5.56	8.43	116956	5134.368
16	8.45	5.32	8.42	190750	8335.775
17	9.47	5.49	8.40	160082	8596.403
18	8.50	5.53	8.49	160945	7226.431
19	9.53	5.57	8.46	158199	8685.125
20	9.56	5.50	8.47	173618	9601.075
21	9.59	5.53	8.45	167092	9256.897
Rata-rata	**8.94**	**5.51**	**8.46**	**194229**	**9528.00**

2. Check Sheet or Sheet Inspection

Check Sheet or Sheet Inspection is a means of collecting and analyzing data to facilitate the data collection process so that it can present the data to be more communicative and can be converted into information [12,13]. The data collected through the of defective products check sheet can be seen in Table 2.

Table 2. Check Sheet PKM Production.

No	Production per	Defective	The Product is not

IOP Conf. Series: Journal of Physics: Conf. Series **1007** (2018) 012024 doi:10.1088/1742-6596/1007/1/012024

	Day (kg)	products per Day (kg)	Defective (kg)
1	218963	12108.65	206854.35
2	228779	11850.75	216928.25

Table 2 (continued)

No	Production per Day (kg)	Defective products per Day (kg)	The Product is not Defective (kg)
3	217252	11384.00	205868.00
4	226180	12055.39	214124.61
5	212608	11650.92	200957.08
6	219401	12132.88	207268.12
7	212207	9421.99	202785.01
8	166555	7428.35	159126.65
9	178450	7976.72	170473.29
10	201274	8936.57	192337.43
11	218258	9821.61	208436.39
12	228308	10273.86	218034.14
13	222911	9852.67	213058.33
14	200022	8880.98	191141.02
15	116956	5181.15	111774.85
16	190750	8431.15	182318.85
17	160082	8596.40	151485.60
18	160945	7226.43	153718.57
19	158199	8685.13	149513.87
20	173618	9601.08	164016.92
21	167092	9256.90	157835.10
Jumlah	**4078810**	**200079.1**	**200753.57**

3. Histogram

Histogram for the period PKM production data made a bar graph (histogram) that shows the composition of the number of defects of each type of disability that can be seen in Figure 2.

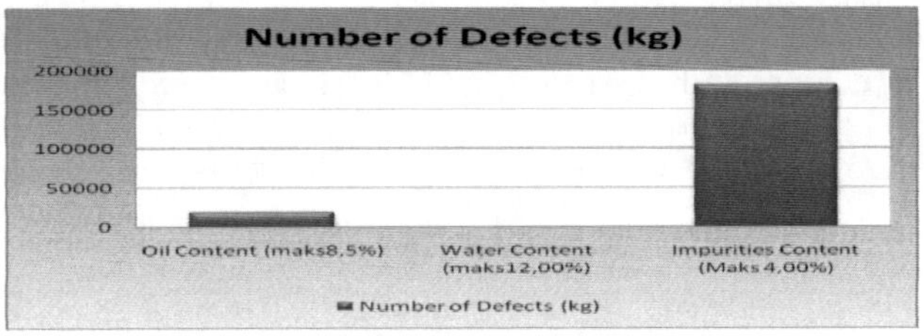

Figure 2. Histogram PKM Production.

IOP Conf. Series: Journal of Physics: Conf. Series **1007** (2018) 012024 doi:10.1088/1742-6596/1007/1/012024

Of The histogram above shows that PKM production defects occur most commonly on the oil content of more than 8.5% and the content of impurities exceeds 4%, so this research is focused on both these factors

4. Scatter Diagram

Scatter diagram used to see the relationship between the amount of PKM production by the number of defects. Data disability oil content of more than 8.5% and impurity content exceeding 4%, for further presented in the scatter diagram in Figure 3. and Figure 4.

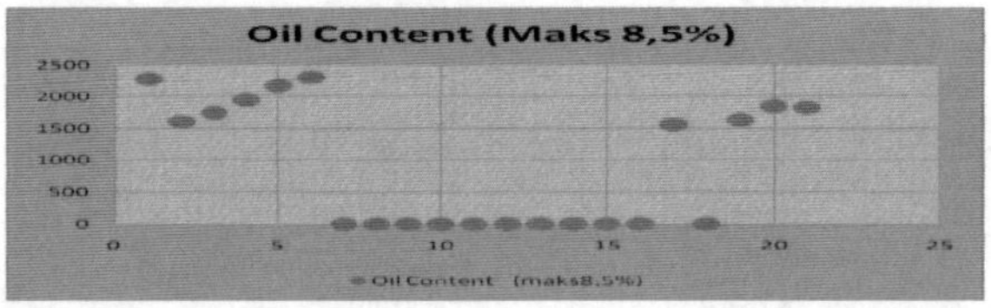

Figure 3. Scatter diagram Defects Type Oil Content (> 8.5%)

Based on Figure 2. known that defects of production at most of the oil content (> 8.5%) occurred in the first week and last. Images of a scatter diagram for the production of disability resulting impurity content (Maks > 4%) can be seen in Figure 4.

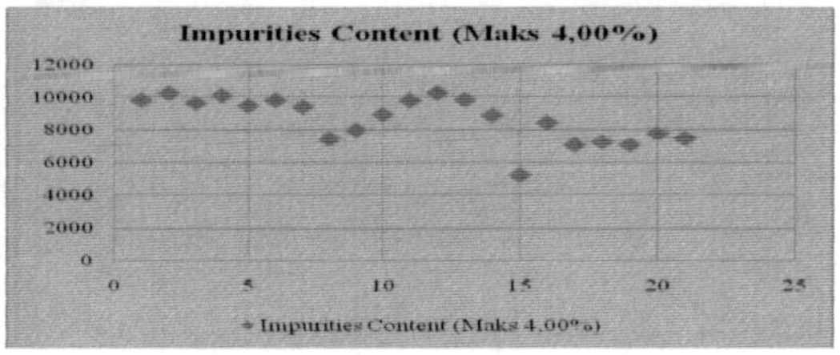

Figure 4. Scatter diagram type of defects Impurities Content (> 4%)

Based on Figure 4. The known that production defects resulting impurity content (Maks > 4%) had a uniform distribution of the data.

5. Map Control (Control Chart)

Map control is used to see whether the data is the percentage of oil content and impurity content in the production of PKM homogeneous or not [14]. If not then do repairs on the data out of control.If homogeneous then repair remains to be done, but as a whole without priority. In this study, used a map c, and to see the percentage of irregularities in the data content of oil and dirt on the production of PKM used several variables, namely: Central Line (CL), Upper Control Limit (UCL), Lower Control Limit (LCL).

The calculation of the value of CL, UCL and LCL conducted based on data from Table 1. Map control c percentage of oil content in the production of PKM can be seen in Figure 5.

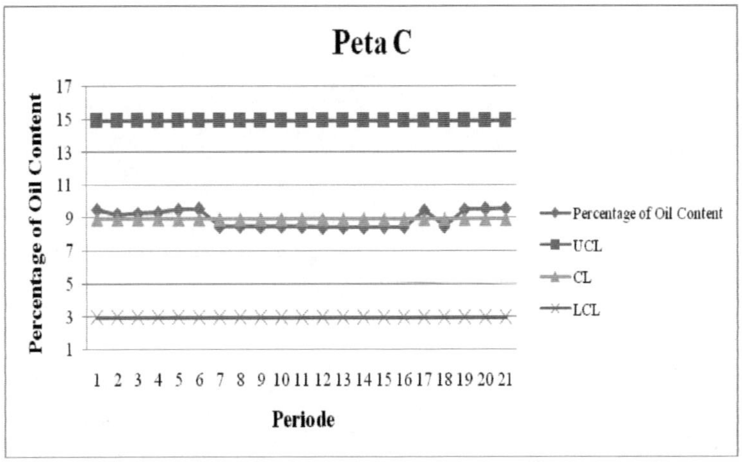

Figure 5. Map Control C Percentage of Oil Content at PKM Production.

From the graph note that all the data, the percentage of oil content in the production of PKM are at in control.This shows that the data does not require revision because no data is out of control.Image map control c percentage of impurity content on the production of PKM can be seen in Figure 6.

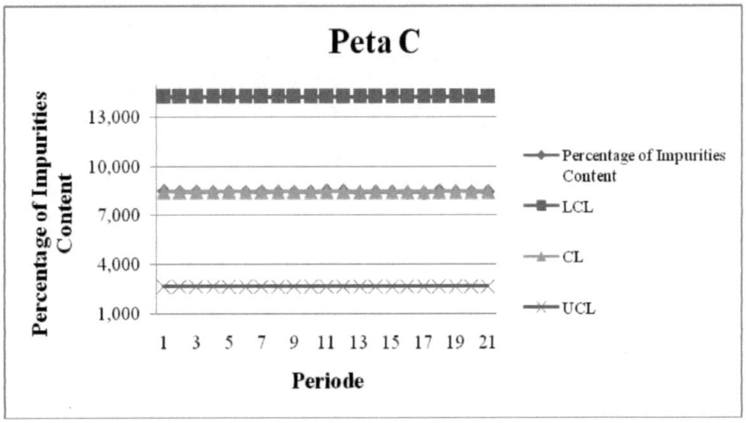

Figure 6. Map Control C Percentage of Impurities Content at PKM Production.

From the graph note that all data on the percentage of impurity content is the production PKM in control.This shows that the data does not require revision because no data is out of control.

6. Cause and Effect Diagram

This diagram is used to determine the cause of the high percentage of oil content and degree of dirt on the production of PKM in detail [15]. The cause of disability levels of oil and dirt levels caused by several factors, such as man, machines, methods, environment, and material. Causal diagram defects oil content can be seen in Figure 7.

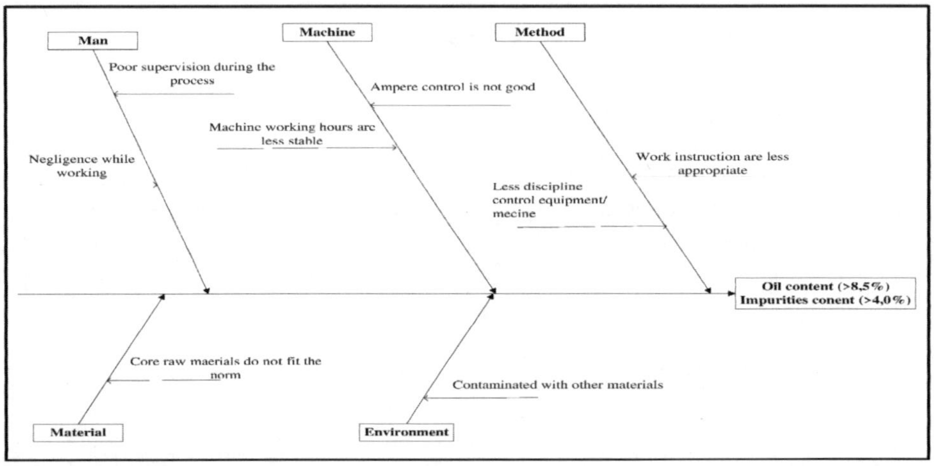

Figure 7. Cause and Effect Diagram of Oil Content (> 8.5%) and Impurities Content (> 4.0%)

Based on interviews conducted by the company that the main factors causing the high percentage of oil content and impurity content directly are Ampere second press oil expeller and hours second press oil expeller.

4. Conclusion.

Based on the results of the application identified by using Statistical Quality Control (SQC), it can be concluded as follows: that the processing operation PKM average oil content was 8.94%, while the maximum permissible percentage of only 8.50%. These conditions indicate oil levels had 0.44% excess of a predetermined maximum value. With an excess of the oil content, it will cause a loss of yield of 0.44% x 194 229 kg = 854.6078 kg PKM. The same thing happens at levels of dirt with an average percentage of 8.45%. Rated maximum percentage allowed is 4.00%, 4.50% means that there are excessive levels of dirt. Thus, the losses suffered by the company due to excessive levels of dirt amounted to 4.50% x 194 229 kg = 8643.193 PKM kg. Analysis of the results of cause and effect diagram and SQC, the factors that lead to poor quality of PKM is Ampere second press oil expeller and hours second press oil expeller.

References

[1] Amin Widjaja Tunggal. 2013. Introduction to Quality Management. Jakarta: Harvarindo
[2] Besterfield, Dale H.. 1998. Quality Control. Fifth Edition. New Jersey: Prentice Hall.
[3] H Subiyakto, Lukmandono, Rony Prabowo. 2017. Analysis of Product Quality Improvement of Precast Concrete with Statistical Process Control Approach and Quality Function Deployment. Surabaya.
[4] M Nazaruddin, S Ikhsan, J E M Napitupulu. 2015. Analysis of Disability and Quality Improvement of Water Tank at PT. Sabang Subur. Medan.
[5] S M Tryana, S Rahmi M, G Jeffrey. 2016. Analysis of The Cause of Defects of Sack Products with Taguchi and Annova Method on PT. Sri Intan Karplas Industry. Medan
[6] M A Rahim, Anizar, P Alexander. 2012. Design of Quality Improvement on Palm Oil Processing by Statistical Quality Control and PDCA Method in PT. PP London Sumatera Tbk. Medan.
[7] Ginting, Rosnani. 2007.Production Systems, Yogyakarta: Graha Science.
[8] Gusniar, Iwan Nugraha. 2012. Analysis of Motorcycle Rear Brake Quality in Industry Small in PT. X by Taguchi Method. Scientific Magazine Solutions Unsika ISSN 1412-86676 Vol. 11 No. 24 Ed.Sep - Nov 2012. Karawang.
[9] Sinulingga, Sukaria.2015.Research Methods, 3rd Edition.USU Press. Medan

IOP Conf. Series: Journal of Physics: Conf. Series **1007** (2018) 012024 doi:10.1088/1742-6596/1007/1/012024

[10] S Khawarita, T Ukurta, K Harris. 2016. Quality Control with Statistical Quality Control and FMEA Method in PT. Prima Indah Saniton. Medan.
[11] T Ukurta, S Erwin, P N Dita. 2015. Proposed Improvement of Product Quality of Caplet Medicine with Statistical Quality Control Method and Fault Tree Analysis on PT. Mutiara Mukti Farma. Medan
[12] B A Wicaksono, W Sulistiyowati. 2017. The Determination of Factors Influences The Quality of The Compressive Strength of Lightweight Brick by Statistical Process Control Method and Taguchi Method. Journal. Prozima, Vol 1, No.1, Juni 2017.
[13] Ahmad, L Laricha, F Wijaya. 2015. Expert System Design for Quality Control on Sparepart with DIE Casting System. Journal. Vol. 3 No. 3.
[14] A B Dewantara, M Kholil. 2015. Automation System as an Effort to Improve Quality with SPC Method on Line Finishing. Journal. Vol. 3 No. 3. 141-149. Jakarta.
[15] Mufidin, C I Parwati, J Susetyo. 2016. Proposed Improvement of Product Quality Ep Yst Pro with Statistical Process Control Method and Failure Mode and Effect Analysis on PT. Mitra Rekatama Mandiri. Journal. Vol. 4, No. 2, December 2016.

Developing soft skill training for salespersons to increase total sales

A Mardatillah, I Budiman*, U P P Tarigan, A C Sembiring, Hendi

Faculty of Technology and Computer Science, Department of Industrial Engineering, Universitas Prima Indonesia, Indonesia

E-mail: *irwanb01@gmail.com

Abstract. This research was conducted in the multilevel marketing industry. Unprofessional salespersons behavior and responsibility can ruin the image of the multilevel marketing industry and distrust to the multilevel marketing industry. This leads to decreased company revenue due to lack of public interest in multilevel marketing products. Seeing these conditions, researcher develop training programs to improve the competence of salespersons in making sales. It was done by looking at factors that affect the level of salespersons sales. The research analyzes several factors that influence the salesperson's sales level: presentation skills, questioning ability, adaptability, technical knowledge, self-control, interaction involvement, sales environment, and intrapersonal skills. Through the analysis of these factors with One Sample T-Test and Multiple Linear Regression methods, researchers design a training program for salespersons to increase their sales. The developed training for salespersons is basic training and special training and before training was given, salespersons need to be assessed for the effectivity and efficiency reasons.

1. Introduction

Today, many people in Indonesia feel saturated with a product offered in the market. The saturation customer level of the product in the market is not only due to the marginal utility factor. It is also caused by the influence of salespersons factors in communicating to the customer[1]. Indonesian society is saturated because of the behavior of the salespersons to the customer. In presenting product or prospects, salespersons mostly looks like trying to win arguments with the customer, lack of active communication, offering products aggressively, and so on.[2] This will create a reluctant effect for the customer to the offered products or even switch to rival products. It will also lead to a mindset that company simply looking for profit at the expense of customer money [1, 3].

Some previous research by Bayuaji, Helena, Yanti, and Barton stated that so many variables affect personal selling skill. Usually, researchers only state only soft skill needed by salespersons such as presentation, communication, adaptation, and technical knowledge skill giving positive impact and significant to the sales [3-8]. The current development in multilevel marketing industry plays an important role in building the country's economy. Due to unprofessional salespersons and irresponsible, it makes the image of multilevel marketing industry to be bad that resulted in the community to distrust multilevel marketing industry so that this resulted in decreased company revenue due to Lack of public interest in multilevel marketing products. [2] Seeing these conditions,

IOP Conf. Series: Journal of Physics: Conf. Series **1007** (2018) 012025 doi:10.1088/1742-6596/1007/1/012025

the researchers created a training design to improve the competence of salespersons in making sales by looking at factors that affect the level of salespersons sales.

2. Research methods

In this research, survey research was used because the study was conducted with the aim of obtaining facts from the existing symptoms and seek information factually to get the information. The ability of salespersons in presenting products influenced by several factors such as operational variables are Presentation Skill, Ability to Ask, Adaptability, Technical Knowledge, Self-Control, Interaction Engagement, Sales Environment, and Intrapersonal Skill [3-6]. Data were collected through questionnaire from the respondents (salespersons), Central Bureau of Statistics, and from some surveys held previously. This research was conducted in Medan, North Sumatra, Indonesia from various multilevel marketing companies registered in Direct Selling Association of Indonesia. Object observed in this study are salespersons (trainers, marketing, sales, and agents) of each multilevel marketing company that conducts presentations to customers where every single salesperson represents their respective companies.

The sample of this research was collected through Cluster Sampling. This research use 100 respondents as samples. Questionnaire given using the semantic differential scale for the respondent's answer consists of two poles of the left statement and the right statement while the result of respondents states in seven scales [10, 13, 15]. Collected information through questionnaire then analyzed through One-Sample T-Test, Linear Multiple Regression, and Ordinary Least Square to ensure the result is not biased. Through the questionnaire, a researcher can understand in which part of skill that the salespersons need and develop training programs that suitable for many salespersons.

Detail steps of determining the soft skill training program can be seen in Figure 1.

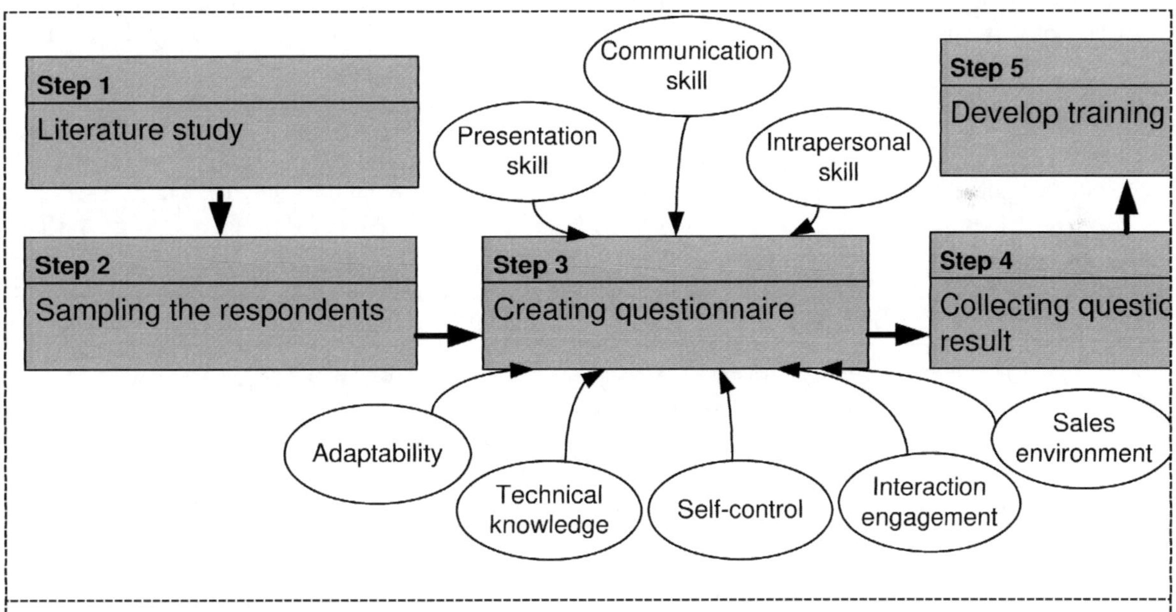

Figure 1. Block Diagram for designing soft skill training program.

3. Result and Discussion

In this research, the basic assumption is tested to determine how much influence of each independent variable individually to the dependent variable with One-Sample T-Test method. It was done by considering the t value of the critical area in the normal distribution curve. The use of One Sample T-Test method for each respondent who gives answers to the questionnaire comes from one group of the same sample and analysis of regression was done by using Linear Multiple Regression which aims to

IOP Conf. Series: Journal of Physics: Conf. Series **1007** (2018) 012025 doi:10.1088/1742-6596/1007/1/012025

determine the linear relationship between two or more independent variables (X_1 , X_2, ...Xn) with the dependent variable (Y).[9] After regression analysis, then continued by testing the classical assumption with Ordinary Least Square (OLS) approach as a requirement to form multiple linear regression models. If the model formed does not meet the classical assumptions required, then the modification/ transformation/ healing of data or multiple linear regression models is required.

From the processing result, it is found that the interaction involvement variable with the value of calculated t (1,283) < t table (1,98422) and the intrapersonal skill variable with t value (1,870) < t table (1,98422) is stated not to have influence to salesperson's sales level in offering products or business opportunities. Based on the value of unstandardized coefficients B test results with SPSS, it can be formed multiple linear regression models which can be seen in the equation below.

$$TPD = 20{,}754 + 1{,}125SP + 0{,}946KBt + 0{,}963KBa + 1{,}015PT + 0{,}845PD + 1{,}401LP + e$$

From ANOVA, it was obtained a value of F calculate (108,968) > F table (2,31) and Sig value. (0.000) < α (0,05) hence reject Ho and accept Ha which means that independent variables simultaneously have the influence to salesperson's sales level in offering the product or business opportunity. From table summary model of test result with SPSS, obtained value of R equal to 0,936 lies between 0,800 < 0,936 < 1,00. Based on the interpretation of the value of R, this shows that there is a very strong relationship between independent variables simultaneously to the dependent variable. From the table of coefficients on the collinearity statistics of the test results with SPSS, it is found that the Tolerance value of each independent variable is greater than 0.01 and the VIF of any independent variable is no greater than 10. Based on the classical linear regression assumption with OLS (Ordinary Least Square), then the linear regression model does not occur multicollinearity.

From the scatterplot graph of heteroscedasticity test, it is seen that the dot distribution docs not form a certain pattern, so it can be concluded that heteroscedasticity is not happening. The classical assumption about heteroscedasticity in this model is fulfilled, ie free of heteroscedasticity. From the diagram of Cartesian horizontal axis DW count, it is seen that the DW count value of 2.102 is in the area there is no autocorrelation so it can be concluded that in multiple linear regression models there is no autocorrelation.

The result was then analyzed and giving result in creating training programs. Generally, they were Basic Training Programs and Specialized Training Programs. Basic training programs for salespersons that can be applied by the company as Personal Development System can be seen in Table 1.

Table 1. Basic Training Programs.

No.	Training Program	Program Objectives	Related Variables
1.	Induction Training	Introduce the organization to newly appointed employees. This is a brief and informative training given shortly after joining the organization.	a. Technical Knowledge b. Sales Environment
2.	Job Training	Provide appropriate information and guidance to employees to enable them to perform the work systematically, appropriately, efficiently, and ultimately with confidence.	a. Presentation Skill b. Ability to Ask c. Ability to Adapt d. Self-control e. Sales Environment
3.	Training for Promotion	Allows employees to adjust themselves to higher-level job assignments.	a. Ability to Adapt b. Self-control
4.	Refresher Training	Updating professional skills, information, and experiences of someone who occupies an important executive position.	a. Presentation Skill b. Ability to Ask c. Ability to Adapt d. Technical Knowledge

			e. Self-control f. Sales Environment
5.	Training for Managerial Development	Given to managers to improve their efficiency and thus enable them to accept higher positions.	a. Presentation Skill b. Ability to Ask c. Ability to Adapt d. Technical Knowledge e. Self-control f. Sales Environment

Whether specialized training programs developed for salespersons who are communicating directly with the customer can be seen in Table 2.

Table 2. Specialized training programs.

No.	Training Program	Information	Related Variables
1.	Sales Marketing Training	In the sales marketing training category, the focus of this training is prioritized to employees and employees as well as any elements that interact directly or indirectly with sales activities or sales marketing. In the Sales Marketing Training category we present, there are various types of training that have various goals and benefits tailored to the needs of sales marketing companies or agencies.	a. Presentation Skill b. Ability to Ask c. Ability to Adapt d. Technical Knowledge e. Self-control f. Sales Environment
2.	Personal Development Training	In the self-development training category, the focus of human resource training is prioritized to all elements within a company, from the top to the bottom level. Activity work not only involves physical ability alone, but mental ability and mindset also affect, and that's the main purpose of this self-development training so that every element of the company can continue to improve motivation and self-development.	a. Presentation Skill b. Ability to Adapt c. Self-control d. Sales Environment
3.	Motivation Training	In the motivation training category, the focus of HR training is prioritized to all elements within a company, whether from top to bottom level. Everyone needs motivation in order to achieve the best performance. Similarly, an employee and head of a company would also require motivation to help companies to achieve the desired target. This is because everyone who works continuously would one day feel	a. Presentation Skill b. Ability to Ask c. Ability to Adapt d. Self-control

		saturated and in the end, it can actually decrease work productivity. Therefore motivation training is needed as a refresher media in order not to feel bored working in the office.	
4.	NLP & Hypnotherapy Training	In the NLP & Hypnotherapy training category, the focus of this training is prioritized to every person who wants to get things better and more from within himself so as to have a broad impact on his life. Because through NLP & Hypnotherapy Training we can understand more about the subconscious mind that is actually a tool to facilitate humans in living his life.	a. Presentation Skill b. Ability to Ask c. Ability to Adapt d. Technical Knowledge e. Self-control f. Sales Environment
5.	Outbound Packages	Outbound Package is one of the systematic training methods that is directly implemented to all human resources in the wild. The outbound package is based on modification techniques: Mind (Cognition), Feelings (affection) and Behavior (behavior) with based on ethics, love, and Spirituality. These components are packed in a choice of Outbound Packages that are implemented in a short time. The outbound package is done in the wild through management games that will lead participants into the experience Experience Experience fun, memorable, and applicative in personal life and organization.	a. Presentation Skill b. Ability to Ask c. Ability to Adapt d. Technical Knowledge e. Self-control f. Sales Environment
6.	HRD Training	In the HRD training category, the focus of HRD training is addressed to every element involved in Human Resources Development, but HRD training can also be followed by leaders and those who need more knowledge and skills in employee development.	a. Presentation Skill b. Ability to Ask c. Ability to Adapt d. Technical Knowledge e. Self-control f. Sales Environment
7.	Leadership Training	In this leadership training category, the focus of this training is how to shape and develop and hone and improve one's leadership skills. Leadership is one of the keys to personal and business success. Careers of someone with strong leadership skills will be more brilliant. Similarly, employees with superior leadership business spirit will bring the company to reach the peak of glory. Various management	a. Presentation Skill b. Ability to Ask c. Ability to Adapt d. Technical Knowledge e. Self-control f. Sales Environment

	studies have proven that leadership is an important key to organizational success.	

Basic training and specialized training programs are the programs given to salespersons to develop their soft skill in communicating with the customer. However, this soft skill must be re-assess biennial to assure that salespersons have the required skill before communicating directly with the customer. For the efficiency reason in running the programs, some programs with a small number of participants are recommended for the third party running. They are Leadership training, NLP & Hypnotherapy training, Training for Promotion and Training for Managerial skills.

4. Conclusions

The result of this research in developing soft skill training in multilevel marketing industry can be concluded as:

1. The most significant factors in making sales simultaneously are presentation skill, communication skill, ability to adapt, technical knowledge, self-control, and sales environment.
2. Before training programs were given, it is recommended to do an assessment to ensure the effectivity and efficiency.
3. The basic training developed for new salespersons is Induction Training, Job Training, Training for Promotion, Refresher Training, and Training for Managerial Development (Training for Managerial Development). While specialized training is given to experienced salespersons, they are Sales Marketing Training, Self Development Training, Motivation Training, NLP & Hypnotherapy Training, Outbound Packages, HRD Training, and Leadership Training.
4. Training recommended being run by the third party were: Leadership training, NLP & Hypnotherapy training, Training for Promotion and Training for Managerial skills.
5. Through specialized training, salespersons can contribute more to increase total sales.

References

[1] Ahmad and Eva 2013 *Evaluasi Pelatihan Selling Skill Karyawan Bagian Sales Promotion Girl PT. Guna Layan Kuasa Jakarta Pusat*(Universitas Indonesia: Jurnal Ilmu Sosial)

[2] Albertus 2010 *Pelaksanaan Personal Selling dalam Pengaruhnya Terhadap Keputusan Pembelian Konsumen Sepeda Motor Honda Pada CV. Aceh Honda Motor Bandung*(Bandung: Jurnal Manajemen)

[3] Bayuaji 2007*Analisis Faktor-Faktor yang Mempengaruhi Selling Skill Terhadap Kinerja Tenaga Penjualan* (Semarang: Tesis Magister Manajemen)

[4] Endiwanto and M Parnawa2015 *Faktor-Faktor Sukses Dalam Proses Personal Selling*(Yogyakarta: Jurnal Ekonomi, Vol. 2)

[5] Helena, Laurence, and Wisang2012 *Pengaruh Softskill (Intrapersonal Skills) terhadap Kinerja Penjualan Pemasar Properti* (Tangerang: Jurnal Teknik Industri, Vol. 12)

[6] Kaswan 2011 *Pelatihan dan Pengembangan untuk Meningkatkan Kinerja SDM* (Bandung: Alfabeta)

[7] Yanti 2006 *Analisis Faktor-Faktor yang Mempengaruhi Adaptif dalam Meningkatkan Kinerja Tenaga Penjual* (Semarang: Tesis Magister Manajemen)

[8] Barton A Weitz 2009 *Personal Selling and Sales Management: A Relationship Marketing Perspective* (Spring: Academy of Marketing Science Journal, Vol. 2)

[9] John Z Sonmez 2015 *Soft Skill: Special Sales Department* (Shelter Island New York: Manning Publications Co)

[10] Morissan 2012 *Metodologi Penelitian Survei* (Jakarta: Kencana)

[11] Sri andAna 2005 *Membangun Kepercayaan Konsumen dengan Strategi Softskill dan Motivasi Tenaga Penjual* (Semarang: Jurnal Management Dynamics Conference, Vol. 1)

[12] Sudjana 1992 *Teknik Analisis Regresi dan Korelasi Bagi Para Peneliti* (Bandung: Tarsito)

[13] Sukaria Sinulingga 2011 *Metode Penelitian* (Medan: USU Press)
[14] Syafaruddin Siregar2005 *Statistik Terapan untuk Penelitian* (Jakarta: PT. Grasindo)
[15] William G Cochran 1991 *Teknik Penarikan Sampel* (Jakarta: UI-Press)

Acknowledgement
In this research, the authors would like to thank profusely to the industry which has been willing to be a place of research for the writing of this paper and highly support researcher for a better result. Authors also want to thank Universitas Prima Indonesia that gives chances to publish and develop author in the process.

IOP Conf. Series: Journal of Physics: Conf. Series **1007** (2018) 012026 doi:10.1088/1742-6596/1007/1/012026

An application of corelap algoritm to improve the utilization space of the classroom

A C Sembiring*, I Budiman, A Mardhatillah, U P Tarigan, and A Jawira

Industrial Engineering Department, Faculty of Technology and Computer Sciences, Universitas Prima Indonesia, Indonesia

E-mail: *anitakembaren@unprimdn.ac.id

Abstract. The high demand of the room due to the increasing number of students requires the addition of the room. The limited number of rooms, the price of land and the cost of building expensive infrastructure requires effective and efficient use of the space. The facility layout redesign is done using the Computerized Relationship Planning (CORELAP) algorithm based on total closeness rating (TCR). By calculating the square distance between the departments based on the coordinates of the central point of the department. The distance obtained is multiplied by the material current from the From-to chart matrix. The analysis is done by comparing the total distance between the initial layout and the proposed layout and then viewing the activities performed in each room. The results of CORELAP algorithm processing gives an increase of room usage efficiency equal to 14 , 98% from previous activity.

1. Introduction

The high demand of the room due to the increasing number of students requires the addition of the room. The limited number of rooms, the price of land and the cost of building expensive infrastructure requires effective and efficient use of the space. The facility layout redesign is done using the Computerized Relationship Planning (CORELAP) algorithm based on total closeness rating (TCR). This study was conducted for improving good space utilization associated with the facility layout [1-5]. Based on the needs of the faculty room by the computer science faculty, then optimized the layout of the building in order to maintain the balance of land based on the comparison between open space and buildings where required by law in at least 30% open space. Therefore, it is necessary to improve the layout of private universities that become the object of research.

Redesigning the layout of an efficient space facility in overcoming material handling problems has been done in previous research using the CORELAP method as the best method in manufacturing companies [6-12]. The purpose of this improvement is to minimize the distance between departments and improve the effectiveness of labor by using the CORELAP method because it locates the most related activities and then adds other activities based on the desired distance and desired size. Proper room layout in building design should be considered as a functional specification and personal needs of users and fulfillment of final production cost calculations [13-16].

In this research, the authors will evaluate the facility's layout using the CORELAP method as it locates the most related and then progressively activities by adding other activities based on the desired proximity and the required size. The purpose of this improvement is to minimize the distance between departments and increase the effectiveness of private university workforce by using CORELAP analysis [17-22].

2. Methodology/Experimental

The algorithm was introduced by Robert C. Lee and Moore in 1967 with the foundation of Muther's Systematic Layout Planning (SLP). The procedure there are three that is problem analysis, search stage, and selection stage. The input data of the algorithm include: (a) a relationship chart, (b) the area of each department, (c) the number of departments, (d) the closeness rating. CORELAP's initial step is to calculate the total closeness rating (TCR) of each department. TCR facility 1 is the sum of the numerical values that state the relationship between the facilities with each other. These values are obtained from the relationship diagram shown by the degree of proximity: A (given a value of 5), E (given a value of 4), I (given a value of 3), O (rated 2), U (given value 1), X (given a value of 0). CORELAP is also a construction algorithm with activity relationship a major consideration. It is designed to accommodate situations when constantly changing conditions prohibit the collection of precise numerical data [21].

CORELAP constructs layouts by locating rectangular-shaped departments when the departmental area and layout scale permit a rectangular representation of the departmental area . It is based on REL chart and the numerical weighted rating assigned to the closeness ratings. The evaluation phase employs a placing rating and a boundary length. Total closeness rating for each department is calculated as:

$$TCR_i = \sum_{j-1}^{m} \mathbf{V(rij)} \quad ; m = \text{no of departement}$$

CORELAP is a deterministic approach that provides an unique solution. Running the program second time with the same data produces the same final layout. The rectangular shape of the departments is divided into the predefined number of squar e blocks. In order to make department rectangular the area of the department which results in over utilization of the floor space. Once the relevant input data and information is entered, the entire layout of the development work is fully performed by the program. There is a high possibly that layout generated by CORELAP is not practically implementedable.

Troubleshooting steps with the OCRELAP algorithm: (a) select one room with the maximum TCR, (b) second allocated space, select the room that has relationship A with the selected space, (c) the third allocated space, select the room which has an A relationship with the first selected chamber, (d) for the next selected chamber having an A, E, I, O, U relationship with the second selected chamber, or the third and the last one being placed if there is an X relationship with the first selected chamber.

This research was conducted at a private university in a medan city requiring a faculty room for lecturer performance improvement. Limited space and the number of students who are highly correlated with high land prices require the effective and efficient use of space. Because the university needs a new layout. Research is descriptive research with method used to describe or analyze research result can't be used to make wider conclusion. The object of research observed is the actual condition of private higher education such as area space, spatial, and environmental management where the goal to look more tidy. The data collected for this study is the measurement of each initial layout and department, space capacity, and measured directly by using tape measure and visually.

3. Results and Discussion

In this section will discuss the point of coordinates of the room, the distance between rooms, the formation of Activity Relationship Chart (ARC), and calculation of CORELAP.

3.1. Drawing on room Layout

Private higher education layout has 13 sections or spaces used by universities to facilitate students and lecturers on the move. Each room consists of one or more work elements. Each layout has an existing size and location. In this layout, drawing depicted space at the university. Determination of coordinate points for each room can be seen in Figure 1.

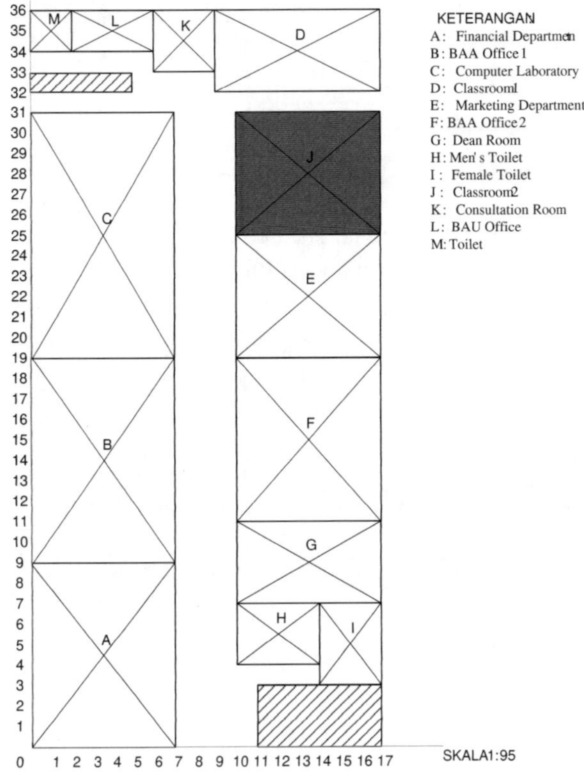

Figure 1. Actual layout of higher private education.

3.2. Coordinate Point Determination

Coordinate points can be measured using the formula:

Coordinate X = $X_0 + \frac{X_1 - X_0}{2}$

Coordinate Y = $Y_0 + \frac{Y_1 - Y_0}{2}$

The diagonal intersection that occurs for room A is at the point:

Coordinate X = $X_0 + \frac{X_1 - X_0}{2} = 0 + \frac{7-0}{2} = 3,5$

Coordinate Y = $Y_0 + \frac{Y_1 - Y_0}{2} = 0 + \frac{9-0}{2} = 4,5$

Coordinate point of room A = (x,y) = (3,5 ; 4,5)

Similarly, for the other room, the calculation of coordinate points in the same way. Data Work Station, Encoding and Data Size Each Room can be seen in Table 1.

Table 1. Work station, encoding and data size each room.

Department	**Code**	**Dimension (m)**	**Area (m^2)**	**X**	**Y**
Financial Department	A	6,745 x 8,925	60,199125	3,5	4,5
BAA Office 1	B	6,745 x 10,4	70,148	3,5	14
Computer Laboratory	C	6,745 x 11,9	80,2655	3,5	25
Classroom 1	D	6,745 x 5,95	40,13275	13,5	28
Marketing Department	E	6,745 x 5,95	40,13275	13,5	22
BAA Office 2	F	6,745 x 8,33	56,18585	13,5	15

Dean Office	G	6,745 x 3,67	24,75415	13,5	9
Male Toilet	H	4,2 x 2,7	11,34	12	5,5
Female Toilet	I	2,55 x 3,7	9,435	15,5	5
Classroom 2	J	9,75 x 3,75	36,5625	13	34
Consultation Room	K	2,9 x 2,55	7,395	7,5	34,5
BAU office	L	3,79 x 1,675	6,34825	4	35
Toilet	M	2 x 1,675	3,35	1	35

3.3. Inter Distance.Measurement

The distance between rooms is measured using a rectilinear distance, where the distance is measured following a perpendicular path. Room distance is calculated by taking the center point of the room. In rectilinear distance measurement used the following formula:

$$d_{ij} = |x-a| + |y-b|$$

The distance between stations is calculated using the Rectilinear distance formula:
Coordinate A (3.5 ; 4.5) and B (3.5 ;14), then the distance A to B is:
$d_{ij} = |x_2-x_1| + |y_2-y_1|$
A-B = |3.5 – 3.5| + |14 – 4.5| = 9.5
Likewise, for other distances are calculated in the same way. The results of calculating the distance between stations as a whole for the initial layout can be seen in Table 2.

Table 2. Distance between rooms.

i/j	A	B	C	D	E	F	G	H	I	J	K	L	M
A		9,5	20,5	33,5	27,5	20,5	14,5	9,5	12,5	39	34	31	33
B	9,5		11	24	18	11	15	17	21	29,5	24,5	21,5	23,5
C	20,5	11		13	13	20	26	28	32	18,5	13,5	10,5	12,5
D	33,5	24	13		6	13	19	24	25	6,5	12,5	16,5	19,5
E	27,5	18	13	6		7	13	18	19	12,5	18,5	22,5	25,5
F	20,5	11	20	13	7		6	11	12	19,5	25,5	29,5	32,5
G	14,5	15	26	19	13	6		5	6	25,5	31,5	35,5	38,5
H	9,5	17	28	24	18	11	5		4	29,5	33,5	37,5	40,5
I	12,5	21	32	25	19	12	6	4		31,5	37,5	41,5	44,5
J	39	29,5	18,5	6,5	12,5	19,5	25,5	29,5	31,5		6	10	13
K	34	24,5	13,5	12,5	18,5	25,5	31,5	33,5	37,5	6		4	7
L	31	21,5	10,5	16,5	22,5	29,5	35,5	37,5	41,5	10	4		3
M	33	23,5	12,5	19,5	25,5	32,5	38,5	40,5	44,5	13	7	3	

3.4. Determination of Total Moving Distance

Total displacement distance is obtained by:
Total distance of displacement = frequency x distance of room
Example: Total distance of displacement = frequency x distance of room = 10 x 9.5 = 95 m
Likewise for the total distance of other displacements is calculated in the same way.

3.5. Activity Relationship Chart

ARC is made on the basis of the frequency of the flow of the room's displacement, the frequency of carrier/labor displacement, the similarity of material handling tools used and also the factors of comfort and safety at work. The ARC between rooms can be seen in Figure 2.

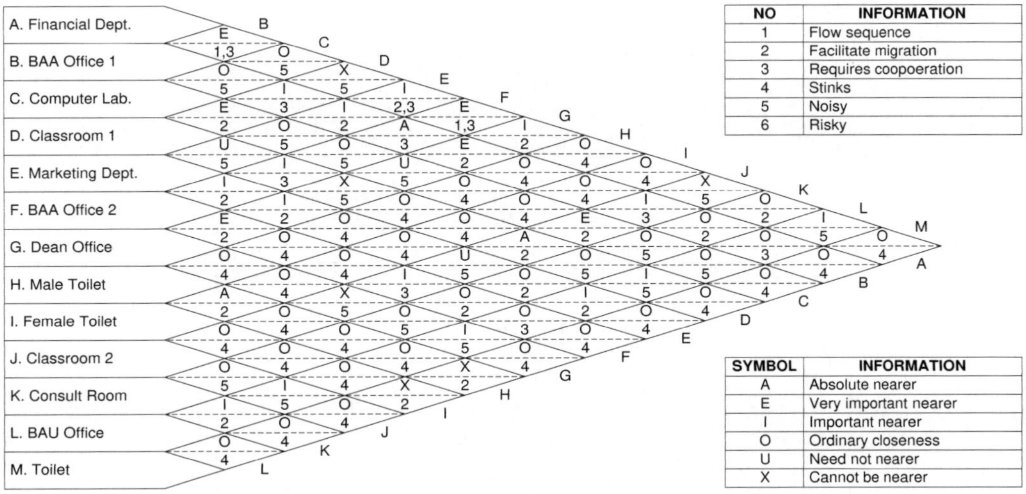

Figure 2. Activity relationship chart.

3.6. Data Processing With Corelap Algorithm

Calculations with the CORELAP Algorithm use a significant proximity rating in the Proximity Rating Total (TCR) in the selection of workplace placements. The TCR calculations are performed based on the qualitative data of ARC in Figure 4, which is converted into figures, in Table 3.

Table 3. Scale of ARC score.

Kode	Value
A	5
E	4
I	3
O	2
U	1
X	0

Step-by-step data processing with CORELAP: Select one of the room that has the value with the highest Total Closeness Rating (TCR) and the largest area of the room as the first room, then select the second room that has relationship A with the first room, then the third room that has the relationship of the value of A to the first and subsequently selected chambers is chosen which has the relationship A, E, I, O, U with the second and third elected.

So using the same way all the rooms are allocated so as to generate allocations from the entire room so that all the rooms are correctly allocated according to the CORELAP algorithm. The layout image of CORELAP algorithm design can be seen in Figure 3.

IOP Conf. Series: Journal of Physics: Conf. Series **1007** (2018) 012026 doi:10.1088/1742-6596/1007/1/012026

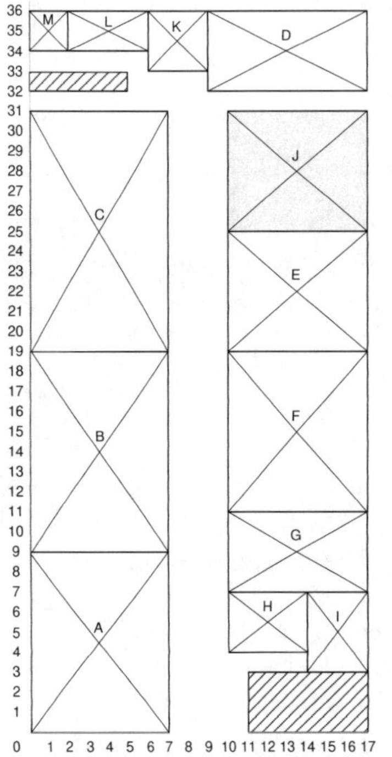

Figure 3. Final result of CORELAP algorithm.

3.7. Initial Condition Analysis

In the initial conditions, the arrangement of the layout at the private University has been seen optimally. This can be seen from the location of the room and the frequency of displacement in the University of Prima Indonesia. However, there are some rooms that are not proportional to the frequency of movement resulting in the total distance travelled to be longer.

3.8. Design Results Analysis

To compare the initial layout and proposed design, total displacement distance is used as a reference. The total distance of the displacements occurring in accordance with the initial conditions of the enterprise can be calculated by multiplying the frequency of displacements by their displacement distance

$$Correction = \frac{Total\ initial\ distance - Total\ final\ distance}{Total\ initial\ distance} \times 100\%$$

$$= 14{,}98\%$$

From this calculation, it can be seen that the design with CORELAP algorithm gives an efficiency of 14,98 %.

4. Conclusion.

Based on the results of the study, the total initial distance of 2062.5 meters from the initial layout was reduced to 1753.5 meters. The result of CORELAP algorithm processing gives 14.98% efficiency and minimum class activity is converted into lecturer room. So that space J which was originally classroom converted into lecturer room.

References

[1] Apple, James M. 1990. Tataletak Pabrik dan Pemindahan Bahan. Bandung: Penerbit ITB.
[2] Inaki Maulida Hakim1*, Vidyahningtyas Istiyanti., Improvement Of Layout Production

Facilities For A Secondary Packaging Area Of A Pharmaceutical Company In Indonesia Using The Corelap Method, International Journal of Technology 2015.
[3] Dwianto, Qodri Azis, dkk. 2016. Usulan Rancangan Tata Letak Fasilitas dengan Menggunakan Metode CORELAP di Perusahaan Konveksi. Jurusan Teknik Industri Itenas No.1, Vol.4
[4] Naik B.S., Kallurkar S., A literature review on efficient plant layout design, International Journal of Industrial Engineering Research and Development 2016, 7, 2, 43-51.
[5] Drira, A., Pierreval, H., Gabouj, S.H., 2007. Facility Layout Problems: A Survey. Annual Reviews in Control, Volume 31, pp. 255-267
[6] Khan, A.J., Tidke, D.J., 2013. Designing Facilities Layout for Small and Medium Enterprises. International Journal of Engineering Resesarch and General Science, Volume 1(2), pp. 1-8
[7] Matalycki M., Kiturko O., Basiewicz N., Application of HM-networks in problems of transport logistics, Scientific Research of the Institute of Mathematics and Computer Science 2010, 9, 2, 153-172.
[8] Tompkins J.A., White J.A., Bozer Y.A., Tanchoco J.M.A., Facilities Planning, 4th ed., John Wiley & Sons Inc., New York 2010.
[9] Setiawati, L., 2012. Production Facility Layout Improvements using Blocplan Algorithm. Industrial Engineering Journal Bung Hatta University, Volume 1(2), pp. 206-216
[10] Kriel, M., 2010. Optimizing Facility Layout through Simulation. University of Pretoria. Pretoria
[11] Dr. Surender Kumar and Dr.A.K. Jha, Technology of computer-aided design and Manufacturing CAD/CAM,Dhanpat Rai & Co (P), 1st ed 1993.
[12] Afridasani, Nuri. 2015. Perancangan Ulang Tata Letak Fasilitas Produksi Dengan Menerapkan Travel Chart, Algoritma BLOCPLAN dan CORELAP di PT. Cahaya Bintang Medan. Medan: Universitas Sumatera Utara
[13] Siregar, R.M., Sukatendel, D., Tarigan, U., 2013. *Re-layout Design of Production Facility by Implementing BLOCPLAN Algorithm and CORELAP Algorithm at PT. XYZ. Industrial Engineering E-Journal FT USU*, Volume 1(1), pp. 35-44
[14] Langgihadi, Dhanudwitri, dkk. *Usulan Rancangan Ulang Tata Letak Fasilitas dengan Menggunakan Metode CORELAP di Perusahaan Distribusi Bahan Bakar Pesawat Udara*. Jurusan Teknik Industri Itenas No.1, Vol.4.2016.
[15] Suhendar, Dede, dkk. *Usulan Perancangan Ulang Tata Letak Fasilitas Produksi Dengan Menggunakan Metode Algorotma CORELAP*. Teknik Industri Universitas Bina Darma. 2012.
[16] Apple, J.M., 1990. *Plant Layout and Material Handling* (3rd ed). Bandung: ITB
[17] M. M. Hassan and G. L. Hogg, *A review of graph theory application to the facilities layout problem, Omega*, vol. 15, no. 4, pp. 291–300, 1987.
[18] Rubianto, Christina Natalia, dkk. *Penentuan Lokasi dan Perancangan Tata Letak Fasilitas Tempat Packaging PT ABC.* Titra No.2, Vol.2.2014.
[19] Siregar, Renata Maywanto, dkk. Perancangan Ulang Tata Letak Fasilitas Produksi dengan Menerapkan Algoritma Blocplan dan Algoritma CORELAP Pada PT XYZ. Teknik Industri FT USU No.1, Vol.1. 2013.
[20] Sembiring, Anita Christine. *Perancangan Ulang Tata Letak Pabrik untuk meminimalisasi material handling di PT. Atmindo.* Medan: Universitas Sumatera Utara. 2012.
[21] Wignjosoebroto, Sritomo. *Tata Letak Pabrik dan Pemindahan Bahan*. Jakarta: PT Guna Widya. 1996.
[22] R. Muther and A. H. Mogensen, "*Systematic layout planning*," 1973.

Acknowledgments
In this research, the authors would like to thank the leaders of private universities who are willing to be the object of research . The authors also thanked the Universitas Prima Indonesia that has given the writerthe opportunity to continue to work and conduct research.

IOP Conf. Series: Journal of Physics: Conf. Series **1007** (2018) 012027 doi:10.1088/1742-6596/1007/1/012027

Feasibility Study Analysis of Establishing Archery Sports Special Place

U P P Tarigan*, A Mardhatillah, I Budiman, A C Sembiring, M S Ramadhan

Industrial Engineering Department, Faculty of Technology and Computer Science, Universitas Prima Indonesia

E-mails: *unipratama@gmail.com, ellalapink@gmail.com, irwanb01@gmail.com, anitakembaren22@gmail.com, ramadhans909@gmail.com

Abstract. This study essentially discusses the basic concepts related to the decision and process of selecting a sports archery business project in order to provide economic and social benefits over time. Archery sports is a sport that requires special skills, both accuracy, coordination mental training and improves the physical condition of the prime. In Medan, it's still a bit of practice archery, but many people love this sport. From the data collected, there are three locations where archery in the city of the field of the Sunggal Knights is located at Jl Tunggul Hitam no.1A, Kec. Medan Sunggal, Avros Educational Park Aval address at Avros Street No.60 Medan, and the Knights of Johor Archery Club which is located at Jl. The work of Darma Gg. Ujung Family, Medan Johor. Therefore a feasibility study was undertaken in establishing an archery sports place in the field. The aspects to be considered in making feasibility studies are market and marketing aspects, technical and equipment aspects, management aspects, legal aspects, economic and social aspects, and financial aspects. Financial analysis using own capital got Payback Period 3 years 5,8 months, Net Present Value Rp 11.684.026 and obtained Internal Rate of Return of 17%.

1. Introduction

Given the impending conditions of uncertainty, certain considerations in an enterprise are required, where the basis of these considerations can be generated through a study of various aspects of the viability of a business to be carried out, so that the results of the study can be used to decide whether or not to have or not. The above mentioned is shown in the business feasibility study will be a lot of teams from various experts in accordance with their respective fields or aspects such as economics, law, psychologists, accountants, technological engineers and so forth.

Archery sports is a sport that requires special skills, both accuracy, coordination and mental training and improves the physical condition of the prime. In Medan, it's still a bit of practice archery, but many people love this sport. From the data collected in the field of wisata.com.com in 2016, there are three locations where archery in the city of the field of the Sunggal Knights is located at Jl Tunggul Hitam no.1A, Kec.Medan Sunggal Medan, Avros Educational Park Aval address at Avros Street No.60 Medan, and the Knights of Johor Archery Club which is located at Jl. The work of Darma Gg. Ujung Family, Medan Johor.

From the results of observation and supported from the data obtained from medanwisata.com The author wants to try to establish a business where archery sports, which in which can provide income for writers and also can help the community economy around the business location to be run, but the current problem is the financial condition is very minimal, the existing funds from own capital of Rp. 10,000,000. With a little capital need to seek additional capital, one of them by seeking additional capital to the investors who want to invest money in order to benefit from the business of this archery sporting place.

A business feasibility study is an activity that studies in depth about an activity or business or business to be run, in order to determine whether or not the business is run [1]. The aspects to be considered in making feasibility studies are those studied are market and marketing aspects, technical and equipment aspects, management aspects, legal aspects, economic and social aspects, and financial aspects [2,3] Understanding is feasible in this assessment is the possibility of business ideas or projects that will be implemented to provide benefits (benefits), both in the sense of financial benefit and social benefit. To find out the feasibility of a business plan required financial and nonfinancial analysis. Financial analysis is the analysis used to determine the feasibility of a business plan from the financial side by using the measurement criteria investment.[4]. Small businesses play their role in tackling big business in Indonesia by increasing national product of the country, in terms of diversifying sources of income, increasing the labor force, lowering unemployment rate, and providing stimulation for technical and vocational education [5,6] The feasibility used along with its advantages and disadvantages associated with public sector in Arab countries. The goal is to help the community the project manager in the decision-making process and assess the overall performance [7]. Feasibility studies are also common in the culinary business. An example is the feasibility study of the establishment of Cozy Chocolate Home Industry [8,9].

2. Methodology/Experimental

The type of research used is descriptive. Descriptive is a method used to describe or analyze a research result but not used to make wider conclusions. [10] This research focuses on six aspects which consist of six independent variables and two dependent variables.
The criteria for the business feasibility of this sports venue are as follows [11,12,13].

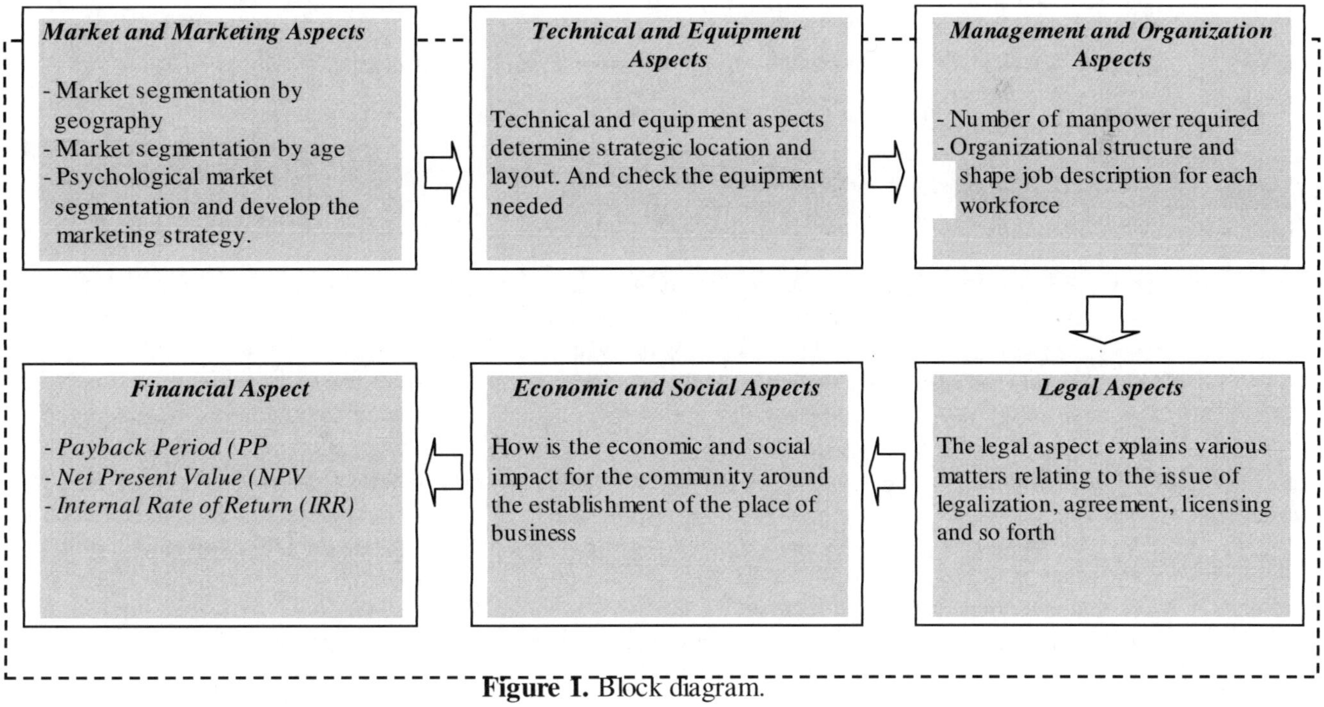

Figure I. Block diagram.

IOP Conf. Series: Journal of Physics: Conf. Series **1007** (2018) 012027 doi:10.1088/1742-6596/1007/1/012027

- Market and Marketing Aspects
  On the market aspect will set the price to be charged to each user, and divide the market based on 3 groups,
  - Market segmentation by geography
  - Market segmentation by age
  - Psychological market segmentation and develop the marketing strategy.
- Technical aspects and equipment
  Technical and equipment aspects determine strategic location and layout. And check the equipment needed
- Aspects of Management and Organization
  Aspects of Management is a process to make preparations for it to run again properly. Aspects range management
  - Number of manpower required
  - Organizational structure and shape job description for each workforce
- Legal Aspects
  The legal aspect explains various matters relating to the issue of legalization, agreement, licensing and so forth.
- Economic and social aspects
  How is the economic and social impact for the community around the establishment of the place of business
- Financial aspect
  This aspect discusses how to calculate funding needs, determine the cost of each user's rent, the length of the capital recovery period, and the feasibility rating, with the criteria used among others:
  - Payback Period (PP)
  - Net Present Value (NPV)
  - Internal Rate of Return (IRR)

  In pricing is used profit-based pricing method. Profit pricing targets are generally in the form of determination of the number of annual profit targets expressed specifically [14,15].

3. Results and Discussion

3.1. Analysis of Market Aspects and Marketing
Based on the results of data processing of market and marketing aspects, the results obtained based on feasibility

- 53% of respondents are interested in archery and want to try archery.
- The target market starts at the age of 14 years both men and women
- The price applied can cover the operational cost.

So according to the above aspects of the business where the archery sport worth to run

3.2. Analysis of Technical and Equipment Aspects
Based on the results of data processing of technical aspect and equipment, the result of feasibility is based on

- Land area contained in the fishing line 1 Medan Deli sub-district for 3,500 m^2
- Range of firing up to 60 m
- Can accommodate 14 players/hour.
- The layout design of the archery sports venue can be seen in Figure 1 as follows.

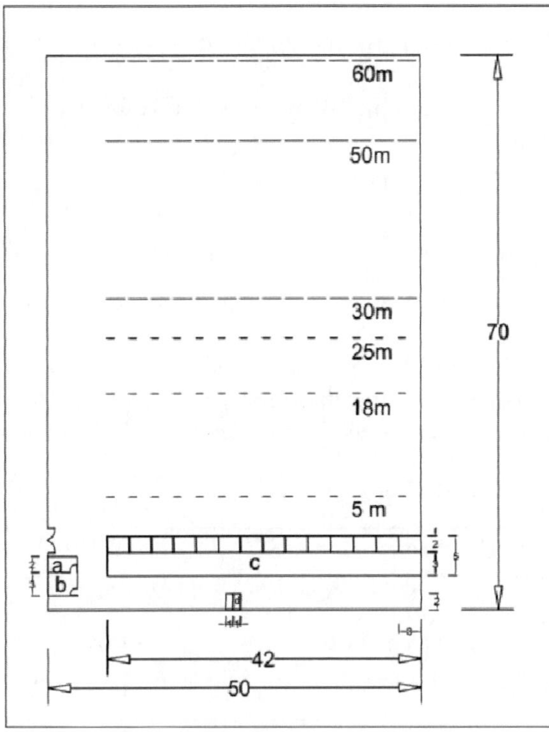

No	Description
a	counter
b	the office and the room maintenance
c	Place archery
d	Toilet

Figure 2. Layout design place sports archery.

3.3. Analysis of Management Aspects

Based on the results of data processing of management aspect, the results obtained based on feasibility

- Number of manpower required 6
- The minimum education background of graduates from SMA / SMK to S1.
- For employees' day off can be done by way of scheduling.

Working hours in this archery business for 8 hours, starting from 8:00 to 12:00 and will be resumed at 14:00 to 18:00. But for 2 hours of rest, there should be kept at least 2 people just in case if there are visitors who come.

This archery sports place is open every day, so for the day of employees are done by turns every person is only given a holiday once a week but with a record of the task in everyday, there must be a replacement. The organizational structure used is the line. Where the delegation of authority directly vertically and completely from the leadership against his subordinates

3.4. Legal Aspect Analysis

Based on the results of legal aspect data processing, a feasibility result is based on the business entity in this business is Commanditaire Vennootschap (CV) where the permission can be handled by the notary from the beginning until the issuance of the business license. So according to the above aspects of the business where the archery sport worth to run.

3.5. Analysis of Economic and Social Aspects

Based on the results of data processing of economic and social aspects, Feasibility results are based on The economic impact will be felt by the community around business location establishment where the archery is among others

- Increased family income
- There is a double income pattern

- Open employment opportunities for the community

While the social impacts that will be in the natural by the community around the location of the establishment of the archery sports venue other

- The existence of a means to exercise archery other than football field foot
- District deli field is known by many people.
- The change of mindset of people around the location

3.6. Financial Aspects Analysis

Based on the results of data processing.

- Payback Period that already exists like the re-invitation period faster than the economic age (5 years)
- Net Present Value value if in accordance with its positive NPV
- Appropriate Internal Return Rate if IRR is greater than bank credit interest (15% and 17%)

Then the results of data processing on the financial aspects can be seen in Table 1 as follows.

Table 1. Financial aspects analysis.

No	Method	Results	Information
1	Payback Period	3 years 5.8 Months	Worthy
2	Net Present Value	Rp 11.684.026	Worthy
3	Internal Rate of Return	17% > 15%	Worthy

4. Conclusion

The initial capital to start the archery business is Rp 162,823,000, based on RI law number 20 the year 2008 this business is included in the small business category. The price of archery sport that will be established in the District of Medan Deli cost Rp 30.000 / jam. Investment repayment period is for 3 years 5 months 8 days. The business of archery sports place is said to be feasible based on the results of the feasibility of six aspects of research and calculation of financial aspects which has been done then got the conclusion that the business where the archery sport is feasible to run.

References

[1] K and J 2009 Studi Kelayakan Bisnis Jakarta: Kencana 2
[2] A Atheem, A Muttaleb Feasibility Study for Investment Decision Taking Egypt: Dar Al-Ma'arifah Al-Jamiah
[3] M Porter 1980 Competitive Strategy: Technique for Analyzing Industries and Competitors Newyork: Free Press
[4] S Danang 2014 Studi Kelayakan Bisnis Yogyakarta: Center of Academic Publishing Service
[5] A Azmi, G Mustafa 2012 Market Analysis and The Feasibility of Esthablishing Small Business European Scientific Journal 8 1857- 7431
[6] T Yousuf, A Rahim 2002 Management of Small Businesses Amman: Dar Safa'a for Publishing and Distribution
[7] A Abou, A Bushraa, M Ezzat 2007 Overview of Feasibility Study Procedures for Public Construction Projects in Arab Countries Cairo Egypt: Structural Engineering Department 18 19-34
[8] A Abidatul, S Muhammad, Dwiatmanto 2015 Analisis Studi Kelayakan Usaha Pendirian Home Industry Studi Kasus Pada Home Industri Cokelat Cozy Kademangan Blitar Jurnal Administrasi Bisnis 35
[9] S Sinulingga 2011 Metodologi Penelitian Medan: USU Press
[10] D H Kim 2015 An Economic Feasibility Study on the Research Infrastructure Project for the

High-speed Big-Data Processing Devices Journal of The Society of Mobile Technology 12
[11] D Ashfa, S Muhammad, F Devi 2016 Analisis Kelayakan Usaha dalam Rangka Rencana Pengembangan Usaha Studi Kasus Pada PO. Zenav Pariwisata Malang. Jurnal Administrasi Bisnis 35.
[12] G William 2000 Business Research Methodology USA: Harcourt College Publisher 6
[13] P Miller, T OLeary 1997 Capital Budgeting Practices and Complementarity Relations in The Transition to Modern Manufacture: A Field- Based Analysis Journal of Accounting Research 35 257-271
[14] P Kotler 1964 Marketing Mix Decisions for New Products Journal of Marketing Research 43-49
[15] W Robert 2001 A Systematic Approach to Market Opportunity Analysis Business Horizons 17 25-35

IOP Conf. Series: Journal of Physics: Conf. Series **1007** (2018) 012028 doi:10.1088/1742-6596/1007/1/012028

Analysis of Optimal Transport Route Determination of Oil Palm Fresh Fruit Bunches from Plantation to Processing Factory

U Tarigan[1]*, R F Sidabutar[1], U P P Tarigan[2], A Chen[2]

[1]Industrial Engineering Department, University of Sumatera Utara, Almamater St USU Campus Medan 20155, Indonesia
[2]Industrial Engineering Department, Faculty of Technology and Computer Science, Universitas Prima Indonesia, Indonesia

E-mails: *ukurta.tarigan@yahoo.com, ronaldfrans.22@gmail.com, unipratama@gmail.com, andreechen96@naver.com

Abstract. Manufacturers engaged in the business, producing CPO and kernels whose raw materials are oil palm fresh fruit bunches taken from their own plantation, generally face problems of transporting from plantation to factory where there is often a change of distance traveled by the truck the carrier of FFB is due to non-specific transport instructions. The research was conducted to determine the optimal transportation route in terms of distance, time and route number. The determination of this transportation route is solved using Nearest Neighbours and Clarke & Wright Savings methods. Based on the calculations performed then found in area I with method Nearest Neighbours has a distance of 200.78 Km while Clarke & Wright Savings as with a result of 214.09 Km. As for the harvest area, II obtained results with Nearest Neighbours method of 264.37 Km and Clarke & Wright Savings method with a total distance of 264.33 Km. Based on the calculation of the time to do all the activities of transporting FFB juxtaposed with the work time of the driver got the reduction of conveyance from 8 units to 5 units. There is also improvement of fuel efficiency by 0.8%.

1. Introduction

The manufacturing company as a whole will always be closely linked to the transportation and distribution process. Starting from the procurement of raw materials to the marketing of the final product are all very closely related to transportation. It is essential that a manufacturing plant minimize the costs sacrificed for transportation. One of the steps that can be done to make the transportation process better is to plan the travel route appropriately so that the product can reach the customer with timely and low cost. Similarly, in the transportation of raw materials. There are four things that are targeted for the smooth transportation of FFB that is maintaining free fatty acids (FFC) related to final product quality, capacity or smooth processing in a factory, FFB security in the field, and minimum transportation cost [1].

XYZ company is one of the companies that produce CPO and kernel produced by obtaining raw materials from the plantation owned by the factory concerned. Transport of raw materials by means of transportation provided by the manufacturer. Trucks used for branded Hino with a capacity of 10 tons of 8 units. The process of transporting FFB from plantation to factory still has not applied the standard system, causing some problems in the transportation process. The uncertainty of this transportation lane makes the cost increase and also increases the waiting time for FFB to be processed immediately.

Distribution route problems can be analyzed using transportation management approach. The Nearest Neighbours and Clarke & Wright Savings methods are a method used in determining the proper transportation path by trying to sequentially enter the place to be visited into the vehicle route by using the closest distance criteria so that it can solve the problem of vehicle scheduling route (vehicle routing and scheduling problem) [2,3] Clark and Wright (C-W) algorithm was developed by entering G. Clarke and J. W. Wright. The basis for the C-W algorithm is savings concept where these savings are realized by linking pairs of delivery points served by a single depot in the network, while Nearest Neighbours Procedure (NNP) builds a tour based on the cost or distance of traveling from the last-visited node to the closest node in the network.

2. Methodology and Experimental

- Data collection methods used in this study are:

The steps on the Clarke & Wright Savings method are as the following:
Specify the node as the central node

- Observation technique, which is to observe directly in the company to get data related to research.
- Interview technique, which is conducting question and answer (interview) to the parties involved directly in about transportation of FFB
- or called the depot and destination node.
- Creating a distance matrix is the intermediate distance matrix depot with node and distance between nodes.
- Calculate the savings value ($S_{i,j}$) in the form of distance travel from one node to another node.
- Create an austerity matrix, then process it is repeatedly driven from the largest matrix to a matrix of small value to each, the austerity matrix was evaluated for further route repair

The steps in the Nearest Neighbours method are as the following:

- Determine the parameter K (the number of nearest neighbors).
- Calculate the quadratic distance of the Euclid (query instance) of each object against the given sample data
- Sort objects into groups that have the smallest Euclidean spacing.
- Collecting category Y (nearest neighbours classification).
- Using the nearest neighbor category of the majority, it can be predicted the value of the query instance that has been calculated [4,5,6].

3. Results and Discussion

3.1. Processing Clarke & Wright Savings Method

In harvest area I there are 26 Point of Result Collection (PRC), then total iteration did as many as 13 iterations with recapitulation as follows in Table 1and while for harvest area II there are 24 Point of Result Collection (PRC).

Table 1. Route recapitulation with clarke & wright savings methods area harvest I.

NO	PRC	Amount of FFB/PRC (ton)	Total of FFB (ton)	Truck Capacity (ton)	Mileage (Km)
1	PP25	4.0	8.2	10.0	15.12
10	PP18	5.5	10.0	10.0	12.56

NO	PRC	Amount of FFB/PRC (ton)	Total of FFB (ton)	Truck Capacity (ton)	Mileage (Km)
	PP26	4.2			
2	PP23 PP22	6.0 4.0	10.0	10.0	13.32
3	PP22 PP21	3.7 6.3	10.0	10.0	12.63
4	PP21 PP24	1.2 7.1	8.3	10.0	13.34
5	PP2 PP1	6.7 3.3	10.0	10.0	12.41
6	PP1 PP19	3.3 6.7	10.0	10.0	13.83
7	PP19 PP20	0.6 7.6	8.2	10.0	12.22
8	PP6 PP5	5.9 4.1	10.0	10.0	11.27
9	PP5 PP15	1.7 7.0	8.7	10.0	12.44
	PP8	4.5			
11	PP8 PP9	4.0 4.6	8.6	10.0	11.52
12	PP18 PP8	4.8 5.2	10.0	10.0	11.25
13	PP36 PP35	2.0 6.4	8.4	10.0	10.58
14	PP17 PP7	5.3 4.7	10.0	10.0	11.59
15	PP7 PP4	2.9 7,1	10.0	10.0	10.92
16	PP13 PP11	7.0 3.0	10.0	10.0	9.85
17	PP11 PP3	4.5 5.1	10.0	10.0	10.42
18	PP12	7.7	7.7	10	8.4
		TOTAL			**214.09**

Minimum Total Mileage of with Clarke & Wright Savings Methods' iteration in harvest I is 214.09 Km. The similar method did recapitulation as many as 12 interations in Harvest II and got the total 264,33 Km.

3.2. Processing with Nearest Neighbours Method

This method uses the simple principle of making the factory as the starting point of departure to the point of result collection (PRC) and to be addressed is the PRC which has the closest distance to the factory then to the nearest PRC with the last visited PRC, and so on until all the PRC has been visited. Recapitulation results can be seen in Table 2.

Minimum Total Mileage of with Nearest Neighbours Method iteration in harvest I is 200.78 Km. The similar method did recapitulation in Harvest II and got the total 264,37 Km.

Table 2. Recapitulation of routes with the nearest neighbours area method of harvest I.

NO	PRC	Amount of FFB/PRC (ton)	Total of FFB (ton)	Truck Capacity (ton)	Mileage (Km)
1	PP12 PP8	7.7 2.2	10.0	10.0	10.05
2	PP8 PP13	6.2 5.1	10.0	10.0	10.61
3	PP13 PP15	3.2 6.8	10.2	10.0	9.98
4	PP16 PP17	6.4 3.6	10.0	10.0	10.58
5	PP17 PP14 PP11	3.4 5.1 1.6	10.0	10.0	11.53
6	PP11 PP10	4.2 5.9	10.0	10.0	10.53
7	PP10 PP9 PP7	6.7 3.3 3.3	10.0	10.0	10.49
8	PP7 PP4 PP5	5.1 3.4 5.1	10.1	10.0	11.35
9	PP5 PP6	4.2 5.9	10.1	10.0	11.27
10	PP45 PP46	0.8 7.6	10.0	10.0	18.06
11	PP1 PP3	3.3 5.1	8.4	10.0	11.81
12	PP20 PP21	7.6 2.4	10.0	10.0	12.17
13	PP21 PP22	5.1 4.9	10.0	10.0	12.63
14	PP22 PP23 PP26	2.8 6.0 1.2	10.0	10.0	14.64
15	PP26 PP25 PP24	3.0 4.0 3.0	10.0	10.0	14.67
16	PP24 PP18	4.1 5.5	9.6	10.0	13.77
17	PP19	7.3	7.3	10.0	11.60
		TOTAL			200.78

IOP Conf. Series: Journal of Physics: Conf. Series **1007** (2018) 012028 doi:10.1088/1742-6596/1007/1/012028

3.3. Calculation of Truck Amount

After calculating the route and the amount of mileage each truck will take in transporting FFB from the garden, calculating loading and unloading times, it can be calculated how many trucks it needs. With a working time of 8 hours per day, this becomes the limitation for consideration in determining the optimal number of trucks.

Calculation time can be calculated by using the largest distance from the calculation route selected above that is by Clarke & Wright Saving method on harvest area II as many as 15 routes with a total distance of 264.33 Km.

$$T = \frac{S}{V}$$

$$T = \frac{264{,}33}{25}$$

$$T = 10.57 \text{ hours}$$

$$T = 634{,}2 \text{ minutes} \qquad \text{...(1)}$$

$$T = 10.57 \text{ hours}$$

$Where \ = Travel\ time\ (hours)$

S = Mileage (km)

V = Speed (km/hour)

Activity data carried out in transporting FFB from plantation to factory for the 15 selected routes can be seen in Table 3.

Table 3. FFB and Time Transportation Activities.

Activity	Amounts (route)	Times (minutes)
Traveling	15	634.20
Loading	15	553.95
Unloading	15	27.49
Weighting	15	30.0
Total		1245.64

From Table 3, we get the amount of time required, but for the calculation of the number of trucks is also affected with the system that is applied to the factory that is completed every 1 trip transporting FFB then the driver and helper are given 1 hour rest period. So for the calculation of the number of trucks performed calculations with assumption, as follows: [7,8,9]

1. With 4 truck units, the following results are obtained:

$$\text{Working hours} = \frac{\text{Total of Time (hours)}}{\text{Amount of Trucks}} + \left[\frac{\text{JAmount of rutes}}{\text{Amount of Trucks}}\right] [\text{resting time (hour)}] \ \ .... (2)$$

$$= \frac{1245{,}64/60}{4} + \left[\frac{15}{4}\right][1] = 8{,}94 \text{ hours}$$

Due to the time result obtained > working hours, then 4 trucks is not possible.

2. With the number of 5 units of trucks obtained the following results:

$$\text{Working hours } = \frac{\text{Total of Time (hours)}}{\text{Amount of Trucks}} + \left[\frac{\text{JAmount of rutes}}{\text{Amount of Trucks}}\right][\text{resting time (hour)}] \text{........(3)}$$

$$= \frac{1245{,}64/60}{5} + \left[\frac{15}{5}\right][1] = 7.15 \text{ hours}$$

Due to the time obtained < working hours, 5 trucks can be transported.

3.4. Allocation of Trucks

For the allocation of trucks, the carrier is within the authority of the transportation foreman, by following the harvest data obtained from the plantation division. The gardens are divided into 2 harvest areas where the goal is to arrange the harvesting schedule once every 2 weeks. Then, the harvested area will be harvested for the next 2 weeks. In the previous chapter has been calculated the number of trucks required by the time of transportation that is a number of 5 units. As for the number of tours that will

be done has been set in the area of harvest I as much as 17 tours and for the harvest area II as many as 15 tours. So for the allocation of trucks can be seen in Table 4.

Table 4. Allocation of truck area of harvest I.

TOUR	TRUCK NO	ROUTE	TOUR	TRUCK NO	ROUTE
1	DT 1	PP0-PP12-PP8-PP0	9	DT 3	PP0-PP5-PP6-PP0
2	DT 2	PP0-PP8-PP13-PP0	10	DT 5	PP0-PP2-PP1-PP0
3	DT 3	PP0-PP13-PP15-PP0	11	DT 4	PP0-PP1-PP3-PP0
4	DT 4	PP0-PP16-PP17-PP0	12	DT 5	PP0-PP20-PP21-PP0
5	DT 5	PP0-PP17-PP14-PP11-PP0	13	DT 2	PP0-PP21-PP22-PP0
6	DT 2	PP0-PP11-PP10-PP0	14	DT 3	PP0-PP22-PP23-PP26-PP0
7	DT 1	PP0-PP10-PP9-PP7-PP0	15	DT 1	PP0-PP26-PP25-PP24-PP0
8	DT 4	PP0-PP7-PP4-PP5-PP0			

FFB TRANSPORTING ROUTE IN THE HARVEST AREA I (NEAREST NEIGHBOUR)

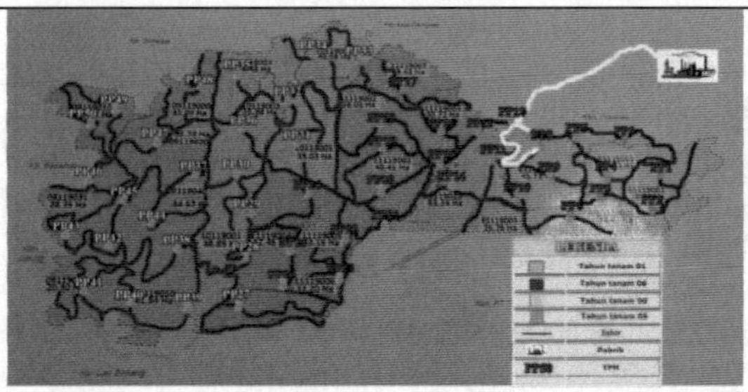

Figure 1. The Line of Harvesting Area 1.

FFB TRANSPORTING ROUTE IN THE HARVEST AREA II *(CLARKE &WRIGHT SAVINGS)*

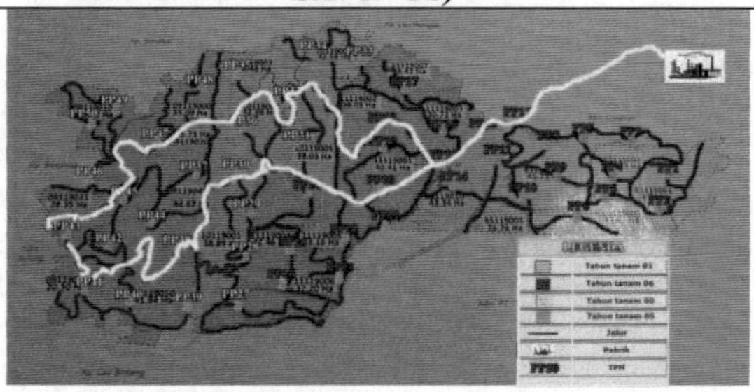

Figure 2. The Line of Harvesting Area II.

For the second harvest area, PRC PP27-PP50 gets the next harvest turn after harvest area I, with the allocation to the 5 planned trucks which can be seen in Table 5.

Table 5. Allocation of truck area of harvest II.

TOUR	TRUCK NO	ROUTE	TOUR	TRUCK NO	ROUTE
1	DT 5	PP0-PP41-PP43-PP0	9	DT 3	PP0-PP29-PP30-PP37-PP0
2	DT 3	PP0-PP50-PP49-PP0	10	DT 5	PP0-PP37-PP47-PP0
3	DT 2	PP0-PP49-PP40-PP42-PP0	11	DT 4	PP0-PP47-PP48-PP0
4	DT 4	PP0-PP42-PP45-PP46-PP0	12	DT 3	PP0-PP36-PP35-PP0
5	DT 1	PP0-PP46-PP39-PP0	13	DT 1	PP0-PP35-PP32-PP0
6	DT 2	PP0-PP39-PP27-PP0	14	DT 2	PP0-PP32-PP31-PP34-PP0
7	DT 1	PP0-PP38-PP44-PP0	15	DT 5	PP0-PP34-PP33-PP0
8	DT 4	PP0-PP28-PP29-PP0			

3.5. Calculation of Total Fuel Usage

In the calculation of the previous FFB transport route, it is known that the chosen route is based on the smallest mileage number which can be seen in Table 6 [10,11].

Table 6. Selected distance data.

Harvesting Area	Choose Route	Mileage (Km)
Harvesting Area I	*Nearest Neighbour*	200.78
Harvesting Area II	*Clarke & wright Saving*	264.33
Total		465.11

Based on the clearing result done by the factory, it is found that the carrying capacity to the fuel usage in the ratio of 1: 2,50 with 1 Liter definition can travel 2,50 Km. So for the calculation of total fuel is calculated as follows: [12, 13]

$$\text{Total Fuel} = \text{Total Distance x Trucking Capacity} \quad (4)$$
$$= 465.11 \text{ x } \frac{1}{2{,}50}$$
$$= 186.05 \text{ Liters}$$

So when compared to actual fuel use with a ratio of 1: 2.48, then obtained efficiency improvements for:

$$\frac{2{,}50-2{,}48}{2{,}48} X\ 100\% = 0{,}8\% \quad (5)$$

IOP Conf. Series: Journal of Physics: Conf. Series **1007** (2018) 012028 doi:10.1088/1742-6596/1007/1/012028

4. Conclusion

The conclusions obtained based on the results of data processing and analysis that has been done is in the area of harvest I Nearest Neighbours method has less mileage compared to Clarke & Wright Savings method. The total distance of Nearest Neighbours method and Clarke & Wright Savings method are 200,78 Km and 214.09 Km respectively. As for the area of harvest II obtained different results, with Nearest Neighbours method obtained more mileage compared with Clarke & Wright Savings method. The total distance of Nearest Neighbor method and Clarke & Wright Savings method are 264.37 Km and 264.33 Km respectively.

The number of FFB transport tours that formed both methods in the first harvest area with Clarke & Wright Savings and Nearest Neighbours method were 18 tours and 17 tours respectively, while in the harvest area II were 15 tours in both methods. In accordance with the time calculation carried out all the activities of transporting FFB, recommend the number of trucks used is reduced from 8 units to 5 units. The average fuel usage ratio of 1: 2.5 makes an increase of 1: 2.48. Thus, the improvement of fuel consumption refers to the achievement of the standard sets by the company, with an efficiency improvement of 0.8%.

References

[1] M Hudori 2017 *Penentuan Rute Angkutan Tandan Buah Segar Kelapa Sawit yang Optimal dengan Metode Saving Matrix* Bekasi

[2] M Rafael 2016 *Heuristic algorithms and Variants of the Vehicle Routing Problem for a Distribution Company: A Case Study* Universidade Nova de Lisboa

[3] I N Pujawan 2005 *Supply Chain Management* Surabaya: Guna Widya

[4] N Medi 2017 *Analisis Kinerja Operasional (Waktu Tempuh, Kecepatan Perjalanan, Dan Load Factor) Angkutan Umum (Pete-Pete) Kota Kendari Pada Trayek R01 Dan R02* Kendari: Halu Oleo University

[5] C Abadi 2014 *Penentuan Rute Kendaraan Distribusi Produk Roti Menggunakan Mentode Nearest Neighbour dan Metode Sequential Insertion* Bandung : Itenas

[6] D J Bowersox 1996 *Manajemen Logistik* Jakarta: Bumi Aksara

[7] R H Ballou 1999 *Business Logistic Management* New Jersey: Prentice Hall, Inc

[8] S Sinulingga 2015 *Metode Penelitian* Medan:USU Press **3**

[9] A V Prasmoro 2014 *Optimasi Produksi Dump Truck Volvo Fm 440 Dengan Metode Kapasitas Produksi Dan Teori Antrian Di Lokasi Pertambangan Batubara* Samarinda: Universitas Mercu Buana

[10] P Kucera 2012 *Different Versions Of The Savings Method For The Time Limited Vehicle Routing Problem* Acta Universitatis Agriculturae Et Silviculturae Mendelianae Brunensis

[11] A K Huda 2014 *Penentuan Rute Distribusi Untuk Meminimalkan Biaya Transportasi Menggunakan Metode Saving Matrix, Nearest Insert Dan Nearest Neighbour di PT XYZ.* Surakarta: Universitas Muhammadiyah

[12] E Volna and M Kotyrba 2016 *Unconventional Heuristics for Vehicle Routing Problems* Ostrava: University of Ostrava

[13] J Jairo 2017 *A Heuristic Algorithm Based On Tabu Search For Vehicle Routing Problems With Backhauls* Columbia: Pontificia Universidad Javeriana Cali

[14] K A Wahyu 2016 *Pengaruh Tingkat Kerusakan Jalan Perkebunan dan Posisi Tandan Buah Segar di Bak Truk Terhadap Kinerja Angkutan Kelapa Sawit* Yogyakarta: Universitas Gadjah Mada

[15] I S Kurniawan 2014 Usulan Rute Pendistribusian Air Mineral dalam Kemasan Menggunakan Metode Nearest Neighbour dan *Clarke & Wright Savings* Bandung: Institut Teknologi Nasional

IOP Conf. Series: Journal of Physics: Conf. Series **1007** (2018) 012029 doi:10.1088/1742-6596/1007/1/012029

Performance analysis of Supply Chain Management with Supply Chain Operation reference model

Abdurrozzaq Hasibuan*, Mahrani Arfah, Luthfi Parinduri, Tri Hernawati, Suliawati, Bonar Harahap, Siti Rahmah Sibuea, Oris Krianto Sulaiman, Adi purwadi

Faculty of Engineering, Universitas Islam Sumatera Utara, Medan, Indonesia

E-mails: *rozzaq@uisu.ac.id, mahrani.arfah@ft.uisu.ac.id, luthfi@ft.uisu.ac.id, trihernawati@ft.uisu.ac.id, suliawati@ft.uisu.ac.id, bonarhrp@ft.uisu.ac.id, sitirahmah@ft.uisu.ac.id, oris.ks@ft.uisu.ac.id, adiurwa@icloud.com

Abstract. This research was conducted at PT. Shamrock Manufacturing Corpora, the company is required to think creatively to implement competition strategy by producing goods/services that are more qualified, cheaper. Therefore, it is necessary to measure the performance of Supply Chain Management in order to improve the competitiveness. Therefore, the company is required to optimize its production output to meet the export quality standard. This research begins with the creation of initial dimensions based on Supply Chain Management process, ie Plan, Source, Make, Delivery, and Return with hierarchy based on Supply Chain Reference Operation that is Reliability, Responsiveness, Agility, Cost, and Asset. Key Performance Indicator identification becomes a benchmark in performance measurement whereas Snorm De Boer normalization serves to equalize Key Performance Indicator value. Analiytical Hierarchy Process is done to assist in determining priority criteria. Measurement of Supply Chain Management performance at PT. Shamrock Manufacturing Corpora produces SC. Responsiveness (0.649) has higher weight (priority) than other alternatives. The result of performance analysis using Supply Chain Reference Operation model of Supply Chain Management performance at PT. Shamrock Manufacturing Corpora looks good because its monitoring system between 50-100 is good.

1. Introduction

Higher consumer needs, and more intelligent in choosing their needs. Starting from the middle to upper class always demand the best quality and economical price. The Company will make every effort to improve productivity, efficiency, prompt service, and continue to create new innovations to stay ahead and survive in the marketplace. In addition to productivity and efficiency that need to be improved, the company must also understand and know what is needed by consumers. That the importance of the role of all parties ranging from suppliers, manufacturers, distributors, retailers, and customers in creating a cheap, quality, and fast products is what then gave birth to a new concept of Supply Chain Management (SCM).

IOP Conf. Series: Journal of Physics: Conf. Series **1007** (2018) 012029 doi:10.1088/1742-6596/1007/1/012029

Supply chain is a network of companies that work together to create and deliver a product into the hands of end users. These companies usually include suppliers, manufacturers, distributors, or retailers, as well as support companies such as logistics services companies. Measurement of suppply chain performance is a performance measurement system that aims to help monitor the running of SCM application to run well. Therefore, the performance indicators used are more specific and relatively different from the organizational performance measurement system. This system is more integrative with the work area that includes suppliers, factories, and distributors that aims to achieve successful supply chain implementation. PT. Shamrock Manufacturing Corpora (PT SMC), continuously strives to optimize glove production to meet export quality standards. One of the efforts that can be done is to conduct performance measurement analysis of SCM.

2. Theoretical basis

2.1. Understanding Supply Chain and SCM

Supply chain is a network of companies that work together to create and deliver a product into the hands of the end user. These companies usually include suppliers, manufacturers, distributors, stores, or retailers, as well as support companies such as logistics services. Supply chain is the physical network, the companies involved in supplying raw materials, producing goods, or sending to end users, SCM is a method, tool, or management approach.

Supply chain performance measurement is a performance measurement system that aims to help monitor the running of SCM applications to run properly. Therefore, the performance indicators used are more specific and relatively different from the organizational performance measurement system. This system is more integrative with the work area that includes suppliers, factories, and distributors that aims to achieve successful supply chain implementation.

2.2. Supply Chain Operation Reference (SCOR)

One way to measure supply chain performance is to use the SCOR method. This method was introduced by the Supply Chain Council (SCC) as a model of supply chain performance measurement across industries. The SCOR model is a process reference model for supply chain operations developed by SCC, Pittsburgh, PA. SCOR splits supply chain processes into five processes including Plan (planning process), Source (procurement process), Make (production process), Deliver (delivery process), and Return (return process).

2.3. Normalization Process

There are various ways of measuring the performance ever performed by the company. That the performance fulfillment rate is defined by the normalization of the performance indicator. Each indicator has different weights with different size scales. Therefore, it is necessary to process the equation of parameters that is by way of normalization. Here normalization plays a significant role in achieving the final value of performance measurement. The normalization process is done by Snorm De Boer normalization formula, namely:

$$S(norm) = \frac{(Si-Smin)}{(Smax-Smin)} \; x\, 100 \quad \textbf{or} \quad \frac{(Si-Smin)}{(Smax-Smin)} = \frac{skor-0}{100-0} \qquad (1)$$

Where :

Si	=	Indicator actual achieved.
Smin	=	The value of achieving the worst performance of the performance indicator.
Smax	=	Best performance achievement value of performance indicator.

In this measurement, each indicator weight is converted to a certain value interval of 0-100. Zero (0) is interpreted at its worst and one hundred (100) is best interpreted. Thus the parameters of each indicator are the same, after that obtained a result that can be analyzed. The table shows the monitoring system of performance indicators.

Table 1. Monitoring system of work indicators.

Monitoring System	Work Indicators
< 40	Poor
40 – 50	Marginal
50 – 70	Average
70 – 90	Good
> 90	Exelent

2.4. Analytical Hierarchy Process (AHP)

Analytical Hierarchy Process, hereinafter called AHP, is a decision support model developed by Thomas L. Saaty. This decision support model will describe complex multi-factor or multi-criteria problems into a hierarchy. AHP has an advantage because it can combine objective and subjective elements of a problem.

Below is an example of pairwise matrix matrix using A1, A2, A3, ... An.

C	A1	A2	A3	...	An
A1	a11	a12	a13	...	An
A2	a21	a22	a23	...	An
A3	a31	a32	a33	...	An
...	...	...	...	...	...
An	an1	an2	an3	...	Ann

Figure 1. Matched comparison matrices.

To begin this pairwise comparison process, start at the top of the hierarchy to select the C criteria, or properties, to be used for the first comparison. Then from the level just below it, take the elements to be compared: A1, A2, A3, and so on. In this matrix, compare the elements A1 in the column to the left with elements A1, A2, A3, and so on in the upper row with respect to the C property in the upper left corner. Then repeat with column element A2 and so on. The goal is to get a single value that represents a respondent. The average geometry formula is as follows:

IOP Conf. Series: Journal of Physics: Conf. Series **1007** (2018) 012029 doi:10.1088/1742-6596/1007/1/012029

$$G = \sqrt[n]{X_1 . X_2 . X_3 \dots Xn} \tag{2}$$

Where :

G	=	Geometry average
Xn	=	Assessment to 1,2,3 ... n
n	=	Number of ratings

The notion of consistency is a type of measurement that can not just happen or have certain conditions. The formula of consistency index (CI / Consistency Index) is:

$$CI = \frac{(\lambda\, maks - n\,)}{(n-1)} \tag{3}$$

Where :

CI	=	Consistency Index
λ	=	Eigenvalue
n	=	The size of the matrix.

The maximum eigenvalue of a matrix will not be less than n so that there can be no negative CI value. The formula of consistency ratio (CR/Consistency Ratio) is written as follows:

$$CR = \frac{CI}{RI} \tag{4}$$

Where :

CR	=	Consistency Ratio
CI	=	Consistency Index
RI	=	Random Index

If CR is greater than 0.10 it means that there is a 10% chance that each element is not worth comparatively. In this case, the decision maker should review the comparative process that has been done.

Table 2. Random cosistency index.

N	1	2	3	4	5	6	7	8	9	10
RI	0	0	0,52	0,89	1,11	1,25	1,35	1,4	1,45	1,49

3. Research Methodology

Research methodology is a problem-solving process used to solve emerging problems, based on the background and objectives to be achieved by using supporting theories in problem solving, and collecting data either through literature or through field studies, processing data until conclusion studied. in the application of the preferred AHP method is the quality of the respondent's data, and does not depend on the quantity. Therefore, AHP assessment requires experts as respondents in decision making in alternative selection. The experts here are competent people who really master, influence the decision-making or really know the information needed. For the number of respondents in the AHP method does not have a certain formulation, but there is only a minimum limit of 2 respondents.

The SCOR method is a supply chain language, which can be used in a variety of contexts to design, describe, and reconfigure various types of commercial/business activities. Normalization plays a significant role in achieving the final value of performance measurement. The normalization process is done by the normalization formula of Snorm De Boer.

4. Data Processing

4.1. Weighting with AHP

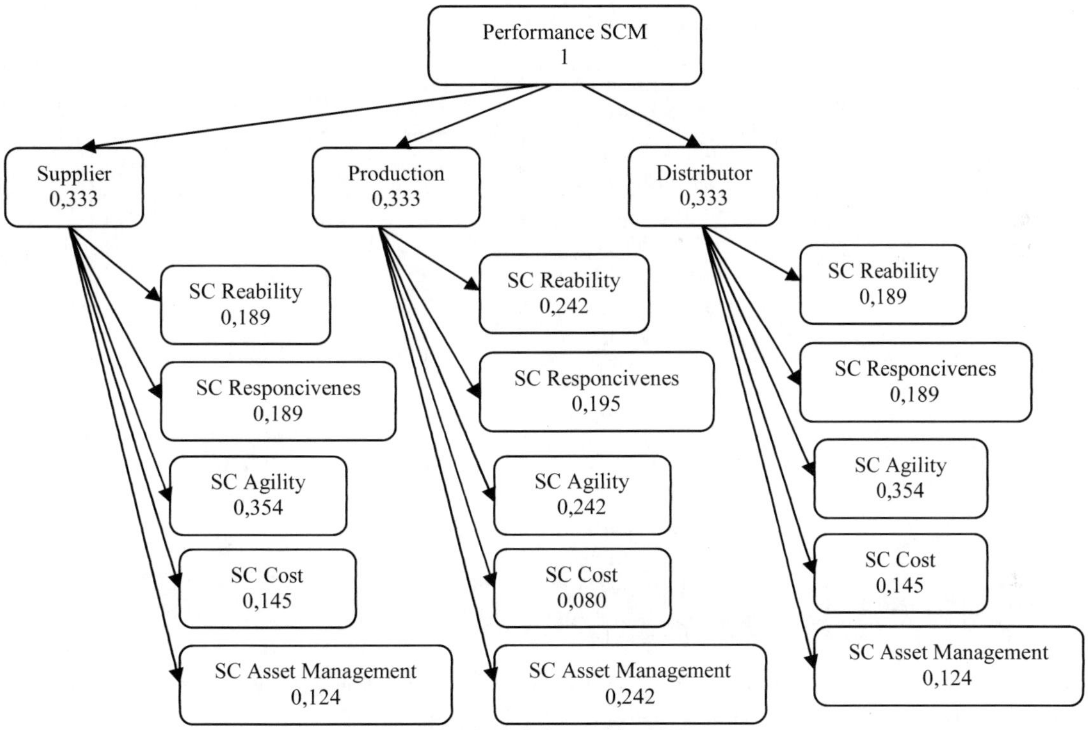

Figure 2. Weighting with AHP

Table 3. CR value (Consistency Ratio)

	CR	Information
Criteria	0	Consistency
Supplier	0,07	Consistency
Production	0,08	Consistency
Distributor	0,08	Consistency

4.2. Performance Analysis based on SCOR

Table 4. Normalization Value.

Process core Level 1	**Dimention Level 2**	**Key Performance Indicator Level 3**	**Si**	**Smax**	**Score**	**Monitoring**
Plan	Reability	Meeting with suppliers	1	2	50	Average
	Responsiveness	Percentage of raw material needs with production target	15 tons	30 tons	50	Average
	Asset	Cash to cash cycle time	3 months	3 months	100	Excellent
Source	Reability	Fulfillment of raw materials	98860,7 Kg	100.000 Kg	98,86	Excellent
	Responsiveness	Lead time of raw materials	4 days	4 days	100	Excellent
	Agility	Supplier availability	9	9	100	Excellent
Make	Reability	Achieving production targets	1.848.000 pieces	2.520.0000 pieces	73,3	Good
	Responsiveness	Time of manufacture of products to achieve production targets	24 hours	24 hours	100	Excellent
	Asset	Type of mold	5	5	100	Excellent
Delivery	Reability	Level of fulfillment of finished product	4.800.000 pieces	9.000.000 pieces	53,3	Average
	Responsiveness	Timely delivery of products	1 weeks	1 weeks	100	Excellent
Return	Reability	The level of customer complaints	-	-	-	-

From the table can be seen that the performance of supply chain management in PT. Shamrock Manufacturing Corpora looks good because of the monitoring system is between 50 – 100 which means good.

5. Analysis and Evaluation

5.1 Analysis

The results of the performance measurement by weighting the AHP can be concluded SC.Responsiveness (0.649) has higher weight (priority) than other alternatives. This can be seen in Figure 3. the result of the AHP weighting. The results of questionnaires filled by respondents prioritize the speed of the supply chain in providing products for consumers (SC Responsivensss).

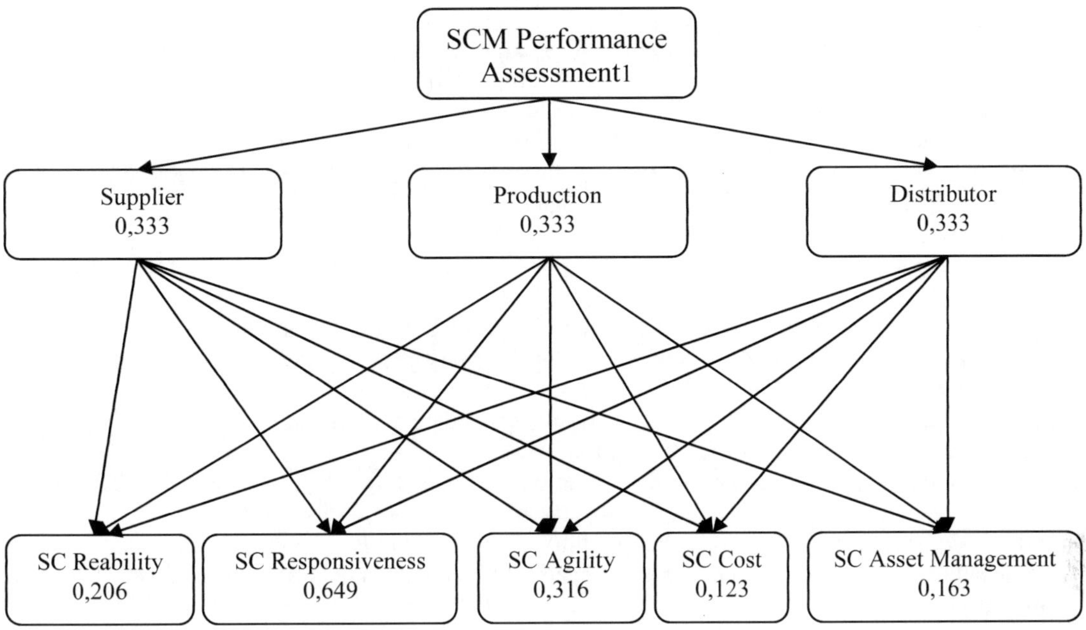

Figure 3. Weight of AHP.

5.2. Evaluation

Identify the products produced per day to see the ability of the company in transforming raw materials into products, production target PT.Shamrock Manufacturing Corpora is 2,000,000 pieces but on three times the production target did not reach 2.000.000 pieces that can be seen in table 5.

Table 5. Recapitulation of Unreachable Total Production

Date	Total Production (Pieces)	Target (Pieces)
18-Juli-2017	1.936.000	2.000.000
24-Juli-2017	1.848.000	2.000.000
29-Juli-2017	1.980.000	2.000.000

The company's ability to achieve production target is still not 100% because on 18,24 and 29 have not reached production target, based on table 5 above. Equalization of the value scale with the normalization process of Snorm De Boer, shows that the value of the received score is at the lowest value of 50. Which is the ability of the company is in the medium value (regular).

6. Conclusions

1) Weighting with AHP shows CR value of criteria and subcriteria below or <0.1 means consistent CR value, ie criteria with CR 0 value, supplier with CR value 0.07, production with CR value 0.08, distributor with value CR 0.08.
2) The results of the performance measurement by weighting the AHP can be concluded SC. Responsiveness (0.649) has higher weight (priority) than other alternatives.
3) The result of performance analysis using SCOR model of SCM performance in PT. SMC looks good because of the monitoring system is between 50-100 which means good.

IOP Conf. Series: Journal of Physics: Conf. Series **1007** (2018) 012029 doi:10.1088/1742-6596/1007/1/012029

References

[1] Apte, P.R, 5 Day Course on Taguchi Method for Quality cost Optimization, http :/www.tifr.res.in/-apte/CU_PRA_TAGUCHI. htm/file : TAGUCHI Method – Abstract and course contents (10 Oktober 2000)
[2] Amrullah, Hasbi Amar. 2011. Pengukuran Performansi Supply Chain Dengan Menggunakan Metode SCOR (Supply Chain Operation Reference) dan AHP (Analytical Hierarchy Process) untuk meningkatkan kinerja perusahaan. Fakultas Teknologi Industri : Universitas Islam Indonesia
[3] David, Simchi Levi, 2007, Designing and Managing Supply Chain, McGraw Hill Book Company, New York
[4] Frankell, Per Axelsson (2014). Performance Measurement System of Warehouse Based on SCOR Model, International Journal of Integrated Supply Management (IJISM), Volume 9, Issue 1, Pages 56-72, January 2014
[5] Heizer, Jay, dan Render, Barry, 2014, Operations Management : Sustainability and Supply Chain Management, 11th Edition,Pearson Education, Inc
[6] Jameh, Shooran, (2015). Assessing Supply Chain Performance through Applying the SCOR Model, Journal of Volcanology and Geothermal Research 98, ISSN:127–15
[7] Latifah Dinar Wigaringtyas, 2013, Pengukuran Kinerja Supply Chain Management Dengan Pendekatan Supply Chain Operation Reference (SCOR), Universitas Muhammadiyah Surakarta, Surakarta
[8] Manahan P. Tampubolon, 2014, Manajemen Operasi dan Rantai Pasok, Mitra Wacana Media
[9] Pujawan, I Nyoman dan ER, Mahendrawati. 2010. Supply Chain Management. Penerbit Gunawidya : Surabaya
[10] Perdana, Y. R., 2014. Perbaiakan Kinerja dengan Pendekatan Supply Chain Operation Reference (SCOR) dan analytical hierarchy process (AHP). Yogyakarta
[11] Reni Rahmayanti, 2010, Pemilihan Supplier Menggunakan Metode Analytical Hierarchy Process (AHP), Universitas Sebelas Maret, Surakarta
[12] Riko Ervil, Patdono Suwignjo, Ahmad Rusdiansyah, 2010. Pengembangan Model Pengukuran Kinerja Supply Chain berbasis Balanced Scorecard (Studi Kasus PT. XYZ). Surabaya
[13] Rosnani Ginting, 2015, Kuesioner, Medan, Indonesia
[14] Thomas L.Saaty, 1993, Pengambilan Keputusan Bagi Para Pemimpin, PT Pustaka Binaman Pressindo
[15] Williem Sahaya, 2016, Sukses Supply Chain Management Akses Demand Chain Management, Penerbit IN MEDIA
[16] Zakiya Muallifa Rahman, 2016, Pengukuran Kinerja Supply Chain Pada PT.Mataram Tunggal Garment, Universitas Islam Negeri Sunan Kalijaga, Yogyakarta
[17] https://bushido02.wordpress.com/2010/06/08/konsep-supply-chain/. Diakses pada 22 Juli 2017
[18] http://ranazura.blogspot.co.id/p/supply-chain-management.html. Diakses pada 22 Juli 2017
[19] https://sriwijayanti.wordpress.com/prinsip-dasar-supply-chain-management/Diakses pada 22 Juli 2017
[20] https://informatika.web.id/langkah-langkah-metode-ahp.htm. Dikases pada 22 Juli 2017

A Framework for Text Mining in Scientometric Study: A Case Study in Biomedicine Publications

V M M Silalahi, R Hardiyati, I M Nadhiroh, T Handayani, R Rahmaida, and M Amelia

Centre for S & T Development, Indonesian Institute of Scinces, Jl.Jend. Gatot Subroto 10, Jakarta, Indonesia

E-mail: mesnans@yahoo.com

Abstract. The data of Indonesians research publications in the domain of biomedicine has been collected to be text mined for the purpose of a scientometric study. The goal is to build a predictive model that provides a classification of research publications on the potency for downstreaming. The model is based on the drug development processes adapted from the literatures. An effort is described to build the conceptual model and the development of a corpus on the research publications in the domain of Indonesian biomedicine. Then an investigation is conducted relating to the problems associated with building a corpus and validating the model. Based on our experience, a framework is proposed to manage the scientometric study based on text mining. Our method shows the effectiveness of conducting a scientometric study based on text mining in order to get a valid classification model. This valid model is mainly supported by the iterative and close interactions with the domain experts starting from identifying the issues, building a conceptual model, to the labelling, validation and results interpretation.

1. Introduction

The data of Indonesians scholarly research publications in the domain of biomedicine has grown in significant quantities. This amount of data can serve as a resource for an analysis to get a picture of the development of science and technology in the domain of biomedicine in Indonesia. The efforts to analyse biomedicine scientific publications in Indonesia have been conducted in the form of bibliometric studies as well as co-authorship network analysis. However, the rich information stored in the full text form has so far not been explored. For this purpose, a scientometric study was conducted through text mining the research publications. One of the goal in the scientometric study is to get a view of how much of the research output that lead to commercialization. To this end a predictive model is built that provides a classification of documents as a proxy to the potency for downstreaming of the research outputs in the direction to commercialization. In addition a framework is proposed which is suited in conducting a scientometric study based on text mining. An investigation is conducted relating to the problems associated with building a corpus and the effort to produce a valid training data through labelling of the document. Most of the existing frameworks in text mining focused on the technicality and management whereby in our framework we put the focus on the close interaction with the domain expert. In general, the issues in biomedical text processing is the problems with large volume, high dimension, and heterogeneous data sources. Moreover there is also issue related with inconsistent and incomplete data, as well as with data which is rich in context. This is a challenge in the field of biomedical informatics given the data existence such as unbalanced data sets, structured or weakly structured data, noisy and ambiguous labelling [6]. For the purpose mentioned above, in this paper, an effort is described to build a special corpus of research data in the domain of Indonesian biomedicine, with the primary objective of supporting research in the field of scientometrics in Indonesia. The corpus is constructed with a good design in mind by investigating some common conditions such as: the representation of the collected sample, the homogeneity according to the selection criteria, and the balance of the class, as well as the careful documents labelling in order to avoid ambiguous results.

2. Literature Review

According to [10] scientometrics can be defined as the application of quantitative methods related to the analysis of science in the context of information dissemination. Scientometrics can also be defined as a quantitative study of science, communication in science, and science policy [5]. The definition of scientometrics research is also given by [14] as a quantitative study of science and technology. Scientometrics is associated with bibliometrics and has partially the same area as bibliometrics [12]. As far as the authors know, there is no text mining based scientometric research has been found on biomedicine which focuses on the condition of Indonesia as a mega biodiversity country. Meanwhile, some researchers, among others, [8] used a co-word approach to map the results of research produced by Biology Research Centre, Indonesian Institute of Sciences. Ref. [9] conducted a scientometric study using social network analysis methods to investigate central actors, institutions involved, and popular topics. Meanwhile Ref. [2] conducted a foresight study using text mining methods to investigate the many tasks involved in the process. The process involves a qualitative method later in the process after the quantitative method. The case study for the foresight study was conducted in the domain of nanotechnology in Brazil.

3. Framework

The classification model is built within a five step processes starting from conceptual modelling, data searching and selection, data cleaning and pre-processing, models building and selection, and finally evaluation and results interpretation. Fig 1. Shows the text data analytics process for the purpose of a scientometric study.

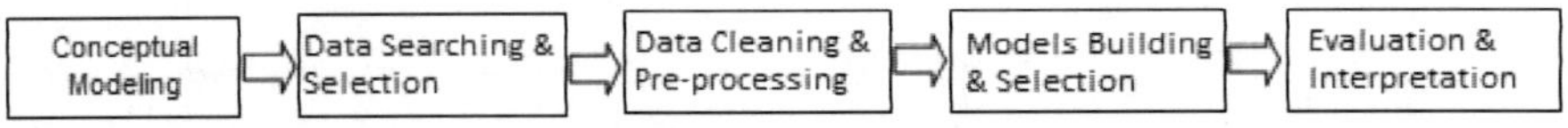

Figure 1. Text mining process for the scientometric study.

There are frameworks proposed for research study based on text mining. However most of the proposed frameworks have a focus on the technicality in a text mining process. The CRISP-DM is a generic model for the data mining based project. It can be used as a generic check-lists useful for planning the project [15]. Some are very specific in purpose such as in [11]. This paper proposes a framework in doing a text mining in a scientometric study which has a focus on the role of the domain experts. Fig. 2 shows a framework built to perform text-based scientometric study. For this purpose a conceptual model of document classification is constructed in accordance with the objective of the study. The objective is to classify document according to drug design process as a proxy to show the potency to commercialization. The basis used is the theoretical approach or best practices contained in official documents that can be used as a reference. For the purpose of building this conceptual model, domain experts in biomedicine are involved. In addition to this, the roles of a domain expert range from validating the conceptual model, labelling the documents, and interpreting results. The conceptual model building activity is supported by a literature study in which key features or predictory terms identification is conducted to support the conceptual model. The corpus development includes the searching, selection and the pre-processing of the document. Furthermore, in the proposed framework there are also activities carried out such as named entity mining, relation mining and event mining for broader purposes in scientometric study of. A named entity (NE) is a phrase or combination of phrases in a document that indicates a particular object or group of objects, such genes, proteins, cells, drugs, chemicals, and diseases in the biomedical literature. An attempts to obtain data on this NE will provide links into descriptions to semantic layers e.g. by connecting annotations from NE in the text with ontologies to gain structured knowledge.

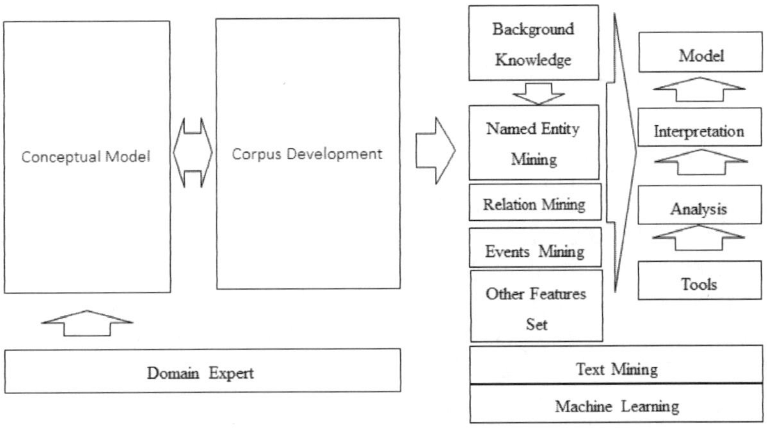

Figure 2. A framework for text mining based scientometric study.

3.1. Conceptual Modeling

Conceptual model constructs are built with reference to the existing official model such as by the FDA [3] and then through the synthesis of some of the models present in the literature such as in [7]. In an effort to develop a conceptual model suitable for the research question in this scientometric study, there will be several in-depth interviews with domain experts and resource persons from research centres, industry as well as from academics (universities). In-depth interviews are also conducted to obtain facts in the practice of promoting biomedicine research results to the industry for its downstreaming purposes, as well as to get an overview of the issues, problems faced and how the solutions are taken to go downstream. From the data analysis perspective, the conceptual model is built by taking into account the amount of publication data that can be collected, the estimated size of the distribution of documents or publications according to the classification model, and to a more detailed level with regard to mentioned predictory terms in the text. The predictory terms obtained were analysed to get input in constructing a conceptual model and then grouped in a list of terms that became predictors for the purpose of the initial classification. The list of terms obtained is then used as the material in the discussion with the domain expert in an in-depth interview. Table 1 shows a sample from the results of the features identification for the classification model.

Table 1. Sample from the result of feature selection for the classification model.

	Stage in new medicine process discovery	Stage in new medicine process discovery
1	Basic Research	potential compound; inhibition; activation
2	Drug Discovery	inhibition; activation; therapeutic effect; promising compound; lead compound; sequence homology studies; DNA microarray; structural genomics; high-throughput screening

3	Pre-Clinical studies	antioxidant; assay; anti-proliferation effect; cytotoxic effect; cultured cell; efficacy, bioactive protein fraction; antithrombotic; thrombolytic agents; toxicity study; dose; mice; male; female; fasted; observation period; major organs; rat; male; female; adult; body weight; pregnant; prenatal developmental study; efficacy; safety; absorption; distribution; metabolism; excretion; bioequivalence; bioavailability; dosage level
4	Clinical Trials	effectiveness; tumor necrosis factor-α (TNF-α) levels; dengue hemorrhagic fever (DHF); double-blind; randomized; placebo-controlled trial; patients; human, treatment group; placebo group; significant; volunteers; safety; efficacy

The initial conceptual model is then used to label the collected documents which will then be validated by the expert. The results of this analysis form the basis for building a computational model by text mining method. After processing the data analysis, the next step is to confirm the results that has been obtained through Focus Group Discussion (FGD). In the FGD, domain experts are expected to build consensus of the results and possibly refining the labelling. The results of these discussions are expected to support in better research policy recommendations in the future.

3.2. Data Searching and Selection

The data used in this study is a collection of international scientific publications from Scopus, Science Direct, DBLP, PubMed Central, Google Scholar as well as other accessible online data sources. The process of collecting data is done by using search keywords that are built in accordance with the research goals. The data of international scientific publications collected are data derived from authors with Indonesian affiliates up to 2017. Due to limited amount of data collected from qualified publication sources, time limitation and other resources, then downloading text from websites and Google Scholar proved to be a viable solution. The search also investigates the research sites and academic institution sites in Indonesia. Some university websites found to be very helpful, because it contains a list of Indonesian biomedical journal. The biomedical documents collected for the corpus are full-text papers in the form of journals, workshop reports, or conference proceedings. Most of them are text in pdf format which is then converted into plain text format encoded in UTF8 format.

Data collection is done in several stages. The first step is to determine the search terms used in collecting data on the basis of the research questions. The search terms used to capture biomedicine-related articles are such as biomedicine, drugs, and herbal medicine, medical biology combined with the word "Indonesia". The result of the searches through the Scopus database is 209 articles which is still quite small. Therefore, the addition of data related to Indonesia biomedicine articles is captured through Google Scholar, university repository, and the like. This addition generated a database of 662 articles. The steps used in the data selection are as follows: the types of documents are only in the form of articles in English, articles are produced by Indonesian authors and articles are related to Indonesian biomedicine, which uses Indonesian resources. Data selection starts from the title, abstract, and then the full-text of an article. In this step, all data is aggregated to be checked for duplication. Based on the results of data integration, we obtained data of Indonesian international scientific publications in biomedicine as much as 543 articles.

3.3. Data Cleaning and Data Preprocessing

Text conversion is largely based on the conversion from pdf format to plain text format. Documents are collected either in pdf image format or pdf text but mostly in pdf text format. To meet the needs with common format documents in the annotation and analysis process, the format will be a plain text. Therefore, two types of document converter tools are used: pdf-to-txt converter and OCR tool which is available online. In preparing our corpus we emphasize the importance to anticipate the process of text annotation. To this end we plan to build a specialized purpose web application to ensure the effectiveness and efficiency of the process. The category of semantic annotations with an identifier of concepts and semantic categories derived from ontology fall outside this study. In building the corpus our objective is to consider that the corpus should be useful for the broader research community and other scientometric related research through the development of practical and useful text mining resources. For the first purpose, four characteristics are considered in the corpus i.e. to make them easy to use through a good documentation, a balance in representation, and the availability of data on inter-annotator agreements [1]. Two major issues arise as common problems are the large feature dimensions and the small sample sizes. To overcome this problem pre-processing is carried out which includes tokenization, word frequency filtering, stop-words removal and stemming. Then a feature selection is designed that includes the use of the TF/IDF method as well as knowledge engineering using domain experts. The reduced number of features will bring benefits in speeding up the algorithms, improving learning accuracy, and a better model understanding. Our goal is to have a highly correlated feature subset with the predictive classes, but not correlated with each other.

3.4. Models Building and Selection
Models building and selection were performed on the basis of lexical analysis based on the bag of words method to quantify a term using TF/IDF measurement. For the text classification some models of the supervised machine learning methods are investigated such as k-NN (Nearest Neighbours) which is a simple model suitable for small dimensional data to complex models such as SVM (Support Vector Machines). In this study, three classifiers are used: Naive Bayes classifier which is often used as a baseline study, a k-NN classifier, and a SVM classifier. The Bayesian approach was used as a basis in our study because of the simplicity and the speed of training. Bayesian classification performance is in fact the best in our overall metric evaluation. On the other hand, SVM produces the least accurate classification in our investigation. Previous research has shown that no single learning approach is clearly superior to all cases, and in fact, different learning algorithms often produce similar results [4] especially after careful optimization process. One factor that can have a major impact on the success of learning algorithms is the nature of the data used to characterize the tasks to be learned. If data fails to show the regularity of the exploited statistics of computer learning algorithms, then learning will fail.

3.5. Results Evaluation and Interpretation
The output from the document classification model is a useful resource for a scientometric study. Combining the results with the bibliometric analysis and a visual presentation tool such as VOSviewer will give many perspectives in the analysis. The results obtained from the development of the document classification model will be validated qualitatively using domain expert in a FGD session. For that purpose the data details of the features which is a variable in a classification is provided. Domain experts can provide interpretation of feature-based relationships and presentation-based classification results. The domain experts are put in the position of interpreting the results of a document classification model, whereby they should certainly trust the evidence they have learned about the texts and the features. At the same time they should cultivate open-mindedness about how the results of the model might challenge conventional thinking. Researcher's domain knowledge is also invaluable for interpreting and contextualizing results in the first place. For the quantitative evaluation of the model, a confusion matrix is used to evaluate a classifier's performance. Two popular performance measures are precision, which measures if a document is correctly classified, and recall, which measures how well the classifier detects documents in a given category. For the overall evaluation purpose we used the aggregated F-score as a single metric whereby the score achieved in the Naive Bayes model is 98% in classifying BR (Basic Research), 97% in classifying DD (Drug Design), 97.8% in classifying PC (Pre-Clinical), and 100% in classifying CT (Clinical Trial). Because of the limitation of our research, the reported results are only from text mining the abstract of the publications. Some studies show that articles in full-text form give different results to those in abstract form whereby large amounts of information or data are not found in abstracts, but only in the full-text

IOP Conf. Series: Journal of Physics: Conf. Series **1007** (2018) 012030 doi:10.1088/1742-6596/1007/1/012030

of f, or even in tables and captions [13]. For a broader purpose various other methods such as clustering and topic modelling can be applied to support a broader perspective in scientometrics and as a tool for evaluating and constructing interpretations of the results. A document clustering model is evaluated by looking at how the cluster model explains variation in the document-term matrix compared to treating the documents as one cluster. Re-clustering is possible as a results of not optimal labelling by the domain expert.

4. Conclusion

This study proposes a framework for doing text mining in a scientometric study. From the experience in developing a document classification model as a proxy to its downstreaming potential, a lot of deviations occurs in the process from the initial planning. This is due to a lack of a guidance that can be used in performing the various tasks required. Therefore a framework is needed to conduct a scientometric study based on text-mining. The experience shows that the development of the corpus requires considerable effort. Based on our experience and the success in developing the document classification model for the purpose of a scientometric study, this framework can guide for the effective conduct. This framework for doing research in scientometrics with text mining relies heavily on the feedback of the domain experts in constructing conceptual models, labelling, validating and interpreting the results. In order to reduce the dependency, large data is required to enable the machine to do the learning, but mostly the amount of data that can be collected is not large so that it becomes a limitation. The framework development will be further enhanced through case study research in the different domain of interests and also through exploring the text mining in the full-text form of document, and the use of semantically richer methods. The analysis used in this study will also provide new knowledge for scientometric research in Indonesia.

References

[1] Cohen B, Fox L, Ogren P and Hunter L 2005 Corpus Design for biomedical natural language processing, ACL-ISMB workshop on Linking Biological Literature Ontologies and Databases.

[2] De Miranda Santo M, Coelho G M, dos Santos D M, and Fellows Filho L 2006 Text mining as a valuable tool in foresight exercises: A study on nanotechnology. Technological Forecasting and Social Change, 73 8 1013-27

[3] FDA 2017 The Drug Development Process http://www.fda.gov/ForPatients/Approvals/Drugs/default.htm. Accesed on 11 January 2017

[4] Hall M A 1999 Correlation-based feature selection for machine learning

[5] Hess D J 1997 Science Studies: An advanced introduction. *New York: New York University Press*

[6] Huang S, Zhou J, Wang Z, Ling Q and Shen Y 2017 Biomedical informatics with optimization and machine learning

[7] Hughes J P, Rees S S, Kalindjian S B and Philpott K L 2011 Principles of early drug discovery *Br. J. Pharmacol.* 162 1239–49

[8] Lukman B 2012 Analisis Bibliometrika Berdasaran Pendekatan Co-words: Pemetaan Laporan Hasil Penelitian Pusat Penelitian Biologi. Bogor: Institut Pertanian Bogor

[9] Nadhiroh I, Amelia M, Handayani T, Hardiyati R and Simamora N 2015 Analisis Jaringan Co-authorship Publikasi Ilmiah Internasional Indonesia dalam Penguatan Iptek Masa Depan (Bidang Energi, Pangan, dan Kelautan) Jakarta: Pusat Penelitian Perkembangan Iptek

[10] Nalimov V V, Mulchenko Z M 1969 Naukometriya. Izuchenie Razvitiya Nauki kak Informatsionnogo Protsessa Scientometrics Study of the Development of Science as an Information Process, Nauka, Moscow, (English translation: 1971. Washington, D.C.: Foreign Technology Division. U.S. Air Force Systems Command Wright-Patterson AFB Ohio NTIS Report No.AD 735- 634

[11] Patumcharoenpol P, Doungpan N, Meechai A, Shen B, Chan J H and Vongsangnak W 2016 An integrated text mining framework for metabolic interaction network reconstruction *PeerJ*. **4** e1811
[12] Saini P K 2014 Application of Scientometric Analysis in Library Net Work: A Comprehensive Study. *International Journal of Emerging Research in Management & Technology* ISSN: 2278-935
[13] Shatkay H and Feldman R 2003 Mining the biomedical literature in the genomic era: an overview. *Journal of computational biology* **10** 6 821-855
[14] Van Raan A 1997 Scientometrics: State of The Art. *Scientometrics* 205-218
[15] Wirth R and Hipp J 2000 CRISP-DM: Towards a standard process model for data mining. In *Proceedings of the 4th international conference on the practical applications of knowledge discovery and data mining* 29-39

IOP Conf. Series: Journal of Physics: Conf. Series **1007** (2018) 012031 doi:10.1088/1742-6596/1007/1/012031

The robust design for improving crude palm oil quality in Indonesian Mill

Siti Maretia Benu[1]*, Sukaria Sinulingga[1], Nazaruddin Matondang[1], Irwan Budiman[2]

[1]Industrial Engineering Department, Faculty of Engineering, Universitas Sumatera Utara, Indonesia
[2]Industrial Engineering Department, Faculty of Technology and Computer Sciences, Universitas Prima Indonesia, Indonesia

E-mails: sitimaretia@gmail.com, sukaria_sinulingga@yahoo.com, nazaruddin_matondang@yahoo.com, irwanb01@gmail.com

Abstract. This research was conducted in palm oil mill in Sumatra Utara Province, Indonesia. Currently, the main product of this mill is Crude Palm Oil (CPO) and hasn't met the expected standard quality. CPO is the raw material for many fat derivative products. The generally stipulated quality criteria are dirt count, free fatty acid, and moisture of CPO. The aim of this study is to obtain the optimal setting for factor's affect the quality of CPO. The optimal setting will result in an improvement of product's quality. In this research, Experimental Design with Taguchi Method is used. Steps of this method are identified influence factors, select the orthogonal array, processed data using ANOVA test and signal to noise ratio, and confirmed the research using Quality Loss Function. The result of this study using Taguchi Method is to suggest to set fruit maturity at 75.4-86.9%, digester temperature at 95°C and press at 21 Ampere to reduce quality deviation until 42.42%.

1. Introduction

This research was conducted in palm oil mill in Sumatra Utara Province, Indonesia. Companies engaged in the processing of palm oil especially palm mill is growing rapidly. This development leads to tough competition on each mill [1,2]. Every company has to achieve standard quality so that the product's quality can meet the customer requirements [1,3,4]. The effort to achieve expected standard quality by using quality control. Quality of CPO is one of the main factors for a customer to make purchasing decisions especially standard stated in Indonesia National Standard SNI 01-2901-2006. The maximum quality requirements of CPO stated are dirt count, free fatty acid, and moisture at 0.5%, and Iodine number at 55 grams Iodine per 100 grams [5].

Several researchers researching about the improvement of CPO's quality through Six Sigma, Seven Tools of Quality improvement or even using Design of Experiment in Laboratory scope. The improvement obtained from that method usually high cost and Improvement of quality using Taguchi method is simpler and lower cost [1,2,5].

Moisture in CPO greatly affects the free fatty acid content. Increased water content causes free fatty acids to increase. This results in the hydrolysis reaction to oils or fats in the presence of water.

The presence of high fatty acid will result in the quality problem [1,2]. Quality is often referred as conformance to specifications. Taguchi, however, proposes a different view of quality, which links it to costs and losses in money value, not only to producers at the time of production but to consumers and society as a whole [2,6,7,8,9].

Currently, the main product of this mill is Crude Palm Oil (CPO) and hasn't met the expected standard quality. CPO is the raw material for many fat derivative products. The generally stipulated quality criteria are dirt count, free fatty acid, and moisture of CPO. The aim of this study is to obtain the optimal setting for factor's affect the quality of CPO. The optimal setting will result in an improvement of product's quality [10,11,12,13,14,15,16].

2. Method and equipment

The Taguchi method uses the Loss Function approach which describes the costs that arise between the producer and the consumer due to the determination of certain quality characteristics of the product [1,2,3,13,14,15]. Loss Function is used in measuring the performance of the quality characteristics in the achievement of the target, i.e how big is the variation around the target. The use of Taguchi Method in this research is expected to know the factors that influence the quality of CPO and perform CPO quality control analysis so that the factory produces CPO with quality in accordance with the expected target.

Taguchi's define that quality is judged from a product loss to society after the product is delivered. The quality objective of the manufacturing process is to take action to minimize loss to society. Quality Loss Function provides the basis for the determination of tolerance. In this case, tolerance is not a deviation between products but is defined as the deviation of the target. Loss due to the deviation L (y) can be formulated as follows

$$L(y) = L(m + y - m) \tag{1}$$

$$L(y) = L(m) + \frac{L'(m)}{1!}(y-m) + \frac{L''}{2!}(y-m)^2 + ... \tag{2}$$

Minimum L value at y = m and Ly (m) = 0 then L (m) = 0. Thus, the third term of Eq. 1 is the main term of the loss function or (y-m) 2 which can then be estimated as follows (assuming L''' (m) and so on is ignored) can be seen in Eq. 3 and the loss function formula can be seen in Eq. 4. Loss function contains a coefficient number k. For products with target value m, to determine the magnitude k, it is necessary to know the magnitude of the losses due to exceeding the tolerance.

$$L(y) = \frac{L''}{2!}(y-m)^2 \tag{3}$$

$$L(y) = kv^2 \tag{4}$$

In this research, Experimental Design with Taguchi Method is used. Steps of this method are identified influence factors, select the orthogonal array, processed data using ANOVA test and signal to noise ratio, and confirmed the research using Quality Loss Function. The conceptual framework build in this research can be seen in Fig. 1.

This research uses associative hypothesis that is a statement about the relation between two variables or more. The character of the relationship between variables is not viewed from the causal point, but only question whether there is a relationship between these variables [16]. The hypothesis build in this research consists of three main hypotheses. They are "H_0: There is no significant effect of the variable to CPO Quality" and "H_a: There is a significant effect of the variable to CPO Quality" for each variable fruit maturity, digester temperature, and ampere press.

There are 9 steps in this research using Taguchi Method as follow:

- Analyze CPO quality level by using seven tools

IOP Conf. Series: Journal of Physics: Conf. Series **1007** (2018) 012031 doi:10.1088/1742-6596/1007/1/012031

- Identification of Quality Factors Against Irregularities
- Determination of Number Level and Level Factor

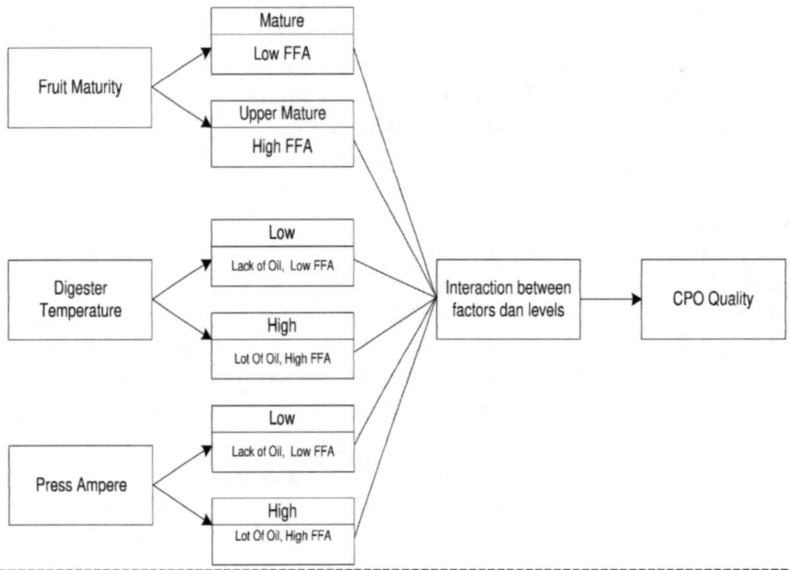

Figure 1. Conceptual framework in the Taguchi Experimental Design.

- Determine freedom degrees to calculate the minimum number of experiments to be performed to investigate the observed factors.
- Selection of appropriate Orthogonal Matrices depends on the value of the factors and expected interactions and the level values of each factor.
- Assignment of Factors to Orthogonal Matrices. Assignment of factors both control factors and interference factors and their interactions on selected orthogonal arrays with respect to linear graphs and triangular tables.
- Calculation of Main Effect
- Data analysis, calculation, and presentation of data using statistical analysis of variance, hypothesis testing and application of empirical formulas on the data result of combinations of factors.
- Interpretation of results is a step taken after the experiment and analysis have been done.

3. Results and Discussions

3.1. Actual Data Analysis

Collected data from current CPO Data can be an indication if the CPO quality problem. Collected CPO Quality for a month can be seen in Fig. 2 till Fig. 4

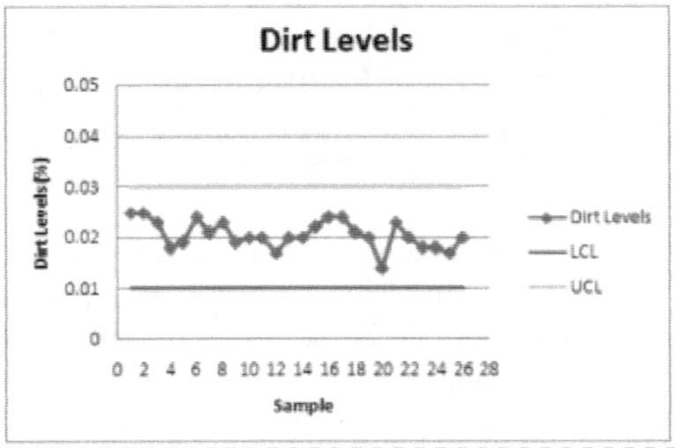

Figure 2. CPO Dirt Count.

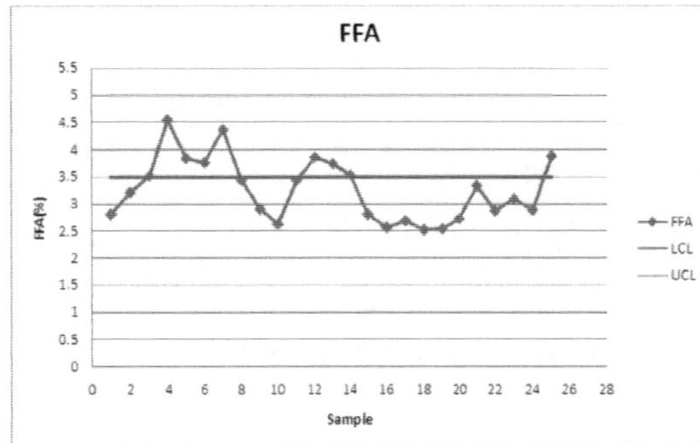

Figure 3. CPO Free Fatty Acid.

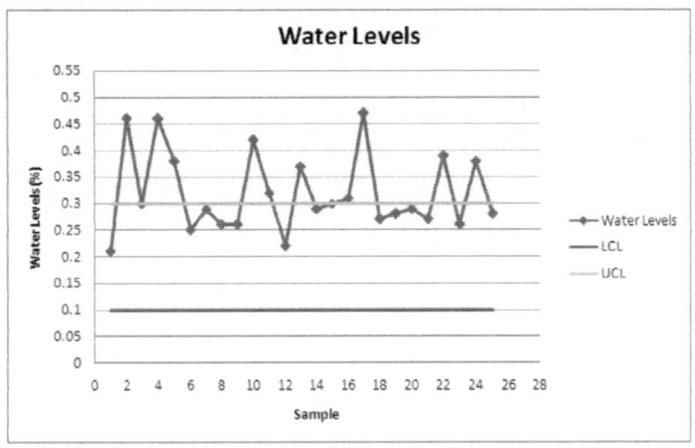

Figure 4. CPO Moisture.

In order to see CPO of production that suffer from accumulative deviation to quality condition, the nonconformities for CPO quality data was taken for 20 months as can be seen in Figure 5.

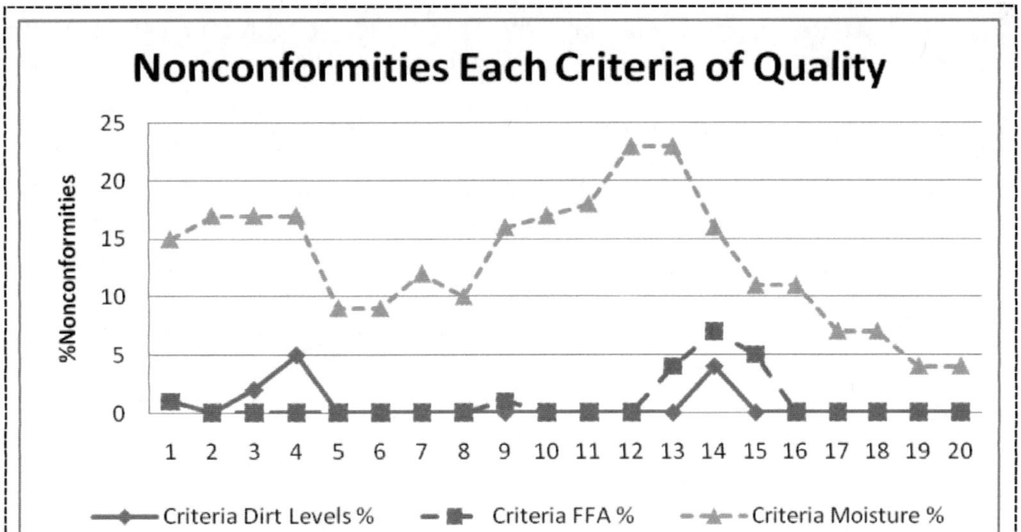

Figure 5. Nonconformities for Criteria Dirt Count, FFA, and Moisture.

The water content in CPO greatly affects the free fatty acid content. Increased water content causes free fatty acids to increase. This is due to the hydrolysis reaction to oils or fats because the presence of a certain amount of water in the oil or fat will change the oil or fat to free fatty acids and glycerol.

3.2. Identification and Discussion

A pooling-up strategy is undertaken to determine which factors have the most significant effect on product quality. To show it in Variant Analysis, the pool and pooled error pools are used. The smallest variant pool is marked Y to indicate the source is pooled into pooled E. Next is calculated F-ratio based on pooled E. Then factor A is the factor with least squares. Pooling factor is done on factor A by marking Y and adding the sum of squares to pooled E. Variance Analysis results for the pooling up strategy can be seen in Table 1.

Table 1. Variance analysis.

Source	Pool	Sq	v	Mq	F-ratio	Sq`	ρ%
A	Y	0.0105	1	0.0105	-	-	-
B		0.0218	1	0.0218	2.6954	0.0137	7.57
C		0.0541	1	0.0541	6.6971	0.0460	25.44
E	Y	0.0944	12	0.0079			
Pool E		0.1049	13	0.0081	1.0000	0.1211	66.99
St		0.1807	15	0.0120	-	0.1807	100.00
Mean		2.0378	1	-	-	-	-
ST		2.2185	16	-	-	-	-

Based on F distribution, with significance level at 5%, degree of freedom for variable 1 is 1 and variable 2 is 13, $F_{(0.05;1.13)} = 4.6672$. Hypothesis testing and conclusion obtained from variance analysis table after pooling of Factor A. For Factor B, F calculated is 2.6954 and receive H_0 because calculated F is less than F table (4.6672). This means there's no impact on Digester temperature to CPO quality. For Factor C, calculated F is 6.6971 more than F table (4.6672), so there's an impact of ampere press to CPO quality. From the contribution percentages in Table 5.17, the significant factors are B and C. These factors amount to 33.01% of the sum of the total squares. The sum of squares pooled error is 66.99%.

Based on Taguchi Experimental Design, the result of this research can be seen in Table 2. From Table 2, we can see that there's a need to set fruit maturity at 75.4-86.9% and press at 21 amperes so that signal to noise ratio can be increased and can reduce quality deviation to 42.42%.

Table 2. Setting recommendation based on Taguchi.

Factor	Level Setting	
	Current Setting	Recommendation setting
Fruit Maturity	Maturity Level 2 87.0-98.5%	Maturity Level 1 75.4-86.9%
Digester Temperature	Level 2 95^{0}C	Level 2 95^{0}C
Ampere	Level 1	Level 2

Press	20 A	21 A
S/N ratio	11.66 dB	20.25 dB

From this research, we can see that fruit maturity is the main factor affecting FFA. Higher digester temperature also gives a better result for chopping, lighten the work of screw press, and reduce broken seed. Changes the electric current also gives impact to press pressure. Pressure higher than standard will increase the amount of oil obtained and the number of broken seeds, while the lower pressure will give impact to the decrease of oil obtained but decrease the number of broken seeds.

4. Conclusions

The result of this study using Taguchi Method is to suggest the setting of fruit maturity at 75.4-86.9%, digester temperature at 95°C and press at 21 Ampere. This setting expected to reduce quality deviation by 42.42%. Implementing Taguchi Method is expected to get suitable CPO criteria based on SNI 01-2901-2006.

References

[1] Iwan Nauli D, Dedy N, Veni A 2013 J. Ekonomi, Analisis Taguchi dalam meningkatkan utilisasi produksi pada industri crude palm oil (CPO), Taguchi analysis in enhancing production utilization on CPO industry 21(2) 34-38

[2] Candra B, Rodhi A 2011 *J. Riset Industri*, Setting Parameter Mesin Press dengan Metode Respon Permukaan Pada Pabrik Kelapa Sawit, Setting Parameter in Press Machine using Surface Response Method in Palm Mill **V** 102-105

[3] Fatemeh A 2013 *Iranica J. Energy & Environment*, Production of Mycophenolic Acid by Penicillium brevicompactum in a Submerged Batch Culture: Optimization of Culture Conditions Using Taguchi Approach **4(4)** 330-335

[4] Elham G, Abdollah A, Elizabeth A C 2015 *Int. J. Quality & Reliability*, Mahalanobis Taguchi System: a review **32(3)** 291-307

[5] National Standardization Bodies of Indonesia 2006 *Crude Palm Oil* SNI 01-2901-2006

[6] Sreeraj P, Kannan T, Subhasis M 2014 *World J. Eng.*, Analysis of flux cored arc welding process parameters by hybrid Taguchi approach **11(6)** 575-588

[7] Periyanan P R, 2014 *Int. J. Quality & Reliability*, Optimization of multiple-quality characteristics in micro-WEDG process using Taguchi technique **31(3)** 205-219

[8] Donny Hidayat A J, Tung K L 2016 *Kybernetes*, Path optimization of CNC PCB drilling using hybrid Taguchi genetic algorithm **45(1)** 107-125

[9] Rahim A, Sharma U, Murugesan K, Sharma A, Arora P 2012 *J. Structural Fire Eng.*, Optimization of Post-Fire Residual Compressive Strength of Concrete by Taguchi Method **3(2)** 169-180

[10] Jenarthanan M P, Jeyapaul R 2013 *Pigment & Resin Technology*, Evaluation of milling characteristics of resin hybrid GFRP laminates using Taguchi approach **42(5)** 288-297

[11] Satyendra Kumar S, Vinod K 2015 *Benchmarking: An Int. J.*, Optimal selection of third-party logistics service providers using quality function deployment and Taguchi loss function **22(7)** 1281-1300

[12] Abbas Khammas H 2017 *Multidiscipline Modeling in Materials and Structures*, Parametric optimization of wt.% Y2O3 modified chromium-aluminide coatings using utility concept-based Taguchi approach **13(3)** 448-463

[13] Roma Mitra D, Ravi S 2014 *The TQM J.*, Emerging trend of customer satisfaction in academic process: An application SPC and Taguchi's robust parameter design **26(1)** 14-29

[14] Naresh N, Jenarthanan M P, Prakash R H 2014 *Multidiscipline Modeling in Materials and Structures*, Multi-objective optimisation of CNC milling process using Grey-Taguchi method in machining of GFRP composites **10(2)** 265-275

IOP Conf. Series: Journal of Physics: Conf. Series **1007** (2018) 012031 doi:10.1088/1742-6596/1007/1/012031

[15] Ali A, Mansour Z, Nima S 2016 *J. Enterprise Information Management*, Supplier selection in closed loop supply chain by an integrated simulation-Taguchi-DEA approach **29(3)** 302-326

[16] Sukaria S 2011 *Metode Penelitian, Research Methodology*, 1st Ed. (Indonesia: USU Press)

Acknowledgments
In this research, the authors would like to thank profusely to the palm mill which has been willing to be a place of research for the writing of this paper and highly support researcher for a better result.

IOP Conf. Series: Journal of Physics: Conf. Series **1007** (2018) 012032 doi:10.1088/1742-6596/1007/1/012032

INA-Rxiv: The Missing Puzzle in Indonesia's Scientific Publishing Workflow

R Rahim[1*], D E Irawan[2], A Zulfikar[3], R Hardi[4], L Arliman S[5], E R Gultom[6], G Ginting[7], S S Wahyuni[8], M Mesran[7], M Mahjudin[9], I Saputra[7], F T Waruwu[7], S Suginam[7], E Buulolo[7] and J Abraham[10]

[1]School of Computer and Communication Engineering, Universiti Malaysia Perlis, Perlis, Malaysia
[2]Institut Teknologi Bandung, Indonesia
[3]Universitas Muhammadiyah Yogyakarta, Indonesia
[4]STMIK Balikpapan, Indonesia
[5]Sekolah Tinggi Ilmu Hukum Padang, Indonesia
[6]Universitas Trisakti, Indonesia
[7]STMIK Budi Darma, Indonesia
[8]Universitas Brawijaya, Indonesia
[9]Universitas 45 Surabaya, Indonesia
[10]Bina Nusantara University, Indonesia

Email: *usurobbi85@zoho.com

Abstract.INA-Rxiv is the first Indonesia preprint server marking the new development initiated by the open science community. This study aimed at describing the development of INA-Rxiv and its conversations. It usedanalyzer of Inarxiv.id, WhatsApp Group Analyzer, and Twitter Analytics as the tools for data analysis complemented with observation.The results showed that INA-Rxiv users are growing because of the numerous discussions in social media, e.g.WhatsApp,as well as some other positive response of writers who have been using INA-Rxiv. The perspective of growth mindset and the implication of INA-Rxiv movement for filling up the gap in accelerating scientific dissemination process are presented at the end of this article.

1. Introduction

Dissemination of research[1], like via INA-Rxiv Preprint Server Indonesia[2]is strategic for speeding up research claims that can be acknowledged and used by other researchers.INA-Rxiv is not the only preprint server available.The paper licenses uploaded to INA-Rxiv are CC-BY (Creative Common Attribution) to guarantee the freedom of usage as long as the source acknowledged.Providing preprint or author original version (AOV) on a community-based preprint server endorses Open Access practice, which means free to re-use, free to remix, and free to redistribute.INA-Rxiv was soft launched on August 17, 2017. It has proven to be in the major league, by hosting 1000+ papers in just three months (August-November 2017). From the initiation of INA-Rxiv by Open Science activistsuntil INA-Rxiv is becoming known today, social media was very helpful. INA-Rxiv is growing with many posts on various social media such as Facebook (Open science Inarxiv),

Twitter(@inarxiv_id), Telegram (e.g. IDRI) and also WhatsApp.WhatsApp wasthe most widely used social media[3] among lecturers or researchers.

INA-Rxivcarrying the concept of Open Access[4][5]indexed in Google Scholar[6]could be a solution to improve publications and citations both for personal and institutional. The Open Science Framework (OSF), a not-for-profit institution, as the backbone of INA-Rxiv, provides a nearly unlimited cloud storage space, where Indonesian and Indonesianist authors can store their papers along with supplementary files and datasets(data, image, and codes)[7][8]. Now it is easy for academics to upload (1) preprint articles (prior peer review), without having to wait for the long winding road of the review process,(2) postprint articles, and even (3) articles that have been published but hasn't online.

Matthias andTennant[9]explained the preprint and postprint process schemes as shown in Figure 1. Nowadays, the most authors know solely the publication version issued by the journal publishers. Preprint and postprint publications[10][11][12]are article versions that have not been widely used to accelerate the results of research dissemination. In the process of publishing the article, there are several stages to be completed, e.g. a review process that takes 3 months or even 1 year. Meanwhile, anarticle which under review process would be useful for other researchers if there is a preprint version that can be used[10]. Another case is a paper published in the Non-Open Access journal requires apaid subscription to the publisher for the article to be read or downloaded. It would be more accessible if there is a postprint version of the article, all free for readers. This study aimed at presenting the development of INA-Rxiv and the use of social media as a forum for the socialization of INA-Rxiv.

2. Methods

This study used descriptive designto analyze the sum and distribution of discussions about INA-Rxiv and openscience in general sense. Data were takenand analyzed with https://inarxiv.idfrequency analyzer, WhatsApp Analyzer of INA-Rxiv Group from middle August2017 to early October2017 as well as Twitter Analytics.

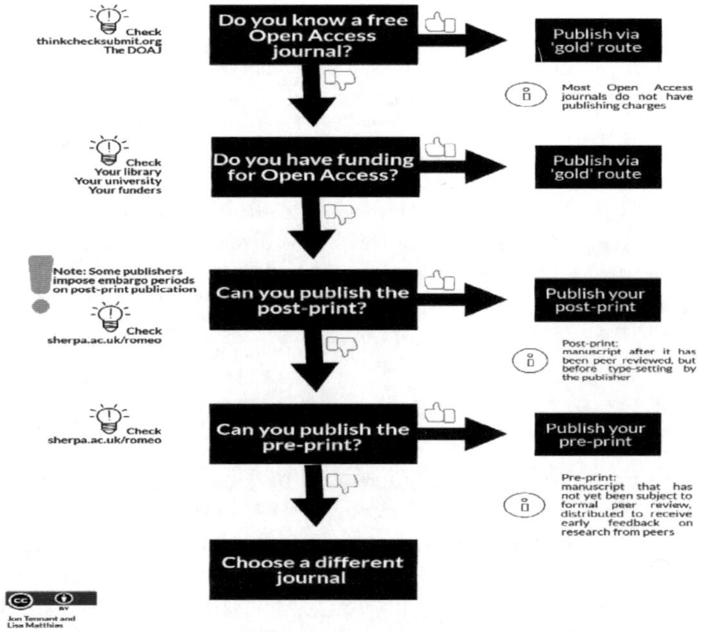

Figure 1. Preprintand postprint open access scheme: how to make your research open access for free and legally.

3. Results and Discussion

One week since its soft launching (August 17, 2017)[13], INA-Rxiv received 77 articles. The majority of the articles came from Physical Sciences and Math (35 articles), Engineering (12 articles), Social and Behavioral Sciences (8 articles), Education Science (8 articles), Medicine and Health Sciences (8 articles). The rest came from Business, Law, as well as Life Science.On October 14, 2017, the increase of articles onINA-Rxiv has reached a total of 638 Articles (more than 800% increasing). It might be caused by massive socialization of questions and testimonials from curious lecturers or researchers about INA-RXIV and its benefits.The majority of the articles came from Physical Science and Math (180 articles), Engineering (160 articles), Education Science (60 articles), Life Sciences (55 articles), Arts and Humanities (55 articles) as well as Social and Behavior Sciences (40 articles). The rest came from Law as well as Medicine and Health Sciences.

Figure 2 shows that the chat activity on the group was quite active at 6pm to 9pm, which is non-working hours. The discussions were about technical bits on INA-Rxivconcepts and uploading techniques, Scopus indexing and how to get published in journals and proceedings. There were also several feedbacks for the national-wide promotion system and positive feedback on how to make the most from INA-Rxivin research workflow.

Figure 2. Chat messages about publication and ina-rxiv.

Twitter analytics shows an overwhelming sum of impressions or engagements gained from tweets which generated from INA-Rxiv twitter handle (@inarxiv_id). Since August 6, 2017, a week before Open Science Framework(OSF) officially launched INA-Rxivand the other five preprint servers, to October 31, 2017, the Twitter handle had earned 120,000 impressions. This was considered as a huge start.Based on the observation on the dialogues that occurred through the WhatsApp group, only a few questioned the mechanism of deletion of files that have been uploaded toINA-Rxiv. Questions about file removal are generally triggered by two subjective things. *First*, the writers'worry that theeditorial board member offuture journal publication targetwill accuse him/her of self-plagiarism, as evidenced by the similarity level of content between the submitted article and the one savedand announced inINA-Rxiv. *Secondly*, the writers'evaluation that the uploaded article is a "rushed" or "error" version containing either substantive, fundamental, and/or redactional error.

The first concern has been eroded with objective fact that "major"publishers can justify the dissemination of manuscripts through preprint servers. SAGE Publishing, for example, clearly states that "*Prior publication of datasets or deposition in a preprinted server do not constitute prior publication*"[14]. Manuscripts posted in the preprint server are not classified as a "publication", so allegations of self-plagiarism or double publication that are unethical can notbe addressed to the author[15]. *The second* concern related to the psychological aspects of a person, i.e.the mindset.

Dweck[16]divided the mindset into two, namely fixed mindset and growth mindset. The characteristics of a person with a *fixed mindset* are (1) assumes that his/her intelligence, character, and personality are permanent entities at all times, (2) wants to constantly indicate or show off their talents, skills, expertise, personality, and character; (3) can not tolerate errors or setbacks, even tend to be depressive because of them, (4) his/her self-confidence is more vulnerable due to errors because the mistake done---though occurring only once---is considered a part of his/her deep personal identity, (5) seeks to nurture and protect his/her self-esteem by proving his/her performances and collecting a lot of recognition and confirmation; if there is a failure they will seek "scapegoat", do "revenge" when there is a way or opportunity (just to feel win and superior above others), or resist to try further.The characteristics of people with *growth mindset* are (1) see their intelligence and personality as entities that can change and evolve over time, (2) give meaning to their mistakes or failures as a means for learning, (3) appreciate processes, not results, and are more able to consider the situational aspects intheir attributions, (4) their confidence is not threatened by failure, and (5) find pleasure in challenges and is not busy attaining and proving perfection.

The growing number of articles in INA-Rxiv coupled with very few complaints regarding the INA-Rxiv policy---stating that "*When you add a new version of a file with the same name to your OSF project, the OSF automatically updates the file and* ***maintains a record of all previous versions*** *stored on OSF Storage"*[17]*(see also its predecessor policy, arXiv, which states "arXiv makes* ***all previous versions of submissions publicly available*** *(since October 1997). Therefore, even though the current version of a paper may be marked as withdrawn,* ***previous versions can still be retrieved***"[18])---can also be viewed from the perspective of the mindset.

Based on the reality, it appears that most academics in Indonesia who uploads their manuscript to INA-Rxiv have growth mindset in the academic field. The academics feel fine that the history or evolutive traces of their manuscripts (version 1, version 2, version 3, and so on) in INA-Rxivstorage can still be accessed and traced transparently. This shows that they *dare to face themselves* in the midst of public presence. The publicis not positioned as the party that will sentence their ability level. In other words, they enjoy the process *of perfecting* their articles from the obtained feedbacks, *not* expecting recognition *of perfection* about how great their articles are. By storing and disseminating manuscripts via INA-Rxivthey are actually very enthusiastic to learn and gain more criticism for their development. Although the Google Scholar indexation of manuscripts stored in INA-Rxiv is a tantalizing potential, the motive of Indonesian academicsjoined in INA-Rxivis clear enough, i.e.involving themselves in a "knowledge crises." Knowledge is always seen in a series of "crisis after crisis" that will drive the advance of that knowledge. The crises coped with ambiguity tolerance [19]are viewed as a site for sharing, correcting self, and sustaining happiness (not pressure and suffering) in the dynamics of knowledge production. The facilitation of INA-Rxivreduces social loafing [20]in the academic virtual teamand it creates optimism for the gait of Indonesian academics.

The dynamics, as indicated by the data, are experiencing proliferation because of appropriation of INA-Rxivin addition to the growth mindset (self/personal aspect) and socialization (social aspect). Appropriation[21]refers to a dynamic activity of which INA-Rxiv does not merely serve as firstly initiated, i.e. as a preprint server for knowledge dissemination and feedback acquisition from a variety of openly and ubiquitously distributed resources (e.g. "peer-as-reviewer"*vis-à-vis*"expert peer reviewer") in one's virtual and social environments;but also,to mark the journey of effort in reaching novelty or originality in a research, and, more essentially, celebrates human freedom.We can view the appropriation in the condition that whenour ideaor even worse, our documents, have been stolen. How do we prove that we are the first to make an idea or draft the document? That's what INA-Rxiv is for. We can say, "*INA-Rxiv---Feel the freedom.*"

4. Conclusion

INA-Rxiv can be a solution for the dissemination process because it receives *preprint and postprint versions* of a manuscript (author version) that can be widely disseminated; both preserve author's original copyright. It is also the missing puzzle in scientific workflow since it allows an immediate public claim of a work and public feedback as well. WhatsApp as a social media has been successfully used to spread the messages on INA-Rxiv, preprint, and postprint to lecturers and researchers.

Further development of this paper may address the effects of INA-Rxiv to improve institutional publications and the number of articles and citations in Indonesia's SINTA (Science and Technology Index; sinta.ristekdikti.go.id) record, and how the Indonesian Ministry of Research, Technology, and Higher Education could use it to improve the quantity and quality of publication in Indonesia.

References

[1] D. E. Irawan, C. N. Rachmi, H. Irawan, J. Abraham, K. Kusno, M. T. Multazam, K. K. Rosada, S. H. Nugroho, G. Kusumah, D. Holidin and N. A. Aziz, "Penerapan Open Science di Indonesia Agar Riset Lebih Terbuka, Mudah Diakses, dan Meningkatkan Dampak Saintifik," Berkala Ilmu Perpustakaan dan Informasi, vol. 13, no. 1, pp. 25-36, 2017.

[2] D. E. Irawan, "INA-Rxiv: The Indonesia preprint server," Medium Corporation, [Online]. Available: https://medium.com/open-science-indonesia/ina-rxiv-2a51aae69d3a.

[3] T. Sutikno, L. Handayani, D. Stiawan, M. A. Riyadi and I. M. I. Subroto, "WhatsApp, Viber and Telegram: which is the Best for Instant Messaging?," International Journal of Electrical and Computer Engineering (IJECE), vol. 6, no. 3, pp. 909-914, 2016.

[4] E. C. McKiernan, P. E. Bourne, C. T. Brown, S. Buck, A. Kenall, J. Lin, D. Dougall, B. A. Nosek, K. Ram, C. K. Soderberg, J. K. Spies, K. Thaney, A. Updegrove, K. H. Woo and T. Yarkoni, "Point of View: How open science helps researchers succeed," eLife, p. 5, 2016.

[5] V. Tracz and R. Lawrence, "Towards an open science publishing platform [version 1; referees: 2 approved]," F1000Research, vol. 5, 2016.

[6] L. Dixon, C. Duncan, J. C. Fagan, M. Mandernach and S. E. Warlick, "Finding articles and journals via Google Scholar, journal portals, and link resolvers: Usability study results," Reference & User Services Quarterly, vol. 50, no. 2, pp. 170-181, 2010.

[7] R. Rahim, N. Nurjamiyah and A. R. Dewi, "Data Collision Prevention with Overflow Hashing Technique in Closed Hash Searching Process," Journal of Physics: Conference Series, vol. 930, no. 1, p. 012012, 2017.

[8] R. Rahim, I. Zulkarnain and H. Jaya, "Double hashing technique in closed hashing search process," IOP Conference Series: Materials Science and Engineering, vol. 237, no. 1, p. 012027, 2017.

[9] L. Matthias and J. Tennant, "How to make your research open access? For free and legally," Figshare, 2017.[Online].Available: https://figshare.com/articles/How_to_make_your_research_open_access_For_free_and_legally_/5285512.

[10] H. W. Inefuku, "Pre-Print, Post-Print or Offprint? A guide to publication versions, permissions and the digital repository," Digital Repository @ Iowa State University, 2013. [Online]. Available: http://lib.dr.iastate.edu/digirep_outreach/2.

[11] B. Swoger, "Understanding your rights: pre-prints, post-prints and publisher versions," Scientific American, [Online]. Available: https://blogs.scientificamerican.com/information-culture/understanding-your-rights-pre-prints-post-prints-and-publisher-versions/.

[12] E. Amsen, "Guide to open science publishing," F1000Research, 2015:21.

[13] D. E. Irawan, "INArxiv: 7 days and 77 preprints," [Online]. Available: dasaptaerwin.net/wp/2017/08/inarxiv-enthusiasts.html.

[14] A. Anonym, "Frequently Asked Questions," Sage Publishing, [Online]. Available: https://insights.sagepub.com/author_resources.php?folder_id=120.

IOP Conf. Series: Journal of Physics: Conf. Series **1007** (2018) 012032 doi:10.1088/1742-6596/1007/1/012032

[15] C. L. Curry, "SHERPA Services and SHERPA/RoMEO," Journal of Electronic Resources in Medical Libraries , pp. 1-4, 2017.
[16] C. S. Dweck, Mindset: The New Psychology of Success, Ballantine Books, 2007.
[17] A. Anonym, "File Revisions and Version Control," Open Science Framework, [Online]. Available: http://help.osf.io/m/files/l/524182-file-revisions-and-version-control.
[18] a. administrators, "To withdraw an article," Cornell University, [Online]. Available: https://arxiv.org/help/withdraw.
[19] J. Abraham and A. Falah, "Is Virtual Marriage Acceptable? a Psychological Study Investigating the Role of Ambiguity Tolerance and Intimacy Illusion in Online Dating among Adolescents and Early Adults," Journal of Psychological and Educational Research, vol. 24, no. 2, pp. 117-143, 2016.
[20] J. Abraham and M. Trimutiasari, "Sociopsychotechnological Predictors of Individual's Social Loafing in Virtual Team," International Journal of Electrical and Computer Engineering (IJECE), vol. 5, no. 6, pp. 1500-1510, 2015.
[21] T. Prayoga and J. Abraham, "A psychological model explaining why we love or hate statistics," Kasetsart Journal of Social Sciences, vol. 38, no. 1, pp. 1-8, 2017.

IOP Conf. Series: Journal of Physics: Conf. Series **1007** (2018) 012033 doi:10.1088/1742-6596/1007/1/012033

MISSIONS: The Mobile-Based Disaster Mitigation System in Indonesia

Rossi Passarella[1], Sarifah Putri Raflesia[1*], Dinda Lestarini[1], Ahmad Rifai[1], Harumi Veny[2]

[1]Faculty of Computer Science, Universitas Sriwijaya, Jl. Raya Palembang-Prabumulih Km. 32, Indralaya, South Sumatra, Indonesia
[2]Faculty of Engineering, UniversitiTeknologi MARA, Shah Alam, Selangor, Malaysia

E-mail: *syarifahpr@gmail.com

Abstract. Disaster mitigation is essential to minimize the effects of disasters. Indonesia is one of the disaster prone areas in Asia and the government explores the usage of Information technology (IT) to aid its mitigation efforts. Currently, there are Indonesian websites which hold information regarding the weather monitoring, climate conditions, and geophysics. But, there is no clear indicator of mitigation efforts or things to do during an emergency. Therefore, this research proposed MISSIONS, a disaster mitigation model using geo-fencing technique to detect the location of the users through their mobile devices. MISSIONS uses mobile-based disaster mitigation system as a way to disseminate critical information to victims during emergency when they are in disaster zones using virtual fences. It aims to help the government to reduce the effects of disaster and aid in the mitigation efforts. The implementation result shows that MISSIONS have a high accuracy in detecting user whereabouts.

1. Introduction

Natural disasters occur at several places around the world and they can happen at any time and at any place. These natural disasters are very common in regions that are claimed as disaster-prone areas. Indonesia is a classic example of a disaster-prone area in the world. The occurrence of natural disasters cannot be predicted accurately[1] and hence the adverse effects on the society and the economy are enormous[2]. Therefore, efforts should be taken in order to mitigate natural disasters. Disaster mitigation could be defined as a set of activities that are carried out to lessen the likelihood of a disaster or to diminish the prevalent risks of disasters if the disaster itself cannot be avoided[3]. Disaster mitigation activities which are done by the government organizations aim to minimize the risk of disasters by decreasing the loss of human and animal lives and loss of property or wealth. The government organization in Indonesia which deals with disasters is Indonesian Agency Disaster Mitigation or BadanNasionalPenanggulanganBencanaAlam (BNPB).

According to BNPB, the most frequent disaster in Indonesia between 1815 and 2016 were flood and tornado but the deadliest disaster was none other than the earthquake. The earthquake has been rightly termed as the deadliest disaster because it has taken the lives of over 57% of the total victims. The main reason for such a high percentage of loss is the lack ofdisaster-related information dissemination [4][5].The people found it difficult to identify any evacuation route and there was no information on safe places at those moments.Hence, there must be a system to accommodate the problem of lack information indisaster mitigation especially during the disasters occur. The system can be used to send alerts and also information about evacuation whenever the disaster occurs. The studies about disaster mitigation which finally generated the innovation to support disaster mitigation have been done all over the world such as the use of drone technology to find survival in the disaster area[6], remote sensing technology to assess the damage caused by the disaster [7], social media to communicate[8], disseminate disaster-related information[9] and map resources[10] in disaster area, data mining to predict the natural disaster[11], and mobile-based technology to evacuate the survival by engaging the geo-fencing technique[12]. In order to develop a system which could be used for disaster mitigation, this research proposes an Information Technology (IT) model by using the geo-fencing technique.

Geo-fencing can be defined as an innovative technique based on telematics and satellite positioning that aims to demarcate a boundary or a virtual fence in a specified geographic area[13]. The geo-fencing technique was developed around the year 2000 in order to track the list of devices entering a particular virtual fence or geographical space[14]. From the time of its development, geo-fencing has been primarily used for tracking purposes, such as tracking trucks, buses, or human beings. Geo-fencing is popularly used in the logistics sector to enable the stakeholders to trace the location of trucks[13]. In the recent times, geo-fencing is being used increasingly in other dimensions, especially in rescue and support operations. In 2016, a model for child safety was proposed where the model enables the parents to monitor the children safety in real time[14] [15].The geo-fencing technique also empowered for health field. It is embedded into wearable device to monitor the persons with Alzheimer's and dementia.

For the first time, in 2015, an application used geo-fencing technique for human protection in disaster management. This application for human protection during disaster management was named HelpMe and it enables the rescuers to discover the locations of disaster victims[16]. According to the previous findings[16], the geo-fencing has been used for supporting IT-based disaster mitigation. Unfortunately, the proposed technology is focused on monitoring the people who have been familiar with the environment. There's no early notification for new people such as tourists, people who arrive for business, etc. Hence, in this research, the improvement of model is adding the early notification for newcomers. In this research, this geo-fencing technique is embedded in the system which is named MISSIONS in order to maximize the efficiency and effectiveness of disaster mitigation activities. MISSIONSprovides disaster mitigation maps, information, regulation, and emergency plans. Meanwhile, the use of geo-fence enabled MISSIONS to track the people trapped in the disaster area. Then, it sends emergency notification to users and instructions for survival such as the nearest safe point and possible evacuation routes.

There are two important lessons that lead to effective mitigation in natural disaster. First, we must consider disaster as unresolved problems which are unpredictable, isolated and independent events. The second lesson is the mitigation process cannot depend only upon technological solutions, it also should be based on others measures, such as engineering devices, land management, social regulation and economic improvement [17]. In order to fulfill the requirements of mitigation processes, the proposed model also includes the IT regulations and governance as a control. It is mandatory to engage IT governance and government regulations so as to ensure that the use of the proposed system aligns with the action plans and strategies formed by the government.

2. Overview of Disaster Mitigation in Indonesia

The importance of disaster mitigation in disaster-prone areas can be seen from those countries' Government Regulations and policies to handle disasters. One such example is the Indonesian Government Regulation No. 33 (2006). It states the availability of information regarding disaster-prone areas and the types of disasters in Indonesia should be made known to the public. The government regulation goes on to encourage socialization among people living in these areas in order to increase the level of social awareness regarding these issues. The people should also know what

IOP Conf. Series: Journal of Physics: Conf. Series **1007** (2018) 012033 doi:10.1088/1742-6596/1007/1/012033

they should do and not during critical situations. Finally, the government regulation also insisted on the importance of governance and the role of Indonesian government in the mitigation of disasters.

Disasters often happen in Indonesia such as earthquake, landslides, flood, and tsunami. It is caused by the geography factors where Indonesia is located. According to Indonesian Government Regulation No. 33 (2006), the keys of disaster mitigation are:

- the availability of disaster-prone areas information and maps;
- the information dissemination about disaster mitigation to enhance the knowledge and awareness of disaster mitigation;
- ensure that all people in disaster-prone areas aware of things to do and things that must be avoided;
- re-organize and rearrangement of disaster-prone areas.

Nowadays, Indonesia spread the information about disaster in the official website Indonesian Agency Disaster Mitigation which means all the people must pro-actively to check the website to get information about disaster. According to the requirement of information dissemination in disaster-prone areas in Indonesia, this paper proposed disaster mitigation model which includes MISSIONS, the mobile-based disaster mitigation

3. The Proposed Model: IT-Based Disaster Mitigation

The proposed disaster mitigation model using geo-fencing technique is applied through the development of MISSIONS which is controlled by well-prepared IT governance and government regulations. The aim of this model is also to support the four keys important of disaster mitigation according to Indonesian regulation. The flow chart of the activities in MISSIONS is depicted in Figure 1.

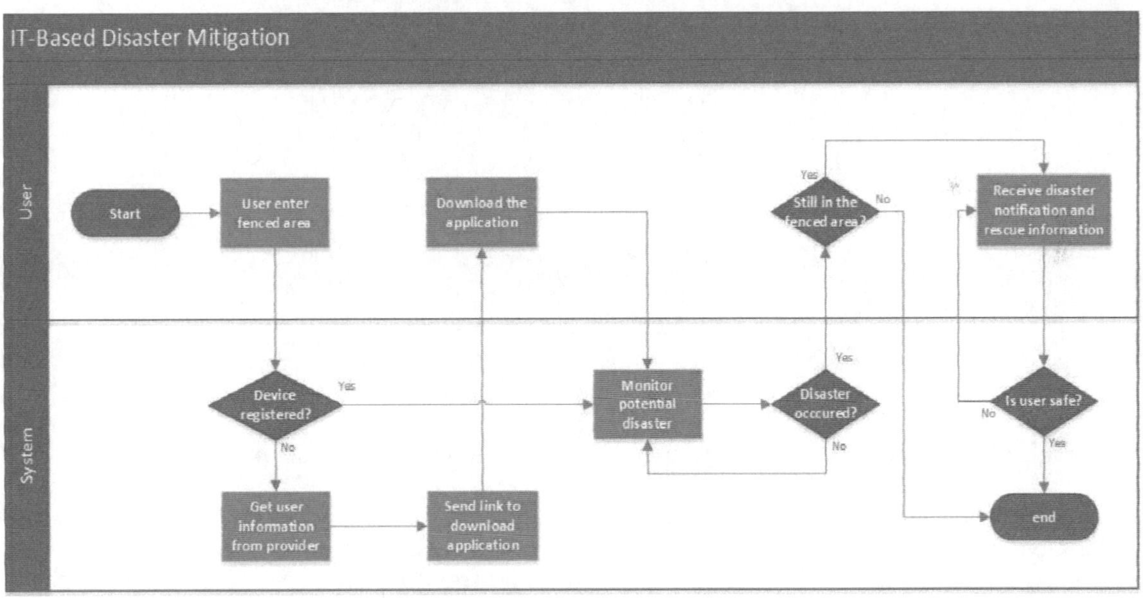

Figure 1. Flowchart of MISSIONS.

First, the geo-fence will be set up around the disaster-prone areas by person in charge. The person in charge can be staff of Indonesian Agency Disaster Mitigation. When users enter the virtual fence, the system will check whether the user is registered or not. If the user is not registered, system sends the short message service (SMS) with a link to download MISSIONS for disaster mitigation and register users' devices. Once the device registered and the users are still in the virtual fence, MISSIONS will be able to send the location of users to the server. It will help the rescuers to trace the location of users.

MISSIONS will proactively monitor potential disaster in the disaster-prone area. Moreover, it also provides the set of standards operation of disaster mitigation, safe points, emergency contacts, important places (hospital, police stations, etc). This feature aims to support the well-spread knowledge about disaster mitigation. Figure 2 shows the MISSIONS presented the current location of user and the nearest safe point.

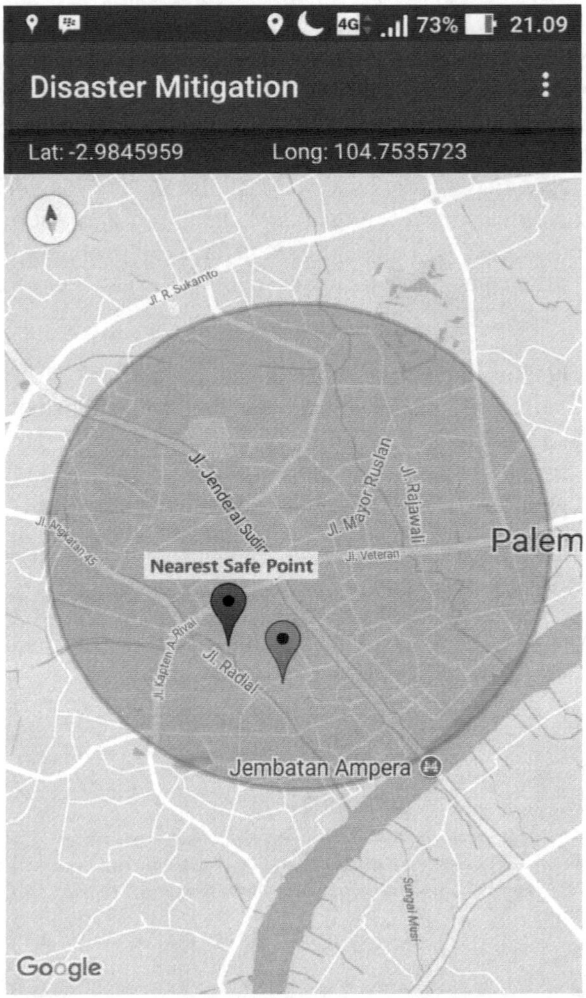

Figure 2.MISSIONS during its operation shows the current location and nearest safe point.

As we mentioned in the previous part that there are four important keys to the disaster mitigation. MISSIONS accommodates directly to the first three keys; as the maps and mitigation routes provider; support the information dissemination and spreading the standardized operations of disaster mitigation. Pertaining the early diffusion of MISSIONS, the model is also supported by traditional approach which means during the process of awareness enhancement by using MISSIONS, the government must also provide the face-to-face meeting to the people in the disaster-prone area, information boards, and spreading information by using mass media.

Meanwhile, when disaster occurs, the system automatically sends notifications to the users in fenced area. The notification will also include rescue plans, such as the nearest evacuation routes and the meeting points. By using the MISSIONS, the users will also be able to send emergency messages. The user can also communicate with rescuers to guide them to the safe zone. In this situation, if the users exit the fence, MISSIONS stops to communicate with the users, understanding that they are in the safe zone. Besides engaging the MISSIONS for disaster mitigation, the proposed model includes the government regulations and IT regulations and governance.

4. Missions Evaluation

This evaluation part aims to show how the MISSIONS works in detecting and monitoring people in disaster-prone area. For the evaluation process, the virtual fence was created as the representation of disaster-prone area. This area was in radius 4 kilometers. In order to evaluate the MISSIONS, the scenario was the user had to enter and exit the disaster-prone area which was previously configured by system administrator. While the user in the disaster-prone area, the user was notified, monitored, and provided with disaster information such as rescue plan, safe points, things to do in emergency situation, etc. Table 1 shows the result of this evaluation.

Table 1. MISSIONS Evaluation Result.

Number of Experiments	Entered Fence (meters)	Exited Fence (meters)
1	32	33
2	37	42
3	25	26
4	42	45
5	33	37
6	32	31
7	42	43
8	38	39
9	36	39
10	33	33

The evaluation result depends on the internet connection and Global Positioning System (GPS) accuracy. According to the evaluation, the MISSIONS sent notification that the users entered a disaster-prone area when they were about 25-42 meters inside the fence. The average distance for entry notification is 35 meters. Users also received notification that they are in safe area when they were about 26-45 meter outside the fence by means the average distance for exit notification is 36.8 meters. According to the evaluation results, MISSIONS has a high accuracy in detecting user location when they entered or exited the fence.

5. Conclusion

Disaster mitigation activities are necessary to protect valuable lives and assets of the country. Indonesia is one of the disaster prone areas in Asia where it is prone to earthquakes, floods, landslides and so on. With the increasing use of IT in all fields, this research explores the use of IT in disaster mitigation. Consequently, a disaster mitigation model using geo-fencing technique has been proposed in this research. This model includes the proposal for development of a MISSIONS using geo-fencing technique with adherence to IT governance and regulations of the country. This proposed model would magnify the efforts of disaster mitigation and when operationalized, the performance measurements will help in the betterment of the model. The implementation of MISSIONS shows that users will entry notification when they are about 35 meters inside the fence. The users will also get the notification when they are about 36.8 meters outside the fence. This result shows that MISSIONS have a high accuracy in detecting the location of users.

References

[1] Arceneaux K, and Stein R.M. 2006 Who is held responsible when disaster strikes? the atribution of responsibility for a natural disaster in urban election. *Int. J. of Urban Affairs* 28 pp. 43-53

[2] Butt S 2014 *Asia-Pacific Disaster Management: Comparative and Socio-Legal Perspectives* ed S Butt, HNasu and L Nottage (Berlin: Springer Berlin Heidelberg) pp 183-196
[3] Whybark D C 2015 Co-creation of improved quality in disaster response and recovery.*Int. J. Qual. Innov.* 13
[4] CFE-DMHA 2015 *Indonesia disaster management reference handbook*
[5] Mileti D.S., and Brien, P.W.O. 2014 Warnings during disaster : normalizing communicated risk. *Soc. Problem* 39 pp 40-57
[6] Lee S, Har D, and Kum D 2016 *Proc. APWC on CSE/APWCE 2016* ed A B M S Ali, M A Khan andK Mamun (Nadi: Conference Pubishing Services) pp 84-89
[7] Radhika S, Tamura Y, and Matsui M 2016 Application of remote sensing images for natural disaster mitigation using wavelet based pattern recognition analysis. *Int. Geosci. Remote Sens. Symp.*, pp. 84–87
[8] Velev D. and Zlateva P 2012 *Int. Proc. Econ. Dev. Res.*, 8 pp. 41–45
[9] Funayama T 2015 *Proc. Int. Con. on ICT and Knowledge Engineering*(Bangkok : Siam University) pp. 42–45
[10] Basu M, Ghosh S, Jana A, Bandyopadhyay S, and Singh R 2017Resource mapping during a natural disaster: A case study on the 2015 Nepal earthquake.*Int. J. Disaster Risk Reduct.,* 24 pp. 24–31
[11] Refonaa J, Lakshmi M, and Vivek V 2015 *Proc. Int. Conf. on Circuit, Power and ComputingTechnologies*(Tamilnadu : the Institute of Electrical and Electronics Engineers, Inc) pp. 1-6
[12] Suyama A and Inoue U 2016 *Proc. Int. Con. on Computer and Information Science*
[13] Reclus F and Drouard K 2009 *Proc. Int. Con. on Intelligent Transport SystemsTelecommunications* pp 353-356
[14] Munson J P and Gupta V K 2002 *Proc. 2nd Int. Work. Mob. Commer.* - WMC 02 40-44
[15] Raflesia S.P, Lestarini D, Taufiqurrahman, and Firdaus, 2017 *Proc. Int. Con. on Electrical Engineering and Computer Science* pp. 160–162
[16] Yelne S and Kapade V 2015 Human protection with the disaster management using an android application. *Int. J. of Scientific Research in Science, Engineering and Technology* 5 pp. 15-19
[17] El-Masri S and Tipple G 2002 Natural disaster, mitigation and sustainability: the case of developing countries. *Int. Plan. Stud.*, 7 pp. 157–175

Acknowledgement
We would like to thank Lembaga Penelitian Universitas Sriwijaya for the funding the research by rewarding grant Hibah Kompetitif 2017.

IOP Conf. Series: Journal of Physics: Conf. Series **1007** (2018) 012034 doi:10.1088/1742-6596/1007/1/012034

Comparison AHP and SAW to promotion of head major department SMK Muhammadiyah 04 Medan

M Saputra, O S Sitompul, and P Sihombing

Department of Information Technology, Faculty of Computer Science and Information Technology, Universitas Sumatera Utara, Medan 20155, Indonesia

E-mail: shot.jos1@gmail.com

Abstract: Decision Support System (DSS) is a system that can help a person to make informed decisions about various types of choices that are performed accurately and in accordance with the desired goals. Many problems can be solved by using decision support systems. In this journal the decision support system is used to assist the Chief of Muhammadiyah Medan branch in the selection of the department chief. The criteria used for the election of department chiefs are: Loyalty, Job Performance, Responsibility, Obedience, Honesty, Cooperation, Education, and Leadership. The selection promotion process consists of Analytical Hierarchy Process (AHP) and Simple Additive Weighting (SAW) methods. The data were obtained through teacher assessment questionnaires by principals and colleagues. The results of this study used a comparison with two decision methods namely SAW method and AHP method so that the decision maker (principal) is more appropriate in the determination of candidates who will be elected head of department at school. The final result of this research is the first rank obtained by muhammad musa with weight value on AHP method (0.274) and weight value on SAW method (0.993), alvin syahrin with weight value on AHP method (0.241) and weight value on SAW method (0.883), noviyanti with weight value on AHP method (0.193) and weight value on SAW method (0.707). So the conclusion on the research that is by using SAW method the value of weight produced more accurate.

1. Introduction

Branch Chief Muhammadiyah Medan has prioritized efforts to improve the quality of education of students of SMK Muhammadiyah Medan, in anticipating the challenges ahead towards the desired concept, it is necessary to continuously develop opportunities and innovations so that the goals can be achieved maximally. Based on the description of the problem, it is necessary to build a decision support system of the promotion of the position of head of department of SMK Muhammadiyah 04 Medan as a solution to the problem of determining the selection process or promotion of existing positions. To overcome these problems with the AHP (Analitycal Hierarchy Process) method solves a complex problem where aspects or criteria are taken quite a lot and the method of SAW (Simple Additive Weighting) is often also known term weighted summation. Other researchers usually use one method of completion of the election of a particular candidate. Therefore the author uses two methods in the completion of the election of candidates. With two outcomes from two different methods will raise the confidence of decision makers to complete the election of a particular candidate.

IOP Conf. Series: Journal of Physics: Conf. Series **1007** (2018) 012034 doi:10.1088/1742-6596/1007/1/012034

2. Theoritical Basis

2.1 *Decision Support System*

According to Bonczek in the book [1], Decision support system as a computer-based system consisting of components such as components of the language system, knowledge system components and components of the problem processing system) that interact with each other. According to Alter in the book [2], decision support system is an interactive information system that provides information, modeling, and data manipulation where the system used to assist decision making in semi-structured and unstructured situations. The thing to note here is that the existence of DSS is not to replace the duties of managers, but to be a means of supporting them. SPK is an implementation of theory of decision theory that has been introduced by science such as operations research and management science. The only difference is that in the past, to solve the problems encountered, the calculation of iterations must be done manually [3]. According to decision support system components are [4]: data management, model management, subsystem dialog, knowledge management.

2.2 *Analytic Hieararchy Process*

Analytic Hierarchy Process or AHP developed by Prof. Thomas L. Saaty as the decision-making algorithm for multicriteria problems (Multi Criteria Decision Making or MCDM)[5]. The multicriteria problem in AHP is simplified in the form of hierarchy consisting of 3 main components[6]. Analytic Hierarchy Process (AHP) can simplify complex and unstructured problems, strategies and dynamics become part of it, and make variables in a hierarchy (level). A complex problem can be interpreted as the criteria of a problem that is so much (multi criteria)[7]. That is the goal or goal of decision making, assessment criteria and alternative choice[8]. Multi Criteria Decision Making (MCDM) approaches are more popular and widely acceptable for complex decision making problems. MCDM methods are used to find the best alternative by evaluating the conflict criterion. Analytical Hierarchical Process (AHP) is one of the widely accepted MCDM approaches which are accepted by researchers because of its nature to make optimal and best decisions [9]. The image of the hierarchy is as follows:

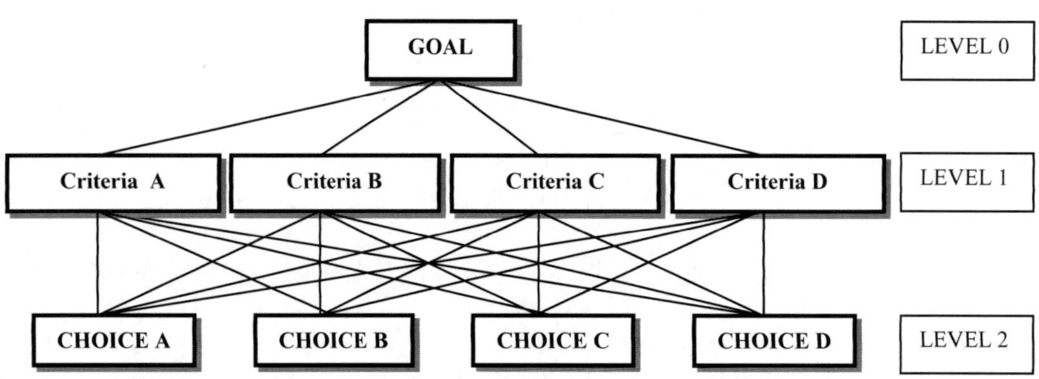

Figure 1. Hierarchy Analityc Process.

Pattern of Ratio Consistency: $CR = \frac{CI}{RI}$ (1)

IOP Conf. Series: Journal of Physics: Conf. Series **1007** (2018) 012034 doi:10.1088/1742-6596/1007/1/012034

CR is a parameter used for check if the pairwise comparison has been done consequently or not. The value of RI is the random value of the index issued by Oakridge laboratory such as which is shown in Table 1 below[10]:

Table 1. Random index (RI).

Ordo	1,2	3	4	5	6	7	8	9	10	11	12	13	14	15
Indeks Random	0,00	0,58	0,90	1,12	1,24	1,32	1,41	1,45	1,49	1,51	1,48	1,56	1,57	1,59

Calculation of consistency index (CI), measurement this is meant to know consistency answers that will affect the validity results. The formula is as follows:

$$CI = \frac{(\lambda_{\max} - n)}{(n-1)} \quad (2)$$

To know if CI is by magnitude certain good enough or not, please note the ratio is considered good, if:

$$\text{CR} \leq 0.1 \quad (3)$$

2.3 *Simple Additive Weighted*

Simple Additive Weighting (SAW) method is often also known as weighted summing method. The basic concept of the SAW method is to find the weighted sum of performance ratings on each alternative on all attributes [11]. Simple Additive Weighting (SAW) method is often also known as weighted summing method. The basic concept of the SAW method is to find the weighted sum of performance ratings on each alternative on all attributes [12]. The SAW method requires the process of normalizing the decision matrix (X) to a scale comparable to all existing alternative ratings [13]. This method is the most famous and most widely used method of dealing with Multiple Attribute Decision Making (MADM) situations. MADM itself is a method used to find the optimal alternative of a number of alternatives with certain criteria (This SAW method requires decision makers to assign weights to each attribute) [14]. The total score for the alternative is obtained by summing all the results of the multiplication between the ratings (which can be compared across attributes) and the weight of each attribute. The rating of each attribute must be dimensionless in the sense that it has passed the normalization process of the previous matrix [15]. Pattern normalization:

$$r_{ij} \begin{cases} \dfrac{X_{ij}}{Max_i \quad X_{ij}} & \text{If } j \text{ is atributte benefit} \quad (4) \\ \dfrac{Min_i \quad X_{ij}}{X_{ij}} & \text{If } j \text{ is atributte cost} \quad (5) \end{cases}$$

Pattern of preferency: $V_i = \sum_{j=1}^{n} W_j \; r_{ij}$

IOP Conf. Series: Journal of Physics: Conf. Series **1007** (2018) 012034 doi:10.1088/1742-6596/1007/1/012034

3. The Result

3.1 Scheme of Analytic Hierarchy Process

The first step in getting a consistency value is to calculate "ʎ" max first. "ʎ" max = 8.470, then calculate Consistency Index (CI) using equation (2): Calculate Consistency Ratio (CR) using Ratio Index equation used CR = (0.0671) /1.41= 0.0476. The Consistency Ratio value obtained shows that ≤ 0.1, which indicates that the weight of the generated criteria has been feasible to use. Counting Weight Results priority of the calculation results are poured in the results of weight using AHP method based on the criteria can be seen in Table 2.

Table 2. Results Weights using AHP Method based on criteria.

Criteria	Loyalty	Job Perfor mance	Responsi bility	Obedi ence	Hone sty	Coope ration	Educa tion	Leade rship
Eigen Vector	0.184	0.155	0.140	0.139	0.110	0.102	0.084	0.086

From the calculation using AHP method based on the criteria, it can be concluded that the highest weight value is in the loyalty criterion with the value 0.184, then the lowest value of weight on the Initiative criteria with the value of 0.084. The results of alternative calculations based on the criteria weighting can be seen in Table 3.

Table 3. Alternative Results Using AHP Methods Based on Criteria.

Criteria / Alter native	Loyalty	Job Perfo rmance	Respon sibility	Obedi ence	Honesty	Coope ration	Edu cation	Leader ship	Final Result	Rangk
Eigen Vector	0.184	0.155	0.140	0.139	0.110	0.102	0.084	0.086		
M. Musa	0.261	0.267	0.254	0.319	0.284	0.286	0.226	0.291	0.274	1
Alvin Syahrin	0.256	0.259	0.255	0.217	0.202	0.249	0.220	0.247	0.241	2
Noviyanti	0.194	0.163	0.218	0.166	0.230	0.194	0.198	0.191	0.193	3
Sofia	0.141	0.189	0.167	0.153	0.142	0.146	0.182	0.150	0.158	4
Syaiful Aswad	0.148	0.123	0.105	0.146	0.142	0.125	0.174	0.121	0.135	5

3.2 Scheme of Simple Additive Weighted

Ranking using SAW method with steps:

1. Decision Matrix

The data used are Candidate List of Head major Department of SMK Muhammadiyah 04 Medan as many as 5 people sample. Seen in Table 3 consisting of criteria such as Loyalty (C1), Job Performance (C2), Responsibility (C3), Obedience (C4), Honesty (C5), Cooperation (C6), Initiatives (C7) and Leadership C8) and alternatives consisting of Muhammad musa (A1), Sofia (A2), Noviyanti (A3), Alvin Syahrin (A4) and Syaiful Aswad (A5). Result of decision matrix can be seen in Table 4.

Table 4. Decision Matrix.

	C1	C2	C3	C4	C5	C6	C7	C8
M. Musa	0.261	0.267	0.254	0.319	0.284	0.286	0.226	0.291
Alvin Syahrin	0.256	0.259	0.255	0.217	0.202	0.249	0.220	0.247
Noviyanti	0.194	0.163	0.218	0.166	0.230	0.194	0.198	0.191
Sofia	0.141	0.189	0.167	0.153	0.142	0.146	0.182	0.150
Syaiful Aswad	0.148	0.123	0.105	0.146	0.142	0.125	0.174	0.121

1. Normalization of Matrices

Normalization of matrices based on equations with reference all attributes is a benefit or benefit, can be seen in Table 5.

Table 5. Normalization of Weighted Matrix Results.

	C1	C2	C3	C4	C5	C6	C7	C8
M. Musa	1.000	1.000	0.995	1.000	1.000	1.000	1.000	1.000
Alvin Syahrin	0.983	0.970	1.000	0.680	0.711	0.872	0.975	0.847
Noviyanti	0.745	0.610	0.854	0.521	0.808	0.679	0.877	0.657
Sofia	0.539	0.708	0.656	0.479	0.500	0.510	0.808	0.514
Syaiful Aswad	0.565	0.462	0.410	0.458	0.500	0.439	0.771	0.416

2. Ranking Process

Determination of ranking process by using weight which have been obtained from Decision Maker of SMK Muhammadiyah 04 Medan then process ranking with SAW method:

W = [0.16; 0.15; 0.14; 0.13; 0.12; 0.11; 0.09; 0.10]

The final result is obtained by calculating multiplication matrix with weighted value. For example, Muhammad musa with initials A1 has the normalization value of each criterion (1,000,1,000, 0.995, 1,000, 1,000, 1,000, 1,000, 1,000) multiplied by weighted (0.16; 0.15; 0,14; 0.13; 0.12; 0.11; 0.09; 0.10) then the result is 0.993. Etc. The results can be seen in Table 6.

Table 6. Results Ranking.

Rangking	Alternative	Value
1	Muhammad musa	0.993
2	Alvin Syahrin	0.883
3	Noviyanti	0.707
4	Sofia	0.578
5	Syaiful Aswad	0.490

4. Conclusions

From the results of research and discussion undertaken conclusions can be drawn as follows:

- Using AHP method makes it easier to make weighted value but not easier to make decision. Then better both of method to combination, not comparison.
- Using two methods as a comparison it will raise the confidence of school leaders in deciding who can become head of department TKJ SMK Muhammadiyah 04 Medan.
- Between the AHP and SAW methods indicates that the 1st rank is occupied by the same alternative but the difference is each of its values. SAW method shows a higher number of AHP methods.
- Selection of Promotion of Candidates Officials based on the criteria in this research has been determined that is Loyalty.

References

[1] Efraim Turban, *et al* . 2005. *Decision Support System and Intelligent Systems Edisi 7 Jilid 1,* Andi Yogyakarta.

[2] Kadarsah, Suryadi. 2003. *Sistem Pendukung Keputusan,* PT.Remaja Rosdakarya Bandung.

[3] Kusrini, 2007. *Konsep dan Aplikasi Sistem Pendukung Keputusan*. Yogyakarta. Penerbit Andi.

IOP Conf. Series: Journal of Physics: Conf. Series **1007** (2018) 012034 doi:10.1088/1742-6596/1007/1/012034

[4] Subakti. I., 2002, *Sistem Pendukung Keputusan,* Jurusan Teknik Informatika, Fakultas Teknologi Informasi, Institut Teknologi Sepuluh Nopember, Surabaya. *http://muhayat.com/downloads/Kuliah%203%20Decision%20Support%20Systems.pdf* 29 April 2011.

[5] Saaty, T.L. 2001. *Decision Making For Leaders. Forth edition, University of Pittsburgh*, RWS Publication.

[6] Anjali Singh, *et. Al.* 2015. *An AHP-PROMETHEE II Method for 2-tuple Linguistic Multicriteria Group Decision Making*. 978-1-4673-7231-2/15/$31.00. IEEE.

[7] Goyal, R.K. & Kaushal, S. 2015. *Effect of Utility Based Functions on Fuzzy-AHP based network selection in heterogenous wireless networks.* RAECS UIET Panjab University Chandigarh 21-22nd :

[8] Balubaid, Mohammed and Alamoudi, Rami. 2015. *Application of the Analytical Hierarchy Process (AHP) to Multi-Criteria Analysis for Contractor Selection*. American Journal of Industrial and Business Management, 5, 581-589.

[9] Brijendra Singh, 2015. *Analytical Hierarchical Process (AHP) and Fuzzy AHP Applications Review Papper.* International Journal of Pharmacy and Technology. ISSN: 0975-766X.

[10] Marimin. 2004. Teknik dan Aplikasi Pengambilan Keputusan Kriteria Majemuk. Penerbit PT Grasindo, Jakarta.

[11] Rahmawati, N.F, Wibawa, H.A. & Bahtiar, N. 2013. *Sistem Pendukung Keputusan Pemilihan Penerima Beasiswa Dengan Metode Simple Additive Weighting (Studi Kasus Di SMA N 1 Karanganyar Kebumen).* Journal of Informatics and Technology 2(3): 59-65.

[12] Kahraman, C., Cebeci, U. & Ulukan, Z. 2003. Multi-Criteria Supplier Selection Using Fuzzy AHP. *Logistics Information Management* 16·(6): 382-394.

[13] Azizollah Memariani. 2009. *Sensitivity Analysis of Simple Additive Weighting Method (SAW): The Results of Change in the Weight of One Attribute on the Final Ranking of Alternatives*. Journal of Industrial Engineering 4: 13 - 18.

[14] Adriyendi. 2015. *Multi-Attribute Decision Making Using Simple Additive Weighting and Weighted Product in Food Choice*. I.J. Information Engineering and Electronic Business, 6, 8-14.

[15] Balaji, N., Sambasivam, G., Murugaiyan, S.R., Basha, M.S.S., Vengattaraman, T. & Dhavachelvan, P. 2013. *Appraisal and Analysis on Diversified Web Service Selection Techniques based on QoS Factors***.** International Journal of Engineering and Technology (IJET*)*, 5(3): 3012 - 3019.

IOP Conf. Series: Journal of Physics: Conf. Series **1007** (2018) 012035 doi:10.1088/1742-6596/1007/1/012035

SQL Collaborative Learning Framework Based on SOA

S Armiati[1*], RM Awangga[2]

[1]Applied Program of Informatics Management, Politeknik Pos Indonesia
[2]Applied Bachelor Program of Informatics Engineering, Politeknik Pos Indonesia

E-mail: *sariarmiati@poltekpos.ac.id

Abstract. The research is focused on designing collaborative learning-oriented framework fulfilment service in teaching SQL Oracle 10g. Framework built a foundation of academic fulfilment service performed by a layer of the working unit in collaboration with *Program Studi Manajemen Informatika*. In the design phase defined what form of collaboration models and information technology proposed for *Program Studi Manajemen Informatika* by using a framework of collaboration inspired by the stages of modelling a Service Oriented Architecture (SOA). Stages begin with analyzing subsystems, this activity is used to determine subsystem involved and reliance as well as workflow between the subsystems. After the service can be identified, the second phase is designing the component specifications, which details the components that are implemented in the service to include the data, rules, services, profiles can be configured, and variations. The third stage is to allocate service, set the service to the subsystems that have been identified, and its components. Implementation framework contributes to the teaching guides and application architecture that can be used as a landing realize an increase in service by applying information technology.

1. Introduction

Learning has four kinds of domains, namely transmission, acquisition, accretion, and emergence[1], and also have a good condition of the body with enough sleep[2] and not in the rehabilitation programs[3]. Transmission is the process of delivering information based on evaluation[4] and analysis[5], knowledge, ideas, and skills to others by way of direct delivery, demonstration and provide guidance. The success of this domain with a domain is 10%. The acquisition provides choice in learning and most relevant for students. This method includes exploring, experimenting, self-instruction, inquiry, and general curiosity [6]. Currently, the acquisition resulted in 20% of what is learned. Accretion has the stages of learning are often gone unnoticed, as the process of learning the language, culture, customs, making assumptions, social rules and behavior. This process can take anywhere and generates 70% success to learn from what is learned [7]. Emergence is the result of the development of ideas and understanding patterned and structured like never before, is the result of a merger between thought and interaction. Forms of learning such as synthesis, creativity, intuition, wisdom, and problem-solving[8]. This method does not rely on in terms of time and can generate new knowledge, are intimately associated with inspiration and authenticity (originality). In the practical context of education today, the success of this domain approximately 1-2% of the learning process[7]. The most important component of collaboration is essential in discussions that do work in the process so that the cognitive benefits would be in collaborative learning demands that must be agreed verbally between students. In learning the reciprocal interactions among the students used to influence student learning [9]. This learning is a form of collaboration and the fact that in conversations that were articulated by the students, as well as the sharing of understanding of the student, the task of cognitive learning potential. Collaboration students in the class formed the thinking process and provide feedback in the form of social behaviour positive interdependence in the form of giving and receiving help, exchange of resources and information, support each other and work together in the process and progress [10].

The interaction in the classroom needs to be managed to build a deep understanding of learning in students. Moore argues that this management involves three types of learning interactions that[11]: the interaction of students with resources, student interaction with teachers, sstudents interaction with students. The research will focus on the design of information systems as a tool for collaborative learning process (Collaborative Learning / CL), because of the nature of collaboration in learning to encourage students to participate. Research in the CL can be seen from seven perspectives: collaborative interaction control, the task of CL, the theory of learning in collaboration, the design of the CL environment, the role of the students, the domain collaboration and learning methodologies that support collaboration. Seven viewpoints are used as an essential dimension of the CL which should be considered when designing, building and using CL system. The output of this research is a model of the CLF for learning SQL Oracle 10g and publication of scientific articles. The contribution of this study is to provide benefits (deliverables) in the form of a framework for collaborative learning guidelines, which will contain: function and task sharing what is being done in a collaborative learning process, document design procedural framework for collaborative learning and basic knowledge and needs of a collaborative learning system.

2. Method

In the phase of literature, used the theory of collaborative learning, how to build the model and the concept of Service Oriented Architecture (SOA) as an approach and the roles and division of tasks involved in the process of learning business[12] in lattice SQL Oracle 10g[13]. In the phase of analysis of business processes of learning, in general, happens in Information Management Prodi modeled using IDEF modeling to determine the rule, entities, roles, inputs and outputs of each process[14]. From the analysis which is accompanied by a document analysis, pattern of learning outcomes and the results of the questionnaire obtained the aspects that can be considered in the design of the learning framework[15]. In the design phase of collaborative learning framework (CLF)[16], designed a framework which is adaptive to the content and interaction is required in the course SQL Oracle 10g.

SOA Foundation has a life cycle phases as shown in Figure 1 [17]. Modelling is the process of designing the depiction of a business based on business needs and objectives.Describing the design of a business that includes the use of supporting equipment can help display the scenario of "what if" with various business parameters. The process that occurs can be simulated using these parameters to estimate the effect on business processes and IT systems.Designing a business used to communicate business goals with IT organizations that will assemble artifacts as a result of the implementation of the information system design phase. Enterprise architects in collaboration with business analysts to change business plan into a set of business process definition, so that it will clarify the services that are needed for business activities. Assembly phase will include the use of a set of policies and rules to oversee the passage of applications in environmental, operational characteristics important, constraint positioning, dependence on the resources, monitoring the integrity and security of access. Deploy phase in the lifecycle includes the process of creating application environments and task declaration of the application. This includes determining the resource dependency of application, operating conditions, integrity and access restrictions. Some things to note with regard to the construction of infrastructure and the environment is a service that has been provided to support the application, then that should be considered is the appropriate platform for user interaction, the flow of business processes, business services, service access and logical flow of information. Phase manage includes task, technology, and software used to manage and supervise the application assets, such as services and business processes that run in a production environment. Manage phase will also include the management of the business model, the placement of the operational environment that meets business goals and to measure the success or failure of your business objectives. SOA Governance is critical for SOA projects, governance helps in the monitoring plan that will be carried out. SOA Governance has four objectives such as make a decision, define the value of business services, managing the life cycle of business assets and measure effectiveness.

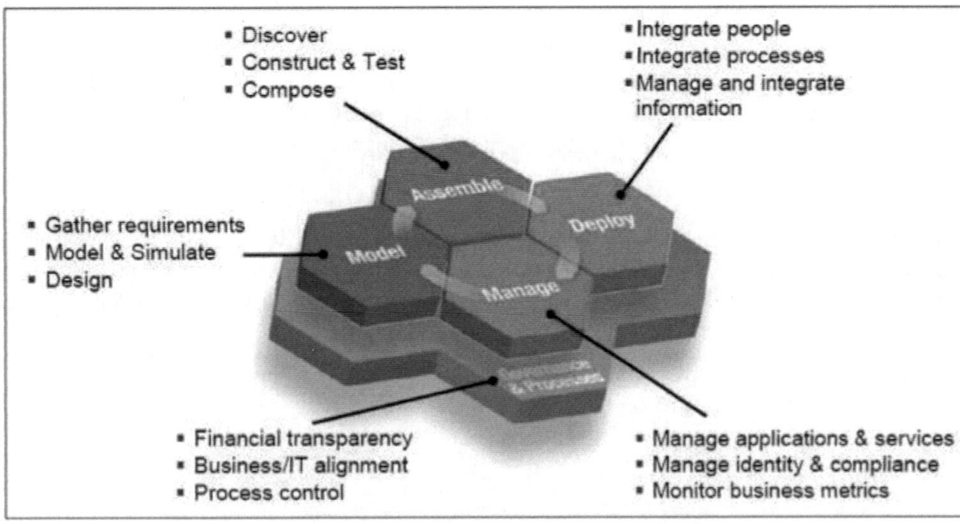

Figure 1. IBM SOA Foundation Lifecycle [17].

3. Experimental setup

It has a theory and practical major. In some subjects that require skills, even more, practice hours than the hours of theory[18]. Most students actively participating in the learning process because of the presence of students is always controlled by the Warning Letter (SP). Students are also encouraged to be active in the process with given tasks independently by lecturers[19]. As control of the material submitted SQL Oracle 10g, professors are required to make *Satuan Acara Perkuliahan* (SAP) and the Outline of Teaching Program (GBPP). For the same courses, but are taught by professors who are different, then SAP and GBPP arranged to be the same. With SAP and GBPP same students are expected to get a relatively uniform material. Lecturers who do not have an academic position will be accompanied by lecturers who already have an academic position which acts as the coordinator of the course. To control the presence of lecturers and the material being taught, after giving the lecture the lecturer is required to put a signature on the attendance sheet. During this time the presence of a lecturer in the Department of Information Management is quite good with the attendance of lecturers mostly close to 100%, or an average of 99%. Evaluation of the learning process applied to both the faculty and the students [20]. The evaluation of the students consists of the provision of initial tests and practice tests every lab meeting, giving a warning letter related to absenteeism, a warning letter related to the Achievement Index (IP) students were substandard, and the provision of Card Study Results at the end of the semester. Evaluation of the lecturer's Database 2 (Oracle 10g) is done by giving questionnaires to students to assess faculty performance distributed at the end of the semester. Once the service is identified then do the classification stages.

Analysis Subsystem activity is used to determine subsystem involved and dependence as well as the flow between the subsystems. Subsystem using use case analysis to identify the domain decomposition service and manufacturing object model that represents internal jobs and design

IOP Conf. Series: Journal of Physics: Conf. Series **1007** (2018) 012035 doi:10.1088/1742-6596/1007/1/012035

service. Use case that is designed to have two domains, the domain administration with staff department of Badan *Administrasi Akademik dan Kemahasiswaan* (BAAK) and domain management majors. The model consists of a collaboration that took solicitation models and team models, they both request and response exchange, solicitation model is a model of time management departments need information that is managed and may have to involve other units to make the needed information. In the subsequent activity carried details of the components that are implemented in the service to include the data, rules, services, profiles can be configured, and variations. Messages and events specifications and management definition occur in Component Specifications phase. The Allocation Service stage is to establish the service of the subsystems that have been identified and its components. Realisation Service stage determines which software can realize the service. This software can be taken from the existing customisation or rebuilt. Other options may include the integration, transformation, subscription or outsourced. At this stage determined the extent to which the modules of existing systems (legacy) will be used to realize the service and the service which will be built from scratch. Realisation service will be proposed as the development of further research.

4. Proposed Framework

The result is designing collaborative learning-oriented framework fulfilment service in teaching SQL Oracle 10g as below contained in figure 2. Figure 3 as a collaboration framework fulfilment service on the organization of learning SQL Oracle has three main components consists of outcome, process and foundation.

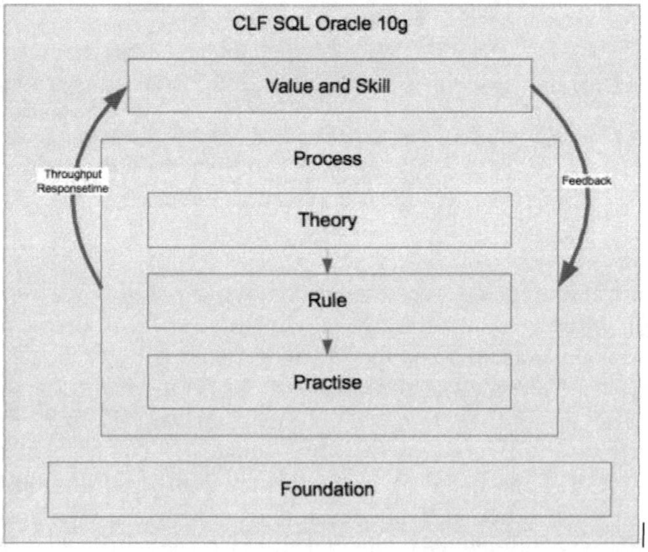

Figure 2. CLF for SQL Oracle 10g.

4.1. Outcome

Outcomes are the conditions desired by the community, in this case, is a student and lecturer SQL Oracle 10g. These results reflect the success in learning and evaluation to achieve the vision of collaboration that produced a quality score of students who achieved at the end of the semester, the value of certification testing SQL database Oracle 10g and value knowledge data use an enterprise-scale database.

4.2. Process

The main activity of the university is organizing academic activities with the aim of producing graduates who can meet the demands of the industry. In order to provide optimum satisfaction to stakeholders, universities need to plan all the activities based on its resources. In Oracle 10g SQL teaching needs to be mastered a theoretical understanding of the Oracle database architecture as work environment, rules regarding data, security, privilege, and syntax as well as the ability of the students practice and execute SQL commands in Oracle 10g.

4.3. Foundation
Foundation is an underlying component of all business processes that occur in academic activities. Each college has a vision and mission that will be used as guidance for the preparation of the vision and mission at the department level to set a specific goal achievement majors. Interest will then be declared measures achievement in the strategic planning department. Strategic planning can be expressed again in a more functional as *Satuan Acara Perkuliahan* (SAP) and *Acara Praktikum* (AP) in the course SQL Oracle 10g, to guide instructional objectives that must be accomplished students. They are composed of the fundamental SQL learning plan.

5. Conclusions
This study has proposed a model that can define the functional requirements and the need for technology on a number of the identification service that will be met. Application of collaborative learning framework as a result of the research have the opportunity contributions. Development of collaborative learning framework that is based on the concept of service-oriented and collaborative propose a model service system, application architecture and formal documentation of functional requirements for the identification services required. Implementation framework contributes to the teaching guides and application architecture that can be used as a landing realize an increase in service by applying information technology and in the development of information systems for internal management majors. Further results from the study can be used as a basis for planning for the development of database universities for the purpose of evaluation and accreditation.

References
[1] Khan, S.R. 2017 Problems and Challenges for the First Generation Learners *International Journal of Applied Linguistics and English Literature.* **6** p 318-332
[2] Yulita, I.N., Purwani, S., Rosadi, R. and Awangga, R.M. 2017 A quantization of deep belief networks for long short-term memory in sleep stage detection *2017 Advanced Informatics, Concepts, Theory, and Applications (ICAICTA), 2017 International Conference on (IEEE)* p 1-5
[3] Awangga, R.M., Fathonah, N.S. and Hasanudin, T.I. 2017 Colenak: GPS tracking model for post-stroke rehabilitation program using AES-CBC URL encryption and QR-Code *2017 2nd International conference on Information Technology, Information Systems and Electrical Engineering (ICITISEE) (IEEE)* p 255-260
[4] Pane, S.F., Awangga, R.M. and Azhari, B.R. 2018 Qualitative Evaluation of RFID Implementation on Warehouse Management System *TELKOMNIKA (Telecommunication Computing Electronics and Control)* **16**
[5] Awangga, R.M., Pane, S.F., Tunnisa, K. and Suwardi, I.S. 2018 K Means Clustering and Meanshift Analysis for Grouping the Data of Coal Term in Puslitbang tekMIRA *TELKOMNIKA (Telecommunication Computing Electronics and Control)* **16**
[6] Passi, A. and Vahtivuori, S. 1998 From cooperative learning towards communalism *Aspects of Media Education* Strategic Imperatives in the Information Age. University of Helsinki. Department of Teacher Education. Media Education Centre. Media Education Publication* **8** p 259-273
[7] Prince, M. 2004 Does active learning work? A review of the research *Journal of engineering education* **93** p 223-231
[8] Sternberg, R.J. 2005 A model of educational leadership: Wisdom, intelligence, and creativity, synthesized *International Journal of leadership in Education* **8** p 347-364

[9] Nelson, J.R. and Roberts, M.L. 2000 Ongoing reciprocal teacher-student interactions involving disruptive behaviors in general education classrooms *Journal of Emotional and Behavioral Disorder* **8** p 27-37

[10] Gokhale, A.A. 1995 Collaborative learning enhances critical thinking *Digital Library and Archives of the Virginia Tech University Libraries*

[11] Smith, C.R. 2003 Learning disabilities: The interaction of students and their environments (Allyn & Bacon)

[12] Marchman, J.F., Jenkinson, L.R. and Page, G.J. 2001 Managing international academic design collaboration *International Conference on Engineering Education*

[13] Laskey, K.B. and Laskey, K 2009 Service oriented architecture *Wiley Interdisciplinary Reviews: Computational Statistics* **1** p 101-105

[14] Indonesia, P.R. 2005 Peraturan Pemerintah Republik Indonesia Nomor 19 tahun 2005 Tentang Standar Nasional Pendidikan (Departemen Pendidikan Nasional Republik Indonesia)

[15] Klemm, W. 1994 JVME v21n1: Using a Formal Collaborative Learning Paradigm for Veterinary Medical Education *JVME* 21

[16] Panian, Z. 2007 Proceesings of the 2007 Collaboration as an imperative of application quality improvement *annual Conference on International Conference on Computer Engineering and Applications* (World Scientific and Engineering Academy and Society(WSEAS)) p 106-111

[17] Keen, M., Ackerman, G., Azaz, I., Haas, M., Johnson, R., Kim, J. and Robertson, P. 2006 Patterns: SOA Foundation-Business Process Management Scenario *IBM WebSphere Software Redbook*

[18] Soehendro, B. 2002 Kerangka Pengembangan Pendidikan Tinggi Jangka Panjang 1996-2005 (Departemen Pendidikan dan Kebudayaan, Direktorat Jenderal Pendidikan Tinggi)

[19] Brodjonegoro, S. 2005 Tanya Jawab Seputar Unit Pengembangan Materi dan Proses Pembelajaran di Perguruan Tinggi *Jakarta: Depdiknas* p 10-37

[20] Walker, H.M. 1997 Collaborative learning: a case study for CS1 at Grinnell College and Austin *ACM SIGCSE Bulletin* **29** p 109-213

IOP Conf. Series: Journal of Physics: Conf. Series **1007** (2018) 012036 doi:10.1088/1742-6596/1007/1/012036

Improving TOGAF ADM 9.1 Migration Planning Phase by ITIL V3 Service Transition

Nisa Hanum Harani[1], Arry Akhmad Arman[2], Rolly Maulana Awangga[1]

[1]Applied Bachelor Program of Informatics Engineering, Politeknik Pos Indonesia
[2]School of Electrical Engineering and Informatics, Bandung Institute of Technology

E-mails: nisahanum@poltekpos.ac.id, arman@stei.itb.ac.id, awangga@poltekpos.ac.id

Abstract. Modification planning of business transformation involving technological utilization required a system of transition and migration planning process. Planning of system migration activity is the most important. The migration process is including complex elements such as business re-engineering, transition scheme mapping, data transformation, application development, individual involvement by computer and trial interaction. TOGAF ADM is the framework and method of enterprise architecture implementation. TOGAF ADM provides a manual refer to the architecture and migration planning. The planning includes an implementation solution, in this case, IT solution, but when the solution becomes an IT operational planning, TOGAF could not handle it. This paper presents a new model framework detail transitions process of integration between TOGAF and ITIL. We evaluated our models in field study inside a private university.

1. Introduction

TOGAF is a framework with detail methodology and beneficial tools in sustain the development and increasing an infrastructure of IT[1]. TOGAF ADM (Architecture Development Method) [2]is a detailed method on how to develop and managing enterprise architecture and information system. In actual, it is designed to sustain the information system (TOGAF 1.0), however since developed it is capable to support business architecture, data, and application architecture.

The latest version of TOGAF is TOGAF 9; according to [3] besides including high detail level in architecture, there is reflection improvement of the utilization and management. It is becoming a disadvantage when TOGAF handling the specific design of domain, the expertise. In this phase, what could be reached by architecture design is being the most concern. By TOGAF 9, the architecture is not final destination or priority of the organization but the suitable strategic planning to the specification of the business demand[4].

The business transformation requirement is triggering comprehensive change and continue to the mainstream, business strategy, culture, and utilization of information technology. In order to adjust to the new business requirement, there is systemic change whether in partial or comprehensively. [5] The planning of system migration activity is the most important. The migration process, involving complex elements such as business re-engineering, transition scheme mapping, data transformation, application development, individual involvement by computer and trial interaction. The most important element in the migration process is human resource involvement. The entailed migration team must focus on the development of migration project as their priority. Therefore, it is required strategic planning to handle the process.

IOP Conf. Series: Journal of Physics: Conf. Series **1007** (2018) 012036 doi:10.1088/1742-6596/1007/1/012036

There are 9 phases in ADM, one of them is migration planning. This phase discusses how to makes suitable implementation and migration planning according to the business requirement. The output of phase is migration projects planning that will be implemented by the organization. The migration planning involves value establishment, business analysis, budgeting of migration project and risk identification. However, [3] arguing that TOGAF only refer to the architecture and migration designs. The migration planning involves implementation project or solution, in this case, the IT solution, but when the solution being the operational design of IT, TOGAF does not handle it.

The Information Technology Infrastructure Library (ITIL) is the most preferred to the IT management in the world [6].ITIL is a the de facto standard to IT Service Management in a collection of process oriented best practices , based on experience in data centres running big mainframes [7].It facilitates the transmission of IT service with high quality and disperses the management procedures in sustain the business of IT operation. The latest version is ITIL V3 consisting of 5 volumes.

From the results of previous studies [8], the integration between TOGAF and ITIL make sense since both frameworks cover and related common subjects. The integration between two framework been demonstrated and validated in the âĂIJ Centro de Dados de DefesaâĂİ CDD, the IT Departement oh the Portuguese Defence Ministry (MDN). This study discusses the adoption of ITIL into IT service and respective support, difficulties found when each technical area contributed to identified the elements they managed and that area were identified with different levels of granularity, with only a small part of the area were identified. After analyzing the causes of failure, the first improvements were made by applying a TOGAF-based approach for EA development assisted together with ITIL as main concepts in the IT service orientation. All information stored in the EA repository will be modeled with ArchiMates as the language that allows modeling systematically alongside different EA layer [9].

This research is more focused on the process of system migration in the Maranatha Christian University using the concept of TOGAF and ITIL framework integration refers to the migration method [10].

2. Methods

2.1. TOGAF (The Open Group Architecture Framework)

TOGAF is a framework and method of enterprise architecture providing a methodology in the analysis of business architecture comprehensively [11]. It is used to the banking, manufacture industry, and education. There are detail method and tools to implementation process; it is differentiated to the other EA framework. The advantages of TOGAF are flexibility and open source. In general, TOGAF is EA of industrial standard. It is the framework with detail methodology and sustaining tools to develop and increasing IT infrastructure for business. TOGAF provides an approach to planning, design, implementation, and regulation for business. It could be defined as framework focused on whole business in the world under the Open Group. A method used in the framework is design, plan, implementation, and management of information and IT architecture for a business that in concrete the framework of TOGAF splits EA into four categories: Business architecture, application architecture, data architecture, and technical architecture [11]. The advantage of TOGAF as follows:

1. IT operation of the business will be more efficient. It is referred to the fact that there is low charge in development, support, and maintenance of software as TOGAF free in used by all framework.
2. The risk of future investment lower as the infrastructure of IT is simple.
3. The decision in correlation with infrastructure supply does not complex relatively as the information framework obtained easily.

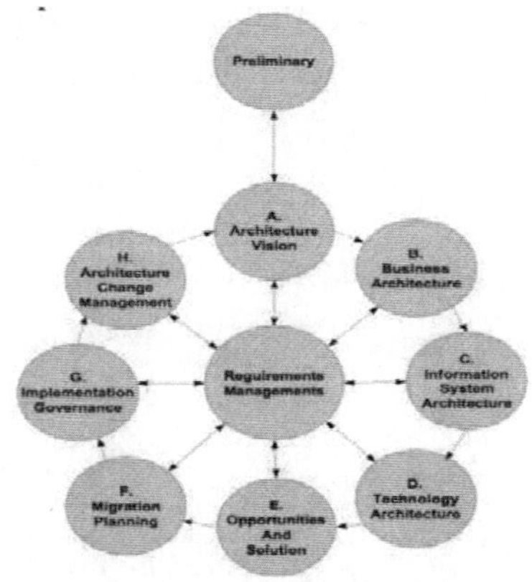

Figure 1. Structure of TOGAF ADM[12].

2.2. ITIL(The Information Technology Library)

ITIL is the most familiar IT service management method in the world. It is a cluster of best practice of IT management service inconsistent and comprehensively to preserve the qualified service approach to attaining the effectiveness and efficiency in business of information system use. Moreover, ITIL is a framework that could be developed and adapted to the system development [6]. It provides the sets of best practice from the public and private sector internationally. It is sustained by comprehensive qualification scheme, accredited training, and implementation assessment. ITIL facilitates the transmission of IT service in high quality. It is described in comprehensive series of management procedure to sustain the business to achieve the quality and value of financial and operation of IT. ITIL is the best practices document providing a manual on development and implementation of the process the present and future technological challenges. The advantages of implementing ITIL as follows:

- Improving customer satisfaction (by service transmission)
- Increase the productivity
- IT service based on process (service level management)
- Containing useful skills and standard experiences as the reference.

The latest version of ITIL is ITIL 3. The principal change is in the viewpoint of IT management; the version 2 provide the services as a cluster of the process and temporary function, while in the latest version the management service as the lifecycle(figure 2).

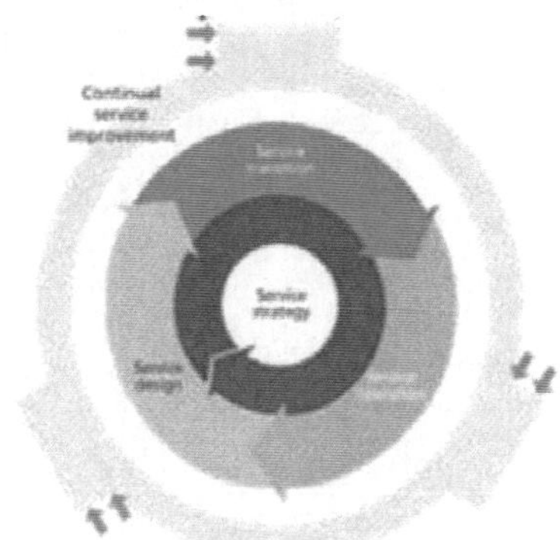

Figure 2. Service lifecycle[6].

3. Experiment and Results

3.1. Analysis of weakness of TOGAF ADM Migration Planning

IOP Conf. Series: Journal of Physics: Conf. Series **1007** (2018) 012036 doi:10.1088/1742-6596/1007/1/012036

The Phase Migration Planning in the framework of TOGAF ADM had the capability to design of migration(figure 3) process architecture. The phase consists of instruction cluster in the design of migration activity and implementation. There is a figure of migration planning process in TOGAF ADM in below.

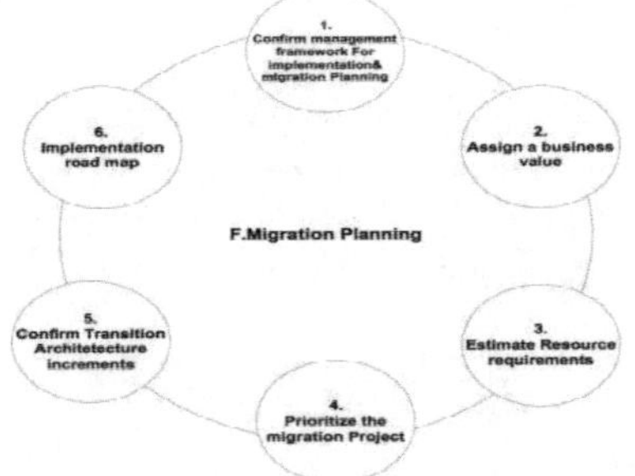

Figure 3. Migration Planning TOGAF ADM[12].

The focus of phase is how to design, create the migration design and implementation, cooperation with a portfolio of project manager. The activity involves dependence, cost, and benefit of any migration projects in correlation with migration demand that must suitable to the business demand. According to [5] the process design of migration must involve a key process to the customer (in this case, the party will experiences a benefit of migration result), the insurance of project migration and control process does not discuss in the framework of TOGAF with Migration Planning phase. A successful transmission of migration project to the ideal situation as was agreed and designed by many interests of business must be identifiable. The control process migration is also being a starting point that must be designed structurally. The migration process oriented to the customer will handle the process control problem in migration phase and could give the guarantee insignificant to the project successful. The migration process oriented to the customer will change the migration design concept to migration service process.

3.2. Mapping TOGAF and ITIL

Synchronization of business and IT is being a challenge and main priority to the CIO continuously. The synchronization occurred there is an ineffective collaboration. The collaborative setting assured that business demand is understandable by business and IT. Synchronization of business/IT(Figure 4) by effective collaboration required conforming to the three principles (had a neutral facilitator to establish the transparency and integrity), communication (assure that all party is understanding on involved terminology), context (preserve and presenting correlation between a component in the project)[6].

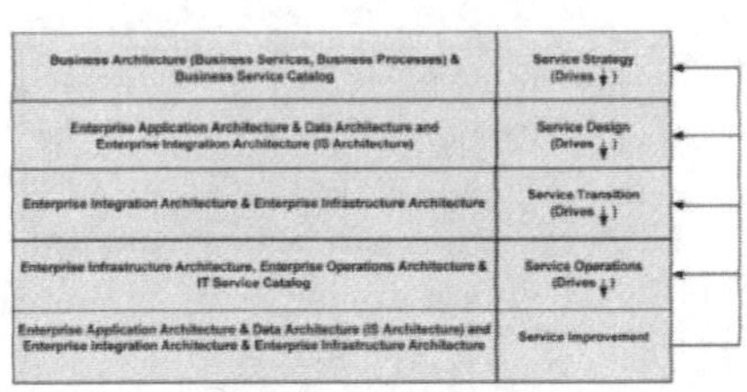

Figure 4. Enterprise Architecture and Service Lifecycle [13].

TOGAF version 9 and ITIL version 3 provides an effective method to transmit the integrated service for all business and IT synchronization[14]. TOGAF and ITIL had focused on IT service integration and business process.

Both of TOGAF and ITIL are frameworks, but handling the problem differentially. ITIL focus on preserving the IT service, while TOGAF is methodology and instrument to develop the enterprise architecture(Figure 5). TOGAF must be considered over the ITIL as it involves the lifecycle concept of product and ITIL being the product management method for the user and customer [15].

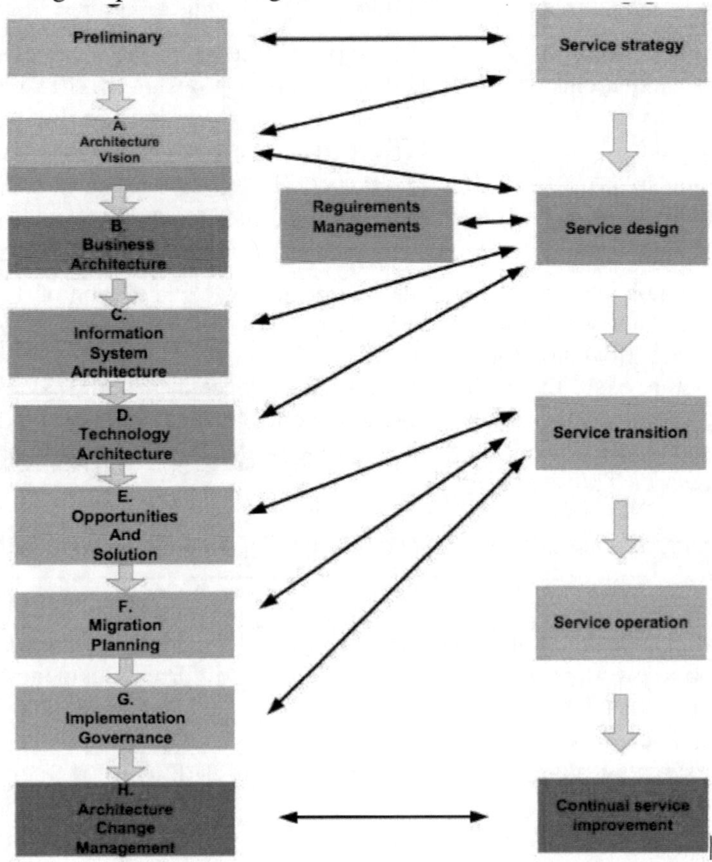

Figure 5. Correlation of TOGAF and ITIL[16].

ITIL defined of service as "value transmission to the customer by facilitating the required result by the customer". The model of architecture enterprise of TOGAF focused on optimizing.

IOP Conf. Series: Journal of Physics: Conf. Series **1007** (2018) 012036 doi:10.1088/1742-6596/1007/1/012036

Table 1. Mapping of TOGAF ADM and ITIL Service Lifecycle [5].

No.	TOGAF	ITIL
1.	Business architecture	ITIL service strategy-business strategy and IT strategy
2.	Information system architecture	ITIL service design-design coordination
3.	Technology architecture	ITIL service design-design coordination
4.	Migration planning	ITIL service transition-change management
5.	Implementation goverment	ITIL service strategy-governance
6.	Architecture change management	ITIL service transition-change management

user, process, and technology to accomplish the business target. Integration of TOGAF and ITIL can provide the structured framework to build the strategy and organization design, process and required infrastructure for service providing [9].

The migration methodology that will be used as a reference in making model mapping based on [10]. The approach written in this book has several important points that must be met in the success of the system migration activities . Following the mapping of the TOGAF ADM phase of the Migration Planning and ITIL Service Transition.

Table 2. Mapping Migration Planning Service according to the Migration Methodology[10].

No.	Migration Integration	Migration Methodology
1.	Planning Migration with transition management and policy frameworks	Identify the stakeholders and Pre-assesment
2.	Determine business value and asset management on each migration project	Define success criteria
3.	Resource Estimation along with management Configuration	Finalize the new environmental
4.	Migration planning and implementation with transition policy	Proof of Concept
5.	Prioritize migration projects along with risk identification and stakeholder decision making	Decision to migrate
6.	Resource Estimation along with management configuration	Resource estimation
7.	Stages of transitional architecture by coordinating activities on the migration process	Actual migration
8.	Roadmap implementation by providing support for the transition team	Actual migration

3.3. Integrated Model Design of TOGAF ADM Phase Migration Planning With ITIL Service Transition

The business demand is developing rapidly triggering strategic design involving a technological requirement to accomplish mutually. However, the implementation design of system does not suitable to the business requirement and result in suffering for business financial. Therefore, the system change process from previous to the latest one should pass through the documented procedure. In this research, TOGAF will provide the method to design and strategic planning of organization business to obtain and documenting a vision, mission, a strategy of the business. Moreover, TOGAF will document the artifacts of business supporter such as application architecture, data, technology, and process. Also, TOGAF will provide the potential strategy to present and being a roadmap to the targeted setting.

After design, the organization business strategy, carried out an identification of change and solution of architecture, that this phase will begin in implementation planning and identification of architecture realization as was defined previously. The phase will be the planning base of architecture migration phase. The transition processes are including the main design of organization, investigation controlling process, organization design process, and lifecycle methodology of the information system.

Perceiving the problem of system change as business change, the framework of TOGAF will bring a recommendation of system change according to the organization strategic design and ITIL will bring the comprehensive manual in transition process involves a process of transition services exist in ITIL Service Transition Manual.The analysis phase of enterprise design of architecture according to the TOGAF ADM as Figure6.

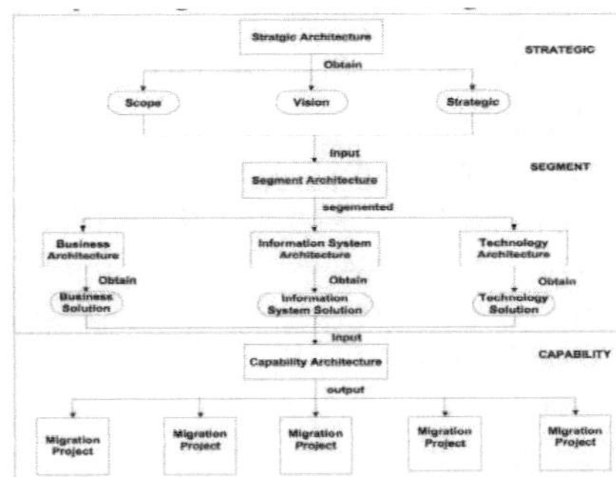

Figure 6. Integration model analysis.

Subsequent to find the architecture target design of TOGAF framework, the migration process will begin. The migration project phase will be synchronized with the demand and organization competence(Figure 7). ITIL will provide the direction to implement the transition service process by IT service oriented.

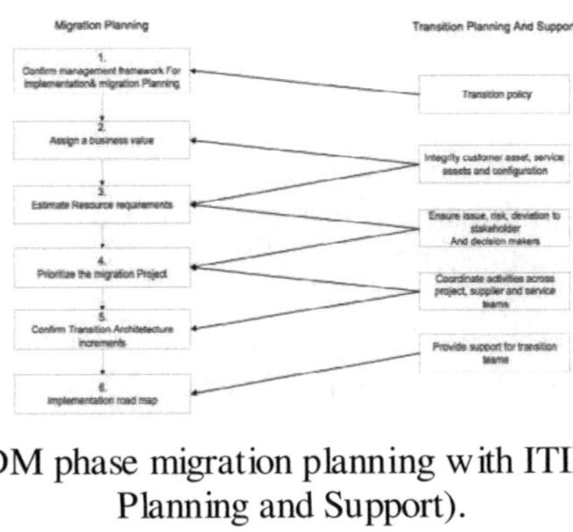

Figure 7. Mapping TOGAF ADM phase migration planning with ITIL Service Transition (Transition Planning and Support).

The phase of Migration Planning in TOGAF involves assessment and order the priority to be the base of the implementation plan. Moreover, the phase involves modeling of assessment matrix and the decision of main demand and sustaining of the organization to the implementation of the information system. By implement ITIL, implementation planning will be sustained by understanding on utility and service insurance to the transition process. ITIL will build the formal policy to implement whole required change and assure that all of them suitable for the organization demand.

Planning and transmission support of ITIL will bring knowledge, decision, and assure the process, system and another element in the transition process. The concept of transition service in ITIL will ensure the availability of knowledge of prior process and can be the direction to the similar situation in future. Migration planning integrity with the migration service could anticipate and manage the restored channel in proactive and could determine of maintenance precondition when the elements of information system service must be synchronized to the specific situation.

4. Conclusion

In this paper suggested planning method of migration process model by using of TOGAF ADM 9.1 Migration Planning and ITIL Service Transition frameworks. It is expected that the model could be used by the organization in carrying out the migration process of the information system. Addition implementation of ITIL Service Transition can assure the precondition and transition process applied logically and documented comprehensively. The result of integration model will bring a complete migration process by operational phase in the standard of migration policy according to the method of IT service.

References

[1] Taleb M and Cherkaoui O 2012 Designing Enterprise Architecture Frameworks 99
[2] Blevins T J, Spencer J and Waskiewicz F 2004 The Open Group and OMG
[3] Radhakrishnan R 2008 The Open Group
[4] Lankhorst Metal. 2009 Enterprise architecture at work vol 352 (Springer)
[5] Aghalatifi O, Tavoosain P and Farahmand M 2011 Computer Research and Development (ICCRD), 2011 3rd International Conference on vol 4 (IEEE) pp 395–399
[6] Sanatdoust A N, Hosseiniun P and Rostam G G 2011 2011 International Conference on Management and Artificial Intelligence IPEDR vol. 6 (2011)
[7] Hochstein A, Zarnekow R and Brenner W 2005 e-Technology, e-Commerce and e-Service, 2005. EEE'05. Proceedings. The 2005 IEEE International Conference on (IEEE) pp 704–710
[8] Gama N, Sousa P and da Silva M M
[9] Gama N, Sousa P and da Silva M M 2013 Building Sustainable Information Systems (Springer) pp 153–165
[10] White B, Fries W, Jorna D, Kordmann H, Nesbitt J, Packheiser F and Palacio E 2008 IBM System z connectivity handbook (Citeseer)
[11] Raynard B 2008 TOGAF The Open Group Architecture framework 100 success secrets-100 most asked questions: the missing TOGAF guide on how to achieve and then sustain superior enterprise architecture execution (Emereo Pty Ltd)
[12] GROUP O et al. 2009 Togaf version 9. the open group architecture framework
[13] Ali S M, Soomro T R and Brohi M N 2013 Journal of Computer Science **9** 1190
[14] Sante T and Ermers J 2009 Getronics Consulting, OGC
[15] van Sante T, Ermers J and Solutions K 2013 White paper, August
[16] Sall´e M 2004 Hewlett-Packard Company 8–17

Acknowledgment

The author would like to thank Dr. Ir. Arry Akhmad Arman, MT., for his most support and encouragement.

IOP Conf. Series: Journal of Physics: Conf. Series **1007** (2018) 012037 doi:10.1088/1742-6596/1007/1/012037

SMS Security System on Mobile Devices Using Tiny Encryption Algorithm

M S Novelan , A M Husein , M Harahap and S Aisyah

Faculty of Technology and Computer Science, Universitas Prima Indonesia, Indonesia

E-mails: teknisiputra@gmail.com, amirmahmud@unprimdn.ac.id, mawaddah@unprimdn.ac.id, sitiaisyah@unprimdn.ac.id

Abstract. The development of telecommunications technology is so rapid has given such great benefits. With the telecommunication technology, distance and time no longer be a significant obstacle. One of the results of telecommunications technology that is well known is the Short Message Service. In this study developed an application on the mobile phone to modify the SMS message into ciphertext so that the information content of the SMS is not known by others. SMS delivery system for encrypting messages into ciphertext using a key that is entered by the sender then sends to the destination number. SMS reception system to decrypt it to others via SMS without the fear of information from these messages will be known by others. The method used in the system encrypt and decrypt the message is the algorithm Tiny Encryption Algorithm and implemented using the Java programming language. JDK 1.7 as the Java programming language ciphertext into plaintext using the key entered by the receiver and displays the original message to the recipient. This application can be used by someone who wants to send a confidential information and the Java compiler. Eclipse, a Java SDK and the Android SDK as a Java source code editor.

1. Introduction

Encryption is the process of converting an original message called plaintext into an unreadable code or code called ciphertext and can not be understood. To return a message to the original form as it needed a process called decryption. Encryption is intended to protect and conceal information so as not to be seen by parties or non-intended persons. Inside the computer security is known a cryptographic technique, which is used for message encoding. Based on the importance and confidentiality of a message it requires a way to secure a message or information using cryptographic techniques. One of the cryptographic techniques is using Tiny Encryption Algorithm (TEA) algorithm. TEA is a block cipher algorithm created by David J. Wheeler and Roger M. Needham from Cambridge University in 1994. The most prominent of the TEA is the simplicity of implementation, the absence of S-Box and P-Box and high speed.

2. Theoritical Basis

2.1 Security

The security of an android device is one of the hallmarks of this device, as it is well known that android devices have security designed for the convenience of users, but some services on certain features that do not have a method of security on the data it stores, one of which is the SMS service

IOP Conf. Series: Journal of Physics: Conf. Series **1007** (2018) 012037 doi:10.1088/1742-6596/1007/1/012037

(Short Message Service) which on the android device by default does not have a security method at all. So it is needed to overcome the third-party software applications in securing the SMS service.

2.2 Cryptography
Cryptography comes from the Greek word from Crypto and Graphia which means secret writing. Cryptography is a science that studies the writing of a secret, in which the definition of cryptography according to (Schneier, 1996) is the science as well as the arts to maintain the security of messages (message) Cryptography is part of a branch of mathematical science called Cryptology. Cryptography aims to maintain the confidentiality of information contained in the data so that the information can not be known by unauthorized parties.

2.3 Encryption
Encryption is the process of securing an information by making the information unreadable without the help of knowledge and or special tools. The way used to perform encryption is to change a code from an understandable into a code that can not be understood. A coding system uses a table or dictionary that has been defined to replace the word of the information or piece of information sent. Encryption can be interpreted as a code or cipher. A cipher uses an algorithm that can encode all stream data streams from a message into an unintelligible cryptogram.

2.4 *Description*
Decryption is an algorithm or a way that can be used to read encrypted information for readability. In other words the description is the process of giving the result given from the encryption process into the initial form before the encryption.

2.5 *Algoritma TEA (Tiny Encryption Algorithm)*
TEA is a block cipher algorithm created by David J. Wheeler and Roger M. Needham from Cambridge University in 1994. The most prominent of the TEA is the simplicity of implementation, the absence of S-Box and P-Box and high speed. TEA operates in 64 bit block size and 128 bit key length. TEA is based on Feistel network and has 32 rounds. The K key will first be divided into 4 internal keys ie K [0..3] each length of 32 bits.

3. The Scheme Of System
The TEA encryption system uses a feistel network process by adding a mathematical function of addition and subtraction as an inverting operator other than XOR illustrated in the figure 1.

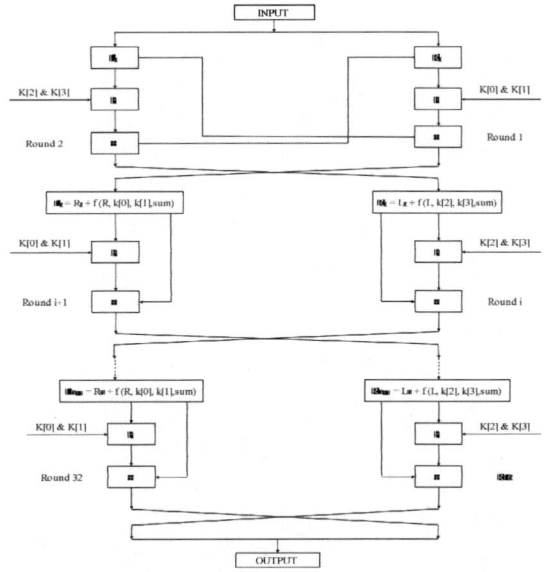

Figure 1. The structure of the Tiny Encryption Algorithm (TEA).

3.1 *Cycle TEA (Tiny Encryption Algorithm)*
The structure of the encoding with the algorithm for one cycle (two rounds) is illustrated in the figure 2:

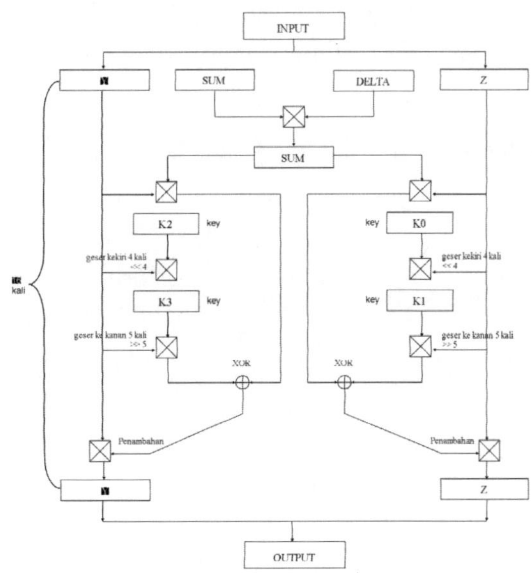

Figure 2. One Cycle TEA (Two Rounds).

3.2 *Flowchart Encryption*
Flowchart for the encryption process illustrated in the figure 3:

IOP Conf. Series: Journal of Physics: Conf. Series **1007** (2018) 012037 doi:10.1088/1742-6596/1007/1/012037

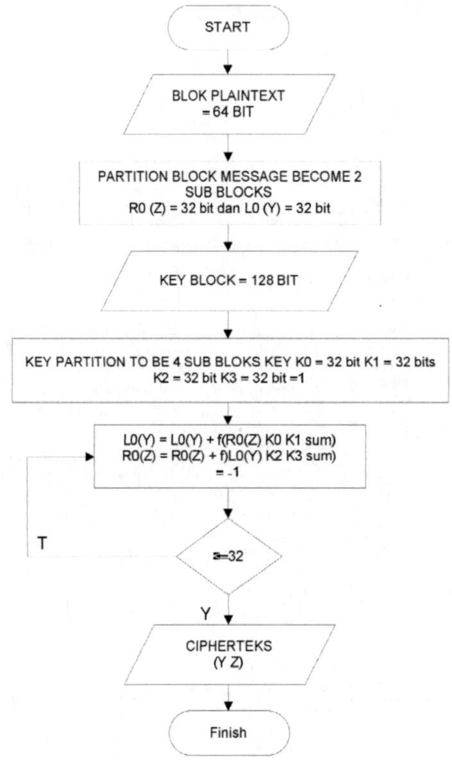

Figure 3. Flowchart of the *Encryption* Process.

3.3 Flowchart Decryption

Flowchart for the encryption process illustrated in the figure 4.

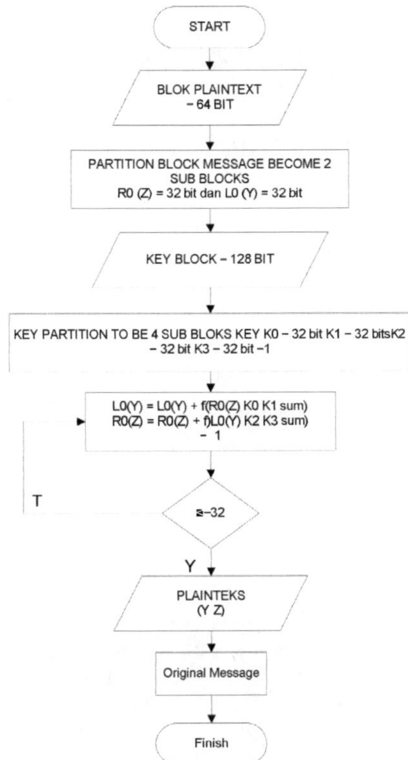

Figure 4. Flowchart of the *Decryption* Process.

4. User Interfaces

4.1 See the preview

Before the operation, the program will display a page like this:

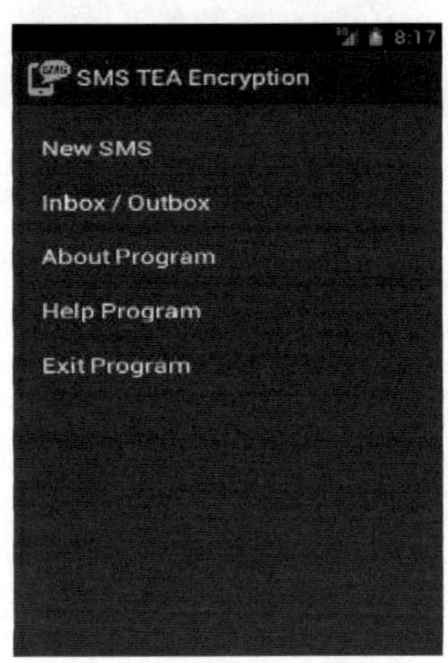

Figure 5. The main menu view.

4.2 Display new message

Access view to write sms and send messages

IOP Conf. Series: Journal of Physics: Conf. Series **1007** (2018) 012037 doi:10.1088/1742-6596/1007/1/012037

Figure 6. New SMS Menu Display.

4.3 List Message

The view a list of incoming messages, sent messages and saved messages

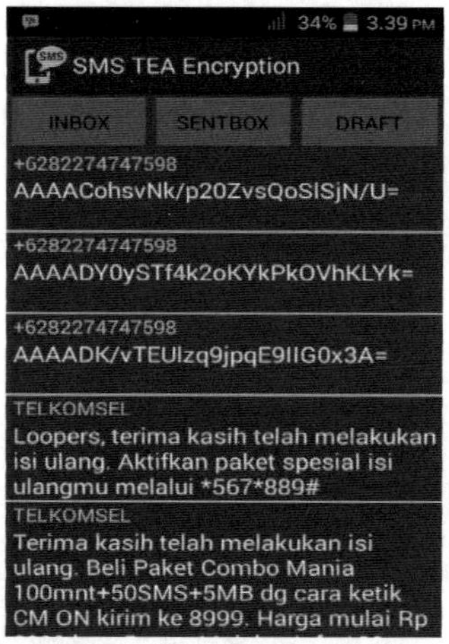

Figure 7. Display List Massage.

5. Conclusions

From the results of research and discussion undertaken conclusions can be drawn as follows:

1. The software designed can be used to process files that have been encrypted to mobile phone, can be done file decryption process to see the actual file.
2. The software can only perform encryption and decryption, done by using one time pad algorithm.
3. The key to decrypt an encrypted file must be the same as the key when the file was sent.
4. The speed of the encryption / decryption process depends on the size of the file, the larger the file size the more time it takes for encryption / decryption.

References

[1] M. Feldhofer, J. Wolkerstorfer, and V. Rijmen, ''AES Implementation on a Grain of Sand,'' IEE Proc, vol. 152, no. 1, Oct. 2005, pp. 13-20.

[2] G. Leander et al., ''New Lightweight DES Variants,'' Proc. Fast Software Encryption (FSE 07), LNCS 4593, Springer-Verlag, 2007, pp. 196-210.

IOP Conf. Series: Journal of Physics: Conf. Series **1007** (2018) 012037 doi:10.1088/1742-6596/1007/1/012037

[3] S. Kumar et al., ''Breaking Ciphers with COPACOBANA—A Cost-Optimized Parallel Code Breaker,'' Proc. Workshop Cryptographic Hardware and Embedded Systems (CHES 06), LNCS 4249, Springer, 2006, pp. 101-118.

[4] T. Good and M. Benaissa, ''Hardware Results for Selected Stream Cipher Candidates,'' Workshop Record State of the Art of Stream Ciphers (SASC 07), 2007, http://www.ecrypt.eu.org/stream/papersdir/2007/023.pdf.

[5] S. Kumar, ''Elliptic Curve Cryptography for Constrained Devices,'' doctoral dissertation, Electrical Engineering and Information Sciences, Ruhr University Bochum: Germany, 2006.

[6] L. Uhsadel, A. Poschmann, and C. Paar, ''Enabling FullSize Public-Key Algorithms on 8-bit Sensor Nodes,''Proc. 4th European Workshop Security and Privacy in Ad hoc and Sensor Networks (ESAS 07), LNCS 4572, Springer-Verlag, 2007, pp. 73-86.

IOP Conf. Series: Journal of Physics: Conf. Series **1007** (2018) 012038 doi:10.1088/1742-6596/1007/1/012038

Text Mining of UU-ITE Implementation in Indonesia

[1]Lukmanul Hakim, Tien F. Kusumasari[2] and Muharman Lubis[2*]

[1]International Islamic University Malaysia, Jalan Gombak,Kuala Lumpur, 50728
[2]Telkom University, Jalan Telekomunikasi No. 1, Bandung, 40257

E-mail: *muharmanlubis@telkomuniversity.ac.id

Abstract. At present, social media and networks act as one of the main platforms for sharing information, idea, thought and opinions. Many people share their knowledge and express their views on the specific topics or current hot issues that interest them. The social media texts have rich information about the complaints, comments, recommendation and suggestion as the automatic reaction or respond to government initiative or policy in order to overcome certain issues.This study examines the sentiment from netizensas part of citizen who has vocal sound about the implementation of UU ITE as the first cyberlaw in Indonesia as a means to identify the current tendency of citizen perception. To perform text mining techniques, this study used Twitter Rest API while R programming was utilized for the purpose of classification analysis based on hierarchical cluster.

1. Introduction

Recently, the rapid growth of social media has led many users to generate web-based contents for various subjects such as sharing information, online chatting, and expressing their opinions, thereby causing the amount of text information become abundance. Due to this phenomenon, the need to analyse unstructured text in social media cannot be ignored. Indeed, analysing social media has become a compulsion for academia and industry [1, 2] in order to discover untapped knowledge. It also supported with a diverse techniques, tools, and platforms that available publicly. The reasons why it is become a major research and business activity today are because of commercial pressures and the potential of social media for analysis.Furthermore, to ensure legal certainty in cyber transaction, encouraging economic growth and as a means to prevent cybercrime for protecting service to the citizen, government regulate UU-ITE in 2008 and has been revised in 2016 involves explanation of insulting and contaminating verse interpretations, decreasing the threat of punishment, setting up intercepts, synchronization with KUHAP, strengthen the role of civil servant investigators, the provision of right to be forgotten or the obligation to remove irrelevant content for e-system providers and the role of government. Meanwhile, there are many controversies regarding the implementation of this Act such as case on defamation involves Prita, Arsyad, Ervani, Florence, Fadhli and other hundred victims, threats to BBM from minister of informatics and SARA (ethnic, religion, race, group) website blocking to prevent radicalism.

In general, the UU-ITE (Act No.14/2008) has main objective to manage information and transaction electronically as well to stipulate the prohibited acts without specifying the principle of privacy. The role of this regulation is to tackle the growing of cybercrime, so the evidence, statement and proof in the digital form or from device became legitimate and be allowed in the court. Despite the lack of agreement between legal experts in interpreting this regulation [5], this act has the vital role as the basis for the direction of personal data protection and cybercrime prevention in Indonesia. The most important thing in the e-commerce transaction is the protection of personal data in every activity involves accessing, browsing, comparing or storing, which can be transformed into secured and trusted e-voting.However, many people considered the revision of the Act did not alter the substance of the potential problem. The provision of defamation or defamation in article 27or so-called "rubber articles" becomes the core of the objection. Some weaknesses regarding UU-ITE has been reported such as digital evidence, jurisdiction procedure, locus delicti and tempus, issues on context, content and concept [6]. This study want to identify the netizens perception on the implementation of UU-ITE based on text mining methods to see how does certain ideas or concepts has been transferred through social media and become a means to encourage government to revise and improve the regulation.

2. Literature Review

Indonesia is one of populous country with highest internet penetration has been struggling to cover all activities in cyberspace with a comprehensive cyber law, recognized as the Electronic Information and Transaction Bill (UU ITE). It is intentionally to give legal certainty in cyberspace which is the principle that should include in national and international law. However, the implementation of ITE getting various responses from Indonesian citizens, particularly from internet users since it was signed at 2008 and after the recent amendment in November, 2016. The presence of social media like Twitter easier for them to involve in a discussion, to share political contents base on their interest, or to create an online forum to criticize government policies with many participants involves. Moreover, the large number of Twitter users also impacted socio-political transition such as Arab Spring, and the Occupy Wall Street movement [16]. With simple features, short, and interactive comments, Twitter attracts many users to join in their network. In 2010, Indonesia categorized as the highest Twitter penetration, and ranked as the most tweeting country in 2014 [13]. This is due to the fact that, there are many cases convicted unprecedented numbers of citizens in Indonesia who interacting in the cyberspace. Such cases unfolded the intention of internet users, particularly in the Twitter that day by day became a trend. In majority, the social media users are the stakeholder who are the most susceptible to be criminalized with the articles of UU ITE rather than the journalisms. According to Southeast Asia Freedom of Expression Network (SAFEnet) noted that there are more than 200 reports about the libels, desecration of religions and threats based on the UU ITE [8].

Text mining could be used to discover useful information, patterns, and knowledge in social media [4] although it still needs human intervention to interpret the output of text mining. It gives new possibility to researchers from wide range of disciplines to use text mining as vehicle of analysis [7], and also constitutes as an effective way to meet user's diverse information needs [10]. Some scholars from multiple studies have exploited the text analytics in social media for their research development, such as political communication [7], security and defence[9], policy implementation [11, 12]. In another hand, some study tried to compare between the official poll or surveys and text mining analysis, they found that the social media data (e.g. twitter) could be used for social study because of strong correlation between them [7]. However, some criticism derived from other studies justifying that social media like Twitter is not a representative sample of any population [12]. Furthermore, the richness of text mining features could also utilize by government agencies to understand public responses, feedbacks, and reactions upon a policy implementation or initiative [15]. In order to achieve the transparency and engagement in government agencies, it is compulsory for them to get closer to the citizens by initiating to glean 'citizen sourcing'. Most of netizens express their opinions in social media because of efficient channel for creation and exchange of political content [11].

Actually, many ways to represent the data in text mining; one of them is regarding of bag-of-words [10, 14]. This text mining approach has been used mostly in text analysis, Natural Language Processing (NLP), and Information Retrieval (IR) [1] rather than an algorithmic point of view regardless of its drawbacks [3]. The features of bag-of-words (BOW) also used by researchers in their study for some purposes, for example, the feature of BOW designed to detect location expression of Twitter users [9]. Therefore, this study engaged a bag-of-words model to retrieve an interesting topics,

IOP Conf. Series: Journal of Physics: Conf. Series **1007** (2018) 012038 doi:10.1088/1742-6596/1007/1/012038

patterns, and terms from social media Twitter in relation with ITE implementation in Indonesia.The intention of havingsecurity policy was not to persuade users but to convince them, by letting the users reflect, on their ownterms, on why information security is important and on how to react in certain circumstances [18].

3. Research Methodology

The text messages (tweets) collected from the availability of web-based Application Programming Interface (API) provided by Twitter. To retrieve relevant text data, some parameters used to filter the need information such as keywords search, hashtags, phrases, regions, and usernames or user ID. There are 1843 tweets present in the corpus.

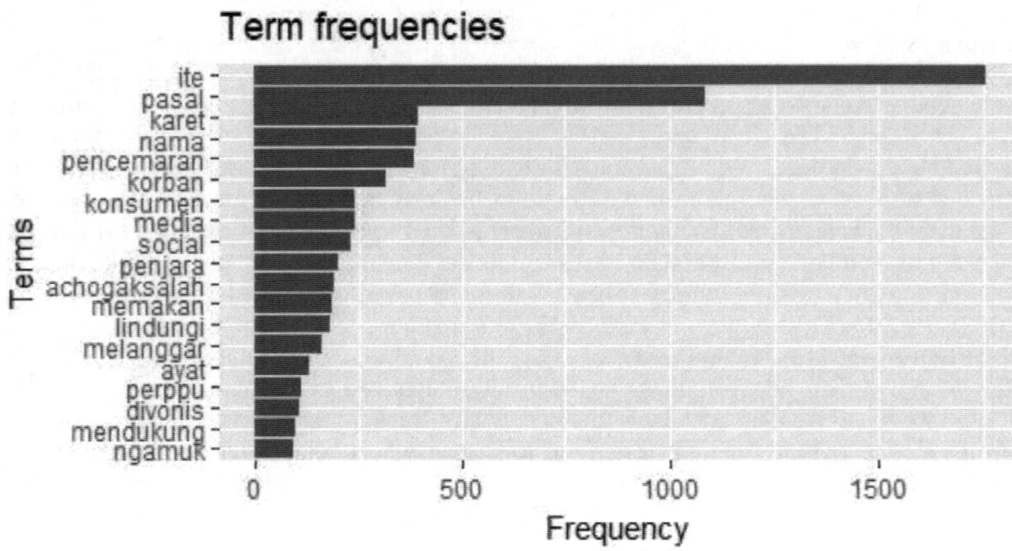

Figure 1. Terms frequency of ITE tweets.

This study applied text mining techniques to discover the perceived issues surrounding ITE implementation in social media (Twitter). The first stages are to scrape tweets in Twitter by using Twitter Rest API, in which R-Programming as the tool that was used in the entire processes of text analysis. The next stage, when the data that was extracted from Twitter APIcompleted, the next stage called pre-processing techniques such as removing stop-word, stemming and TF/IDF algorithm. The motive that stop-words should be removed from a text is that they make the text look heavier and less important for analysts. Removing stop words reduces the dimensionality of term space. The most common words in text documents are articles, prepositions, and pro-nouns that does not give the meaning of the documents. Thus, the identification of root word should be done to remove various suffixes, to reduce the number of words, to have accurately matching stems, to save time and memory space. The third step is to build the term-document matrix where the features of terms were selected in a collection of the documents. In the next section the terms are plotted with word association, which available in R. Finally, the data then stratified with clustering techniques such as hierarchical cluster for further analysis. By building cluster tree or dendrogram to represent data, each node links to two or more successor node, which are nested and organized as a tree, ideally ends up as a meaningful classification scheme. Each node in the cluster tree contains a group of similar data; nodes group on the graph next to other, similar nodes. Clusters at one level join with clusters in the higher level by using a degree of similarity. The process carries on until all nodes are in the tree, which gives a visual snapshot of the data contained in the whole set. The total number of clusters is not predetermined before starting the tree creation.

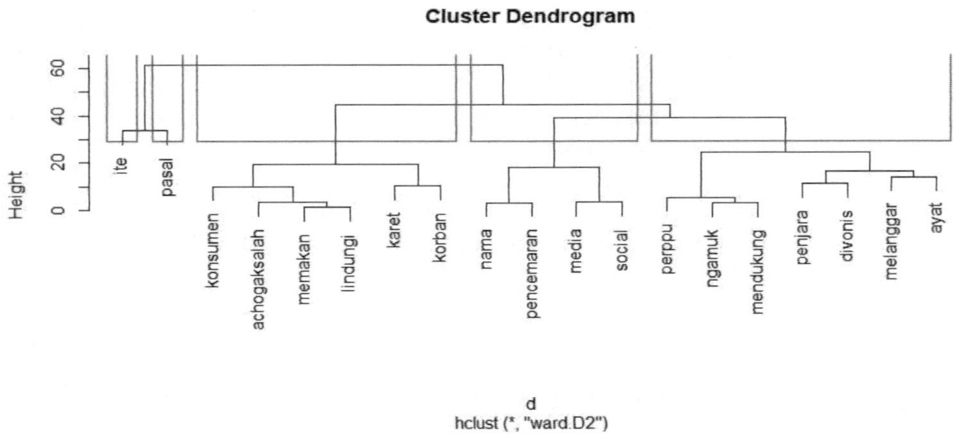

Figure 2. Hierarchical cluster of ITE tweets.

The figure 1 presented obviously the frequency of each term which simply to show what the popular terms that discussed in the Twitter are. The word "ite" is the most used term in the tweets followed by "pasal/article", "karet/elastic", "nama/name", "pencemaran/defamation", and "korban/victim", while the clusters of each term could be seen in the figure 2 which correlated to another term. These words grouped according to the distance of each term based on the method of hierarchical clustering. The distance between terms are calculated with dist()after scaling, then clustered with hclust() and dendrogram is cut into 5 clusters. The agglomeration method is set to ward to indicate the increase in variance when two clusters are merged. Other than that there are many other options can be use such as single linkage, complete linkage, average linkage, median linkage, and centroid. Some of sparse terms are removed to prevent the plot with a crowd words. These two graphics are used to facilitate the researchers easy to interpret the issues surrounding ITE implementation in Indonesia from Twitter users that vocally expressed their opinion, especially the case related to freedom of expression. Based on the figure 2, from the first cluster that the case is occurred involving a person named "Acho", which perceived against consumer right, so that they blamed some articles in the UU ITE has potential to be interpret improperly. The second cluster denoted about the libel cases that still prevailed in Indonesia, particularly in social media that attracted the "netizens" attention. The remains clusters discussed about the issues of infringement related UU ITE that became popular among the netizens in the twitter.

4. Discussion

4.1 Legal Consequence

The Twitter users denoted that they confronting such a fear issue in the internet, whether to communicate, to interact, or to express their opinions by considering the legal effect upon certain conducts. This is due to the fact that particular cases happened in Indonesia that prosecuted some perpetrators, received many responses of the netizens to the articles which charged the suspects with improper articles. Moreover, the defamation of others reputation is still prevailing among the internet users in Indonesia. It is recorded in the cluster dendrogram that the words "pencemaran" and "nama" closed together. The action of libel often involves parties and individuals or vice versa that potential to be misuse by certain parties to disadvantages others. It is because the vague definition of defamation leading to multi-interpretation among citizens, or certain parties. Furthermore, the perception of doubt upon certain articles such as the article 27 section (3) has disclosed public trustworthy while article 28 also became the consideration due to possibility to be misinterpreted. Even though the first amendment has been made, the anxiety is still leaving a big trauma of the citizens. There as suggestion to remove this article and use article 310-311 of the Criminal Code (KUHP). Therefore, it can be threaten freedom of expression in the cyberspace. There are also high number of hateful messages in the social media such as Luna Maya comment's in her twitter account says that the degrees of infotainment is more despicable than prostitutes and murderers. Indeed, the attempts to meddle in private matters should be discourage through regulation to preserve the individual privacy. In short, the issues of UU-ITE based on text mining relate to the importance of defamation article, the uniformity of insulting case, vagueness of "good name" and strange proof attempt in the court. Some problematic verses has been rejected by constitutional court after the Act initiated, so the other judicial review cannot be conducted. Thus, government should take initiative by revise plan or develop new regulation to give clear definition of certain vague term and to remove distrust among citizen on the proper intention of this Act.

4.2 Precedence
There are many cases convicted unprecedented number of netizens who interacting in the cyberspace. Based on the recent cases discussed in the Twitter, some of these issues attracted the internet users, particularly in the Twitter that day by day became a trend. The presumption of innocent went viral among them. It is begin with tweets of someone or news portal in the Twitter somehow shared quickly to other users. The sympathy channelled to the particular person who presumes commits libel based on themselves perceptions. In short, the assumption of arbitrary power abuse by some parties encouraged them to express their opinion in the Twitter. It is aligned with other research [17] found that only a minority understood the UU ITE really well and applied it in their daily operations either in the individual task or at company operation consistently and strictly while the majority of Indonesians knew nothing about it at all. There were so many companies in Indonesia, which had just started operations and were still ignorant of the rules and regulations governing this sector and at the same time did not feel obliged to consult legal experts. The worst part was that there was no proper control or enforcement in cases of fraud. Occasionally, most Indonesians had made bad experiences in using electronic devices, often in the form of spam delivered to mobile phones or personal computers, either to exploit the emotion in the form of cheating or allegation of defamation [6], which might result in the reluctant support of the government policy and regulation implementation to overcome cyber issues. Moreover, since fraud occurred freely without proper response from the government, Indonesians rightfully and understandably doubted in honest, fair and proper handling of cybercrime based on UU-ITE ever to take place. Indeed, government should not be worried to get criticism from citizen while at the same time provide the way to remind every components include social and mass media to consider ethics and norm in sharing and delivering their mature opinion and comprehensive thought. Certain restriction can be implemented in other type of regulation to prevent other citizen's right be violated by the others such as cyberbullying or cyber stalking. The plan to draft criminal code in information technology should be supported instead of commenting on the negative effect on certain regulation implementation. Government should take a note that strict restriction should be imposed on specific sensitive issues only such as SARA, hoax, hateful message, spam and other negative activity, not to whole content of communication from citizen about policy or their experience.

5. Conclusion
To maintain the enforcement of personal data protection, government is encouraged to create the periodical auditory, proper compliance and privacy impact assessment towards the implementation of the policy. Therefore, it is really resourceful to understand the factor that shape the negative sentiment from netizens. At certain extent, educational level, environment location, economic condition and social status play essential role to determine on how people to perceive government attempt in solving national problems. By understanding these issues through indicators, government can anticipate several possibilities to avoid worst case scenarios, in which certain party might exploit for their own interest.Thus, the policy should measure the attribute of the process, data subject behavior and characteristics. It is recommended that the Indonesian government created a special unit with the task

of monitoring the transaction in the digital and handle the cybercrime investigation. Nevertheless, the trust is essential to create good relationship between netizens and government, which focus to the ground issue not the politics as the sign of good intention and concern [19]. For the sake of realizing the ideal and admirable level of society, the presence of the UU-ITE Law and its revisions need to be appreciated. It forms the responsibility of the state to its people to protect its people from the negative things due to freedom in the virtual world, the application of the UU-ITE and its revisions could be used as a tool for the government to silence all forms of criticism.

References

[1] Batrinca, B & Treleaven, P. C. (2015). Social media analystics: A survey of techniques, tools and platforms. AI & Soc , 89-116.
[2] Palitopoulou, V., & Maragoudakis, M. (2013). On Mining Opinion on Social Media. EANN (pp. 474-484). Halkidiki, Greece: Springer-Verlag Berlin Heidelberg.
[3] Aggarwal, C. C., & Zhai, C. X. (2012). An Introduction to Text Mining. In C. C. Aggarwal, & C. X. Zhai, Mining Text Data (pp. 1-10). Springer US.
[4] Guidi, M., Ruiz-Agundez, I., & Canga-Sanchez, I. (2012). Knowledge Mining from the Twitter Social Network: The case of Barack Obama. In A. Abraham, Computational Social Networks: Mining and Visualization (pp. 211-229). London: Springer-Verlag.
[5] E, Kusumasari TF, Lubis M and Lubis AR. Study to the Current Protection of Personal Data in the Educational Sector in Indonesia. ICCAI 2017. IOP Conference Series.
[6] LubisM and Maulana F A. 2010. Information and Electronic Transaction Law Effectiveness (UU-ITE) in Indonesia. IEEE ICT4M, pp. C-13-19.
[7] Ampofo, L., Collister, S., O'Loughlin, B., & Chadwick, A. (2015). Text mining and social media: When quantitative meets qualitative, and software meets humans. In P. Halfpenny, & R. (. Procter, Innovations in Digital Research Methods (pp. 161-192). London: Sage.
[8] Southeast Asia Freedom of Expression Network (SAFEnet) (2017). Daftar Kasus Netizen yang Terjerat UU ITE – SAFEnet Voice. Retrieved from http://id.safenetvoice.org/daftarkasus/.
[9] Inkpen, D. (2016). Text mining in social media for security threats . In R. Abielmona, R. Falcon, N. Zincir-Heywood, & H. A. Abbass, Recent advances in computational intelligence in defense and security (pp. 491-517). Springer, Cham.
[10] Hu, X., & Liu, H. (2012). Text analytics in social media. In C. C. Aggarwal, & C. X. Zhai, Mining Text Data (pp. 385-414). LLC: Springer Science+Business Media.
[11] Charalabidis, Y., Maragoudakis, M., & Loukis, E. (2015). Opinion mining and sentiment analysis in policy formulation initiative: The EU-community approach. International Conference on Electronic Participation (pp. 147-160). Thessaloniki: Springer, Cham.
[12] Arunachalam, R., & Sarkar, S. (2013). The new eye of government: Citizen sentiment analysis in social media . Workshop on Natural Language Processing for Social media (SocialNLP) (pp. 23-28). Nagoya : IJCNLP.
[13] Carley, K. M., Malik, M., & Kowalchuk, M. (2015). Twitter usage in Indonesia. Pittsburgh: Institute for Software Research, School of Computer Science Canegie Mellon University.
[14] Witten, I. H. (2004). Text Mining. In M. P. Singh, The practical handbook of internet computing (pp. 1-22). Washington: DC: Chapman & Hall/CRC.
[15] Finlay, S. (2014). Predictive analytics, data mining and big data. Steven Finlay.
[16] Kumar, S., Morstatter, F., Liu, H. (2014). Twitter Data Analytics. New York. Springer.
[17] Lubis M, Kartiwi M and Zulhuda S. 2017.Privacy and Personal Data Protection in Electronic Voting: Factors and Measures. Telkomnika, March vol. 15(1) pp. 512-521.
[18] Ahlan A, LubisM and Lubis A R. Information Security Awareness at the Knowledge-Based Institution: Its Antecedents and Measures. Procedia Computer Science, 2015. 72, 361–373.
[19] Lubis M and Kartiwi M. Privacy and Trust in the Islamic Perspective: Implication of the Digital Age. International Conference on Information and Communication Technology for the Muslim World, 2013.

IOP Conf. Series: Journal of Physics: Conf. Series **1007** (2018) 012038 doi:10.1088/1742-6596/1007/1/012038

Acknowledgement
This work would not have been possible without the academic support from my supervisor, Associate Professor Mira Kartiwi and the financial support from Telkom University.

IOP Conf. Series: Journal of Physics: Conf. Series **1007** (2018) 012039 doi:10.1088/1742-6596/1007/1/012039

Increasing Prediction the Original Final Year Project of Student Using Genetic Algorithm

Rijois Iboy Erwin Saragih [1], Mardi Turnip [2], Delima Sitanggang [2], Mendarissan Aritonang [1], Eva Harianja [1]

[1]Universitas Methodist Indonesia, Jl. Hang Tuah No. 8 Medan, 20152, Indonesia
[1]Faculty of Technology and Computer Science, Universitas Prima Indonesia, Indonesia

E-mail: 0105128001@methodist.ac.id

Abstract. Final year project is very important forgraduation study of a student. Unfortunately, many students are not seriouslydidtheir final projects. Many of studentsask for someone to do it for them. In this paper, an application of genetic algorithms to predict the original final year project of a studentis proposed. In the simulation, the data of the final project for the last 5 years is collected. The genetic algorithm has several operators namely population, selection, crossover, and mutation. The result suggest that genetic algorithm can do better prediction than other comparable model. Experimental results of predicting showed that 70% was more accurate than the previous researched.

1. Introduction

Higher Education has the goal of producing qualified graduates [1]. The best graduates are not createdin a night. It required a long process, year by year, course by course to write final Year Project. Student's final year project is one of the most important aspects of his/her engineering degree. Final year project is an important element as a final step of the education journey of a student [2]. A long process has to be passed by a student starting from the first semester until the last semester to get a proposal seminar and the defence time. The real problem which is faced by some universities in Indonesia is plagiarism of final year project of a student. Plagiarism, collusions and other forms of academic misconducts have always been regrettable but unavoidable aspect of academic existence [3].This situation has become a habit among the students; even more it has already become a bad culture.Students are willing to pay someone to make their final year project. This practice has put a bad image on education. Not only student involves in this situation but also some lecturers that offer a service for final year project of a student. This situation has become a big problem for universities.

In refer to previous worked [4] used different method for the prediction original final year project of student. This system used the apriori algorithm that found the correlation between various factors. The output of this system was the accuracy prediction of the original final year. In this study, GPA, presence, practical work, and mentoring were used as input variables. Prediction accuracy using this system revealed about 50%. Therefore, in this research, approaches are done to predict the original final year of student with data collecting techniques. Data collecting is the solution to this serious problem. Data collecting is an essential step in the process of knowledge discovery in databases. Thus,it refers to extracting or finding knowledge from large amounts of data [5]. GPA, presence,

IOP Conf. Series: Journal of Physics: Conf. Series **1007** (2018) 012039 doi:10.1088/1742-6596/1007/1/012039

practical work, and mentoring variables are used to predict the original final year project of student. These variables can be obtained from historical data of the students in the last few years. From the explanantion above, this research obtained the historical data of the students in the last few years by implementing data collecting techniques [6]. The purpose is to get new knowledge or information from those data regarding to the research problems. The same data will be used from previous researched on this paper with different algorithm which is genetic algorithm.

2. Methodology

There are several steps of process to solve optimization problems in genetic algorithm. The steps are: population, selection, crossover and mutation. Those steps are continually repeated until finding the expected result [7]. Prediction approaches are to obtaining the available data and then using genetic algorithm to get new information from the process [8-15]. The population consists of ten chromosomes or populations, the chromosome are ordered by gens which content scores. In the Genetic Algorithm, the problem to be solved is represented by a list of parameters which drives an evaluation procedure, called chromosomes or genomes. Chromosomes can be defined as simple strings of data and instructions. In the first step of the algorithm, such chromosomes are generated randomly or heuristically to form an initial pool of possible solutions called first generation pool. The proposed algorithm can be described as [16]:

1. Firstly suppose we have a population of N size, with chromosomes generated randomly;
2. Apply fitness to each chromosome or genomes of population;
3. Make new chromosomes or genomes through crossings of selected chromosomes of this population. Apply recombination and mutation in these chromosomes;
4. Eliminate old population members, so that there is enough space to insert new chromosomes, keeping the population with the same N chromosomes;

Further explanation, a population of genetic algorithm also called chromosomes that are randomly distributed in the solution space is selected as the starting point of the search. The goal of fitness function is to numerically encode the performance of the chromosomes. Crossover causes to form a new offspring between two randomly selected good parents. In this researched is used one point crossover. The crossover occurs only with some probability. The last step is mutation, this operator play important role to ensures that the probability of reaching any point in the search is never zero [17].

2.1 Population

Genetic algorithm isgenerally started with an initial population that is generated randomly. Some researches have been conducted by using special techniques to produce a higher quality initial population. Thus, an approach is designed to give the GA a good start and speed up the evolutionary process [18].

Table 1. Population.

Population									
10	6	5	4	1	7	3	8	2	9
10	9	7	1	3	8	5	2	4	6
5	8	2	10	3	9	6	1	7	4
4	7	1	9	3	2	6	5	8	10
9	2	10	8	7	3	5	4	6	1

Every scorein the table is the chromosome gen which represents the students. For example number 10 is the student who is chosen before number 10. This method is based on data factor in all variables such as GPA, presence, practical work and mentoring.

IOP Conf. Series: Journal of Physics: Conf. Series **1007** (2018) 012039 doi:10.1088/1742-6596/1007/1/012039

2.2 Fitness

The fitness function makes certain that evolution forward to optimization with accumulating for each individual on population [19 - 21].

$$F = V_1 + V_2 + V_3 + V_4$$

Where,

F = Fitness value
V1 = GPA
V2 = Presence
V3 = Practical Work
V4 = Mentoring

- *2.2.1 GPA*

  GPA is Grade Point Average in term in academic system. GPA plays an important role to evaluate the students.

Table 2. GPA.

No	Scale	Value
1	2.0 – 2.7	4
2	2.7 – 3.0	3
3	3.0 – 3.5	2
4	3.5 – 4.0	1

- *2.2.2 Presence (P)*

  A presence is an evidence which shows the students' presence. This is also important because it shows the attitude of the students.

Table 3. Presence.

No	Scale	Value
1	P < 40%	3
2	40% > P < 70%	2
3	P > 70%	1

- *2.2.3 Practical Work (PW)*

  To know the students in computer study program have skills in computer by practical work. This practical work has to be done by students with good grade.

Table 4. Practical Work.

No	Scale	Value
1	PW<60%	3
2	40% > PW< 70%	2
3	PW> 70%	1

- *2.2.4 Mentoring(M)*

Mentoring time must be followed by student so that lecturer or supervisor can lead them to the right way how to write a good paper, or how to do good research. The student must bring a card of mentoring that show their presence by lecturer signature.

Table 5. Mentoring.

No	Scale	Value
1	PW< 60%	3
2	40% > PW < 70%	2
3	PW > 70%	1

3. Discussion and Result

For this step, it will be discussed about the process and the result of the increasing prediction of the original final year project using genetic algorithm. The parameters and values of genetic algorithm on this research are:

1. Population = 10
2. Generation = 100
3. Crossover probability = 0.7
4. Mutation probability = 0.1

Table 6. 10 Population 100 generation.

Experiment	Generation	Prediction	Reality
1	99	7	5
2	92	9	8
3	87	10	7
4	70	11	7
5	95	9	9
6	90	7	9
7	85	12	5
8	90	7	5
9	89	10	7
10	95	7	5

On Table 6 is seen that predictionand the reality resultover all the experiment are different. This experiment is done in ten times. The result shows that on generation 90 the prediction is7 students but in reality is 9 students.It means that 2 score different compare to other generation which is average above 2 score. Therefore, percentage result between 7 and 9 is about 70% accurate predicted.

Table 7. The testing result.

Previous Researched	**Recent Research**
Implementation of Data Mining Association Rules to know the Original Final year Project of Student	Increasing Prediction the Original Final Year Project of Student Using Genetic Algorithm

Prediction Accuracy 50%	Prediction Accuracy 70%

In the table 7 shows the comparison to previous work and recent work on the same context or data research. The different is seen in the prediction accuracy of 50% and 70%, it means that recent work shows a better result than the last research. This result is based on the prediction accuracy with the same variable and data. It took 10 times experiment with the same genetic algorithm parameter.

4. Conclusions

Based on the discussion and implementation which have been done on this research, therefore the conclusion shows that the implementation of genetic algorithm to predict the original final year project of student shows better result than previous researchin term of time and accuracy. The result describes that generation = 90, prediction = 7, and reality = 9. Based on that result it means 7 to 9 differentiate 2 numbers which is about 70% accuracy predicted. The contribution from this research is that genetic algorithm with the proper parameter setting and formulated the problem can give good result in prediction system.

References

[1] De Corte, E., (2003), Excellence in Higher Education, Portland Press, London.
[2] Wanli X, Petakovic G R and Sean G 2015 Participation-based student final performance prediction model through interpretable Genetic Programming: Integrating learning analytics, educational data mining and theory. Computer in Human Behavior. Vol. 47 Pg. 168-181
[3] Carroll, and Appleton, J (2001) "Plagiarism: A Good Practice Guide". JISC publication.
[4] Saragih R I E and Sembiring H 2015. Penerapan Data Mining Association Rules. MajalahIlmiahMethoda. Vol. 5 No. 2
[5] Sivanandam S.M, and Deepa, S.N. 2008. Introduction to Genetic Algorithms, Springer, Berlin:Germany.
[6] Guo X, Shao andHua N. 2009. A hierarchical Text clustering algorithm with cognitive situation Dimensions. Second International Workshop on Knowledge discovery and Data Mining
[7] David de la Fuente, Alejandro Garrido, Jaime Laviada, Alberto Gomez,(2006) "Genetic Algorithms to Optimize the Time toMake Stock Market Investment". GECCO 2006, Vol.2, pp.1857- 1858.
[8] Naik L R, Ramesh, D, Manjula B and Govardhan, A2012. Prediction of Stock Market Index Using Genetic Algorithm. Computer Engineering and Intelligent System. Vol. 3 No. 7
[9] Ratnakar S, Rajeswari K and Jacob R 2013Prediction Of Heart Disease Using Genetic Algorithm For Selection of Optimal Reduced Set Of Attribute. International Journal of Advance Computational Engineering and Networking. Vol. 1 Issue 2
[10] Ahmadi F, Amjadian S andPardegi, K. 2012. New Approach to Bankruptcy Prediction Using Genetic Algorithm. International Journal of Computer Application. Vol. 44 No. 4
[11] Guo X, Shao Z and Hua N. 2009. A hierarchical Text clustering algorithm with cognitive situation Dimensions. Second International Workshop on Knowledge discovery and Data Mining
[12] Hu J Xiong and Shu J Zhou X., Zhu J. 2009. A novel Text clustering method based on TGSOM and Fuzzy K-means ",First International Workshop on Educationa Technology and Computer Science
[13] Zang Z and Cheng H. 2008. Clustering Aggregation based on genetic algorithm for Document clustering. IEEE Congress on Evolutionary Computation.
[14] Cyril Schoreels, Brian Logan, Jonathan M.Garibaldi. 2004"Agent based Genetic Algorithm Employing FinancialTechnical Analysis for Making Trading Decisions Using Historical Equity Market Data", IAT04.

IOP Conf. Series: Journal of Physics: Conf. Series **1007** (2018) 012039 doi:10.1088/1742-6596/1007/1/012039

[15] L.S. Duke and Long, J., (1993) "Neural network futures trading - A feasibility study," Adaptive Intelligent Systems, ElsevierScience Publishers.
[16] Reeves C, danRowe, J. (2002). Genetic Algorithms Principles and Perspectives. Kluwer, Norwell, MA.
[17] Naik R. L, Ramesh D, Manjula B, and Govardhan A. 2012 " Prediction of Stock Market Index Using Genetic Algorithm" Computer Engineering and Intelligent System. Vol. 3 No. 7
[18] K F Man, K S and Tang, S. Kwong, "Genetic Algorithms: Concept and Designs", Springer, Chapter 1-10, Pg 1-348.
[19] Coley D A (1999). An introduction to genetic algorithms for scientists and engineers. World scientific.
[20] Smith J E 2002. Handbook of Global Optimization. Volume 2, 275-362
[21] Reeves C, and Rowe, J. (2002). Genetic Algorithms Principles and Perspectives. Kluwer, Norwell, MA. Coley, D. A. (1999). An introduction to genetic algorithms for scientists and engineers. World scientific.

Acknowledgments
Thanks to all the authors and the supervisors for all the times and supports.

IOP Conf. Series: Journal of Physics: Conf. Series **1007** (2018) 012040 doi:10.1088/1742-6596/1007/1/012040

Analysis of Tourists Preferences on Smart Tourism in Yogyakarta (Case: Vredeburg Fort Museum)

Rima Amanda*, PInsap Santosa, M Nur Rizal

Department of Information Technology and Electrical Engineering, Universitas Gadjah Mada, Yogyakarta, Indonesia

E-mails: *rima.amanda@mail.ugm.ac.id

Abstract. Smart tourism is a supporting system of an individual tourism in the terms of a comprehensive and integrated information and technology service. An educational tourist destination such as a museum is expected to present an informative and interactive atmosphere. Vredeburg Fort Museum as one of the tourist destinations in Yogyakarta begins to lose its visitors. The lack of interest of public towards the museum and the assumption that the museum is an ancient, less well maintained, and boring place become main obstacles in attracting tourists. This research aims to find the important factors becoming the preferences of tourists to visit the Vredeburg Museum in Yogyakarta. The research method used is the Principal Component Analysis. The analysis shows there are four main factors with eigenvalue more than 1, i.e. the first factor of 8,623, the second factor of 1,920, the third factor of 1,175, and the fourth factor of 1.082. Those four factors are the result of the grouping of 20 preference determinant variables.

1. Introduction

The large number of users of gadget and internet lead to an opportunity for the government to improve the quality of public services and economic improvement of the tourism sector[1]. The tourism industry itself has implemented the Information and Communication Technology (ICT) to reduce costs, improve service quality and customer quality. The tourism industry has adopted the Information and Communication Technology (ICT) to reduce costs, improve the service quality and customer quality[2]. Due to the economic progress, the ordinary tourism potentially turns into smart tourism, which makes use of intelligent systems [3]. The main objective of smart tourism is to focus on the needs of tourists by integrating ICT with the tourism industry to promote, to improve the quality of services and tourism management, and to enlarge the industrial scale into a wider scope [4]. With the implementation of ICT in tourist destinations, there can be more tourism experience and enhance the competitiveness of the place [5].

Yogyakarta, as the second national tourist destination after Bali, relies on the same tourism destinations such as KeratonYogyakarta , Taman Sari, Malioboro, Vredeburg Fort Museum, Taman Pintar and GembiraLoka Zoo. Most of the tourism destinations in Yogyakarta is a cultural and historical place. It is dominated by the museum (14 museums) [6]. Among the 14 museums in Yogyakarta, Vredeburg Fort Museum is situated in the most strategic location, around JalanMalioboro which is the center of tourism in Yogyakarta. Based on the field observation, the writer chose the Vredeburg Museum of Yogyakarta as the object of empirical research since Vredeburg Fort Museum has shown progress towards smart tourism. From the data obtained, it is known that Vredeburg Fort Museum has improved its technology system such as the use of a visualization room dioramas with QR code technology to enable visitors to access information about the objects displayed (Figure 1). A screen facility is also provided for visitors as media information and educational games about the history of the struggle for Independence of Indonesia. Audiovisual room for playing movie of the struggle is also available every Friday and Sunday.

Figure 1. Diorama room of Vredeburg Fort Museum.

The city of Yogyakarta has a vision to be a city with quality, character and inclusive education, culture-based tourism, and center of environmentally minded public services and people-based economy[7]. To realize the vision, Yogyakarta government has made smart city roadmap in 2016. In the roadmap, it is mentioned that there are three focuses of smart city development in Yogyakarta : smart culture, smart tourism, and smart education[8]. In smart tourism, Yogyakarta has three variables to be achieved, the number of tourists, length of stay, and the amount of money spent. The number of tourists visiting Yogyakarta during 2016 is 5.5 million people, 2 percents lower than the one in 2015.

A museum is a permanent and non-profit institution, serving the needs of public openly by collecting, conserving, researching, communicating, and exhibiting real objects to the community for study, education, and pleasure[9]. Museum development can be done by cooperation in the educational, social, science and technology, cultural, and tourism field [10]. Vredeburg Fort Museum has covered the three aspects of smart city of Yogyakarta; smart culture to lift the culture, smart education as a means of education improvement, as well as smart tourism as a cultural and historical tourist destinations.

This research will summarize the important factors supporting the implementation of smart tourism in Yogyakarta, especially in BentengVredeburg Museum. Existing technology must be available because providing integrated technology is a requirement to make a smart tourism object. Analysis of the preferences of tourists on Museum Vredeburg Fort needs to be done to determine the strengths and weaknesses of this tourist destination in order to facilitate managers in preparing strategies for the realization of smart tourism.

2. Literature Review

2.1. Smart Tourism

The concept of smart tourism is a supporting system of an individual tourism within the terms of a comprehensive and integrated information and technology services[11]. Smart tourism involves several components and elements supported by ICT, as shown in Figure 2[12]. Component of smart experience aims to enhance the experience of the tour by getting updated information, sharing experiences, and doing tourism activities through the media technology[13]. Smart business components have characteristics such as dynamic stakeholder interconnection, core business that have shifted toward digital process, and friendliness to investors[5]. The last component is the smart

IOP Conf. Series: Journal of Physics: Conf. Series **1007** (2018) 012040 doi:10.1088/1742-6596/1007/1/012040

destination which is the integration of ICT with physical infrastructure in tourism destinations offering more value than other destinations.

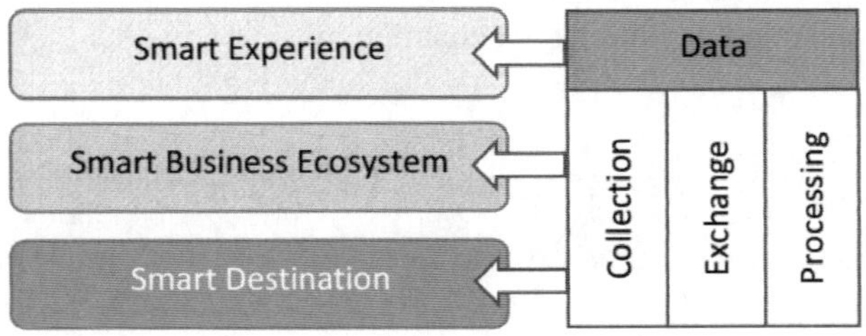

Figure 2. Components dan Elements of smart tourism.

The most important thing of smart tourism is to offer personal service to the tourists by considering some aspect such as access to real-time information to collect user data. According to Zhang et al. [14].There are three technologies supporting smart tourism system: cloud computing, Internet of Things (IoT), and end-user internet service system.

2.2. Vredeburg Fort Museum

Vredeburg Fort Museum is a museum of the Indonesian's people struggle for independence in Vredeburg Fort Museum is a museum of struggle for independence in Yogyakarta. Vredeburg Fort Museum of Yogyakarta is located in the zero kilometer of the center of Yogyakarta city or at the southern end of Malioborostreet. In its service to the community, Vredeburg Castle museum can not be separated from the elements of education because basically the museum has two functions, a recreation and a place of education. Information about the history, culture and noble values of the independence struggle is presented to the young generation in nuances of edutainment, the term derived from the word education and entertainment.

Attractions presented at the Vredeburg Fort Museum are in the form of dioramas or miniatures of the history of the struggle for independence. The diorama of historical events of independence such as the diorama in room I depicts historical events in the Diponegoro war period until the Japanese colonization in Yogyakarta (1825-1942). Diorama room II depicts the events from the proclamation or the beginning of independence until the Military Aggression of Netherland (1945-1947). The diorama room III illustrates the event of the Renville Agreement until the recognition of RIS sovereignty (1948-1949). Diorama room IV describes the history from the period of the establishment of Republic of Indonesia until the New Order (1950-1974).

2.3. Tourist Preferences

Preference is the choices made by tourists regarding the tourism destinations visited. The strength of the tourist preferences will determine what facilities they like in the tourism destinations. Tourist preference is closely related to personal information. Each person has different personal characteristics. In the field of tourism, the data of tourist preference can be used to highlight the advantages possessed by these tourism destinations so that the tourists have the desire to have another visit and recommend it to others.

Garcia et al. in Baggio et al. [15] set up a tourism recommendation system based on tourist preferences on Point of Interest (POI). Preference is used as a tool in making decisions for other tourists to visit a tourist destination.

2.4. Principal Component Analysis (PCA)

Principal Component Analysis (PCA) is a technique for constructing new factors that are linear combinations of the original variables[16]. The maximum number of these new factors will be equal to the sum of the old variables, and these new factors are not correlated one another. Calculating the analysis using the PCA method is about solving the eigen equation where the eigenvalues should be more than 1. The PCA algorithm steps in general are: calculating the covariance matrix, calculating the eigenvalues, calculating the eigenvectors, and defining the new variable (principal component) by multiplying the original variable with an eigenvector matrix.

3. Research Methodology

Quantitative method is used to collect data from respondents. The number of tourist populations visiting the museum during the data collection period is unknown (infinite) so that the minimum sample count is calculated using the Lemeshow formula[17], wherein the minimum sample used is 96 respondents. The method of analysis is used is Principal Component Analysis because this research uses multi variable. This research uses questionnaires in the form of a set of formal questions to obtain information from respondents. The purpose of the questionnaire is to translate the information needs of the researcher into a set of questions that make the respondent willing and able to answer. To measure the respondent's perception of the questions given, Likert scale with 6 point response is used. According to Chomeya[18], the scale with 6 points of response has a better degree of discrimination and reliability than a scale with a response point of 5 or any other odd number.

This research uses 20 variables to determine the preferences of tourists towards smart tourism in Yogyakarta. The variables used will be adopted from the results of research of Wang et al. [19]and have been adapted to the situation at the Vredeburg Museum of Yogyakarta as shown in Table 1.

Table 1. Smart tourism variables and operational definition.

No.	Variables	Operational Definition
1.	Website of Tourism destination	The tourists make use of digital sources such as websites in searching information and communicating. Their travel intention is reinforced by the website features.
2.	Mobile Apps	Mobile apps can improve the quality of customer service at the desired tourism destination by pointing out the right locations, booking places, and estimating waiting time at various tourist attractions
3.	Free Wifi	It allows tourists to connect their gadget devices to the internet via wireless connections that are widely available in hotels, airports, restaurants.
4.	QR code	Dapat mengakses informasi tentang tempat menarik melalui perangkat mobile dengan scan kode quick response The tourists can access information about interesting places through mobile devices by scanning quick response code
5.	Touch Screen Devices	Touch screen technology can provide services for tourist who needs information, and is used as an interactive medium of educational games
6.	Signal Coverage	Tourists have no difficulty in accessing mobile phone signals and can use the internet connection from a service provider during visit in the tourism area
7.	Service center call	Service center provides useful and timely information for the travel plan of the tourists
8.	Smart Guidance System	It can provide experience and more or different information for tourists
9.	Information Guidance Services	The system combines Geographic Information Systems (GIS) with location-based services to provide a better travel experience for tourists and a deeper understanding of the importance of precious views
10.	Smart Card	It contains personal information attached to an RFID chip.

11.	System of Electronic Entrance Security	A good information system should consist of several IoT information systems, including security systems and ticketing systems with RFID technology
12.	Smart environment	Smart environment is one of the most important dimensions of smart tourism destination, it is related to energy optimization leading to sustainable resource management
13.	Smart education	Smart destinations focus not only on utilizing the new technologies, but also on educating tourists on how to use them.
14.	E-complain	Tourism organizations should have an e-complain handling system to handle complaints and give feedback from travelers
15.	E-Commerce	Buying tourism products through websites and apps is a growing market nowadays since more tourists have mobile devices.
16.	E-Ticketing	Tourism entrance system with modern information system must apply RFID technology
17.	E-parking system	RFID can be used for parking systems, so parking fee is no longer required
18.	Emergency system	It is an IoT-based tourism security platform and cloud computing. This platform connects security equipment with service facilities at the tourism destination via IoT and will detect warnings via wireless sensors
19.	Augmented Reality	Informasi interaktif multi media dengan bantuan smartphone untuk meningkatkan pengalaman berwisata It is a multimedia interactive information by means of smartphone to enhance the travel experience
20.	Tourism Destination Blog	Tourists trust the website with reviews more rather than professional guides and travel agents

4. Findings and Discussion

Demographic data collected from 110 questionnaires that can be analyzed include age, gender, education, occupation, experience of visiting, and the area of origin of the respondents. Most visitors coming to Vredeburg Fort Museum of Yogyakarta are 21-24 years old that (43%). This is because the visitors are mostly students who are studying at universities in Yogyakarta.

Validity and reliability tests have been done towards questionnaire items with 30 samples. The validity test is carried out to know how good the questionnaire is to precisely be used as survey instrument in this research. This test is done by comparing r obtained and r table. If r obtained is higher than r table, so the item is said to be valid. On the contrary, if r count is lower than r table, so the item is considered invalid. Based on the result of validity test, it is known that the 20 questions tested are considered valid with r table above 0,361. Therefore, the questionnaire can be used as research instrument. The test reliability is done by using Cronbach's Alpha. The score of Cronbach's Alpha is assumed to be reliable if it is higher than 0.7 and can be considered good if it is higher than 0.8 [20]. The reliability test result shows that Cronbach's Alpha score is 0.953, so it can be used as instrument in this research.The assumptions to be met in the PCA analysis are the correlation matrix score, Kaiser Mayer Olkin (KMO), Anti Image Matrix, and the score of Communalities.

4.1. Matriks Korelasi

The correlation matrix table contains correlation scores among variables to be analyzed. The correlation among independent variables must be strong enough (higher than 0.5). One of matrix correlation result of tourism destination website variable with mobile apps variables shows a score of 0.667 indicating that there is a strong correlation. The correlation between the variable of tourism

IOP Conf. Series: Journal of Physics: Conf. Series **1007** (2018) 012040 doi:10.1088/1742-6596/1007/1/012040

destination website and mobile application is significant. It can be seen from the p-value score of 0 (<0.005) meaning that there is a correlation between the variables.

4.2. Uji Kaiser Mayer Olkin (KMO)

In this research, Kaiser-Meyer Olkin (KMO) test is a method to see the adequacy of data for factor analysis. This KMO method measures the adequacy of sampling thoroughly and the adequacy of sampling for each variable. The score of KMO obtained from this test result is 0.867 which can be categorized as good data.

4.3. Anti Image Matriks

Besides checking the KMO, the Anti-Image Matrices is also checked to determine whether the variables are partially feasible to be analyzed and not issued in the test. Based on Anti Image Matrices, it is known that all 20 variables to be analyzed have Measure Sampling Adequacy (MSA). In this research, all variables are higher than 0.5, so the variables are considered eligible to be included in the factor analysis.

4.4. Communalities

Communalities are used to see the closeness of the correlation between variables and factors formed. From the total scores in communalities table, it is found that 19 variables have large communalities (> 0.5). One variable has a communalities score of 0.379 (emergency response system variable) meaning that it must be taken out for further analysis.

4.5. Total Variance Explain

Total Variance Explained shows the percentage of total diversity that can be explained by the diversity of factors formed. To determine how many factors are used in order to explain the total diversity, it is done by considering how large the eigenvalue. The component with eigenvalue > 1 is the component to use. The result of analysis shows that there are 4 variables with eigenvalue value> 1. In the table, there is also an eigenvalue value of each factor formed. It is known Factor 1 with eigenvalue of 8.355, Factor 2 with eigenvalue of 1.876, Factor 3 with eigenvalue of 1.175, and Factor 4 with eigenvalue of 1.079.

4.6. Interpretaion of Final Result

The factors generated from factor rotation are seen from the loading values of each variable as shown in Table 2.

Table 2. Rotated component matrix.

	Component			
	1	2	3	4
V1	0,112	0,324	0,739	0,107
V2	0,081	0,440	0,682	0,093
V3	0,134	0,061	0,766	0,136
V4	0,359	0,134	0,648	0,436
V5	0,545	0,105	0,315	0,428
V6	0,055	0,584	0,186	0,433
V7	0,083	0,191	0,200	0,850
V8	0,462	0,223	0,263	0,604
V9	0,358	0,639	-0,036	0,390
V10	0,619	0,121	0,051	0,499
V11	0,650	0,343	0,180	0,243

V12	0,310	0,571	0,257	0,335
V13	0,206	0,775	0,274	0,013
V14	0,335	0,700	0,286	0,032
V15	0,828	0,240	0,142	0,050
V16	0,758	0,094	0,147	0,055
V17	0,729	0,231	0,106	0,130
V18	0,587	0,048	0,343	0,240
V19	0,743	0,225	-0,014	0,068

The result of grouping of factors formed determines the main priority of the tourists' preference towards smart tourism according to the predetermined variable.The four factors formed can be named according to the variables contained in it. The naming of each factor formed is according to the name of the forming variable as shown in Figure 3.

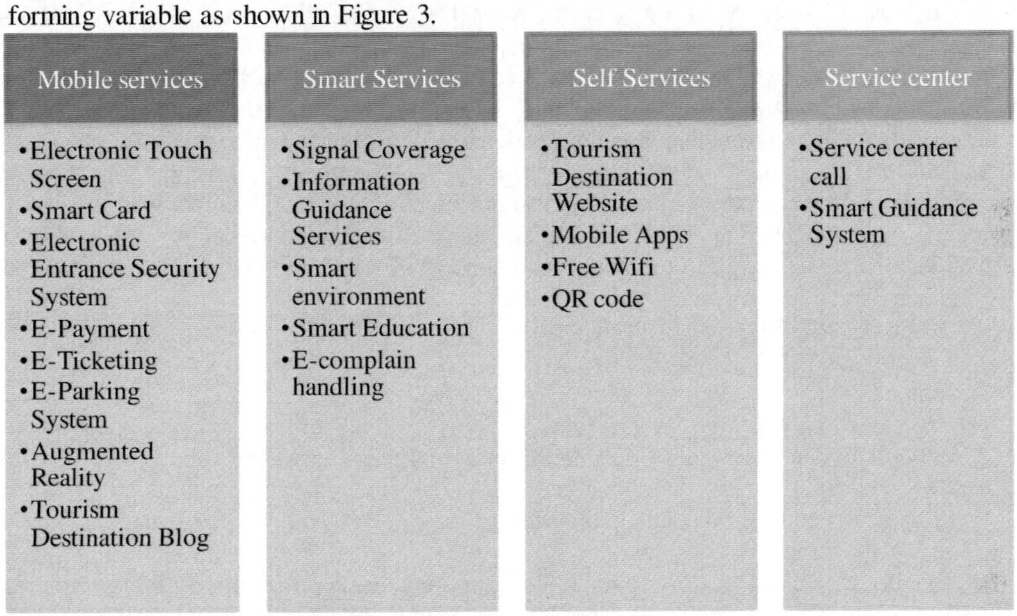

Figure 3. Preferences factors oftourists.

5. Conclusion

Based on the results of the research on Tourist Preferences towards Smart tourism at Vredeburgfort Museum some conclusions can be drawn as follows:

1. There are 20 assessment variables used to measure the preferences of tourists at Vredeburg Fort Museum Yogyakarta. These variables have been tested through the field observation and the adoption of previous research conducted by Wang et al. [19];
2. From the process of clustering factors by using Principal Component Analysis, 4 main factors are formed which determine the preferences of tourists towards Smart tourism in Vredeburg Fort Museum. Those factors formed have eigenvalue higher than 1. Factor 1 has eigenvalue of 8.623, Factor 2 has eigenvalue of 1.920, Factor 3 has eigenvalue of 1.175, and Factor 4 has eigenvalue of 1.082. Based on the grouping, a name of the factor formed is given based on the forming variables. These factors are mobile service (consisting of 8 variables), smart service (consisting of 5 variables), self service (consisting of 4 variables), and service center (consisting of 2 variables).

IOP Conf. Series: Journal of Physics: Conf. Series **1007** (2018) 012040 doi:10.1088/1742-6596/1007/1/012040

References

[1] H. Werthner dan F. Ricci, "E-commerce and tourism," *Commun. ACM*, vol. 47, no. 12, hal. 101–105, Des 2004.

[2] R. Law, R. Leung, dan D. Buhalis, "Information Technology Applications in Hospitality and Tourism: A Review of Publications from 2005 to 2007," *J. Travel Tour. Mark.*, vol. 26, no. 5–6, hal. 599–623, Sep 2009.

[3] U. Gretzel, "Intelligent systems in tourism: A social science perspective," *Ann. Tour. Res.*, vol. 38, no. 3, hal. 757–779, 2011.

[4] X. Huang, J. Yuan, dan M. Shi, "Condition and key issues analysis on the smarter tourism construction in China," in *Multimedia and signal processing*, Springer, 2012, hal. 444–450.

[5] D. Buhalis dan A. Amaranggana, "Smart Tourism Destinations," in *Information and Communication Technologies in Tourism 2014*, vol. 58, no. 12, Z. Xiang dan I. Tussyadiah, Ed. Cham: Springer International Publishing, 2013, hal. 553–564.

[6] Badan Pusat Statistik Kota Yogyakarta, *Kota Yogyakarta Dalam Angka 2017*. Yogyakarta, 2017.

[7] Pemerintah Kota Yogyakarta, "Website Resmi Kota Yogyakarta," 2017. [Daring]. Tersedia pada: https://www.jogjakota.go.id/about/visi-dan-misi.

[8] Tim PSPPR UGM, "Roadmap Kota Yogyakarta Menuju Smart City," no. 1, hal. 1–27, 2016.

[9] The World Museum Community, "Museum Definition." [Daring]. Tersedia pada: http://icom.museum/the-vision/museum-definition/. [Diakses: 03-Jan-2018].

[10] Kementerian Hukum dan HAM, "PP No 66 Tahun 2015 tentang Museum." hal. 1–36, 2015.

[11] Y. Li, C. Hu, C. Huang, dan L. Duan, "The concept of smart tourism in the context of tourism information," *Tour. Manag.*, hal. 1–8, 2016.

[12] U. Gretzel, M. Sigala, Z. Xiang, dan C. Koo, "Smart tourism : foundations and developments," hal. 179–188, 2015.

[13] D. Buhalis dan A. Amaranggana, "Smart Tourism Destinations Enhancing Tourism Experience Through Personalisation of Services," in *Information and Communication Technologies in Tourism 2015*, Cham: Springer International Publishing, 2015, hal. 377–389.

[14] Z. Ling, "On the Basic Concept of Smarter Tourism and Its Theoretical System," *Tour. Trib.*, vol. 27, no. 5, hal. 66–73.

[15] R. Baggio, M. Sigala, A. Inversini, J. Pesonen, dan Eds., *Information and Communication Technologies in Tourism 2014*. Cham: Springer International Publishing, 2013.

[16] H. Abdi dan L. J. Williams, "Principal component analysis," *Wiley Interdiscip. Rev. Comput. Stat.*, vol. 2, no. 4, hal. 433–459, 2010.

[17] S. Lemeshow, D. W. H. Jr, J. Klar, dan S. K. Lwanga, "Adequacy of Sample Size in Health Studies," *World Heal. Organ.*, hal. 11–14, 1990.

[18] R. Chomeya, "Quality of Psychology Test Between Likert Scale 5 and 6 Points," *J. Soc. Sci.*, vol. 6, no. 3, hal. 399–403, 2010.

[19] X. Wang, X. R. Li, F. Zhen, dan J. H. Zhang, "How smart is your tourist attraction?: Measuring tourist preferences of smart tourism attractions via a FCEM-AHP and IPA approach," *Tour. Manag.*, vol. 54, hal. 309–320, 2016.

[20] U. Sekaran, *Research methods for business*, vol. 65, no. 3. 2013.

IOP Conf. Series: Journal of Physics: Conf. Series **1007** (2018) 012041 doi:10.1088/1742-6596/1007/1/012041

Analysis of e-learning implementation readiness based on integrated elr model

K Adiyarta[1], D Napitupulu[2], R Rahim[3], D. Abdullah[4], MI Setiawan[5]

[1]Program of Master of Computer Scocience, Universitas Budi Luhur, Indonesia
[2]Research Center for Quality System & Testing Technology, Indonesian Institute of Sciences, Tangerang Selatan, Banten, Indonesia
[3]School of Computer and Communication Engineering, Universiti Malaysia Perlis, Malaysia
[4]Department of Informatics, Universitas Malikussaleh, Aceh, Indonesia
[5]Department of Civil Engineering, Narotama University, Surabaya, Indonesia

E-mails: krisna.adiyarta@gmail.com, darm003@lipi.go.id, usurobbi85@zoho.com, dahlan@unimal.ac.id, ikhsan.setiawan@narotama.ac.id

Abstract. E-learning nowadays has become a requirement for institutions to support their learning activities. To adopt e-learning, an institution requires a large strategy and resources for optimal application. Unfortunately, not all institutions that have used e-learning got the desired results or expectations. This study aims to identify the extent of the level of readiness of e-learning implementation in institution X. The degree of institutional readiness will determine the success of future e-learning utilization. In addition, institutional readiness measurement are needed to evaluate the effectiveness of strategies in e-learning development. The research method used is survey with questionnaire designed based on integration of 8 best practice ELR (e-learning readiness) model. The results showed that from 13 factors of integrated ELR model being measured, there are 3 readiness factors included in the category of not ready and needs a lot of work. They are human resource (2.57), technology skill (2.38) and content factors (2.41). In general, e-learning implementation in institutions is in the category of not ready but needs some of work (3.27). Therefore, the institution should consider which factors or areas of ELR factors are considered still not ready and needs improvement in the future.

1. Introduction

The rapid development of Information and Communication Technology (ICT) has encouraged various institutions compete to use it in order to provide added value to the products or services offered while enhancing the competitiveness of organizations in a very competitive era. It is sooner or later that an educational institution will be "forced" by a community which demands to adopt ICT in its learning activities. This is because the process of learning is a major business process in educational institutions. It is proven that more than a thousand institutions in 50 countries used e-learning to support their learning activities [1]. Many studies have identified the advantages of implementing e-learning, but in reality not all institutions that have used e-learning got the desired results [2]. Implementation of e-learning is not just uploading teaching material or doing learning content but more to change the paradigm of learning process. The change or paradigm shift of the learning system begins to appear in the transfer of science. The current learning process tends to emphasize more on the teaching process, based on the content base, abstract and only for certain groups. Along in the development of ICT, the process in learning began to shift in learning, based on the problem (case base), contextual and not limited only to certain groups. To adopt e-learning, organization required well strategy and plan to give optimal result as expected. In addition, an institution also need to evaluate e-learning readiness level. Measurement of ELR could be conducted at the analysis stage or evaluation stage. In the analysis phase, the ELR is used to compile the baseline requirement document for the next design phase. While at the evaluation stage, ELR is used to measure success and determine recycling decision for further improvement process. In other words, an evaluation is needed to determine the level of achievement of desired goals [3]. Institution X has utilized ICT to improve the learning process through e-learning. In this study, the measurement of ELR was conducted in the evaluation phase so the organization could know what factors or areas that are considered not ready or still far from expected.

2. Literature Review

Measurement of readiness of e-learning implementation is needed to know the condition and suitability of institutional strategy compared with ideal condition expected. E-learning readiness (ELR) is defined as the mental or physical readiness of an organization for a learning experience [4]. ELR becomes important because in the implementation of e-learning often encountered a variety of obstacles such as resistance, computer literacy, infrastructure to organizational culture [5]. The ELR model is designed to simplify the process of obtaining the basic information necessary to develop e-learning. Many researchers have developed both individual and organizational ELR models such as Kirkpatrick's model that the e-learning evaluation consists of 4 (four) levels of reaction, knowledge, behaviour and result. These four levels further illustrate the evaluation of the results obtained by an institution after implementing e-learning. At each level there are a number of assessments that can be used as information for each characteristic [6]. ELR models developed by other researchers such as the proposed Swatman & So (2005) model measure the readiness of e-learning implementation based on six assessment components: Student's Preparedness, Teacher's Preparedness, IT infrastructure, Management Support, School Culture and Preface to meet face-to- face [7]. The ELR model proposed by Akaslan & Effie (2011) specifically measures the level of e-learning readiness of the faculty aspect based on three components: Technology, People, Content and Institution [8]. Besides, the proposed ELR model of Seakow & Samson (2011) related to e-learning readiness in higher education in Thailand consists of five dimensions of measurement namely Policy, Technology, Financial, Human Resources and Infrastructure [9]. While the ELR model proposed by Kaur & Abas (2004) that measures e-learning readiness in Malaysia has several components of assessment that are Learner, Management, Personnel, Content, Technical, Environment, Culture and Financial [10]. The ELR model proposed by Chapnick classified the readiness of e-learning implementation into the following eight factors [11], they are psychological readiness, sociological, environmental, human resource, financial, technological skill, equipment and content readiness.

In addition, other ELR best practice models, such as the proposed Psycharis (2005), consist of 8 factors of readiness: technology, human resource, economic, content, educational, entrepreneural, culture and leadership [12]. Engholm (2001) also developed an ELR model consisting of five dimensions: organizational & industry, learning content, individual learner, organization culture and ICT [13]. Haney (2002) uses seven categories in the ELR model: human resources, learning management systems, learners, content, information technology, costs and vendors [14]. ELR models will produce scores that can determine the e-learning readiness rating of an institution. The ELR model at the evaluation stage is useful for measuring the success of e-learning implementation and determining the policies and strategies for the improvement process over the next period based on the recommendations and recommendations provided.

IOP Conf. Series: Journal of Physics: Conf. Series **1007** (2018) 012041 doi:10.1088/1742-6596/1007/1/012041

3. Research Methodology

The research conducted descriptively by using questionnaire-based survey method to know the level of organizational readiness in the implementation of e-learning system. The questionnaire was design based on several studies of ELR best practice model. The Chapnick ELR model (2000) as well-known model in this study was integrated with other best practices models such as Engholm (2001) model, Haney model (2002), ELR Kaur & Abas model (2004), Swatman & So (2005), Psycharis model (2005), Akaslan & Effie (2011) model and Seakow & Samson (2011). Those models have been widely used and highly suitable for the case of e-learning implementation in developing countries including Indonesia [15][16]. The integration of eight ELR best practice models in the form of mapping of e-learning readiness factors could be presented as follows [7] [8] [9] [10] [11] [12] [13] [14]:

Table 1. Mapping E-Learning readiness (ELR) factors.

No	E-learning Readiness Factors	Chapnick (2000)	Engholm (2001)	Haney (2002)	Kaur & Abas (2004)	Swatman & So (2005)	Psycharis (2005)	Akaslan & Effie (2011)	Seakow & Samson (2011)
1.	Psychological	√	√	√		√			
2.	Sociological	√				√			
3.	Environmental	√			√	√			
4.	Human Resource	√		√	√		√	√	√
5.	Financial	√		√	√		√		√
6.	Technological Skill	√	√			√	√		√
7.	Equipment	√						√	√
8.	Content	√	√	√	√		√	√	
9.	Innovation						√		
10.	Institution		√	√	√	√	√	√	√
11.	Leadership						√		
12.	Culture		√	√			√		
13.	Policy			√		√			√

Based on Table 1 above, the integration of eight ELR best practice model resulted 13 readiness factors as the basis for the assessment of e-learning implementation readiness (ELR) in Institution X. The questionnaire in this study adopted Likert scale which consists of 5 measurement scale, where 1 = "Strongly Disagree", 2 = "Disagree", 3 = "Neutral", 4 = "Agree" and 5 = "Strongly Agree"had been distributed to 104 respondents by purposive sampling. The criteria of respondents are lecturers who had conducted e-learning at least three years. This criterion is determined because to obtain valid data from a number of teachers who have experience in the learning process of e-Learning so as to know the problems or conditions that run. Total data of 105 respondents will be tested on validation and reliability criteria to be processed in the next stage. To know or map the degree of readiness of the organization in the implementation of e-learning, this research used Aydin and Tasci scale that divided the readiness in four levels as Figure 1 below [17]:

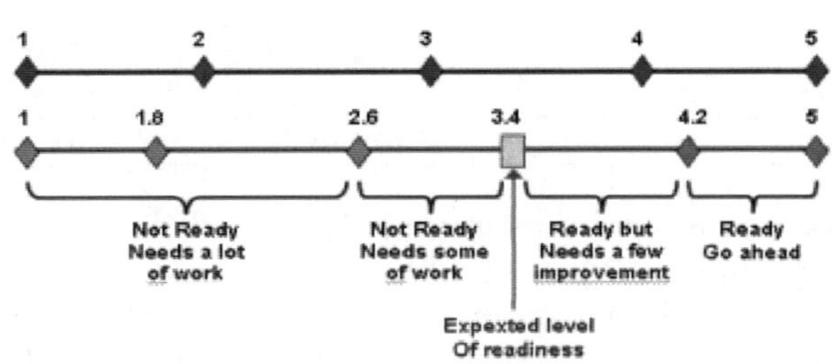

Figure 1. Measurement scale of model ELR Aydin & Tasci [17].

IOP Conf. Series: Journal of Physics: Conf. Series **1007** (2018) 012041 doi:10.1088/1742-6596/1007/1/012041

In Figure 1 above, it could be seen that the average score of 3.41 is the minimum score for the degree of e-learning readiness either for the average score of questions for a factor or the total average score of all questions to be considered or categorized as ready level in e-learning implementation.

4. Result and Discussion

Based on the next approach, the level of readiness of e-learning implementation was analysed. Before the data processing and analysis, the results of questionnaire measurement tested the validity and reliability. For reliability testing is through calculation coefficient reliability Cronbach Alpha where each item is above 0.6 (> 0.6) shows an accurate internal consistency [18]. The results of reliability testing as a whole showed Cronbach Alpha coefficient value of 0.895 (> 0.6), meaning that each item in the questionnaire is considered to be reliable because it meets the minimum requirements. The results of the validity test questionnaire in this study seen from the value of r_{count} (corrected item-total correlation) where if the value of r_{count} greater than r_{table} then the item is valid [18]. From the results of testing the validity of the questionnaire obtained that each measurement item has a value $r_{count} > 0.191$ with 5% significance level where for 104 respondents r_{table} value is 0.191 based on table r product moment. Thus, each item of the questionnaire have met validity and reliability as required.

The next step is analysis the data obtained by descriptively using tools SPSS IBM version 16.0. The statistically mean was computed of each item and also total mean that indicated existing level of e-learning readiness in organization. The result then mapped to Aydin & Tasci (2005) scale as shown which consists of 4 (four) readiness category that are category not ready and needs a lot of work, category not ready but needs some of work, ready category but needs a few improvement as well as the ready category of implementation of e-learning could be continued. The results of ELR could be presented in Table 2 as follows:

Table 2. E-Learning readiness (ELR) and readiness category.

No	ELR Factor	Mean Score	Readiness Category
1.	Psychological	3.58	Ready but needs a few improvement
2.	Sociological	3.62	Ready but needs a few improvement
3.	Environmental	3.45	Ready but needs a few improvement
4.	Human Resource	2.57	Not Ready and needs a lot of work
5.	Financial	3.52	Ready but needs a few improvement
6.	Technological Skill	2.38	Not Ready and needs a lot of work
7.	Equipment	3.79	Ready but needs a few improvement
8.	Content	2.41	Not Ready and needs a lot of work
9.	Innovation	3.56	Ready but needs a few improvement
10.	Institution	3.53	Ready but needs a few improvement
11.	Leadership	3.71	Ready but needs a few improvement
12.	Culture	3.60	Ready but needs a few improvement
13.	Policy	3.48	Ready but needs a few improvement
	Total Mean	**3.27**	**Not Ready but needs some of work**

Based on the results of data processing presented Table 2 above that of 13 e-learning readiness factor obtained total average score is 3.27 which indicates the institution is in the category of not ready but needs some work in the implementation of e- learning if it refers to the Aydin & Tasci value category [17]. Consecutive score of ELR obtained were Psychological 3.58, Sociological 3.62, environmental 3.45, Human Resource 2.57, Financial 3.52, Technological Skill 2.38, Equipment 3.79, Content 2.41, Innovation 3.56, Institution 3.53, Leadership Factor 3.71, Culture of 3.60 and last Policy for 3.48. It could be seen that from 13 factors of readiness of e-learning implementation measured, there are 10 (ten) factors which are considered or included in the ready category ie Psychological,

Sociological, Environmental, Financial, Equipment, Innovation, Institution, Leadership, Culture and Policy. But on the other hand there are 3 (three) ELR factors included in the category of not ready is Human Resource, Technological Skill and Content readiness. Of the 10 (ten) factors that are considered ready, there is one factor that obtained the highest value of Equipment factor of 3.79. Equipment factor is related with infrastructure or technology owned by institutions that support the implementation of e-Learning system both hardware and software. Thus, in terms of ICT infrastructure, institutions have been perceived to be ready for having adequate capacity such as dedicated servers in the form of Web-application Server as well as Database Server, Bandwidth connection large enough, power supply and Moodle-based open source software has also been used so that it is expected to ensure easy and fast accessibility to the e-learning system implemented. But on the other hand, human resource, technology skills and content are still considered not ready or still low of all factors. Of the three factors, technological skill factor has the lowest level of readiness that is equal to 2.38. This suggests that human resource factors need to be a concern and a priority for institutions to further improve their capacity. Infrastructure or technology that is ready without dissertation of qualified human resources then e-learning system will not run well even tend to technology will not be utilized optimally. Therefore, the readiness of human resources becomes very important in determining the success of a technology including e-Learning. The low technology skills factor also shows that users are not familiar with e-learning system being implemented. In other words, the e-learning system is still considered difficult to use which can be caused, among others, interface design, in-site navigation, layouts, precise information, etc. In addition, the content factor is still considered not ready where the learning content provided is mostly still in the form of text only. Whereas in e-Learning need to process the digitalization of learning content to the multimedia in the form of video, audio or animation so as to enrich the learning and improve the understanding of the teaching participants. This is certainly related to the quality of human resources to be able to process and present learning content that is able to meet the learning needs of learners.

5. Conclusion

Based on the results of research that has been done can be drawn some conclusions as follows:

- The contribution of this study compared with previous work was to proposed integrated eight (8) ELR best practice model. This integrated ELR model consisted of 13 factors which useful to measure the readiness of e-learning implementation.
- The degree of institutional readiness in the implementation of e-learning included in the category of not ready but still requires a small increase shown ELR score obtained by 3.27. Thus institutions need to pay attention to which factors that considered not ready then the one could be improved.
- There are 10 ELR factors that are considered as ready level: Psychological, Sociological, Environmental, Financial, Equipment, Innovation, Institution, Leadership, Culture and Policy.
- There are 3 ELR factors included in the category of not ready level: Human Resource, Technological skill and Content readiness.
- Research in the future could be conducted is in depth analysis of any critical factors that influence the readiness of e-learning. Additionally, the addition of number and kind of respondents could be extended to student and management.

References

[1] Bhuasiri W, Xaymoungkhoun O, Zo H and Jeung J 2012 Critical success factors for e-learning in developing countries: A comparative analysis between ICT experts and faculty *Computers & Education* 58.

[2] Sun, P C, Tsai, R J, Finger G, Chen Y Y and Yeh D 2008 What drives a successful eLearning? An empirical investigation of the critical factors influencing learner satisfaction *Computer & Education* p. 1183-1202.

[3] Rosenberg M J 2006 What Lies *Beyond* E-Learning? Source from: http://www.learninddgcircuits.org/2006/March/
[4] Borotis S and Poulymenakou A 2004 E-Learning Readiness Components: Key Issues to Consider Before Adopting e-Learning Interventions Source from http://www.eltrun.gr/papers/eLReadiness_ELEARN2004.pdf
[5] Mungania P 2003 The seven e-learning barriers facing employees Research final report of the masie center of e-learning consortium University of Louisville USA.
[6] Prayudi Y 2009 Kajian Awal: ELearning Readiness Index (ELRI) sebagai Model bagi Evaluasi ELearning pada Sebuah Institusi *Seminar Nasional Aplikasi Teknologi Informasi 2009 Universitas Islam Indonesia, Yogyakarta.*
[7] Swatman T and So T 2005 E-Learning Readiness in the classroom: a study of Hongkong primary and secondary teachers *Fifth IEEE International Conferene on Advanced Learning Technologies.*
[8] Akaslah D and Effie L 2011 Measuring teacher's readiness for e-learning in higher education institutions associated with the subject electricity in turkey *Global engineering education conference – Learning environments and ecosystems in engineering education, IEEE.*
[9] Seakow and Samson 2011 A study of e-learning readiness of Thailand's higher education comparing to the United states of America (USA)'s case IEEE.
[10] Kaur K and Abas Z 2004 An assessment of e-learning readiness at the Open University Malaysia International Conference on Computers in Education Malbourne.
[11] Chapnick S Elearning 2000 Readiness TM Assessment by Samanta Chapnick (2001) *Are You Ready for E-Learning?* Source from: http://www.researchdog.com
[12] Psycharis S 2005 Presumptions and actions affecting an e-learning adoption by the educational system Implementation using virtual private networks" *University of the Aegean*
[13] Begicevic N and Blaženka D 2006 Validation of Theoretical Model For Decision Making About E-learning Implementation *Journal of information and organizational sciences* 30(2) Sumber : http://www.projekti.hr/articles/1_EDEN_v5_FINAL.pdf.
[14] Haney D 2002 Assesing Organiztional Readiness for E-learning: 70 Questionsto Ask. Performance Improvement 41(4) pp. 10-15. Sumber: http://www.qou.edu/english/scientificResearch/eLearningResearchs/assessingOrganizational.pdf
[15] Economist Intelligence Unit & IBM Corporation 2003 The E-learning Readiness Rankings. Newyork: Economist Intelligence Unit.
[16] Priyanto 2008 Model E-Learning Readiness Sebagai Strategi Pengembangan E-Learning. Fakultas Teknik Universitas Negeri Yogyakarta.
[17] Aydin C H and Tasci D 2005 Measuring Readiness for eLearning: Reflections from an Emerging Country *Educational Technology & Society*, 8(4), pp. 244-257.
[18] Azwar S 2012 Reliabilitas and Validity Yogyakarta: Pustaka Pelajar.

Acknowledgments
We would like to thank for the institution X and research team involved that had conducted the research successfully.

IOP Conf. Series: Journal of Physics: Conf. Series **1007** (2018) 012042 doi:10.1088/1742-6596/1007/1/012042

Analysis of user readiness toward ICT usage at small medium enterprise in south tangerang

D Napitupulu[1], **M Syafrullah**[2], **R Rahim**[3], **D. Abdullah**[4], **MI Setiawan**[5]

[1]Research Center for Quality System & Testing Technology,
Indonesian Institute of Sciences, Tangerang Selatan, Banten, Indonesia
[2]Program of Master of Computer Science, Universitas Budi Luhur, Indonesia
[3]School of Computer and Communication Engineering, Universiti Malaysia Perlis, Malaysia
[4]Department of Informatics, Universitas Malikussaleh, Aceh, Indonesia
[5]Department of Civil Engineering, Narotama University, Surabaya, Indonesia

E-mail: darm003@lipi.go.id, mohammad.syafrullah@budiluhur.ac.id, usurobbi85@zoho.com, dahlan@unimal.ac.id, ikhsan.setiawan@narotama.ac.id

Abstract. Utilization of Information and Communication Technology (ICT) is still relatively low in the level of SMEs due to various limitations ranging from access to capital, till the marketing network. ICT is present to provide the ability for SMEs in improving the benefits and competitive advantage of the organization. This study aims to determine the level of readiness of SMEs in utilizing technology, especially ICT. The methodology used is a survey to see the technology readiness of 107 SMEs in South Tangerang selected by purposive sampling. The approach used is TRI (Technology Readiness Index) which is the individual perception of technology based on four criteria that is optimism, innovativeness, discomfort and insecurity. The results showed that the optimism and innovativeness variables significantly positively influence the technology readiness while the variables of discomfort and insecurity also significantly positively influence the readiness of ICT.

1. Introduction

The role of small and medium enterprises (SMEs) in Indonesia is very strategic in making a significant contribution to the national economic growth because SMEs are the largest business operators with percentage above 99% of the total national business actors [1]. SMEs are also a buffer of the national economy where SMEs are proven to survive when Indonesia is experiencing a crisis or a prolonged global economic recession. What makes SMEs survive is the characteristic of SMEs themselves. SMEs that most of the perpetrators of local communities also use local resources in their business so that it does not depend on raw materials or import resources. Furthermore, SMEs have the advantage of products based on local resources that exist in their respective areas and create opportunities for export due to the uniqueness or characteristic of the products produced. Therefore, it can be said that the role of SMEs as the spearhead of the progress of the national economy is very important and could

[1] To whom any correspondence should be addressed.

not be underestimated. The number of SMEs in Indonesia according to BPJS (2012) has reached 56.5 million where most of them are micro business actors [2]. SMEs increased by 2.02% over the previous year and have proven to employ 107 million (99.6%) of the workforce by 2012 where the SMEs sector contributes substantially to 56.7% of Gross Domestic Product (GDP).

Although the role is very strategic but the SMEs sector in Indonesia can be said in a less favourable conditions. This is due to the high level of business competition, especially with large-scale companies that generally have utilized modern technology in developing and running its business. In addition, low capital access is due to the lack of alignment of banks or financial institutions to SMEs or the lack of government policies in support of this small and medium sized business sector. In addition, there is an open free market era like Asean Free Trade Area (AFTA) that will threaten the existence of SMEs if their products are unable to compete with foreign products. The low competitiveness of SMEs is generally due to limited capital, human resources and access to information to network marketing. To overcome the above mentioned, SMEs require an innovation to the whole series of activities or business processes in order to provide added value or superiority of products / services offered. One way to innovate is with the utilization of Technology, especially ICT. In other words, ICTs can facilitate SMEs in the innovation process. The rapidly growing ICTs provide great potential and opportunities that can help solve SMEs problems caused by various limitations.

Despite the considerable the benefits of ICT, the utilization of one in SMEs is still relatively new, characterized by low technological and managerial capabilities [3]. The use of ICT at the SME level is still lagging behind so that SMEs are considered less competitive in the market. Some studies indicate the adoption of ICT by SMEs is still low especially when compared with large-scale companies [4]. This is one of them can be shown by the results of research conducted by research institute AMI Partners in 2000 that only 20% of SMEs who have computers in Indonesia [5]. Based on his research the reason why SMEs in Indonesia have not used ICT is that SMEs do not need ICT to support their business by 82.2%, SMEs do not have adequate budget for ICT expenditure of 41.1% and SMEs do not have human resources skilled in ICT field of 4.1%. In other words, the main reason for the weakness of ICT penetration in SMEs is the lack of awareness of SMEs about the potential of ICTs in improving excellence and competitiveness [6]. Awareness in this respect is a sufficient understanding of the strategic role that ICTs can provide in order to improve the channels of product / service marketing, customer relationships and product or service development [7][8]. Based on the mentioned problems related to the low utilization of ICT at the SME level so in this study want to know exactly how far the readiness level of SMEs in receiving and using technology especially ICT. This study is very important to help the success of SMEs in implementing ICT to improve competitiveness in this global era.

2. Literature Review

Based on Parasuraman (2000), the definition of technology readiness is the desire of the customers to use and adopt new technology to meet the needs and business goals and life every day. The concept of technological readiness can help organizations and academics to understand different user behaviours in adopting technology-based products and services [9]. Parasuraman (2000) emphasizes individual readiness to be a very important factor and affect the process of adoption of a technology. At a certain level of technological readiness, the individual as a user has a positive or negative view of a technology-based product or service [9]. This user view was evaluated based on four dimensions of personality traits: optimism, innovativeness, discomfort and insecurity [9].

The dimensions of optimism and innovativeness are contributors that contribute positively to improving the user's readiness to technology while the dimensions of discomfort and insecurity are included in the category of inhibitors that negatively contribute to technological readiness. Individuals who are said to be ready to adopt technology generally have high level contributors with low inhibitors. These four dimensions of TRI (optimism, innovativeness, discomfort and insecurity) can be used to predict whether users have a positive perception of technology and eventually adopt the technology. Optimism can be defined as the level of user confidence that technology will be useful in improving productivity in business and life, innovativeness is defined as the desire of the user to always be a leader in the utilization of the latest technology. Discomfort indicates the user's difficulty in adopting technology due to incompetence, etc. While insecurity associated with the user's trust in data security, especially in conducting transactions.

The TRI model is based on user beliefs on a technology. This user's belief is related to user openness and acceptance of technology [10]. In other words, the higher the level of user confidence in technology will drive the level of acceptability and ultimately adopt the technology. Based on the TRI model, users or individuals are classified into five groups of readiness: explorer, pioneer, sceptic, paranoid and laggard [11]. The explorer type is a group of users that had the highest level of readiness, while the laggard one is a group will adopt the technology lately.

3. Research Methodology

The methodology used in this study is a survey with a quantitative approach to determine the extent of the level of readiness of SMEs (Small and Medium Enterprises) in utilizing technology, especially ICT. The survey was conducted on 107 respondents representing SMEs in South Tangerang city where the questionnaire was designed based on TRI (Technology Readiness Index) approach proposed by Parasuraman (2000). Based on Parasuraman (2000), individual readiness is a determining factor in adopting a technology because it concerns the individual's beliefs about the technology product. Therefore respondents are asked to respond to their beliefs about ICTs from 4 dimensions of optimism, innovation, discomfort and insecurity through four Likert scoring scales: STS = "strongly disagree", TS = "disagree", S = "agree" and SS = "strongly agree".

The TRI model has 4 dimensions and 34 research variables where the optimism dimension consists of 9 indicators, innovative dimensions has 7 indicators, discomfort dimensions are 9 indicators and the dimensions of insecurity consists of 9 indicators. The theoretical model to be studied in this research is TRI (Technology Readiness Index) model which consists of five hypotheses as follows:

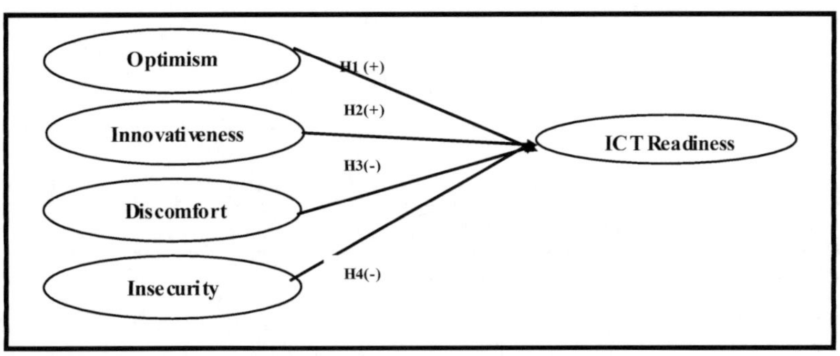

Figure 1. Theoritical research model [9].

H1: Optimism has a significant positive influence on ICT Readiness
H2: Innovativeness has a significant positive influence on ICT Readiness
H3: Discomfort has a significant negative influence on ICT readiness
H4: Insecurity has a significant negative influence on ICT Readiness

Hypothesis testing in this study based on multiple regression analysis using tool of SPSS IBM ver. 2.2 to analyse the data of the questionnaire results. The general equation of multiple linear regression can be presented as follows:

$$Y = a + b1X1 + b2X2 + cX3 + dX4 + e \qquad (1)$$

Where :

IOP Conf. Series: Journal of Physics: Conf. Series **1007** (2018) 012042 doi:10.1088/1742-6596/1007/1/012042

Y = ICT Readiness
a = Direct relation of independent and bound variable
b1 = Optimism Coefficient
b2 = Innovative Coefficient
b3 = Discomfort Coefficient
b4 = Insecurity Coefficient
e = Residue error

4. Result and Discussion

Before the hypothesis testing to see the effect of optimism, innovativeness variables, discomfort and insecurity to technological (ICT) readiness at SMEs level, first tested the validity and reliability of the questionnaire data. Total 250 questionnaires were distributed, the rate of return (response rate) only reached 42.80% ie 107 respondents. Overall data 107 respondents calculated the coefficient of validity and reliability for each research variable. For validity test using Pearson correlation where r_{count} (corrected item-total correlation) must be greater than r_{table} (r_{table} for 107 respondents with significance level 0.0005 is 0.313). Thus items that do not meet the minimum requirements ($r_{count} > r_{table}$) were eliminated from the list of questionnaire items. Total 34 items of the tested study, only the remaining 21 valid items are optimism variables consisting of 9 items (OP1-OP9), innovative variables are 3 items (INO1, INO5 and INO6), the discomfort variables consist of 4 items (DIS2, DIS6, DIS 8 and DIS 9) and the insecurity variables (INS1-INS4 and INS8) because they had fulfilled the minimum requirement ($r_{count} > r_{table}$) with r_{table} is 0.313. Furthermore, reliability testing was conducted by computing the Cronbach Alpha coefficient of each item showed good consistency and safisfaction result.

In the following sections, it will be tested research hypothesis conducted by multiple regression analysis to see how far the relationship and influence of independent variables optimism, innovativeness, discomfort and insecurity to technology readiness (technology readiness). The results of multiple regression analysis can be seen in Table 5 below:

Table 1. Result of regression analysis.

Model		Unstandardized Coefficients		Standardized Coefficients		
		B	Std. Error	Beta	t	Sig.
1	(Constant)	-7.382E-5	.003		-.027	.979
	OPTIMISM	.251	.001	.392	371.635	.000
	INNOVATIVE	.251	.001	.480	459.445	.000
	DISCOMF	.249	.001	.412	413.395	.000
	INSECURE	.250	.000	.534	544.383	.000

a. Dependent Variable: ICT_READINESS

Based on Table 5 above, regression analysis results can be obtained by the following regression equation:

$$Y = a + b1X1 + b2X2 + cX3 + dX4 + e \quad (2)$$

$$Y = -7.382E\text{-}5 + 0.251X1 + 0.251X2 + 0.249X3 + 0.250X4 \quad (3)$$

Through T test (T-Test) can be shown the results of hypothesis testing where H1 is supported in this study because the optimism variable significantly and positively affect the readiness of technology. This is evidenced by a positive regression coefficient (0.251) and a significant probability of 0.000 (<0.05) with t value of 371,635. From the results of this study can be said in general the SMEs have felt confident to use technology, especially ICT because it is considered to increase business productivity. The result of subsequent hypothesis testing is H2 is also supported by this research because innovative variables have been proven to have a significant and positive influence on technology readiness. This can be seen from the magnitude of the regression coefficient is 0.251 and the probability significant 0.000 (<0.05) with t value of 459.445. From the results of this regression analysis can be said the SMEs have a positive view of ICT and tend to want to try and use the latest technology in the market.

Furthermore, the result of hypothesis testing of H3 indicates that the discomfort variable significantly influence the technological readiness. This can be seen from the magnitude of the significant probability 0.000 (<0.05) with a value of t that is 413.395. But from the regression coefficient obtained a positive relationship between the variables of discomfort with technology readiness. This can be due to SMEs in general are already familiar with the use of ICT as a means of communication and marketing or sales, especially social media like Facebook, Instagram and Whatsapp so they do not feel disturbed by the presence of ICT except for more advanced technologies such as data mining, big data, customer behaviour, etc.

H4 hypothesis test results also show that the variables of insecurity significantly affect the readiness of technology This can be seen from the magnitude of the significant probability 0.000 (<0.05) with the value of t is 544.383. However, from the regression coefficient obtained a positive or proportional relationship between the variables of insecurity with technology readiness. This can be due to the fact that SMEs in South Tangerang have not utilized online transactions for payment of products/services offered. Most SMEs still use traditional media such as cash payment or transfer via ATM machines. Thus it is necessary to socialize again to convince SMEs of the advantages and security of the use of online transactions.

The contribution of this study compared to the previous work was Hypothesis H3 and H4 that not supported according to Parasuraman (2000), the discomfort and insecurity are significantly negative respectively but the result showed vice versa. Next part is the testing result of the simultaneous parameters to see whether the parameters or variables optimism, innovative, discomfort and insecurity togetherly influenced the ICT readiness variable which can be presented in Table 6 below:

Table 2. Result of simultaneous testing analysis.

Model		Sum of Squares	df	Mean Square	F	Sig.
1	Regression	7.620	4	1.905	2.709E5	.000a
	Residual	.001	102	.000		
	Total	7.620	106			

Based on the F test (Anova) at Table 6 above, we could see the value of F arithmetic of 2.709E5 with probability significance 0.000. Because probability significance is much less than 0.05 (5%) so it can be concluded that optimism, innovativeness, discomfort and insecurity influenced the ICT readiness simultaneously.

5. Conclusion

The user's attitude towards Information and Communication Technology (ICT) is the key that determines whether the user will use the technology or not. A positive attitude will encourage users to take advantage of ICTs while negative attitudes become a barrier as users. SMEs in the midst of its limitations require innovation of products/services produced in order to create competitive advantage and can survive in the era of increasingly fierce competition. ICT is present as a tool that can be used to increase the added value of products/services offered. The user's attitude in using technology is evaluated based on four variables: optimism, innovativeness, discomfort and insecurity based on Parasuraman & Colby (2001). The results showed that the four variables significantly influence the readiness of ICT which is proven through hypothesis testing with regression technique. SMEs are confident with the potential of ICT in supporting their business activities and not resistant to using new technology as long as it can improve business productivity and marketing of its products. The contribution of this study compared to the previous work was Hypothesis H3 and H4 that not supported according to Parasuraman (2000), the discomfort and insecurity are significantly negative respectively but the result showed vice versa. The SMEs in South Tangerang was familiar with social media but they have not fully utilized ICTs such as online transactions, thus awareness and technical assistance is needed.

References

[1] Dahnil M I 2014 Factors Influencing SMEs Adoption of Social Media Marketing *Procedia - Social and Behavioral Sciences* **148** p 119 – 126
[2] Bagnoli C and Vedovato M 2014 The impact of knowledge management and strategy configuration coherence on SME performance *Journal of Management & Governance* **18(2)** p 615–647
[3] Consoli D 2012 Literature analysis on determinant factors and the impact of ICT in SMEs *Published by Elsevier Ltd, Procedia - Social and Behavioral Sciences* **62** p 93–97
[4] León O A 2016 Relationship between the use of ICT and the degree and type of diversification *Procedia Computer Science* **100** p 1191 – 1199
[5] Naggi R 2011 Understanding the ICT Adoption Process in Small and Medium Enterprises (SMEs), *Springer-Verlag Information Technology and Innovation Trends in Organizations, Berlin Heidelberg*
[6] Bhuasiri W, Xaymoungkhoun O, Zo H and Jeung J 2012 Critical success factors for e-learning in developing countries: A comparative analysis between ICT experts and faculty *Computers & Education* 58
[7] Ramayah T 2016 Factors influencing SMEs website continuance intention in Malaysia *Journal Telematics and Informatics* **33** pages 150–164
[8] Penz D 2016 The Influence of Technology Readiness Index in Entrepreneurial Orientation: A Study with Brazilian Entrepreneurs in The United States of America *International Journal of Innovation* **5(1)** p 66–76
[9] Parasuraman A 2000 Technology Readiness Index (Tri) A Multiple-Item Scale to Measure Readiness to Embrace New Technologies *Journal of Service Research* **2(4)** p 307–320
[10] Corrocher N and Fontana R 2008 Objectives, obstacles and drivers of ICT adoption: What do IT managers perceive? *Journal Information Economics and Policy* **20** p 229–242
[11] Parasuraman A and Colby C L 2001 Techno-Ready Marketing: How and Why Your Customers Adopt Technology *Free Press, New York*

Acknowledgments
We would like to thank the research institution and all SMEs in South Tangerang that had already participated in this survey that enable the research activities was being conducted successfully.

IOP Conf. Series: Journal of Physics: Conf. Series **1007** (2018) 012043 doi:10.1088/1742-6596/1007/1/012043

Testing of technology readiness index model based on exploratory factor analysis approach

AF Ariani[1], DNapitupulu[1], RK Jati[1], JA Kadar[1], M Syafrullah[2]

[1]Research Center for Quality System and Testing Technology,Indonesian Institute of Sciences, Indonesia
[2]Program of Master of Computer Science, Universitas Budi Luhur, Indonesia

E-mails: amelia.febri@gmail.com, darwan.na70@gmail.com, rahmikj@gmail.com, jimmy.ak3@gmail.com, usurobbi85@zoho.com

Abstract. SMEs readiness in using ICT will determine the adoption of ICT in the future. This study aims to evaluate the model of technology readiness in order to apply the technology on SMEs. The model is tested to find if TRI model is relevant to measure ICT adoption, especially for SMEs in Indonesia. The research method used in this paper is survey to a group of SMEs in South Tangerang. The survey measures the readiness to adopt ICT based on four variables which is Optimism, Innovativeness, Discomfort, and Insecurity. Each variable contains several indicators to make sure the variable is measured thoroughly. The data collected through survey is analysed using factor analysis methodwith the help of SPSS software. The result of this study shows that TRI model gives more descendants on some indicators and variables. This result can be caused by SMEs owners' knowledge is not homogeneous about either the technology that they are used, knowledge or the type of their business.

1. Introduction

Small and Medium Enterprises (SMEs) is now a sector of the economy that attracts public attention. In 2013, SMEs controlled 99.99% of business units in Indonesia amounting to approximately 57.8 million [1]. This amount is very big when compared with big business which only amounted to 5,066 units. In addition, SMEs plays an important role in the Indonesian economy, because SMEs are considered quite stable in the face of global economic change compared to big enterprise. This is due to the dominant source of capital and raw materials of SMEs from domestic, so it is not significantly affected if there is a change in global exchange rate or the global economic crisis [2]. This is evidenced by the contribution of SMEs to Gross Domestic Product based on current price and constant price. Figure 1 shows that the value of SMEs greater when compared with big enterprise.

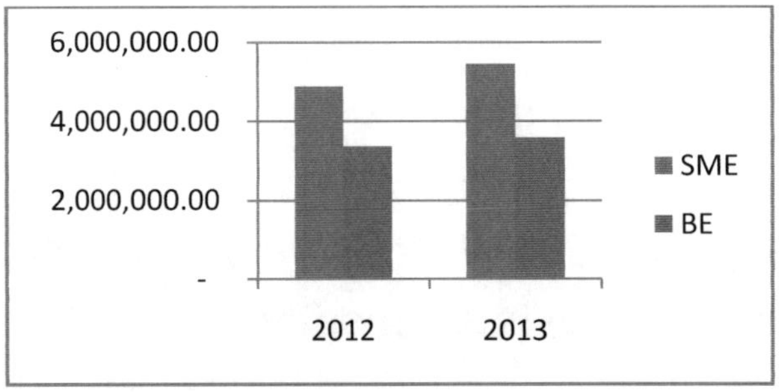

Figure 1. Comparation of gross domestic product based on current price.

Along with the increased global competitiveness in the business and business sectors, now SMEs have to improve themselves in order to have a strong competitiveness. The development of SMEshav been proclaimed by the Government of the Republic of Indonesia as stated in the Long Term Development Plan (RPJP) 2005 - 2025, as a superior economic player and has a competitive spirit [3]. In order to achieve this, it is necessary to strengthen SMEs, and one of its strengthening strategy is through the adoption of innovation and application of technology. The rate of technology application, espescially information technology, is low, which is about 30 to 40% [4].

Implementation of information technology itself can be interpreted as the use of electronic technology-based systems to manage information at all business levels [5]. The positive effect of the application of information technology to SMEs is a significant performance improvement in the areas of marketing, communication, networking, and resource planning, and can improve financial and operational performance if information technology can be utilized appropriately [6]. Bruque and Moyano [7] conducted research to find out the factors that influence the application of information technology to family business and growth cooperation, including organizational growth, management role, and the establishment of new fields related to technology. Some studies discuss the successful application of information technology utilization after the technology has been adopted by SMEs, including supporting and inhibitingfactorsto adopt e-commerce in SMEs in Indonesia [8], measuring e-readiness for SMEs in Iran [5], adoption rate of e- trade for SMEs in Korea [9]. However, the study does not address the readiness of SMEs themselves in applying information technology.

Parasuraman [10] presents an instrument that can be used to measure an organization's readiness in applying information technology, the Technology Readiness Index (TRI). There is not much literature on TRI implementation for SMEs. Parasuraman states that this instrument can give different results when applied in different places because there are other factors that influence. Therefore, in this research, TRI will be analyzed using factor analysis method, to know its suitability in measuring readiness of SMEs to apply information technology. TRI will be analyzed by factor analysis method. This method was once used by Darmawan [11] to analyze e-Government ranking in Indonesia, validate the Nursing Student Competence Isntrument (NSCI) [12], and measure the appropriateness of using Virtual Patient Adoption and Integration in Nursing (VPAIN) [13]. This research contributes to the evaluation of TRI when the method is used on different subject, in this research is SMEs.

2. Literature Review

The study conducted by Parasuraman [10], is a study that clarifies the position of users of technology. Users are the ones who have to learn more deeply to understand the technology system to be used. Correspondingly, the utility of such technology is the primary goal of the user to master the technology system [14].Users in accepting new technology systems will experience various difficulties and ease experienced. Curiosity that generates opinions from users, users will gradually begin to understand with the service module. This opinion which will be adjusted in four factors namely Optimism, Innovativeness, Discomfort and Insecurity.

Understanding these four factors have the understanding that the optimism factor as a positive vision with the confidence of easy to master, flexible and efficient in understanding knowledge, the next factor is innovativeness with the tendency to be a pioneer / pioneer, in the use of technology, discomfort factors with the perceptions of anxiety users of technology and Insecurity is the user's perception of the insecurity of technology.

In technology readiness, the user's point of view can be grouped into five segments: explorers, pioneers, skeptics, paranoid, and sluggish [15]. This type of explorer occupies the highest score in the Contributor factor (optimism, innovativeness) and the lowest score in the dimensions of inhibitors (discomfort, insecurity.). In the other hand, the sluggish type is the type that is not very interested in new technology and has the highest score in the inhibiting dimension and the lowest score in the contributors factor [15].

3. Research Methodology

This research is a survey-based quantitative research with factor analysis approach. The purpose of this factor analysis is to test the validity of the model that is model Technology Readiness Index (TRI). In this model there are variables Optimism, Innovativeness, Discomfort, Insecurity as described in research [10,15] as follows:

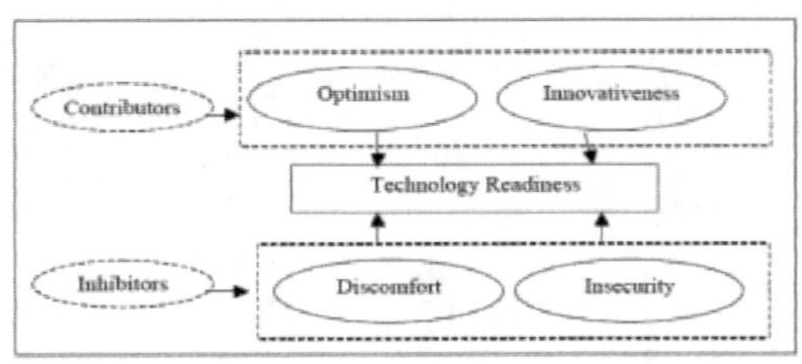

Figure 2. Model Technology Readiness Index (TRI) [10,15].

The data was collected by questionnaire-based survey of 107 SMEs owners in South Tangerang and surrounding areas. The distributed questionnaire consisted of 4 variables from TRI, with 34 indicators. Adequacy of data is complied with the theory of comparison of the number of samples with the number of variables (N/p) that is equal to 3/1 [17]. Data processing in this study using SPSS ranging from reliability test data questionnaire to factor analysis to test the model empirically.

Factor analysis is a unique multivariate statistic by testing the hypothesis of the correspondence between the observed value of the manifest variable [16]. Factor analysis has many uses, among others, can reduce most of the variables into a smaller set of variables as dimension reduction, estimating the underlying dimensions between measured variables and latent constructs, in other words enabling the formation and refinement of theories or hypotheses. and can provide evidence of empirical validity [18,19,20,21].

4. Result and Discussion

Factor analysis is performed after testing the reliability of the data first. The Cronbach's Alpha value obtained from the reliability test for all variables is 0.740. Meanwhile, when tested reliability for each variable Optimism, Innovativeness, Discomfort, Insecurity successively the value of Cronbach's Alpha produced is 0.771; 0.539; 0.658; and 0.738. Arikunto (2013) mentions that when the value of Cronbach's Alpha produced is more than 0.6 then the data has a high degree of reliability, while for values between 0.4 to 0.6 has a moderate degree of reliability [22, 23]. Based on this, then the data is said to be reliable, so that can be done further analysis.

4.1. Variable Optimism

IOP Conf. Series: Journal of Physics: Conf. Series **1007** (2018) 012043 doi:10.1088/1742-6596/1007/1/012043

There are nine indicators in the variable of Optimism. Prior to further analysis, the indicators were analyzed to determine their significance through measurement of Kaiser-Meyer-Olkin (KMO) and Bartlett's Test values. The KMO measurement value for the indicator in the Optimism variable is 0.693, where the value is> 0.5 [24]. The significance value based on Bartlett's Test is 0.000, is <0.05 [25]. Therefore, the data for these variables is feasible for further analysis. After that, each indicator is analyzed correlation with each other in the form of value of Measures of Sampling Adequacy (MSA). The MSA value for each indicator in this Optimism variable is> 0.5, so it can be concluded that each indicator correlates to another indicator. Therefore, the indicators can be analyzed further to find out the number of factors formed by looking at the value of eigenvalues> 1. The result obtained that there are 3 factors that have formed eigenvalues value more than 1, ie 3.283; 1.507; and 1.026.

To know the indicators that can be grouped on certain variables can be seen in loading factor data contained in Table 1. If loading factor> 0.5 is considered very significant, loading factor> 0.4 is more important, and loading factor> 0.3 is significant.

Table1. Rotated component matrix variabel optimism.

	Component		
	1	2	3
OP1	.145	.057	.920
OP2	.213	.327	.769
OP3	.244	.728	-.078
OP4	.066	.671	.259
OP5	.674	.090	.269
OP6	.851	-.180	.061
OP7	.717	.235	.069
OP8	.699	.409	.072
OP9	-.039	.634	.386

Therefore, the grouping of indicators into factors based on the loading factor is as follows:

- Factor 1: OP5, OP6, OP7, OP8
- Factor 2: OP3, OP4, OP9
- Factor 3: OP1, OP2

Based on the result of factor analysis, then nine indicators (OP1-OP9) can not be used to represent one variable that is Optimism, because after analyzed formed three factors that is factor 1, factor 2 and factor 3. Referring to variable and operational indicator in Table 1 it can be said that factor 1 (performance) that reflects that technology can improve the productivity of its users, factor 2 (flexibility) reflects freedom of movement with ICT, and factor 3 (control) can reflect the technology is under user control.

4.2. Variable Innovativeness

In Innovativeness variables there are seven indicators, with Kaiser-Meyer-Olkin measurement value of 0.653, where the value is> 0.5. The significance value based on Bartlett's Test is 0.000, is <0.05. Therefore, the data for these variables is feasible for further analysis. When analyzed to determine the MSA value of each indicator, it appears that the INO2 indicator has a value of 0.398 where the value is <0.5, so the INO2 indicator has no significant correlation with other indicators and the indicator is not done further analysis. After INO2 is deleted and re-analyzed to determine the MSA value, all indicators have a value> 0.5.

Based on the value of eigenvalues, the result obtained are two factors that can be formed because it has eigenvalues value>1, that is equal to 2,173 and 1,055. Table 2 showed the loading factor data so that the indicators can be grouped into two factors as follows:

- Factor 1: INO5, INO6, INO7

- Factor 2: INO1, INO3, INO4

The results of the analysis also showed that seven indicators (INO 1-INO 9) can not be used to represent Inovativeness variables, as two factors can still be formed. Factor 1 can be said as a challenge factor that represents that the latest technology requires an effort to follow it, while factor 2 (leader) reflects that user leadership in using technology. In addition, at the time of the MSA value analysis it was found that the INO2 indicator was not correlated with other indicators so it was not continued in the factor analysis.

Table2. Rotated component matrix variabel innovativeness.

	Component	
	1	2
INO1	.381	.438
INO3	.212	.571
INO4	-.111	.859
INO5	.723	.249
INO6	.566	.467
INO7	.797	-.103

4.3. Variable Discomfort

In the Discomfort variable there are nine indicators, with Kaiser-Meyer-Olkin measurement value of 0.636, where the value is> 0.5. The significance value based on Bartlett's Test is 0.000, is <0.05. Therefore, the data for these variables is feasible for further analysis. The measurement of the MSA value for all indicators in the Discomfort variable is entirely> 0.5, so that each indicator correlates to the other indicators and can be further analyzed.The result obtained according to egien value are 3 factors that are formed because it has eigenvalues more than 1, that is 2,458; 1,233; and 1.139.

Table 3. Rotated component matrix variabel discomfort.

	Component		
	1	2	3
DIS1	.199	-.155	.820
DIS2	.401	.223	.523
DIS3	**-.260**	**.569**	**.594**
DIS4	.234	.704	-.260
DIS5	.096	.720	.142
DIS6	.434	.392	.138
DIS7	.619	-.030	.028
DIS8	.603	.064	.268
DIS9	.668	.260	.001

Based on the data in Table 3, there is a cross loading on the DIS3 variable. Cross loading is a condition where on one indicator there is a loading factor value that has the same high value on two or more of all the factors formed. If such things happen, what can be done is to manipulate the number of factors that are formed. If the previous factors formed a number of three factors, then the next determined by two factors to be formed. Table 4 shows that there is no indicator that has crossloadvalue after determined two factors to be formed, so it can be known the division of indicators in Discomfort variable into the following factors:

- Factor 1: DIS1, DIS2, DIS3, DIS7, DIS8
- Factor 2: DIS4, DIS5, DIS6, DIS9

Table4. Rotated component matrix variabel discomfort (2 Factor).

	Component	
	1	2
DIS1	.789	-.292
DIS2	.664	.200
DIS3	.369	.246
DIS4	-.065	.783
DIS5	.194	.636
DIS6	.367	.469
DIS7	.364	.193
DIS8	.558	.204
DIS9	.378	.476

Based on the results of factor analysis, it can be seen that nine indicators (DIS1-DIS9) can not be used to represent one variable that is Discomfort, because because after analyzed two factors formed. Factor 1 can be said as a pessimism factor that reflects the use of technology can not be directly understood and contains risks in its use. Meanwhile, factor 2 (prejudice) reflects the distrust of the user when utilizing the technology.

4.4. Variable Insecurity

In the Insecurity variable there are nine indicators, with Kaiser-Meyer-Olkin measurement value of 0.653, where the value is> 0.5. The significance value based on Bartlett's Test is 0.000, is <0.05. Therefore, the data for that variable is feasible for further analysis. The MSA value generated by each indicator is> 0.5 so each indicator correlates to the other indicators and can be further analyzed.Based on the value of eigenvalues obtained, it is known that the resulting three factors because it has eigenvalues value>1, that is 3,019; 1,718; and 1.075.

Table5. Rotated component matrix variabel insecurity.

	Component		
	1	2	3
INS1	.782	-.131	.214
INS2	.775	.337	-.115
INS3	.838	.180	.042
INS4	.502	.696	.117
INS5	.106	.494	.229
INS6	.085	.268	.677
INS7	-.304	.331	.724

INS8	.427	-.083	.713
INS9	-.037	.802	.103

Based on the loading factor value in Table 5, it can be seen dividing indicator of Insecurity variable into two factors as follows:

- Factor 1: INS1, INS2, INS3
- Factor 2: INS4, INS5, INS9
- Factor 3: INS6, INS7, INS8

Based on the results of the analysis, nine indicators on Insecurity variables can not be used to represent these variables because after the analysis can form three factors. Based on the description of the indicator in Table 1, it can be said that factor 1 is insecurity, reflecting the potential leakage of confidential information provided with the mediation of technology. Meanwhile, factor 2 (distrust) is the user does not believe fully that the information conveyed through the technology can be up and properly understood by the next party. Factor 3 can be said to be an interaction factor that reflects the user's need to stay in touch with humans, even if they have used technology.

5. Conclusion

Based on the test result of TRI model validation, it can be concluded that the model is irrelevant to be used on the study case of SMEs in South Tangerang. This can be shown in the findings for nine indicators of optimism can not be used to represent one variable that is Optimism, because after the analysis formed three factors. It is also applied on other three variables.ForInovativeness,the variable can create two factors, Discomfort form two factors and Insecurity form three factors. This result can be caused by SMEs owners' knowledge is not homogeneous about the technology that they are used, knowledge, the type of their business, etc. Therefore, the suggestion for the next research is to test the TRI model by increasing the number of samples and compare with the SMEs in other areas. Besides, this research can also be combined with qualitative approach or observation in order to see the effect of knowledge level of the owners, SMEs infrastructure availability, and SMEs business type.

References

[1] Kementerian Koperasi dan UKM 2013 *Perkembangan Data Usaha Mikro, Kecil, Menengah (UMKM) dan Usaha Besar (UB) Tahun 2012-2013* Retrieved at April 18, 2016 fromhttp://www.depkop.go.id/berita-informasi/data-informasi/data-umkm/

[2] Bank Indonesia 2012 *Profil Bisnis Usaha Mikro, Kecil, danMenengah (UMKM)*

[3] Republik Indonesia *Undang-UndangNomor 17 Tahun 2007 Tentang Rencana Pembangunan Jangka Panjang Nasional Tahun 2005-2025*

[4] Tempo 2014 *Baru 40 Persen UKM Manfaatkan Teknolog iInformasi* Retrieved at November 1, 2017 from https://bisnis.tempo.co/read/564637/baru-40-persen-ukm-manfaatkan-teknologi-informasi

[5] Fathian M, Akhavan P and Hoorali M 2008 E-readiness assessment of non-profit ICT SMEs in a developing country: The case of Iran *Technovation* **28** p 578-590

[6] Tarute A and Gatautis R 2014 *Procedia Social and BehavioralSciences*vol 110 (American Elsevier) p 1218-1225

[7] Bruque S and Moyano J 2007 Organisational determinants of information technology adoption and implementation in SMEs: The case of family and cooperative firms *Technovation* **27** p 241-253

[8] As'ad I, Ahmad F and Sentosa I 2012 An empirical study of e-commerce implementation among SME in Indonesia *International Journal of Independent Research and Studies* **1** p 13

[9] Oh K, Cruickshank D and Anderson A R 2009 The adoption of e-trade innovations by Korean small and medium sized firms *Technovation* **29** p 110-121

[10] Parasuraman A 2000 Technology readiness index (TRI): A multiple-item scale to measure readiness to embrace new technologies *Journal of Service Research* **2** p 307-320
[11] Napitupulu D B 2017 Pengujian kerangka kerja pemeringkatan e-government di Indonesia (PeGI): Studi kasus di tingkat kementerian *Jurnal Penelitian Komunikasi* **20** p 15-30
[12] Lin C, Wu C, Hsiao Y, Han C and Hung C 2017 An exploratory factor analysis for developing and validating a scale of Nursing Students Competence Instrument *Nurse Education Today* **50** p 87-91
[13] Kleinheksel A J and RitzhauptA D 2017 Measuring the adoption and integration of virtual patient simulations in nursing education: An exploratory factor analysis *Computers & Education* **108** p 11-29
[14] Erdogmus N and Esen M 2011 *Procedia Social and Behavioral Sciences*vol 24 (American Elsevier) p 487-495
[15] Parasuraman A and Colby C L 2001 *Techno-Ready Marketing: How and Why Your Customers Adopt Technology* (New York: Free Press)
[16] Costello A B and Osbourne J W 2005 Best practices in exploratory factor analysis: four recommendations for getting the most from your analysis *Practical Assessment, Research & Evaluation* **10** p 1-9
[17] Catell R B 1978 *The Scientific Use of Factor Analysis in Behavioral and Life Sciences* (New York: Plenum)
[18] Hair J F, Anderson R E, Tatham R L and Black W C 1998 Multivariate Data Analysis with Readings (New Jersey: Prentice Hill)
[19] Tabachnick B G and Fidell L S 2007 *Using Multivariate Statistics* (London: Pearson)
[20] Thompson B 2004 *Exploratory and ConfirmatoryFactor Analysis: Understanding Concepts and Applications* (American Psychological Association)
[21] Williams B, Onsman A and Brown T 2010 Exploratory factor analysis: A five-step guidefor novices *Journal of Emergency Primary Health Care (JEPHC)* **8**p 1–13
[22] Sekaran U 1992 *Research Methods For Business: A Skill Building Approach*, 2nd ed (New York: John Willey & Sons, Inc)
[23] Arikunto S 2013 *Prosedur Penelitian Suatu Pendekatan Praktik* (Jakarta: RinekaCipta)
[24] Kaiser H F 1974 An index of factorial simplicity *Psychometrika* **39** p 31-36
[25] Bartlett M S 1950 Test of significance in factor analysis *British Journal of Mathematical and Statistical Psychology* **3** p 77-85

IOP Conf. Series: Journal of Physics: Conf. Series **1007** (2018) 012044 doi:10.1088/1742-6596/1007/1/012044

E-Business, the impact of regional growth on the improvement of Information and Communication Development

MI Setiawan[1*], C Hasyim[2], N Kurniasih[3], D Abdullah[4], D Napitupulu[5], R Rahim[6], A Sukoco[7], I Dhaniarti[8], J Suyono[7], IN Sudapet[7], RD Nasihien[1], DAR Wulandari[1], Reswanda[7], SW Mudjanarko[1], Sugeng[9], MBN Wajdi[10]

[1]Narotama University, Department of Civil Engineering, Surabaya, Indonesia
[2]Universitas Darul Ulum, Jombang, Indonesia
[3]Universitas Padjadjaran, Faculty of Communication Science, Bandung, Indonesia
[4]Universitas Malikussaleh, Department of Informatics, Aceh, Indonesia
[5]Research Center for Quality System and Testing Technology, Indonesian Institute of Sciences, Indonesia
[6]School of Computer and Communication Engineering, Universiti Malaysia Perlis, Malaysia
[7]Narotama University, Department of Management, Surabaya, Indonesia
[8]Narotama University, Department of Law, Surabaya, Indonesia
[9]Universitas Islam "45", Bekasi, Indonesia
[10]STAI Miftahul 'Ula, Nglawak Kertosono, Indonesia

E-mail: *ikhsan.setiawan@narotama.ac.id

Abstract. ICT becomes a key element to improve industrial infrastructure efficiency and sustainable economic productivity. This study aims to analysis the impact of regional improvement on information and communication development in Indonesia. This research is a correlational study. Population of this research include 151 regions in Indonesia. By using a total sampling, there were 151 sample regions. The results show there are the strong impact of regional growth on increasing Gross Regional Domestic Product (GRDP) of information and communication. It can be seen from all regional improvement sub variables that have a high correlation in increasing GRDP of Information and Communication in Indonesia. Only two sub-variables that have low correlation to GRDP of Information and Communication variable i.e. GRDP of Agriculture, Forestry and Fishing (0.01) and GRDP of Mining and Quarrying (-0.04). The correlation coefficient (R) is 0.981, means the variable of information and communication GRDP has a very strong correlation with regional growth variable. Thus the value of Adjusted R Square is 95.8%, means there are impact of regional growth variables in increasing GRDPof Information and Communication, while the increase of 4.2% of Information and Communication GRDP is influenced by other factors aside from regional improvement.

IOP Conf. Series: Journal of Physics: Conf. Series **1007** (2018) 012044 doi:10.1088/1742-6596/1007/1/012044

1. Introduction

Information and Communication Technology (ICT) is an important element for sustainable development. ICT becomes a key element for industrial infrastructure efficiency and for increasing productivity as a sustainable and competitive development foundation[1]. On the one hand, ICT can lead pricesincreased but on the other hand an overly high ICT development can lead the prices fall. The price reduction is usually influenced by the provision of free products and services, the ease of adherence to standardization, the Gross Regional Domestic Product (GRDP) which derived from the advancement of the internet that encourages a more free culture, the consumption based on the aspects of usefulness and happiness, but cannot be calculated in the GRDP data that measured income [2]. However, region growth can be measured from the GRDP of the region. To achieve a good GRDP, a region should be able to create valueadded for the production of goods and services. GRDP is one indicators to measure a region performance or as a reflection of a government's success in mobilizing the economic sectors. Information and communication development covers the information production and distribution, cultural products, inventory of tools to transmit or distribute the products, data or communications, information, information technology activities, data processing as well as other information services activities. ICT business fields consists of several industries, namely publishing, production of moving pictures, video, voice and music recording, broadcasting and programming (radio and television), telecommunication, programming, computer and information technology[3]. Regional economic growth reported by Central Bank of Indonesia (Bank Indonesia) shows the highest contribution to the national economy sequentially in Central Kalimantan province (9.5%), Southeast Sulawesi (8.4%), South Sulawesi (7.5%), North Maluku (7 , 5%), West Sulawesi (7.4%), Gorontalo (7.3%), Java (58%), Sumatera (22%) and Eastern Indonesia (KTI) at 20%[4].This study aims to analyze the impact of regional improvement on information and communication development in Indonesia.

2. Methodology

This research used correlational study. Research variables include X Variable (regional growth) and Y Variable (information and communication development). Regression analysis is used to analyze the impacts on both variables. Population on this research include 151 regions in Indonesia. By using a total sampling there were 151 region as research sample. The data of research was analyzed by using Person Correlations, R test and multiple linear regression equations[5][6][7][8][9].

3. Result and Discussion

Analysis of the impact of regional growth on the information and communication development used multiple linear regression formula. The data include 151 of districts/ cities in Indonesia with research variables and sub-variables as follows:

Y = GRDPof Information and Communication
X_1= GRDPof Agriculture, Forestry and Fishing
X_2= GRDP of Mining and Quarrying
X_3= GRDP of Manufacturing
X_4= GRDP of Electricity and Gas
X_5= GRDPof Water Supply, Sewerage, Waste Management and Remediation Activities
X_6= GRDP of Construction
X_7= GRDPof Wholesale and Retail Trade; Repair of Motor Vehicles and Motorcycles
X_8= GRDP of Transportation and Storage
X_9= GRDPof Accommodation and Food Service Activities
X_{10}= GRDPof Financial and Insurance Activities
X_{11}= GRDP of Real Estate Activities
X_{12}= GRDP of Business Activities
X_{13}= GRDP of Public Administration and Defense; Compulsory Social Security

X_{14}= GRDP of Education
X_{15}= GRDPof Human Health and Social Work Activities
X_{16}= GRDP of Other Services Activities

The results of the analysis of 151 regions using Pearson Correlations, R test and multiple linear regression equations can be seen in Table 1, Table 2 and Table 3 below:

Table 1. Pearson Correlations.

GRDP of Information and Communication (Y)	1.000
GRDP of Wholesale and Retail Trade; Repair of Motor Vehicles and Motorcycles (X_7)	0.93
GRDP of Financial and Insurance Activities (X_{10})	0.92
GRDPof Other Services Activities (X_{16})	0.90
GRDPof Water supply, Sewerage, Waste Management and Remediation Activities (X_5)	0.87
GRDP of Human Health and Social Work Activities (X_{15})	0.83
GRDPof Public Administration and Defense; Compulsory Social Security (X_{13})	0.82
GRDPof Construction (X_6)	0.81
GRDPof Accommodation and Food Service Activities (X_9)	0.79
GRDPof Education (X_{14})	0.79
GRDPof Real Estate Activities (X_{11})	0.75
GRDPof Manufacturing (X_3)	0.68
GRDPof Transportation and Storage (X_8)	0.68
GRDPof Business Activities (X_{12})	0.59
GRDPof Electricity and Gas (X_4)	0.46
GRDPof Agriculture, Forestry and Fishing (X_1)	0.01
GRDPof Mining and Quarrying (X_2)	(0.04)

Table 1 shows that almost all regional growth sub-variables have a high correlation with the GDRP of information and communication variable, there are only 2 (two) variables that have low correlation withGRDP of Information and Communication variable ie GRDPof Agriculture, Forestry and Fishing (0.01) and the GRDPof Mining and Quarrying (-0.04).

Table 2. Model Summary.

R	R Square	Adjusted R Square	Std. Error of the Estimate
.981[a]	.963	.958	477.81556

a. Predictors: (Constant), GRDP of Other Services Activities, GRDP Agriculture, Forestry and Fishing, GRDP Mining and Quarrying, GRDPof Electricity and Gas, GRDPof Transportation and Storage, GRDPof Water supply, Sewerage, Waste Management and Remediation Activities, GRDP of Construction, GRDPof Accommodation and Food Service Activities, GRDPof Manufacturing, GRDPof Real Estate Activities, GRDPof Public Administration and Defence; Compulsory Social Security, GRDPof Human Health and Social Work Activities, GRDPof Wholesale and Retail Trade; Repair of Motor Vehicles and Motorcycles, GRDPof Financial and Insurance Activities, GRDP of Business Activities, GRDP of Education .
b. Dependent Variable: GRDPof Information and Communication.

Table 2 shows that the correlation coefficient (R) is 0.981, meaningthat GRDP Information and Communication variable has a very strong correlation with regionalgrowth variable. Thus the value of Adjusted R Square is 95.8%, means there are impact of regional growth variables in increasing

IOP Conf. Series: Journal of Physics: Conf. Series **1007** (2018) 012044 doi:10.1088/1742-6596/1007/1/012044

GRDPof Information and Communication, while the increase of 4.2% of Information and Communication GRDP is influenced by other factors aside from regional improvement.This shows the strong regional growth impact on the increase of Information and CommunicationGRDP.

Table 3. Multiple Linear Regression Equations.

Variable	Unstandardized Coefficients
(Constant)	-94.968
PDRB Agriculture, Forestry and Fishing (X_1)	-0.019
PDRB Mining and Quarrying (X_2)	0.008
PDRB Manufacturing (X_3)	-0.005
PDRB Electricity and Gas (X_4)	-0.892
PDRB Water supply, Sewerage, Waste Management and Remediation Activities (X_5)	8.043
PDRB Construction (X_6)	0.109
PDRB Wholesale and Retail Trade; Repair of Motor Vehicles and Motorcycles (X_7)	0.027
PDRB Transportation and Storage (X_8)	0.125
PDRB Accommodation and Food Service Activities (X_9)	0.073
PDRB Financial and Insurance Activities (X_{10})	0.050
PDRB Real Estate Activities (X_{11})	-0.462
PDRB Business Activities (X_{12})	-0.600
PDRB Public Administration and Defence; Compulsory Social Security (X_{13})	-0.070
PDRB Education (X_{14})	0.695
PDRB Human Health and Social Work Activities (X_{15})	-0.676
PDRB Other Services Activities (X_{16})	1.112

Multiple linear regression equation, as follows:

$$Y = -94.968-0.019 X_1 +0.008 X_2-0.005 X_3-0.892 X_4 +8.043 X_5 +0.109 X_6 +0.027 X_7 +0.125 X_8 +0.073 X_9 +0.050 X_{10}-0.462 X_{11}-0.600 X_{12}-0.070 X_{13} +0.695 X_{14}-0.676 X_{15} +1.112 X_{16}$$

Y = GRDPof Information and Communication
X_1= GRDPof Agriculture, Forestry and Fishing
X_2= GRDP of Mining and Quarrying
X_3= GRDP of Manufacturing
X_4= GRDP of Electricity and Gas
X_5= GRDPof Water supply, Sewerage, Waste Management and Remediation Activities
X_6= GRDP of Construction
X_7= GRDPof Wholesale and Retail Trade; Repair of Motor Vehicles and Motorcycles
X_8= GRDP of Transportation and Storage
X_9= GRDPof Accommodation and Food Service Activities
X_{10}= GRDPof Financial and Insurance Activities
X_{11}= GRDP of Real Estate Activities
X_{12}= GRDP of Business Activities
X_{13}= GRDPof Public Administration and Defense; Compulsory Social Security
X_{14}= GRDP of Education
X_{15}= GRDPof Human Health and Social Work Activities
X_{16}= GRDP of Other Services Activities

Based on the multiple linear regression equation above, it can be seen that the variables X_2, X_5, X_6, X_7, X_8, X_9, X_{10}, X_{14}, X_{16}have a positive impact onincreasing GRDP of Information and Communication, while the variables X_1, X_3, X_4, X_{11}, X_{12}, X_{13}, X_{15}have a negative impact on decreasing GRDP of Information and Communication.

4. Conclusion

There is the strong impact of regional growth on increasing GRDP of information and communication. It can be seen from all regional improvement variables that have a high correlation in increasing GRDP of Information and Communication in Indonesia.Only two sub-variables that have low correlation to GRDP of Information and Communication variable i.e. GRDP of Agriculture, Forestry and Fishing (0.01) and GRDP of Mining and Quarrying (-0.04). The correlation coefficient (R) is 0.981, means the variable of information and communication GRDP has a very strong correlation with regional growth variable. Thus the value of Adjusted R Square is 95.8%, means there are impact of regional growth variables in increasing GRDPof Information and Communication, while the increase of 4.2% of Information and Communication GRDP is influenced by other factors aside from regional improvement.

References

[1] S. Dutta and I. Mia, "The Global Information Technology Report 2009–2010: ICT for Sustainability," 2010.

[2] C. Watanabe, K. Naveed, and W. Zhao, "New paradigm of ICT productivity - Increasing role of un-captured GDP and growing anger of consumers," *Technol. Soc.*, vol. 41, pp. 21–44, 2015.

[3] B. P. Statistik, *Tinjauan Regional berdasarkan PDRB Kabupaten/Kota tahun 2011-2015 pulau Nusa Tenggara, Maluku dan Papua*. BPS, 2016.

[4] C. Watanabe, K. Naveed, and P. Neittaanmäki, "Dependency on un-captured GDP as a source of resilience beyond economic value in countries with advanced ICT infrastructure: Similarities and disparities between Finland and Singapore," *Technol. Soc.*, vol. 42, pp. 104–122, 2015.

[5] Laerd, "Linear regression in SPSS," *Laerd.com*, pp. 1–7, 2013.

[6] S. Weisberg, *Applied Linear Regression*, vol. 528, no. 0. 2005.

[7] N. Altman and M. Krzywinski, "Points of Significance: Simple linear regression," *Nat. Methods*, vol. 12, no. 11, pp. 999–1000, 2015.

[8] X. Su, X. Yan, and C.-L. Tsai, "Linear regression," *Wiley Interdiscip. Rev. Comput. Stat.*, vol. 4, no. 3, pp. 275–294, 2012.

[9] J. Suyono, A. Sukoco, M. I. Setiawan, S. Suhermin, and R. Rahim, "Impact of GDP Information Technology in Developing of Regional Central Business (Case 50 Airports IT City Development in Indonesia)," *J. Phys. Conf. Ser.*, vol. 954, no. 1, p. 12045, Dec. 2017.

Acknowledgment

This paper is a part of a PhD Dissertation titled "Business Centre Development Model of Airport Area in Indonesia" with Prof. Ir. Surjono Surjokusumo, M.Sc.F, Ph.D., Ir. Dadang M. Ma'soem, MSCE, Ph.D. and Dr. Ir. Johny Johan, M.Eng, MM as the counselors.

IOP Conf. Series: Journal of Physics: Conf. Series **1007** (2018) 012045 doi:10.1088/1742-6596/1007/1/012045

E-Business, The impact of the Regional Government Development (APBD) on Information and Communication Development in Indonesia

M I Setiawan[1*], C Hasyim[2], N Kurniasih[3], D Abdullah[4], D Napitupulu[5], R Rahim[6], A Sukoco[7], I Dhaniarti[8], J Suyono[7], IN Sudapet[7], RD Nasihien[1], DAR Wulandari[1], Reswanda[7], SW Mudjanarko[1], Sugeng[9], MBN Wajdi[10]

[1]Narotama University, Department of Civil Engineering, Surabaya, Indonesia
[2]Universitas Darul Ulum, Jombang, Indonesia
[3]Universitas Padjadjaran, Faculty of Communication Science, Bandung, Indonesia
[4]Universitas Malikussaleh, Department of Informatics, Aceh, Indonesia
[5]Research Center for Quality System and Testing Technology, Indonesian Institute of Sciences, Indonesia
[6]School of Computer and Communication Engineering, Universiti Malaysia Perlis, Malaysia
[7]Narotama University, Department of Management, Surabaya, Indonesia
[8]Narotama University, Department of Law, Surabaya, Indonesia
[9]Universitas Islam "45", Bekasi, Indonesia
[10]STAI Miftahul 'Ula, Nglawak Kertosono, Indonesia

E-mail: *ikhsan.setiawan@narotama.ac.id

Abstract. The decline of development delived from natural resources in several regions in Indonesia requires local governments to make a breakthrough in increasing revenues beyond natural resources, including through increased revenue in the field of information technology. The purpose of this study is to analyze the impact of the regional government development (APBD) on information and communication development in Indonesia. By using correlational research, it is known that all of APBD sub variables have a high correlation in increasing Gross Regional Domestic Product (GRDP) of Information and Communication in Indonesia, only sub-variable of surplus/deficit development, unexpected development sub-variable, sub-variable of development aid to province/regency/city and village government, and sub-variable of development special allocations that have not a significant correlation to increase of GRDP of Information and Communication in Indonesia. There is 86,7% of APBD variable influence the increase of GRDP of Information and Communication, while the increase of 15,3% of Information and Communication GRDP influenced by other factors aside from APBD. It shows the strong impact of APBD on increasing GRDP of Information and Communication

1. Introduction

Regional Government Development (APBD) is an effort of local government in Indonesia in optimizing the existing resources, increasing its regional development, and allocating it to the welfare of the community. By looking at the potential of an area in terms of development, it can be formulated appropriate policies, either by the local government itself, the central government, as well as other stakeholders[1]. For example, the 2017 APBD of Sumatra has increased to Rp 227.8 trillion compared to the 2016 APBD of Rp 196.8 trillion. The highest growth of APBD in Sumatra is in the province of Bangka Belitung Islands (6.42%), while the largest increase of APBD is in North Sumatra at 26.57% (yoy). Meanwhile, in some provinces on the island of Sumatra showed a decrease in budget as in Riau as the DBH development of oil and gas decreased. The performance of regional absorption in APBD in Eastern Indonesia (KTI) in the first quarter of 2017 was 21.31%, lower than the achievement of the same quarter of the year in 2016 of 22.41%, the decrease of PAD development and other legitimate development compared with the same quarter last year. In Kalimantan, in particular, the impact of coal exports has not had a direct impact on development performance in the region and is expected to impact and development capacity in the region in the current quarter. Decrease in development in some areas that are sourced from natural resource activities require local governments should be able to make a breakthrough income increase beyond natural resources, one of which increased development in the field of information technology[2].Urban infrastructure development is a promising business. In the midst of urbanization around the world, especially in developing countries, infrastructure investment, especially in the electricity, rail, road, water, sanitation, home, airport, mobile and internet sectors is accelerating due to higher income and middle-class growth[3]. In the field of information and communications technology, the linear impact of Internet penetration and usage is positive and significant for low-income countries. The focus of ICT development is to promote ICT innovations for sustainable development. For that, the use of the Internet is expected to provide great benefits[4].Technological innovation has always been regarded as a major stimulus for development growth. High-speed Internet access through broadband infrastructure has grown rapidly since the late 1990s as a result of the deployment of fixed and mobile technologies, a positive correlation between broadband and development growth, the relevant quantitative relation between broadband and dynamics in the short, medium and long term[5][6]. In the Arab and Middle East regions, especially in high-income Arab and Middle East countries, the internet has become a significant growth determiner [7].

This study aims to see the effect of the size of the regional budget (APBD) on Information and Communication Economic in Indonesia.

2. Methodology

This research is a correlational research. Correlational analysis describes the relationship between independent vaiabel and dependent variable[8]. In this research, APBD become Variable X and Information and Communication development become Variable Y. The population in this research is 151 cities / districts in Indonesia and by using total sampling, all units of population then become sample . Pearson Correlation, R test and regression analysis were used to process the data in this study[9][10][11][12].

3. Result and Discussion

The impact of Information and Communication APBD development, to know the magnitude of the influence of APBD on Information and Communication development used multiple linear regression formula. Table 1 shows the dependent and independent variables in this study.

Table 1. Research Variables

Y = PDRB Information and Communication	X_{16} = Employee Indirect of development
X_1 = Local development	X_{17} = Grant of development
X_2 = Local PAD of development	X_{18} = Social ofdevelopment
X_3 = Local Tax of development	X_{19} = Expenditure ofdevelopment Assistance to the Provincial/District/Municipal and Pemdes
X_4 = Local retribution of development	
X_5 = development of Local assets	

IOP Conf. Series: Journal of Physics: Conf. Series **1007** (2018) 012045 doi:10.1088/1742-6596/1007/1/012045

X_6 = Other valid PAD of development	X_{20} = Unexpected development
X_7 = Balancing Fund of development	X_{21} = Direct development
X_8 = Non-tax and tax share	X_{22} = development of Goods and services
X_9 = General Allocation DAU	X_{23} = Capital of development
X_{10} = Special Allocation DAK	X_{24} = Surplus/Deficit of development
X_{11} = Other valid Local of development	X_{25} = Local Financing of development
X_{12} = Provision of tax development sharing by Province and others	X_{26} = Local Revenue of development X_{27} = SILPA before
X_{13} = Adjustment Funds and Special Autonomy	X_{28} = Spending
X_{14} = Local Expenditure of development	X_{29} = Investment of development
X_{15} = Indirect Expenditure of development	X_{30} = SILPA

Table 2 and Table 3 show the analysis results based on data from 151 cities throughout Indonesia

Table2. Model Summary.

R	R Square	Adjusted R Square	Std. Error of the Estimate
0.933	0.871	0.847	913.72248

Table 2 above shows that the magnitude of the correlation coefficient (R) is 0.933, meaning that the GRDP Information and Communication variable has a very strong correlation with the variables of the APBD.

While the value of Adjusted R Square is 0.847, it means 86,7% influence of APBD to increasing GRDP Information and Communication, while the rest equal to 15.3% increase of GRDP Information and Communication influenced by factors other than APBD. This shows the strong impact of APBD on increasing GRDP Information and Communication.

Table 3. Multiple Linear Regression Equations.

Variable	**Unstandardized Coefficients**
(Constant)	-428.6115365
X_4	-1.13263E-09
X_5	-1.06629E-08
X_6	6.61786E-11
X_8	-5.49497E-09
X_9	-2.48352E-09
X_{10}	-2.51226E-09
X_{11}	-3.64188E-09
X_{12}	1.25897E-08
X_{13}	2.4628E-09
X_{15}	-3.88216E-09
X_{16}	7.06921E-09
X_{17}	6.70682E-09
X_{18}	5.9941E-09
X_{19}	3.3714E-09
X_{20}	-1.16585E-08
X_{21}	-7.46809E-09
X_{22}	1.06337E-08
X_{23}	1.10151E-08
X_{24}	6.49394E-09
X_{27}	4.10982E-09

X_{28}	-6.89825E-11
X_{29}	-1.7214E-08
X_{30}	-4.25754E-09

The Multiple Linear Regression Equation, as follows:
$Y = -428.6115365 - 1.13263E^{-09} X_4 - 1.06629E^{-08} X_5 + 6.61786E^{-11} X_6 - 5.49497E^{-09} X_8 - 2.48352E^{-09} X_9 - 2.51226E^{-09} X_{10} - 3.64188E^{-09} X_{11} + 1.25897E^{-08} X_{12} + 2.4628E^{-09} X_{13} - 3.88216E^{-09} X_{15} + 7.06921E^{-09} X_{16} + 6.70682E^{-09} X_{17} + 5.9941E^{-09} X_{18} + 3.3714E^{-09} X_{19} - 1.16585E^{-08} X_{20} - 7.46809E^{-09} X_{21} + 1.06337E^{-08} X_{22} + 1.10151E^{-08} X_{23} + 6.49394E^{-09} X_{24} + 4.10982E^{-09} X_{27} - 6.89825E^{-11} X_{28} - 1.7214E^{-08} X_{29} - 4.25754E^{-09} X_{30}$

Based on the multiple linear regression equation above, it can be seen that the variables X_6, X_{12}, X_{13}, X_{16}, X_{17}, X_{18}, X_{19}, X_{22}, X_{23}, X_{24}, X_{27} has a positive impact of increasing GRDP Information and Communication, while variable X_4, X_5, X_8, X_9, X_{11}, X_{15}, X_{20}, X_{21}, X_{29}, X_{30} has a negative impact on the reduction of GRDP Information and Communication

4. Conclusion

All APBD variables have a high correlation to increasing GRDP of Information and Communication in Indonesia, only variable of surplus/deficit of development, unexpected expenditure of development, financial expenditure of development to province / regency / city and village government, and special allocation fund of development which have not significant correlation to the increase of GRDP Information and Communication in Indonesia. 86,7% influence of variable of APBD to increase of GRDP Information and Communication, while the rest equal to 15.3% increase of GRDP Information and Communication influenced by factors other than APBD. This shows the strong impact of APBD on increasing GRDP Information and Communication

References

[1] B. P. Statistik, *Tinjauan Regional berdasarkan PDRB Kabupaten/Kota tahun 2011-2015 pulau Nusa Tenggara, Maluku dan Papua*. BPS, 2016.
[2] Bank Indonesia, "Kajian Ekonomi Regional," 2017.
[3] H. Jung and J. D. Son, "Estimation of urban infrastructure demand with consideration of urban characteristics," *ICIC Express Lett. Part B Appl.*, vol. 7, no. 2, pp. 295–300, 2016.
[4] M. M. Albiman and Z. Sulong, "The linear and non-linear impacts of ICT on economic growth, of disaggregate income groups within SSA region," *Telecomm. Policy*, vol. 41, no. 7–8, pp. 555–572, 2017.
[5] A. Castaldo, A. Fiorini, and B. Maggi, "Measuring (in a time of crisis) the impact of broadband connections on economic growth: an OECD panel analysis," *Applied Economics*, pp. 1–17, 2017.
[6] J. Suyono, A. Sukoco, M. I. Setiawan, S. Suhermin, and R. Rahim, "Impact of GDP Information Technology in Developing of Regional Central Business (Case 50 Airports IT City Development in Indonesia)," *J. Phys. Conf. Ser.*, vol. 954, no. 1, p. 12045, Dec. 2017.
[7] M. Badran, "The impact of broadband infrastructure on economic growth in Egypt and some Arab and emerging countries," *... Univ. Fac. Econ. Polit. Sci.*, pp. 1–26, 2011.
[8] B. E. W. Jr and Mary E. Kite, *Principles of Research in Behavioral Science: Third Edition 3rd Edition*. Routledge, 2012.
[9] X. Su, X. Yan, and C.-L. Tsai, "Linear regression," *Wiley Interdiscip. Rev. Comput. Stat.*, vol. 4, no. 3, pp. 275–294, 2012.
[10] N. Altman and M. Krzywinski, "Points of Significance: Simple linear regression," *Nat. Methods*, vol. 12, no. 11, pp. 999–1000, 2015.
[11] M. J. Crawley, "Regression," in *The R Book*, 2007, pp. 387–448.
[12] S. Weisberg, *Applied Linear Regression*, vol. 528, no. 0. 2005.

Acknowledgment
This paper is a part of a PhD Dissertation titled "Business Centre Development Model of Airport Area in Indonesia" with Prof. Ir. Surjono Surjokusumo, M.Sc.F, Ph.D., Ir. Dadang M. Ma'soem, MSCE, Ph.D. and Dr. Ir. Johny Johan, M.Eng, MM as the counselors.

E-Business, Airport Development and Its Impact on the Increasing of Information of Communication Development in Indonesia

MI Setiawan[1*], C Hasyim[2], N Kurniasih[3], D Abdullah[4], D Napitupulu[5], R Rahim[6], A Sukoco[7], I Dhaniarti[8], J Suyono[7], IN Sudapet[7], RD Nasihien[1], DAR Wulandari[1], Reswanda[7], SW Mudjanarko[1], Sugeng[9], MBN Wajdi[10]

[1]Narotama University, Department of Civil Engineering, Surabaya, Indonesia
[2]Universitas Darul Ulum, Jombang, Indonesia
[3]Universitas Padjadjaran, Faculty of Communication Science, Bandung, Indonesia
[4]Universitas Malikussaleh, Department of Informatics, Aceh, Indonesia
[5]Research Center for Quality System and Testing Technology, Indonesian Institute of Sciences, Indonesia
[6]School of Computer and Communication Engineering, Universiti Malaysia Perlis, Malaysia
[7]Narotama University, Department of Management, Surabaya, Indonesia
[8]Narotama University, Department of Law, Surabaya, Indonesia
[9]Universitas Islam "45", Bekasi, Indonesia
[10]STAI Miftahul 'Ula, Nglawak Kertosono, Indonesia

E-mail: *ikhsan.setiawan@narotama.ac.id

Abstract. The increasing number of the internet usage by households have an effect on the tourism sector. On the other hand, the aviation industry is growing as one of the development centers. This study aims to analyze the impact of information and communication development to airport performance in Indonesia. This is a correlation research with 151 of airports in Indonesia as a population. The sampling technique was done by using total sampling. The results of correlation (R) indicates the Gross Regional Domestic Product (GRDP) of Information and Communication has a relatively strong relationship with the Airport Performance. Meanwhile the results of Adjusted R test shows that they are other factors in increasing GRDP of Information and Communication besides Airport Performance. It shows the low impact of Information and Communication GRDP to the Airport Performance.

1. Introduction

Currently Air transport has developed into an economic centre. It is supported by an increase in local travelers and foreign tourists reached 11.519 million tourists, with 2,387 five-star hotel accommodation, with a five-star hotel room occupancy rate reached 54.34%. Wherein, the share of foreign tourists to the total national tourists reached almost 23%, while the share of domestic tourists reached 16%. Increased tourism competitiveness is achieved through improved quality of public

IOP Conf. Series: Journal of Physics: Conf. Series **1007** (2018) 012046 doi:10.1088/1742-6596/1007/1/012046

facilities infrastructure, IT technology, promotion of tourism, conducive business climate, integration connectivity and accessibility. For example there are many tourism objects in Lembang West Java Indonesia use website and social media as their promotion. It shows how potential social media to spread information widely, including information regarding tourism objects[1]. Bank Indonesia estimate shows that any increase in the number of foreign and domestic tourists by 10% will be able to increase the regional growth of 0.31% and encourage increasing the number of workforce of around 0.68%[2].

Electronic era changing industry competition and regional development emphasizing on information management and supply chain network. To be able to compete in regional and global markets required speed and agility. Modern logistics technology and infrastructure based on low to be important for many companies in the new development era. Companies need to manage e-commerce strategy B2C (Business to Consumer) B2B (Business to Business) and B2B appropriately. Industry profits earned by companies that respond flexibly and quickly to their domestic and global customers by delivering products quickly and efficiently.

Flights, digitalization, globalization and competition time-based architecture interact to create new competition while positioning the airport as an international logistics gateway. Airports became the center of commercial activity concentration as a new urban form aero city[3]. The developmental advantages gained by countries that combine digital technology and trade of air through the logistics infrastructure to optimize their position in the global network of information flow and product.

Indonesia has 296 official airport that most ASEAN, covering 13 airports managed by PT Angkasa Pura I, 13 airports managed by PT Angkasa Pura II, 53 airports managed by TNI, 178 airports managed by the Ministry of Transportation Unit Operator Airport and 39 airports managed by the Local Government Unit[4], The use of the airport consists of an international airport and the domestic airport. International Airport is the airport that serves the domestic flights and routes from and to foreign countries. Determination international airport is to consider: (a) The master plan of the national airport; (B) defense and security; (C) The growth and development of tourism; (D) Interest and ability of national air transport; (E) The development of national economy and foreign trade. Determination airports inernasional determined by the Minister after coordination with the minister duties and responsibilities in the field of immigration, customs and kekarantinaan in the framework of work units and personnel placement. Meanwhile, the domestic airport is an airport that serves the domestic flights[5].

Other air transport role in supporting development growth not only in serving the passengers but also cargo services. There are three factors that can increase the positive impact of air cargo, namely liberalization of air services, improve the quality of customs, and reduce corruption[6], This study aims at analyzing the development effects of information and communication on airport performance in Indonesia.

2. Methodology

This is a quantitative research with correlational approach. Correlational research is used to describe the relationship between two or more variables that occur naturally among these variables[7][8]. In this study, Airport Performance is X variable and information and communication development is Y variable. The population of this research include 151 of airports in Indonesia. By using a total sampling, there are 151 of airports as sample. The data of 151 of airports was analyzed by using Person Correlations, R test and multiple linear regression equations.

3. Result and Discussion

The increasing number of internet usage by households has an effect on tourism. The role of information technology in an area can be seen in the increasing investment realization at home and abroad in 2016, where transportation, storage and communication reach Rp. 27.5 trillion[9]. The impact of increasing internet usage is to facilitate the public to access information so that the impact on the growth of development.

Analysis of the impact ofGRDP of information and communication to the airport development Performance used multiple linear regression formula. The data include 151 of districts/ cities in Indonesia with research variables and sub-variables as follows:
Two variables in this study are Y variable (the GRDP of information and communication) and X variable (airport performance). Sub-variable X consists of:
X1 = Aircraft Arrival

X2 = Aircraft Departure

X3 = Passenger Departure

X4 = Passenger Arrival

X5 = Baggage Unloaded

X6 = Baggage Loaded

X7 = Cargo Unloaded

X8 = Cargo Loaded

The results of the analysis of 151 regions using Pearson Correlations, R test and multiple linear regression equations can be seen in Table 1, Table 2 and Table 3 below:

Table 1. Pearson Correlation.

No	Variable	Value
1	Aircraft Arrival (X_1)	0.401
2	Aircraft Departure (X_2)	0.405
3	Passenger Departure (X_3)	0.379
4	Passenger Arrival (X_4)	0.375
5	Baggage Unloaded (X_5)	0.378
6	Baggage Loaded (X_6)	0.365
7	Unloaded Cargo (X_7)	0.217
8	Cargo Loaded (X_8)	0.211

Table 1 shows a sub-variable of Departure Arrival Aircraft and Aircraft have the highest impact to increase of Information and Communication GRDP in Indonesia, while sub-variablesof Cargo Loaded and cargo unloaded do not have a significant impact to increase of Information and Communication GRDP in Indonesia.

Table2. Model Summary.

Model	R	R Square Adjusted	R Square	Std. Error of the Estimate
1	0.572a	0.327	0.288	1989.21673

Predictors: (Constant), Cargo Loaded, Unloaded Cargo, Baggage Unloaded, Aircraft Departure, Passenger Arrival, Baggage Loaded, Passenger Departure, Arrival Aircraft.
Dependent Variable: GRDP of Information and Communication.
Table 2 shows that the correlation coefficient (R) is 0327, which means that the GRDPof Information and Communication variable has relatively strong correlation with variable of Airport Performance.Adjusted R Square is 28.8%, which means the impact of Airport Performance variable in increasingGRDPof Information and Communication, The increased of 71.2% in the GRDPof Information and Communication influenced by others factors aside from Airport Performance variable. It shows the low impact of Information and Communication GRDP to the Airport Performance.

IOP Conf. Series: Journal of Physics: Conf. Series **1007** (2018) 012046 doi:10.1088/1742-6596/1007/1/012046

Table 3. Regression Equations

(Constant)	84.159
Aircraft Arrival (X_1)	-0.888
Aircraft Departure (X_2)	1.199
Passenger Departure (X_3)	0.005
Passenger Arrival (X_4)	-0.003
Baggage Unloaded (X_5)	2.563E-05
Baggage Loaded (X_6)	0.000
Unloaded Cargo (X_7)	-5.661E-05
Cargo Loaded (X_8)	-4.724E-05

a. Dependent Variable: PDRB of Information and Communication

Table 3 shows the Multiple Linear Regression Equations with the following details:

$Y = 84.159 -0.888X_1 + 1{,}199\ X_2 + 0.005\ X_3 -0.003\ X_4 + 2.563E\text{-}05\ X_5\text{-}5.661E\text{-}05X_7$

$-4.724E\text{-}05\ X_8$

with:

Y = the GRDP of Information and Communication

X_1 = Aircraft Arrival

X_2 = Aircraft Departure

X_3 = Passenger Departure

X_4 = Passenger Arrival

X_5 = Baggage Unloaded

X_6 = Baggage Loaded

X_7 = Cargo Unloaded

X_8 = Cargo Loaded

Based on the multiple linear regression equation above, it appears that the X_2, X_3, X_5, X_6 (aircraft departure, passenger departure,Unloaded baggage, baggage loaded) Have a positive impact in increasing GRDP of Information and Communication, Meanwhile the variable X_1, X_4, X_7 and X_8 (aircraft arrival, passenger arrival, cargo Unloaded and cargo loaded) Have a negative impact to the GRDP of Information and Communication.

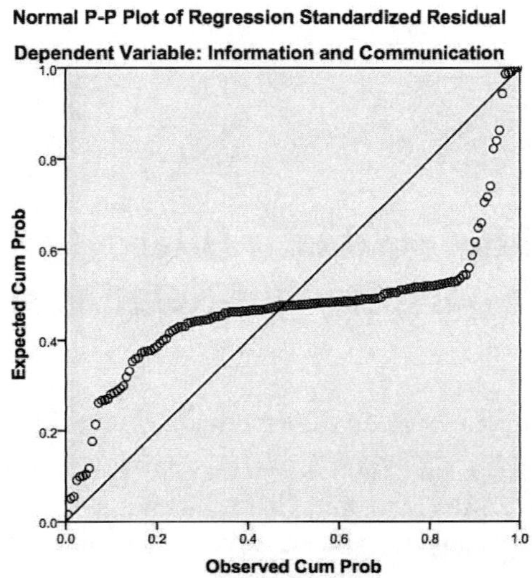

Figure 1. SPSS Analysis-Normal P-P Plot of Regression.

4. Conclusion

The results of correlation (R) indicates the Gross Regional Domestic Product (GRDP) of Information and Communication has a relatively strong relationship with the Airport Performance. Meanwhile the results of Adjusted R test shows that they are other factors in increasing GRDP of Information and Communication besides Airport Performance. It shows the low impact of Information and Communication GRDP to the Airport Performance.

References

[1] Nuning Kurniasih and Pawit M Yusup, "Social Media Utilization to Promote Tourism Industry In Lembang: A Content Analysis Study Of Social Media Used By Management of Tourism Objects In Lembang, Indonesia," in *The 2016 International Conference on Media: Mass Media at Crossroads*, 2016, pp. 489–500.

[2] Bank Indonesia, "Report of the Archipelago in May 2017, Regional Economic and Financial Study," 2017.

[3] J. D. Kasarda, "New Logistics Technologies and Infrastructure for the Digitized Economy."

[4] K. P. R. Indonesia, "Airport Indonesia." [Online]. Available: http://hubud.dephub.go.id/?id/bandara/index.

[5] M. of Transportation, "regulations PM 69 in 2013 on the order of the national airport affairs," 2013.

[6] J. D. Kasarda and J. D. Green, "Air cargo as an economic development engine: A note on opportunities and constraints," *J. Air Transp. Manag.*, vol. 11, no. 6, pp. 459–462, 2005.

[7] B. E. W. Jr and Mary E. Kite, *Principles of Research in Behavioral Science: Third Edition 3rd Edition*. Routledge, 2012.

[8] J. Suyono, A. Sukoco, M. I. Setiawan, S. Suhermin, and R. Rahim, "Impact of GDP Information Technology in Developing of Regional Central Business (Case 50 Airports IT City Development in Indonesia)," *J. Phys. Conf. Ser.*, vol. 954, no. 1, p. 12045, Dec. 2017.

[9] B. P. Statistik, "Statistical Yearbook of Indonesia," 2017.

Acknowledgment

This paper is a part of a PhD Dissertation titled "Business Centre Development Model of Airport Area in Indonesia" with Prof. Ir. Surjono Surjokusumo, M.Sc.F, Ph.D., Ir. Dadang M. Ma'soem, MSCE, Ph.D. and Dr. Ir. Johny Johan, M.Eng, MM as the counselors.

IOP Conf. Series: Journal of Physics: Conf. Series **1007** (2018) 012047 doi:10.1088/1742-6596/1007/1/012047

Distribute Off-Time Office Internet bandwidth Using Topology Mesh For Sorrounding Neighbour

Niskarto Zendrato[1*], Oloan Sihombing[2], Yonata Laia[2], Ertina Sabarita Barus[3]

[1]Program Studi Doktor Ilmu Komputer, Fakultas Ilmu Komputer dan Teknologi Informasi, Universitas Sumatera Utara, Indonesia
[2]Faculty of Technology and Computer Science, Universitas Prima Indonesia, Indonesia
[3]STMIK Kristen Neumann Indonesia, Indonesia

Email: *niskarto@stmikneumann.ac.id

Abstract. The Internet as one of the very rapidly growing information technology can provide data and information with wide world, complete, and up to date. Users can download and upload data such as the application file, multimedia and text through the Internet network. But for the Internet availability is still less equal access because of the lack of availability of adequate infrastructure, therefore the author make the utilization of bandwidth that can be establish Internet balancing although still on a small scale. By this research the authors use bandwidth from PT. Deltauli Home Teknikarya that where bandwidth necessity on when time off-time unused office, where the office always pay full for Internet connection even though at the time of the off-time. It's many of the available bandwidth, so that the author is trying to take advantage of the bandwidth at the time of the off-time the office to be used by the community using radio connection link and use the radius server as user management and server to send sms and user and password to the users who want to enjoy free internet connection.

1. Introduction

The development of the current age demands of human life is not removed from the internet technology. The Internet as a source of information can simplify and speed up a work good in education, literacy field, health, commerce, government and other fields all take advantage of the internet as the summer information. For this reason, the government in demand can provide sufficient infrastructure to provide internet distribution throughout the territory of Indonesia.

According to Rudiantara MenkomInfo that internet network in Indonesia have not spread to all across the archipelago, one is a problem is the telecommunication infrastructure yet to reach all the area, there is a gap between the capacity in rural areas and large cities [1]. According to Niskarto Zendrato in a thesis entitled Analysis of the distribution of bandwidth off-time has held various using mikrotik and radius server [2]. Explain that the bandwidth utilization on off-the-shelf the office by the radio connection link and use the RADIUS server as management and SMS server to send the user and password to the users who want to enjoy the connection Internet.

In general all government offices and industry has been using the internet as a means of supporting the work. Each of the office usually take advantage of unlimited package that is provided by the provider. Office hours are at 8 am up to at 5pm where on the hours of the internet will be used with maximum by all employees in the work. After the hours of the office finished usually internet usage will be low and of course this situation can be used for distributed internet usage for the surroundings of the community.

1.1. Formulation Of Problem
The formulation of the problem in this research is how to build a system and distribute bandwidth in Off-Time Office to spread to the general public so that the public can enjoy internet fast and qualitySome text.

1.2. Limitation Of Problem
Now the limitations of the problem in this research is : 1. The Bandwidth used in this research is from the office on the off-time 2. Using Mikrotik RouterOS as management bandwidth 3. Using WIFI Network 2.4 Ghz to spread to the community 4. Use the 5 Ghz Frequency for mesh relationship between offices 5. Use FreeRADIUS Server as user management and control 6. Research areas in the area surrounding Pancurbatu.

1.3. Research Aims
The purpose of this research is to analyze the unused internet usage on the office to be used by the community that has not been coverage internet network a fast and cheap.

1.4. Benefit Research
Now the benefits that can be obtained from this research is as follows : 1. Help the government especially bent on in some small matter in equitable access to the internet. 2. Add insights and documentary captures the knowledge and can become a means of entertainment with the presence of the internet is fast and cheap. 3. Growing public interest in selling or buying goods with discounted rates online.

2. Literatur Review

2.1. Network Defenition
The network is a collection of hardware that is connected to each other through a medium such as copper cables, optic cable, micro waves and the infrared so that allows for the exchange of data. The hardware can be computer, PDA, printers and other hardware that is able to send and receive data [3] The computer network is a single computer model that serve all computing tasks of an organization that is replaced by a collection of a computer that is separate settlements but are related to each other in carrying out their tasks [4]. The computer network is two or more computers that are connected and can share data, applications, computer equipment and internet connection or some combination of it . The benefits of the computer network to fill needs by reassignment is Resource, it sharing the aim of the whole program, equipment, especially data can be used by everyone who is without affected the location of the resource and users in other words resource sharing is an effort to remove obstacles to the distance.

2.2. Mikrotik
Mikrotik originally a small company founded by John Trully and Arnis Riektins in Latvia. John Trully is a American nationals who emigrated to Latvia while Arnis is a graduate of physics and mechanics both met in 1995. Mikrotik began to be used in Moldova around 1996 using Linux system and MS-DOS which combined with the technology of Wireless LAN (WLAN) Aeronet 2 Mbps speedAlthough using wireless technology basically mikrotik holding the principle of building a reliable router program can be used throughout the world. This time Latvia became the center of research development Mikrotik because John and rot away when they start to help other countries including Sri

IOP Conf. Series: Journal of Physics: Conf. Series **1007** (2018) 012047 doi:10.1088/1742-6596/1007/1/012047

Lanka where they serve around 400 user. The first Linux is used in the development of the kernel Mikrotik 2.2 developed with the help of 5-15 staff Research and Development (R&D) [5]. According to Arnis, in addition to rely on the staff of the R&D in his own Mikrotik environment, they also recruit release and a third party develop a Mikrotik protocol [6].

2.3. Bandwidth
Bandwidth is the number of bits that can be ditranmisikan in a network on a specific period of time. As an example of a network that has a bandwidth 10 Mbps means that network capable of delivering ten million bits per second. Bandwidth also determine how long the time required for every bit transmitted data on 10 Mbps, 0.1 microsecond time required to transmit every bit [7]

Bandwidth is a counting the consumption data available on a telecommunication. Calculated in units of bits per seconds (bits per second).[8] Note that the bandwidth that indicated wireless communication, modem data transmission, digital communication, electronics is bandwidth that refers to the analog signal that is measured in units of bits per second.

2.4. Radius Server
RADIUS (Remote Access Dial-in User Service) Server or in this research is called as authenticator server is an access network model which separates the three kinds of control function, namely Authentication, Authorization, and Accounting, for independent processed. In this research, RADIUS server that is used is the freeradius which is application from RADIUS Server.[9] AAA protocol (Authentication, Authorization, Accounting) set the mechanism how the procedures to communicate both between client to domain network and between the client with a different domain to ensure the security of data exchange.

2.5. Mesh Topology
The mesh topology is a computer network topology that connect all of the computers in full, this topology topology is most complex compared with the other network topology. This type of topology many used by your internet service provider (ISP). [10] The concept of this topology is each computer in your network are connected to each other so that if the failure occurs on one of the computer has no effect on another computer or effect on the network. This mesh network topology apply the relationship between central full. The multitude of this channel must be prepared in order to form a network of mesh topology namely central number of minus 1 (n-1, with n is the number of central).

Type of Mesh Network Topology Connection . Topology connection type on the mesh network topology consists of 2 types of both the mesh topology include: 1. The Mesh Topology Fully Connected Mesh Topology Fully Connected has main characteristics where each computer in your network are connected to each other in full. For example if there are 5 computers in the network then one computer will be connected to 4 other computers. 2. The mesh topology partial connected in mesh topology this type has characteristics that each computer in the network is not all computers will be connected with other computers so that there are some computers that are connected to each other and some of the computer does not relate to each other [11-12].

3. Research Metodology
In this study, the authors use research methods as Analyze the effectiveness of the network, the implementation of network topology and testing the application of network topology. 1) Analysis of network effectiveness Analysis of network effectiveness applied; 2) direct field observations with questionnaires to users. Direct field observations with questionnaires to the user make a number or multiple questions and then distributed to the user, the user, in this case, is the community; 3) Direct Field observation (test case / technical). Direct field observation (test case / technical) Perform observation of patients in network management Local Area Network (LAN) PT Deltauli This technique is managed by PT Deltauli NOC Engineering; 4) in Figure 1 is part of the Network Topology implementation. Implementation of network topology design topology and perform a

number of topology testing; 5) in Figure 2 is part of the implementation of Network Topology Apply topology and see results through a number of Tests.

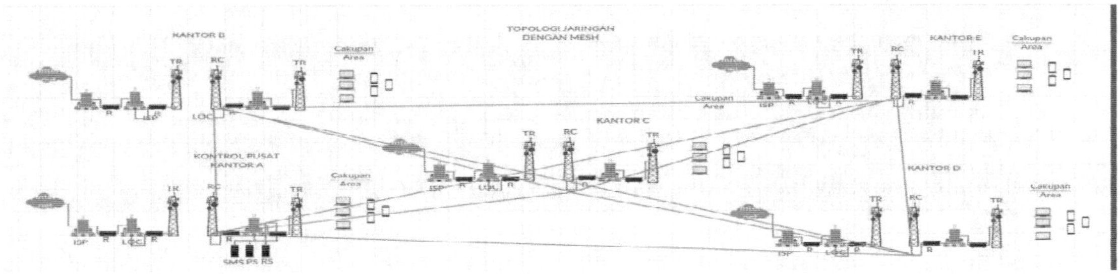

Figure 1. The mesh network topology design.

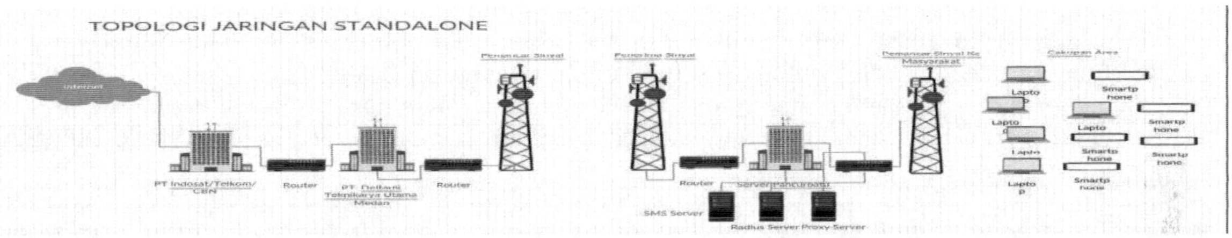

Figure 2. Standalone network topology design.

4. Result

4.1.1 Results of the use of bandwidth with Management

Bandwidth Usage with management also can perform these settings capacity restrictions or speed that evenly on each user or user in the use of the internet facilities bandwidth in off-time the office so that the bandwidth capacity of the internet connection is divided on average each user or user (user). Below can be seen the results of the division of bandwidth internet connection that evenly on each user or user (user) as below :

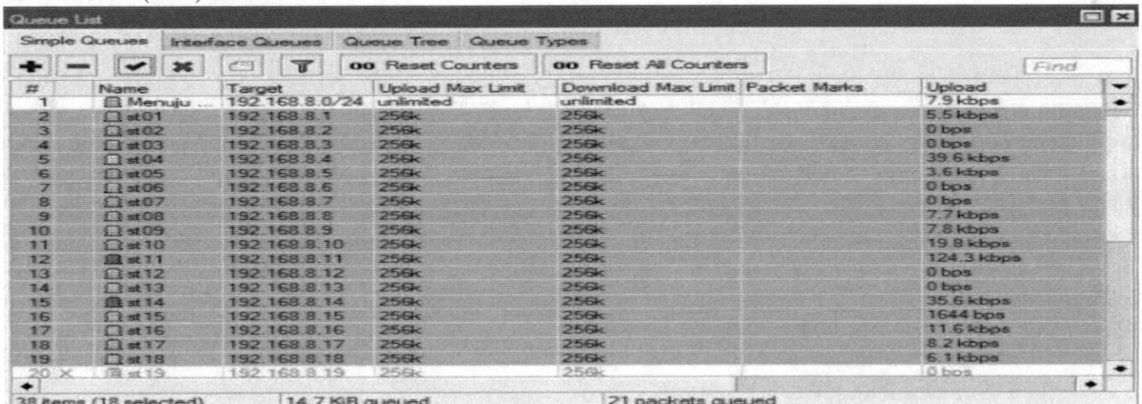

#	Name	Target	Upload Max Limit	Download Max Limit	Packet Marks	Upload
1	Menuju ...	192.168.8.0/24	unlimited	unlimited		7.9 kbps
2	st01	192.168.8.1	256k	256k		5.5 kbps
3	st02	192.168.8.2	256k	256k		0 bps
4	st03	192.168.8.3	256k	256k		0 bps
5	st04	192.168.8.4	256k	256k		39.6 kbps
6	st05	192.168.8.5	256k	256k		3.6 kbps
7	st06	192.168.8.6	256k	256k		0 bps
8	st07	192.168.8.7	256k	256k		0 bps
9	st08	192.168.8.8	256k	256k		7.7 kbps
10	st09	192.168.8.9	256k	256k		7.8 kbps
11	st10	192.168.8.10	256k	256k		19.8 kbps
12	st11	192.168.8.11	256k	256k		124.3 kbps
13	st12	192.168.8.12	256k	256k		0 bps
14	st13	192.168.8.13	256k	256k		0 bps
15	st14	192.168.8.14	256k	256k		35.6 kbps
16	st15	192.168.8.15	256k	256k		1644 bps
17	st16	192.168.8.16	256k	256k		11.6 kbps
18	st17	192.168.8.17	256k	256k		8.2 kbps
19	st18	192.168.8.18	256k	256k		6.1 kbps
20 X	st19	192.168.8.19	256k	256k		0 bps

38 items (18 selected) | 14.7 KiB queued | 21 packets queued

Figure 3. Results of the distribution of bandwidth capacity with the management.

4.1.2 The location of the radio transmitter

On this research, location used as a radio transmitter located at PT Deltauli teknikarya that will be transmitted to the location of the pancurbatu with distance or wide coverage. The process of the spread of the internet facility will diserbarkan bandwidth through POP (Point Of Presence) which is located at the location of the Pancurbatu. Below can be seen the results of the distance between the POP Pancurbatu to PT Deltauliteknikarya Home through google eart as below:

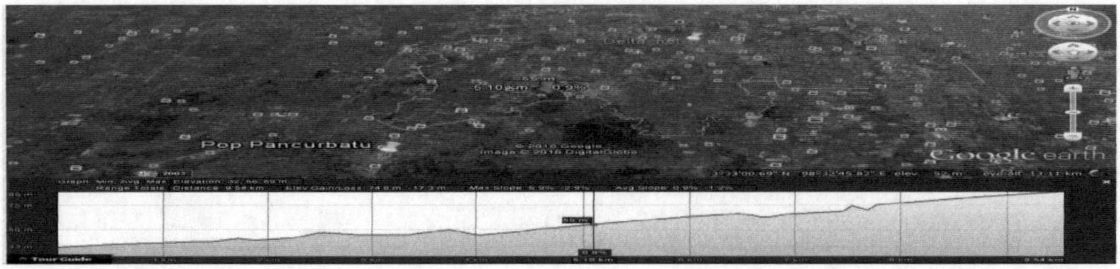

Figure 4. The results of the distance between the pop pancurbatu to pt deltauliteknikarya home.

The location of the signal coverage Internet connection that can be accessed the community. The use of the internet facility bandwidth that spread to the community has a very wide coverage so that the community can enjoy the internet facility on the radius ± 1 Km marked on the green line. Below can be seen the results of signal coverage internet location that can be accessed by the community through google earth.

Figure 5. The results of the location signal coverage internet connection.

5. Conclusion

After the research is done then the conclusion that I can deduce that internet distribution is unused in the offtime computer runs fine, the public can access the internet at the Office during the offtime. On a standalone network topology if the internet disconnected so no backup of internet lines while if using mesh network topology then if any one or more of the Office network is cut off then it could be replaced by other office network so the internet can continue to be accessed by the public.

References

[1] Tiga Tantangan Utama Pemerataan Internet di Indonesia "http://www.cnnindonesia.com/teknologi/20150925141413-213-80931/tiga-tantanganutama-pemerataan-internet-di-indonesia/.

[2] Zendrato Niskarto, 2016, *Analisis Distribusi Bandwidth Off-Time Kantor dengan Menggunakan Mikrotik dan Radius Server* , Tesis, Universitas Sumatera Utara, Medan. p 42.

[3] Dusit Niyato, Ekram Hossain, Dong In Kim, Vijay Bhargava, Lotfollah Shafai, 2016, *Wireless-Powered Communication Network: Architeures, Protocols and Applications*, Cambridge University Press. p.

[4] Harry Prihanto,Juni 2011, *Membangun Jaringan Komputer*, Ilmukomputer.com.

[5] Dwi Febrian Hadriyanto, 2009, *Kajian Penggunaan Mikrotik Router Sebagai Router Pada Jaringan Komputer*.
[6] Saputro, Daniel T, Kustanto. 2008. Membangun Server Internet dengan Mikrotik OS. Gava.
[7] Imam, Riadi.2010.*Optimasi Bandwidth Menggunakan Traffic Shapping*, Jogjapress.
[8] Ekram Hossain, Mehdi Rasti, Long Bao Le, 2017, *Radio Resource Management in Wireless Networks: An Engineering Approah,* Cambridge Uninversity Press. p.
[9] Lingyang Song, Dusit Niyato, Zhu Han, Ekram Hossain, 2015, *Wireless Device to Device Communications and Networks*, camridge University Press, p.
[10] Prastyaningsih, M. Irfan dan Hariyady, 2012, *Sistem Kontrol server Untuk OptimasiPembagian Bandwith Jaringan Mennggunakan Metode HTM (Hierarchical Token Bucket)*, Jurnal Teknologi & Industri.
[11] R. Tornero, J. M. Orduña, A. Mejia, J. Flich, J. Duato, 2010, *A Communication-Driven Routing Technique for Application-Specific NoCs.International Journal of Parallel Programming*, p 357-374.
[12] Savo G.Glisic, 2016, *Advanced Wireless Network: Technology and Business Models*, Jhon Wiley & Sons. p.

IOP Conf. Series: Journal of Physics: Conf. Series **1007** (2018) 012048 doi:10.1088/1742-6596/1007/1/012048

Application of forwardchaining method to diagnosis of onion plant diseases

Delima Sitanggang[1*], Saut D. Siregar[1], Suryani M. F. Situmeang[2], Evta Indra[1], Ayu R. Sagala[1], Oloan Sihombing[1], Marlince Nababan[1], Hendra Pasaribu[1], Rudolf R. Damanik[1], Mardi Turnip[1], Rijois I. E. Saragih[3]

[1]Faculty of Technology and Computer Science, Universitas Prima Indonesia, Indonesia
[2]Poltekkes Kemenkes Department of Health Analyst, Indonesia
[3]Methodist University of Indonesia, Indonesia

Email: *Djoshlimasitanggang@gmail.com

Abstract. Red Onion is a tuber plant that is widely used by the people of Indonesia, both as herbs and herbal medicines. Onion farmers have limitations in identifying diseases that attack their crops.This disease can cause crop failure against the onion.This design begins with the creation of a knowledge base up to input-output design with forward chaining method. The results of this design can assist farmers in identifying their plant diseases. Based on diagnostic results of several methods that have been done testing can diagnose diseases contained in onion plants. With symptoms data that has been determined by the expert with the value of each symptom is different. As for the symptoms that have been determined that the leaves contain patches with a value of 0.3, White leaf spots value 0.4, Leaf spots form a purple zone if it is severe 0.5, Leaf tip of 0.2, Tubers rot 0.4. Based on the above diagnostic results then get the value of diagnosis 67% forward chaining with trotol disease type, Purple spotting.

1.Introduction

Horticulture is a potential agricultural commodity to be developed in Indonesia and has a bright prospect in the future as well as a source of foreign exchange for Indonesia. Horticulture crops are widely cultivated in Indonesia, but when viewed from the results still not satisfactory.This is due to various factors, among them is the cultivation technique, environmental conditions as well as pests and diseases. Pests and diseases are major constraints in the cultivation of agricultural crops that can reduce production by up to 40% [1]. Limitations of an expert sometimes become an obstacle for farmers who will consult to solve a problem to get the best solution.In this case an expert system for the diagnosis of pests and diseases of horticultural crops made to serve as a means for consultation, a means of learning in an Agency of Agriculture or Agricultural laboratory and can be used as a tool for an expert in diagnosing and socializing the types of pests and diseases.With the application of expert systems also a Farmers Group can easily help farmers who are experiencing problems about pests and diseases of horticultural plants and the best solution to be taken without relying entirely on an expert and can share information or knowledge among farmers based on the system . Comparative techniques of forward and backward chaining in applied because the data and facts in conducting the research process has been obtained, and from the data or facts can be made a system that will provide a conclusion or solution based on a collection of data and facts. Using this inference technique the chances of getting a more specific conclusion can easily be obtained [2].

By using the rule-based reasoning method that has an if-then pattern on the knowledge base in the application of this expert system's diagnosis, then the experts with their performance can solve the problem sequentially [3]. With this approach method, the rules that have been generated can be reviewed by experts to be improved or modified in order to get better results.Besides that, this method is also used whenever an explanation of traces (steps) in obtaining an achievement of a solution or steps regarding the achievement of the results of a diagnosis of pests and diseases of horticultural crops.Expert Systems are applied in several areas of life, for example in the diagnosis of rice disease.
In this research an expert system developed on the agricultural area, explains the design and development of the rule base in the expert system using the ESTA shell (Expert System for Text Animation). This research produces a system that integrates a structured knowledge base along with solutions to the problem.The resulting system has been tested between an existing dataset and an expert in the field of plant disease [4].

Earlier researchers used expert systems to diagnose eye disease [5]. The CLIPS language of this research is used as a tool for designing expert systems. Based on research results, it was found that the performance of frame-based was more reliable than rule-based in terms of the speed of the diagnosis process [7]. In this study expert system using demster Shafer method. The resulting diagnosis of coronary heart disease is positively accepted by the user [6]. Expert systems are used for detection and early treatment of cow disease. In this research expert system is implemented by using mobile based applications. However, with limited computing resources on mobile devices. In this research diagnosed coronary heart disease [8]. Then researchers use expert systems in diagnosing skin diseases.This expert system was built using the backward chaining method for decision making [9].

2. Methodology

Expert systems in [6-10], the expert system is a computer program designed to simulate - expert human problem solving in a narrow or disciplined domain. Expert systems usually consist of a knowledge base (information, it yuristics,etc.), machine conclusion (basic knowledge analysis), and end user interface (receiving input, generating output). [11-15].Forward Chaining is a search technique that begins with known facts, then match those facts with the IF part of the IF-THEN rules. If any facts match the IF section, then the rule is executed. When a rule is executed, then a new fact (part THEN) is added to the database. Each time to match, starting from the top rule.Each rule can only be executed once. The matching process stops when there are no more executable rules. The search method used is Depth-First Search (DFS), Breadth-First Search (BFS) or Best First Search. The approach in tracking starts from the input information and then tries to illustrate the conclusion, future tracking of facts that match the IF section of the IF-THEN rules. With the forward chaining method of the approach and the rules that have been produced can be reviewed by experts to be improved or modified to obtain better results [16-17].

As shown in figure 2. Disease diagnosis process on shallot plant will start from checking G5 symptom, where there are three types of diseases that have symptoms of G5, namely disease P1, P2 and P6.After that, the next symptom check will be done. If G1 symptoms are detected then the likelihood of the disease is P1. Meanwhile, if detected symptoms of G7, then the possibility of the disease is P2. Finally, if G29 symptoms are detected, then the possibility of the disease is P6.

IOP Conf. Series: Journal of Physics: Conf. Series **1007** (2018) 012048 doi:10.1088/1742-6596/1007/1/012048

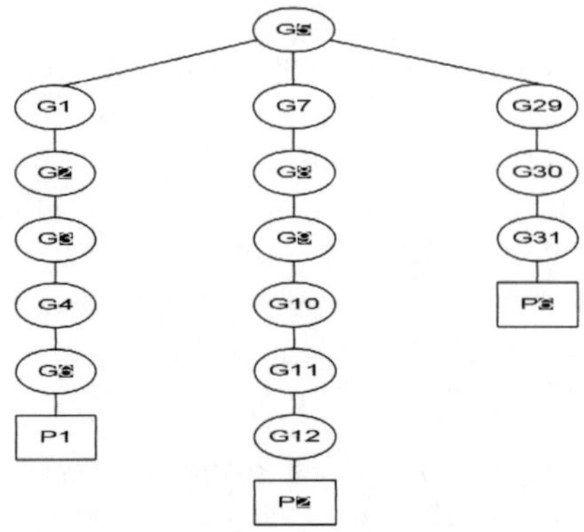

Figure 1. The decision tree of forward chaining method.

Block diagram is used to find out the working order of expert systems to reach a decision and made to model work processes on a computer for easy understanding. The following is a description of the block diagram of the forward chaining expert system:

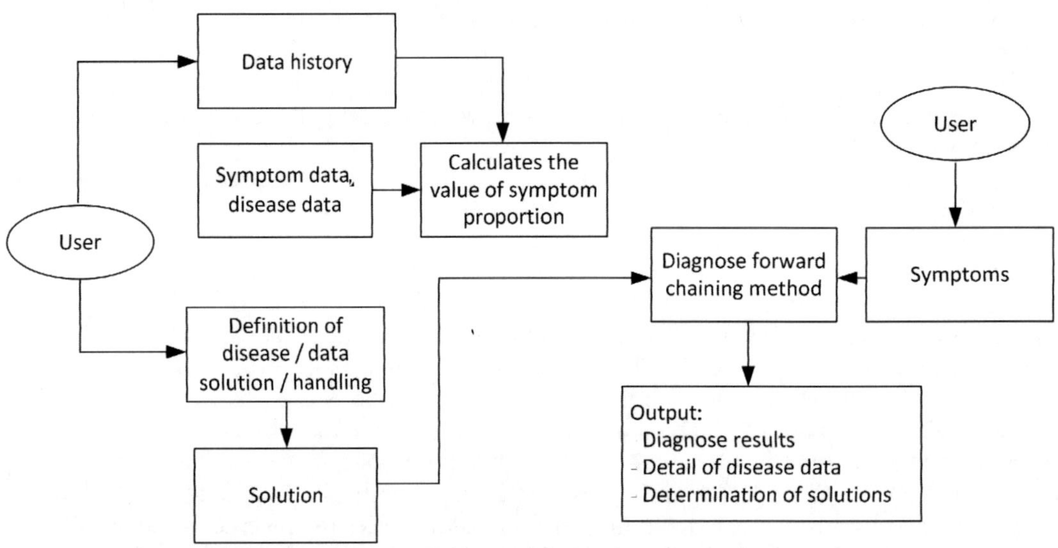

Figure 2. Block diagram of diagnosis of onion diagnosis expert system.

The search process will start from the process of reading the data characteristics of the database. After that, the system will wait for the user's response to the features displayed by the system. If the data entered by the user in accordance with the characteristics of certain onion diseases, then the system will take the next characteristic of the onion fruit disease in question and request a response from the user again. If not, then the system will take the characteristics of other onion diseases and ask for the response from the user. Workflow from the data search process can be seen in Figure 3 as follows:

IOP Conf. Series: Journal of Physics: Conf. Series **1007** (2018) 012048 doi:10.1088/1742-6596/1007/1/012048

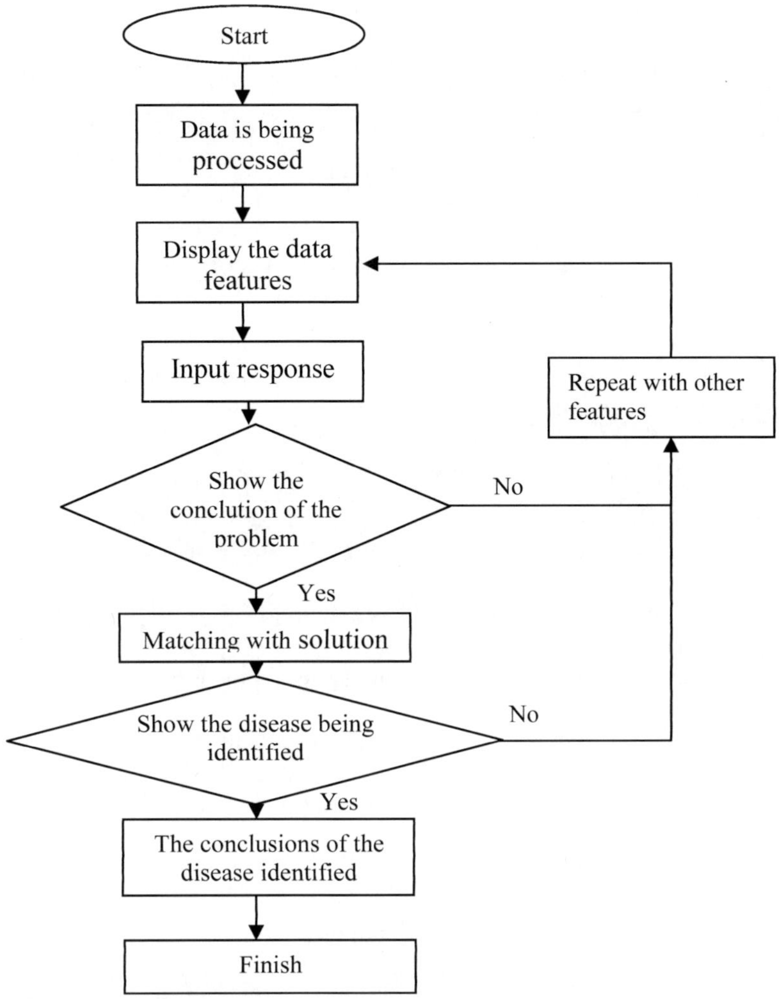

Figure 3. Structure of data search algorithm.

If the system has obtained an onion disease solution or any list of features has been asked, then the system will display the conclusions obtained. The following data of symptoms used in this onion disease expert system amounted to 16 symptoms. The data of these symptoms can be seen in table 1 below.

Table 1. Data on onion disease symptoms.

Codesymptoms	Symptom name	Weight Value
G1	Leaves contain spotted curves	0.3
G2	White leaf spots	0.4
G3	Leaf spots form a purple zone if it is severe	0.5
G4	Leaf tips dry	0.6
G5	Tubers rot	0.4
G6	Tuber tissue dries	0.1
G7	Leaf tips are pale green spots	0.4

IOP Conf. Series: Journal of Physics: Conf. Series **1007** (2018) 012048 doi:10.1088/1742-6596/1007/1/012048

G8	There are mycelium and spores on the leaf spot	0.7
G9	dwarf plant	0.4
G10	user wrinkled	0.6
G11	brownish tubers	0.3
G12	he tuber parts look dry and pale	0.2
G13	Spotted leaf round and elongated	0.5
G14	brown leaf spot	0.3
G15	the largest number of spots on the tip of the leaf	0.2
G16	tissue on dead leaf spot	0.5

The number of diseases processed in this onion disease expert system are 9 kinds of diseases. The data of this disease can be seen in table 2.

Table 2. Onion disease.

Codesymptoms	Symptom name
P1	Tramples, purple blots
P2	Fake flour
P3	Leaf spots cercospora
P4	Automatic, anthracnose
P5	Leaf spots cercospora
P6	Dead shoots
P7	Penicillium reeds
P8	Yellow dwarf virus, mosaic virus
P9	Nematodes root cultivation

3. Results and Discussion

In Figure 4 describes the process done on this system application where when this application is run then that appear first is the login menu when the login menu appears if the user and password is true then go to the main page of the application if wrong then appear your password message wrong. When you enter there are some menus that contain some links to certain pages for example on the menu of symptoms then this menu displays the form of selection of symptoms when it has been selected symptoms into the final diagnosis.Figure 4. Login Process Go to Main menu and Selection of Symptoms.

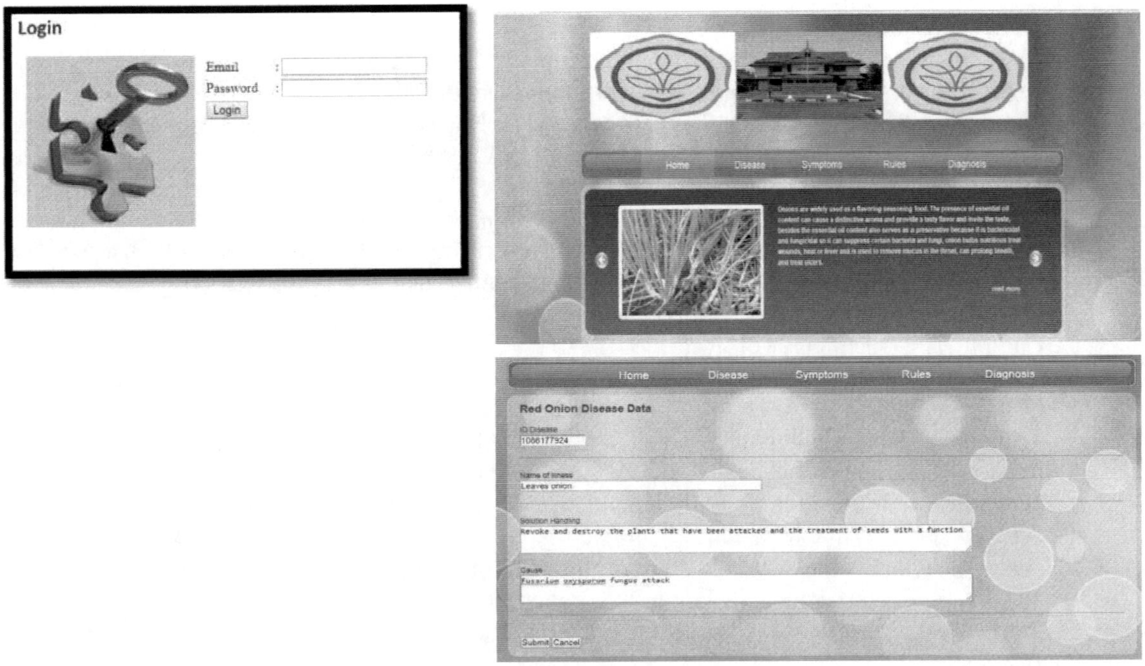

Figure 4. Login process go to main menu and selection of symptoms.

In table 3 it shows the results of onion diagnosis with forward chaining

Table 3. Report on the diagnosis of onion disease.

Date	Disease	Solution
2017-11-29:15:15:60	Moler or Fusarium Lumber (P5)	1. Planting resistant varieties, such as philippine and sumenep varieties, long rotation of plants, adding organic fertilizer in the field of 5-10 tons /ha, planting healthy tuber seeds, avoiding the cultivation of bulbs either at planting or harvesting, immediately revoking and destroying affected plants and seed treatment with recommended functions (100kg of seedlings + 100 g of functionalis). 2. Effective soil dyeing and treatment registered and legalized by the Minister of Agriculture such as Anaconda 50 WP and Saco P

4. Conclusions

Based on the diagnostic results of several methods that have been done testing can diagnose the disease contained in shallot plants. With symptoms data that has been determined by the expert with the value of each symptom is different .As for the symptoms that have been determined that Leaves contain spotted curve with a value of 0.3, White leaf spots value 0.4, Leaf spots form a purple zone if it is severe 0.5, Dry leaf tip 0.2,Tubers rot 0.4. Based on the diagnostic results above then get a value of 67% forward chaining diagnosis results with trotol disease type, Chugging.

References

[1] Badan Penelitian dan Pengembangan Pertanian, " Kebijakan Tanggap Ledakan Hama Penting Tanaman Perkebunan", Kementrian Pertanian, 2011
[2] G. R. Baur, and D. V. Pigford, " Expert System For Business : Concept and Implementations", Boyd & Fraser Publishing Company, Boston-USA, 1990
[3] J. P. Ignizio, " Introduction To Expert Systems : The Development and Implementation of Rule Based Expert System",McGraw-Hill International Editions, 1991
[4] S. K. Sharma, K. R. Singh, and A. Singh , " An Expert System for diagnosis of diseases in Rice Plant",International Journal of Artificial Intelligence, vol.1, no. 1, 2010
[5] P. S. K. Patra, " An Expert System for Diagnosis Of Human Diseases", International Journal of Computer Applications, vol. 1, no. 13, 2010
[6] S. S. Naser, and A. Z. A. Ola, "An Expert System For Diagnosing Eye Diseases Using Clips",Journal of Theoretical and Applied Information Technology, vol. 4, no. 10, 2008
[7] W.Ardianto, W, Anggraeni, and A. Mukhlason, "Pembuatan Sistem Pakar Untuk Pendeteksian dan Penanganan Dini Pada Penyakit Sapi Berbasis Mobile Android dengan kajian Kinerja Teknik Knowledge Representation", Jurnal Teknik ITS, vol.1, no. 1, 2012
[8] E. G. Wahyuni, and W. Prijodiprojo, "Prototype Sistem Pakar untuk Mendeteksi Tingkat Resiko Penyakit Jantung Koroner dengan Metode Dempster- Shafer",IJCCS-Indonesian Journal of Computing and Cybernetics Systems , vol.7, no.2, 2013, pp. 133-144
[9] D. Agustina, H. Mustafidah, and M. R. Purbowati, "Sistem Pakar Diagnosa Penyakit Kulit Akibat Infeksi Jamur", JUITA (Jurnal Informatika), vol. 4 Nomor 2, 2016
[10] Hashem Hashemi, Hossein Alizadeh Moghaddam, Pegah Keyvan, Shahram Jafari, 2013, A Decision Support System for Polyuria Patient's Treatment, International Journal of Engineering Science Invention, Volume 2 Issue 1, January 2013, ISSN (Online): 2319 – 6734, ISSN (Print): 2319 – 6726, PP.70-76
[11] P. Santosh Kumar Patra, *et al* (2010). " An expert System for Diagnosis of Human Diseases" *International Journal of Computer Application (0975-8887)* .
[12] Y.A. Nada, "Construction of Powerful Online Search Expert System Base d on Semantic Web," in nternational Journal of Advanced Computer Science and Applications vol.4, 2013, pp.181-187
[13] Dada, J. O., HansD. K., Joerg P., 2010. Web-Based Expert System for Classification of Industrial and Commercial Waste Products, Journal of Emerging Trends in Computing and Information Sciences, Vol. 2 No.6, JUNE 2011 ISSN 2079-8407, pp257-262
[14] Fadlil, Abdul, dan Tuswanto, 2013, Sistem Pakar untuk Mendiagnosa Hama dan Penyakit Tanaman Bawang Merah Menggunakan Metode Certainty Factor, Jurnal Sarjana Teknik Informatika Vol. 1 No. 1, e-ISSN:2338-5197, Universitas Ahmad Dahlan, Yogyakarta
[15] Sasmito, G. W., 2010. Aplikasi Sistem Pakar untuk Simulasi Diagnosa Hama dan Penyakit Tanaman Bawang Merah dan Cabai menggunakan Forward chaining dan pendekatan berbasis aturan, Universitas Diponegoro, Semarang.[16]. Hartati, S., Iswanti, and Sari, 2015. Sistem Pakar dan Pengembangannya, Graha Ilmu: Yogyakarta
[16] Ginanjar Wiro Sasmito, Bayu Surarso dan Aris Sugiharo (2011), "Application Expert System of Forward Chaining And The Rule Based Reasoning For Simulation Diagnose Pest and Disease Red Onion and Chili Plant", Proceedings of the 1 International Conference on Information Systems For Business Competitiveness(ICISBC)
[17] Ivo Randi MS, Zaenal Wafa dan Ruri Hartika Jhon(2013), " Perancangan Sistem Pakar Untuk Mendiagnosa Penyakit Kanker Serviks Dengan Metode Forward Chaining",Jurnal Sarjana Teknik Informatika 09101152630035

The effect of mining data k-means clustering toward students profile model drop out potential

Windania Purba*, Saut Tamba, Jepronel Saragih

Faculty of Technology and Computer Science, Universitas Prima Indonesia, Indonesia

E-mail: *winda.nia04@gmail.com.

Abstract. The high of student success and the low of student failure can reflect the quality of a college. One of the factors of fail students was drop out. To solve the problem, so mining data with K-means Clustering was applied. K-Means Clustering method would be implemented to clustering the drop out students potentially. Firstly the the result data would be clustering to get the information of all students condition. Based on the model taken was found that students who potentially drop out because of the unexciting students in learning, unsupported parents, diffident students and less of students behavior time. The result of process of K-Means Clustering could known that students who more potentially drop out were in Cluster 1 caused Credit Total System, Quality Total, and the lowest Grade Point Average (GPA) compared between cluster 2 and 3.

1. Introduction

The Highest of successful and the lowest of unsuccessful students were reflecting from the quality of the university. The university becomes the main social in scoring successful or unsuccessful in high education. The successful or students achivement study could be seen as successful and superior of universtity. Vise versa, the unsuccessful or the lowest students study was unable of university in doing learning process in high education. One of the problem is students drop out. Drop out was a process of students quit from students status, that caused many things that depend on university itself. The highest number of drop out students in university can be minimize with the university policy to guide and prevent students from drop out to detect students in first education was very important to keep students from drop out. It maybe department organizer eduation to give guidelines to the students needed [1].

The students uderstanding information who potentially drop out was important to known and undrestood. The understanding could be done by find the data to showing the data which had and then clustering to the result of data so showing the pattern or students drop out clustering. This education could be used in helping university to knowing the students situation and become education early in processing of taking a decision to preventive in anticipation of drop out students, to increase the achievement students, to increase curriculum, to increase process of learning and teaching activity and there were another advantages that could be gotten from the result of getting data.

2. Review Literature

Mining data is the using automatic of analysis technique to knowing the previous relationship cannot detected between the item data. Mining data is a process to finding the same data from database/data set in different area likely, financial, retail industry, science, statistic, medicine, intelligece, neurology. The size of data is increase quickly, so the new technology and high speed and algorithm that needed to collect and process of data [2-3]. Clustering is one of data mining that unsupervised. Clustering is one of data mining that unsupervised. Clustering is a process divided data to the cluster based on the similarity. K-Means is a cluster technique data that the point data of existence in cluster is depend on degree of the member [4].

Steps to do clustering with K-Means is: 1) Determining how many cluster that want to shape, where K value is the number of cluster; 2) Determining the centroid cluster first. The begining Centroid was depend on randomly from the data and number of centroid same with the number of cluster; 3) After determined the begining centroid, so very data will find nearby centroid is count every distance of each centroid by using correlations formula between two objects of Euuclidean Distance; 4) After counting the distance of data the cetroid, so the nest steps is clustering of data based on minimum distance. A data will become a member of cluster that has the nearest (small) from the cluster; 5) Based on the clustering, the next is finding the new centroid based on membership from each cluster is counting the cluster average; 6) Back to step 3; 7) Iteration stop means if there is no any moving data. Determine centroid in iteration wil use the formula below: $\bar{v}_{ij} = \frac{1}{N_i} = \sum_{k=0}^{N_i} x_{ki}$ .... (1). Where: 1) V_{ij} Centroid of average cluster k-i to k-j variable; 2) Ni The number of membership of cluster; 3) k-i; i, k index from cluster; 4) j Index from variable; 5) X_{kj} The value of data k variable ke-j in the cluster. Determine the correlation between two objects is using Euclidean Distance below: $d_{Euclidean}(x, y) = \sqrt{\sum_{i=1}^{n}(x_i - y_i)^2}$ ....(2)
Where: 1) $d_{(x,y)}$ the distance of x to cluster y; 2) x_i The data i in data attibute to n; 3) y_i The data to-j in data attribute to n. That is the formula which is used to determine the correlation between two objects with Euclidean Distance formula [5-8].

3. Result and Discussion

The mechanism of test would be done was:
To determining the cluster based on data that had been available, it needed a flowchart to easier in determine the counting plot as a plot to find the result of cluster implementation to processing the data. There were some steps. They were:

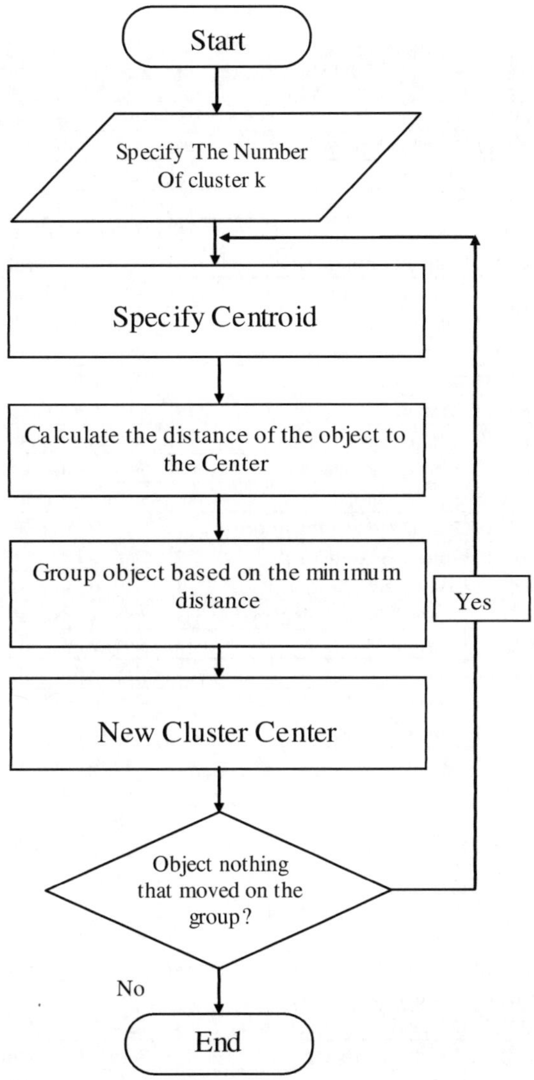

Figure 1. Flowchart Algoritma K-Means Clustering.

There was variable that used in clustering was Semester Credit System, quality and Grade Predicate Academic (GPA), the students data that belong to drop out in last four years, so their score could be seen in table 1.

IOP Conf. Series: Journal of Physics: Conf. Series **1007** (2018) 012049 doi:10.1088/1742-6596/1007/1/012049

Table 1. Student Data.

NO	NIM	NAME	COURSES PROGRAM	CREDITS SKS	QUALITY TOTAL	GPA
M1	8330303014	Vera	S1 - Teknik Informatika	14	47	3.36
M2	8330303015	Caroline Nata Wijaya	S1 - Teknik Informatika	14	56	4.00
M3	8330303016	Silvia	S1 - Teknik Informatika	3	12	4.00
M4	8330303017	Dewi Sandro Maria	S1 - Teknik Informatika	14	24	1.71
M5	9330303001	Dearni Rapita Saragih	S1 - Teknik Informatika	57	154	2.70
M6	9330303002	Hamonangan Nasution	S1 - Teknik Informatika	88	106	1.20
M7	9330303003	Dadang Suhenda	S1 - Teknik Informatika	61	157	2.57
M8	9330303004	Paulus P.H Nababan	S1 - Sistem Infomasi	48	154	3.21
M9	9330303006	Muhammad Tarmizi	S1 - Sistem Infomasi	48	144	3.00
M10	9330303007	Bunga Andriani Lubis	S1 - Sistem Infomasi	57	151	2.65
M11	9330303008	Muhammad Fachri	S1 - Sistem Infomasi	48	148	3.08
M12	9330303009	Rizkina Fitri Fauziah	S1 - Sistem Infomasi	48	121	2.52
M13	9330303011	Melda Riski Harahap	S1 - Sistem Infomasi	57	181	3.18
M14	9330303013	Eka Fitri Siregar	S1 - Sistem Infomasi	57	165	2.89
M15	9330303015	Siskia Lasmana Togatorop	S1 - Sistem Infomasi	57	160	2.81
M16	9330303016	Rohani Tinambunan	S1 - Sistem Infomasi	62	138	2.23
M17	9330303018	Hanova Margareth	S1 - Sistem Infomasi	57	168	2.95
M18	9330303019	Sherly Minarti Nainggolan	S1 - Sistem Infomasi	23	0	0.00
M19	9330303020	Musmulyadi	S1 - Sistem Infomasi	57	164	2.88
M20	9330303022	Robin	S1 - Sistem Infomasi	57	163	2.86
M21	9330303024	Eddy	S1 - Sistem Infomasi	57	176	3.09
M22	9330303025	Feriyanto Tanwijaya	S1 - Sistem Infomasi	57	162	2.84
M23	9330303027	Melianty Sitompul	S1 - Sistem Infomasi	65	173	2.66
M24	9330303028	Shahani Singh	S1 - Sistem Infomasi	71	163	2.30
M25	9330303031	Jovin	S1 - Sistem Infomasi	54	168	3.11
M26	9330303032	Yoko Junior	S1 - Sistem Infomasi	57	136	2.39
M27	9330303037	Ferry Wijaya	S1 - Sistem Infomasi	54	156	2.89
M28	9330303041	Muhammad Ramadhan Lubis	S1 - Sistem Infomasi	54	149	2.76
M29	9330303048	Radius Hia	S1 - Sistem Infomasi	47	128	2.72
M30	9330313002	Winter Pandiangan	S1 - Sistem Infomasi	14	0	0.00
M31	10330303001	Abdul Ghani Abadi	S1 - Teknik Informatika	98	322	3.29
M32	10330303004	Armand Pernando Sitorus	S1 - Teknik Informatika	21	0	0.00
M33	10330303005	Bayakta Sebayang	S1 - Teknik Informatika	159	200	1.26
M34	10330303008	Indra Putra	S1 - Teknik Informatika	104	291	2.80
M35	10330303009	Jeffri Aritonang	S1 - Teknik Informatika	101	302	2.99
M36	10330303010	Jeffry Suherman	S1 - Teknik Informatika	98	338	3.45

To do the cluster data become cluster it was done some steps, there were:

1. Determine the specific cluster. In this research the data would be clustering become three clusters. In cluster 1 had a characteristic data. That was students that potentialy in drop out.
2. In cluster 2 belonged to good characteritic data. In cluster 3 was gotten to characteristic very good students.

Determined the centroid cluster begining. In this research the strting point was determined randomly and gotten in the starting point every cluster. It could be seen in tble 4.3. where cluster point that taken.

Table 2. The Central Point Of The Initial Cluster.

1st Data	Cluster	Credit Total	Quality Total	GPA
4	C1	14	24	1,71

29	C2	47	128	2,72
36	C3	98	338	3,45

After determining the centroid, so every data would found the near centroid by counting the distance every data to each centroid by using correlation formula between two objects of Euclidean Distance. There was a count of manual centroid. It had just count 36 data only.

Table 3. Clustering Result.

Cluster	Member of Cluster	Number of members
1	[1, 2, 3, 4, 18, 30, 32]	7 member
2	[5, 6, 7, 8, 9, 10, 11, 12, 13, 14, 15, 16, 17, 19, 20, 21, 22, 23, 24, 25, 26, 27, 28, 29, 33]	25 member
3	[31, 34, 35, 36]	4 member

From the conclusion above it can be shown that students that have potentially dropped out more belong to cluster 1. It was caused of total od System Credit Semester, Quality Total and the lowest of GPA. It could be compared between cluster 2 and 3.

4. Conclusion

Based on the the discussion and result of test, so it could be concluded:

1. Based on the result of the test so the clustering was taken from students position that potentially drop out was 3 clusters. They were students who very drop out potentially, drop out potentially, and not belong to drop out.
2. Clustering result could be used to conclude in determine students who potentially drop out.
3. To predict the students who potentially drop out, the data which was analyzed consist of scoring students data, quality and GPA.
4. The result of counting for all datas (36 records) based on the test that showed tht cluster 0 consists of 7 members that shown the students who potentially drop out. Then cluster 1 that consist of 25 members shown the well students cluster. Meanwhile cluster 2 shown the very good students with 4 members.

References

[1] Dekker, G.W., 2009, Prediction student drop out: A case study, USA, Academic Press, 2nd international Conference On Educational Data Mining, Cordoba, Spain.

[2] Ramesh Singh Yadava and P.K.Mishra. (2012). "Performance Analysis of High Performance k-Mean Data Mining Algorithm for Multicore Heterogeneous Compute Cluster." Vol. 2 No.4.

[3] Tutik Khotimah 2014. "Pengelompokan Surat Dalam Al Qur'an Menggunakan Algoritma K-Means." Ed. Jurnal Simetris, Vol 5 No 1.

[4] Deka Dwinavinta 2014. "Klasterisasi Judul Buku dengan Menggunakan Metode K-Means." Ed. Seminar Nasional Aplikasi Teknologi Informasi (SNATI).

[5] Bondu Venkateswarlu and Prof G.S.V.Prasad Raju. 2013. "Mine Blood Donors Information through Improved K-Means Clustering." International Journal of Computational Science and Information Technology (IJCSITY) Vol.1, No.3.

[6] Johan Oscar Ong. 2013. "Implementasi Algoritma K-Means Clustering Untuk Menentukan Strategi Marketing Presiden University." Ed. Jurnal Ilmiah Teknik Industri, Vol.12, No.1.

[7] Afrisawati. 2013. "Implementasi Data Mining Pemilihan Pelanggan Potensial Menggunakan Algoritma K-Means." Ed. Pelita Informatika Budi Darma, Volume : V,No. 3.

[8] Budanis Dwi Meilani dan Nofi Susanti, 2014, "Aplikasi Data Mining Untuk Menghasilkan Pola Kelulusan Siswa Dengan Metode Naïve Bayes." Ed. Jurnal LINK Vol 21/No.2.

[9] Dr. M.P.S Bhatia and Deepika Khurana. 2013. "Experimental study of Data clustering using k-Means and modified algorithms." International Journal of Data Mining & Knowlegde Management Process (IJDKP). Vol. 3 No.3.
[10] Er. Nikhil Chaturvedi and Er. Anand Rajavat. 2013. "An Improvement in K-mean Clustering Algorithm Using Better Time and Accuracy." International Journal of Programming Language and Aplications (IJPLA) Vol. 3 No.4.
[11] Fadlina. 2014. "Data Mining untuk Analisa Tingkat Kejahatan Jalanan Dengan Algoritma Association Rule Metode Apriori (Studi Kasus Di Polsekta Medan Sunggal)." Voume.III No.1.
[12] Goldie Gunadi dan Dana Indra Sensuse, 2012. "Penerapan Metode Data Mining Market Basket Analysis Terhadap Data Penjualan Produk Buku Dengan Menggunakan Algoritma Apriori dan Frequent Pattern Growth (FP-Grouth) : Studi Kasus Percetakan PT. Gramedia." Ed. Jurnal Telematika MKom Vol.4 No.1.
[13] Lindawati. 2008. "Data Mining Dengan Teknik Clustering Dalam Pengklasifikasian Data Mahasiswa Studi Kasus Prediksi Lama Studi Mahasiswa Universitas Bina Nusantara." Ed. Seminar Nasional Informatika.
[14] Mujib Ridwan, 2013. "Penerapan Data Mining Untuk Evaluasi Kinerja Akademik Mahasiswa Menggunakan Algoritma Naïve Bayes Classifier." Ed. Jurnal EECCIS Vol.7, No.1.
[15] Oyelade. O. J, 2010. "Application of k-Means Clustering algorithm for prediction of Students' Academic Performance." International Journal of Computer Science and Information Security. Vol. 7 No.1.
[16] Rajashree, 2010. "A hybridized K-means clustering approach for high dimensional dataset." International Journal of Engineering, Science and Technology Vol. 2, No. 2.
[17] Ramzi A. Haraty, 2015. "An Enhancedk-Means Clustering Algorithm for Pattern Discovery in Healthcare Data." Hindawi Publishing Corporation International Journal of Distributed Sensor Networks Volume 2015, Article ID 615740.

IOP Conf. Series: Journal of Physics: Conf. Series **1007** (2018) 012050 doi:10.1088/1742-6596/1007/1/012050

Optimization of training backpropagation algorithm using nguyen widrow for angina ludwig diagnosis

Siti Aisyah*, Mawaddah Harahap, Amir Mahmud Husein Siregar, Mardi Turnip

Faculty of Technology and Computer Science, Universitas prima Indonesia

E-mail: *sitiaisyah@unprimdn.ac.id

Abstract. Tooth and mouth disease is a common disease, with a prevalence of more than 40% (children aged less than 7 years) in milk teeth and about 85% (adults aged 17 years and over) on permanent teeth. Angina Ludwig is one of mouth disease type that occurs due to infection of the tooth root and trauma of the mouth. `In this study back propagation algorithm applied to diagnose AnginaLudwig disease (using Nguyen Widrow method in optimization of training time). From the experimental results, it is known that the average BPNN by using Nguyen Widrow is much faster which is about 0.0624 seconds and 0.1019 seconds (without NguyenWidrow). In contrast, for pattern recognition needs, found that back propagation without Nguyen Widrow is much better that is with 90% accuracy (only 70% with NguyenWidrow).

1. Introduction

Tooth and mouth disease is a common disease, with a prevalence of more than 40% (children aged less than 7 years) in milk teeth and about 85% (adults aged 17 years and over) on permanent teeth. Dental and oral diseases can be determined by looking at various symptoms, there are 40 symptoms of dental and mouth disease with 18 types of diseases [1]. Angina Ludwig one of mouth disease type diffusa cellullitis on the base of the mouth that can soul [2]. Lack of public knowledge about this type of disease makes them unaware of the dangers that threaten their lives. In helping doctors to diagnose dental and oral diseases, some researchers propose methods of artificial intelligence, one of which is Neural Network (NN).

NN is one of the most accurate methods in helping doctors analyze, model and understand complex clinical data in various medical fields such as diabetes diagnosis. [3], chest diseases [4], hepatitis [5], and skin diseases [6]. NN is accurate in nonlinear predictions and the ability to tolerate errors [7]. NN's ability is useful for generalizing, identifying nonlinear relationships and deployment in various applications [8]. One of the most widely used architectures by researchers is the Back propagation Neural Network (BPNN) which has the learning process and error correction in reverse. This model is widely used both for introduction and prediction with a fairly good level of accuracy [9]. BPNN is accurate to solve many real-world problems by building a trained model that runs well on some non linear problem, but lacks the need for large training data and requires high computational time to recognize patterns in high dimensions [10]. It's easy to get caught in a local flaw, causing failure to find optimal solutions and relying on initial weights, bias and parameters.

Some researchers apply a heuristic algorithm to improve the accuracy of BNN, such as particle swarm optimization (PSO) [11], ant colony optimization (ACO) [12-113], bee colony algorithm (BCA)

IOP Conf. Series: Journal of Physics: Conf. Series **1007** (2018) 012050 doi:10.1088/1742-6596/1007/1/012050

[14], and others. Nguyen Widrow is a method that can be applied to the back propagation method. This method can accelerate the initialization process of the value of the weight and value of the back propagation architecture bias so as to facilitate the achievement of its convergence value. On Nguyen Widrow initialization, initializing the bias and weights from the input unit to the output unit using the bias and scale weights within a certain range. [15]. In this research apply method of artificial neural network to diagnose dental and mouth disease angina Ludwig which optimized with Nguyen Widrow algorithm. This paper is presented as follows: Methodology Section, Results and Discussion section 3 and Conclusion Section 4.

2. Methodology

Artificial neural networks provide counters that are inspired by the operating structures of the brain and the central nervous system [16]. Back propagation allows the acquisition of exponential knowledge of input-output mapping in multilayer networks [17]. Artificial neural network models require the best combination of network parameters such as training cycles, level of learning, neurons in the input layer (η), hidden and output, momentum term (α), and a good degree of accuracy of the algorithm in the prediction process [18]. The Nguyen Widrow method is the algorithm used to initialize the weights on artificial neural networks in reducing training time. Initialization of Nguyen Widrowis as follows: Set: n = input unit, p = output unit, β = scale factor = 0.7 (p) 1 / n = 0.7 √ (n & p); for each hidden unit (j = 1, ..., p), do step (c) - (f); for i = 1, ..., n (all input units), vij (old) = random numbers between -0.5 to 0.5; calculate value || vj (old) ||; initialization of the weight of the input unit (i = 1, ..., n); use the bias weight as initialization: voj = random number between -β and β. Back propagation has many layers or multi layers, the goal is to train the network to get a balance of the ability to recognize the patterns used during the training and the network's ability to respond correctly to the pattern of inputs with the patterns used during the training.

The data was obtained from one of the private hospitals in Medan which was also used for studying some of the health students at the time. Table 1 Data on symptoms of Angina Ludwig is commonly felt by patients consisting of eight symptoms. Table 2 and table 3 contain the data used in this study (30 sample data) where s1 to s30 are subject and the number of symptoms of variable 8 (x1-x8). Training process and testing process using Mat lab application. The amount of data used for the training process is 20 (to be able to recognize many patterns), whereas the test uses 10 sample data. The value of variable each patient is the initialization of the perceived condition, where 1 means (yes) and 0 (no). The target on the sample data is the goal to be obtained from the data being processed. If the value generated from the testing process is still far from the target error then the iteration is done back to the target value. The sample description of column s1 in table 2 only has symptoms of pain neck (x1) and dazed (x8), so that x1 and x2 is 1 and target value 0. The description in column s2 has only neck pain (x1), difficult chew (x2) fever (x5) and dazed (x8), so that (x1) (x2) (x5) and (x8) has a value of 1 and target 1. For each description in the sample above the rest can be seen in table 2 and table 3.

Table 1. Symptom data.

Symptoms
Neck Pain (x1)
difficult chew (x2)
Neck Swelling (x3)
Crowded (x4)
Fever (x5)
Rash (x6)
Ear Pain (x7)
Dazed (x8)

Table 2. Sample data.

	s1	s2	s3	s4	s5	s6	s7	s8	s9	s10	s11	s12	s13	s14	s15
x1	1	1	1	0	0	1	0	1	0	0	0	0	1	0	1
x2	0	1	0	1	0	0	1	1	0	1	0	1	1	1	1
x3	0	0	0	1	0	1	0	1	1	0	1	0	0	0	1
x4	0	0	1	1	0	0	1	1	0	0	1	0	1	1	1
x5	0	1	0	0	0	0	1	1	0	1	1	0	0	0	0
x6	0	0	1	1	1	0	1	1	1	0	0	1	1	1	1
x7	0	0	1	0	0	1	1	1	0	0	0	1	1	0	1
x8	1	1	0	0	1	0	1	0	1	0	1	1	0	1	1
Target	0	1	1	1	0	0	1	1	0	0	1	1	1	1	1

Tabel 3. Advanced sample data.

	s16	s17	s18	s19	s20	s21	s22	s23	s24	s25	s26	s27	s28	s29	s30
x1	0	0	1	1	0	1	1	0	0	0	1	1	0	1	0
x2	0	1	1	0	0	1	1	0	1	1	1	1	1	0	0
x3	1	1	0	1	0	1	0	0	1	0	1	0	1	1	0
x4	0	0	1	0	1	1	0	1	1	0	0	0	1	1	0
x5	1	0	1	1	1	1	0	0	1	0	0	1	1	1	0
x6	0	1	0	0	0	1	0	0	1	1	0	1	0	0	0
x7	0	0	1	1	1	1	1	1	1	1	0	1	0	1	1
x8	0	0	1	0	1	1	0	1	0	0	1	1	0	1	1
Target	0	0	1	1	1	1	0	0	1	0	1	1	1	1	0

The design architecture in figure 1. of this research is arranged with the pattern of 8-3-1, The variables in the input layer consist of x1, x2, x3, x4, x5, x6, x7, x8 which is a symptom of the disease (8 symptoms). Hidden layers consisting of 3 variables are optional, and the output layer is the output target of the input and hidden layer. The block diagram in figure 2 illustrates how the introduction process of BPNN and Nguyen in processing the data. The dataset to be managed is called raw data (data preprocess). The raw data is divided into two groups (training and test data), where the amount of training data is greater than the test data so that more networks recognize the pattern. The next stage of normalization or scaling is done to transform the data in order to be processed. After transformation for the next process of training on the train data. Based on how to modify the weighting is divided into two things: training with supervision and unsupervision. In this learning process uses supervision training. At each training input provided to the network, when the network will process and output . The difference of between the output of the network and the target (the desired output) is the error that occurs, and the network will modify the weights accordingly. The training process also enters a further stage of propagation. the training phase will explain the steps taken in the BPNN process. Step 0: initialize the weights between neurons using small random numbers (-0.5 to 0.5); each input receives a signal and passes it across the hidden layer unit; Each hidden unit calculates the weight of the input signal; for each unit of output summing the weight of the input signal; the output unit receives a mutually appropriate pattern on the training pattern input and calculates the error information and calculates the bias weight correction; Each hidden layer unit sums the input results multiplied by a derivative of the activation function to calculate error information. From the training process obtained new weight data, this process is done repeatedly to find the appropriate weighting value. For further testing conducted by relying on training data and new weighted data on BPNN algorithm training process. At the last stage is the result of the testing process, whether Angina Ludwig disease can be recognized or not from pattern recognition. While for testing with the algorithm Nguyen Widrow for weighting using scale factor 0.7 (p) 1 / n = 0.7 $\sqrt{p}$, but for the next process has the same stages with

BPNN. Testing with test data both BPNN and Nguyen done by using all sample data, so that can know the ability of network expressed in level of accuracy.

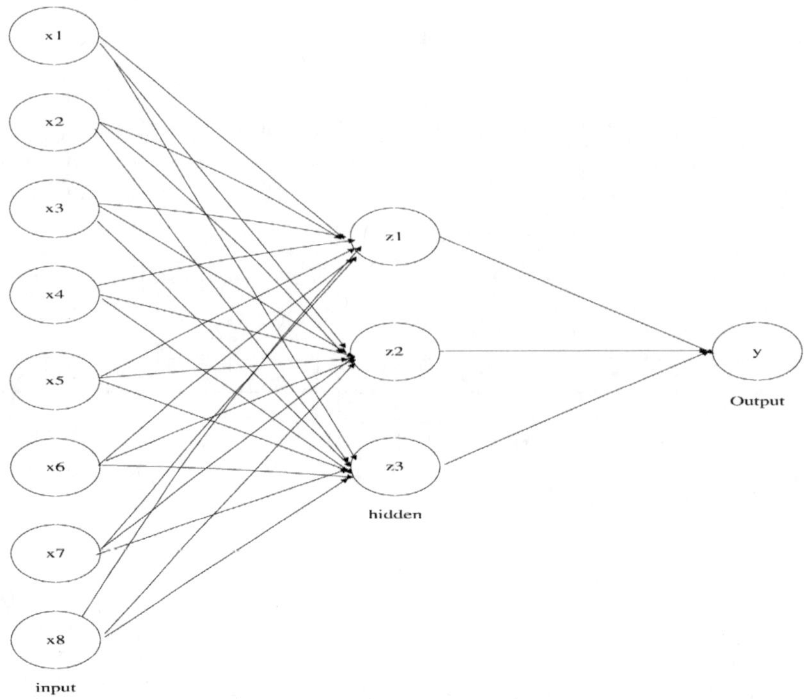

Figure 1. Design architecture.

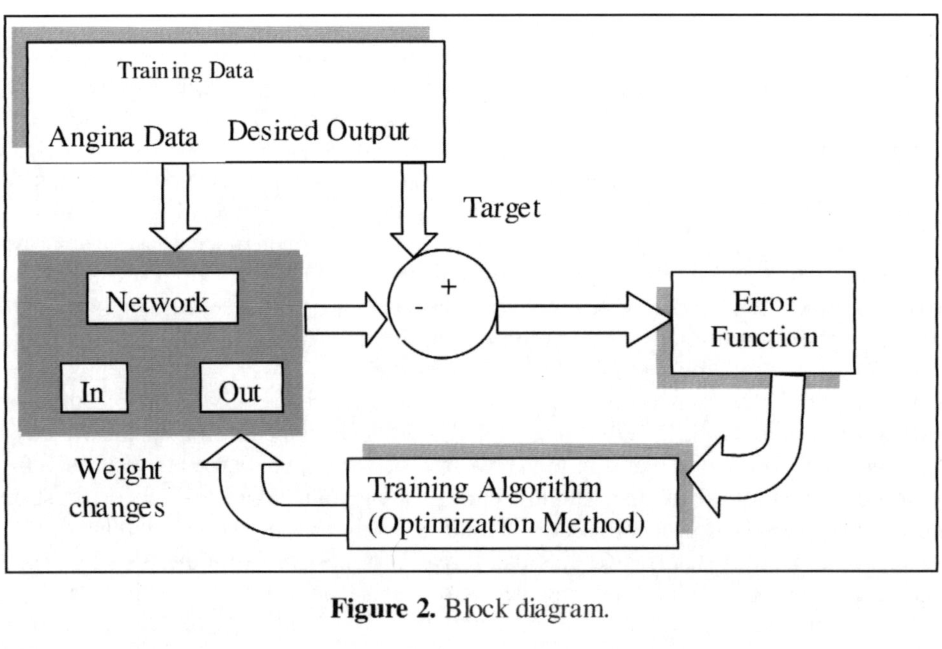

Figure 2. Block diagram.

3. Results and Discussion

This study focuses on the impact of the value of this activation weight during BPNN training using the Nguyen Widrow algorithm for diagnosis. Table 4 and table 5 are training data tables. The initialization given is for each variable having two terms or criterion, where the criterion are worth (Yes) or (No); Variables that contain the criterion(Yes) will get a weight of 1, while for the variable with the criterion (No) is given weight of 0. For the training process using 20 of the 30 sample data (70%), the amount of data on the Training Process is greater than the data used in the testing process, it means that the pattern is recognized by the network.

Table 4. Training data.

	s1	s2	s3	s4	s5	s6	s7	s8	s9	s10
x1	1	1	1	0	0	1	0	1	0	0
x2	0	1	0	1	0	0	1	1	0	1
x3	0	0	0	1	0	1	0	1	1	0
x4	0	0	1	1	0	0	1	1	0	0
x5	0	1	0	0	0	0	1	1	0	1
x6	0	0	1	1	1	0	1	1	1	0
x7	0	0	1	0	0	1	1	1	0	0
x8	1	1	0	0	1	0	1	0	1	0
Target	0	1	1	1	0	0	1	1	0	0

Table 5. Advanced training data.

	s11	s12	s13	s14	s15	s16	s17	s18	s19	s20
x1	0	0	1	0	1	0	0	1	1	0
x2	0	1	1	1	1	0	1	1	0	0
x3	1	0	0	0	1	1	1	0	1	0
x4	1	0	1	1	1	0	0	1	0	1
x5	1	0	0	0	0	1	0	1	1	1
x6	0	1	1	1	1	0	1	0	0	0
x7	0	1	1	0	1	0	0	1	1	1
x8	1	1	0	1	1	0	0	1	0	1
Target	1	1	1	1	1	0	0	1	1	1

Based on the datasets in table 4 and 5, the next step is to normalize the matrix form with the aim to be recognized on the network architecture of BPNN. Normalization of data is done so that the network output in accordance with the activation function used. The dataset is normalized in intervals [0, 1] Binary sigmoid activation function with interval transformation data [0.1; 0.8 with equation x ^ '= (0.8 (x-a)) / (b-a) +0.1 where: a minimum data, b is maximum data, x data normalized, and x' has changed.

Data processing is done in two stages, the first step is done before the training so that the pattern can be recognized, after the training continued with the testing in the second stage. In this research result of training data taken sample with log sig activation function; target error 0.1; learning rate 0.1 then the results obtained in table 6 and 7. The number of epoch given influences the outcome, the higher the epoch given the faster it will approach the target to be achieved. However in table 4 the training target results are still far from the value to be achieved. The value obtained in each sample subject on the epoch 100 is s1 0.0133, s2 0.140, s3 0.0597, s4 0.0135, s5 0.0042, s6 0.0607, s7 0.0589, s8 0.5339, s10 0.265, s11 0.0133, s12 0.1403, s13 0.0597, s14 0.0135, s15 0.0042, s16 0.0607, s17 0.0589, s18 0.5339, s19 0.0308, s20 0.781. On epoch 1000 is s1 0.0496, s2 0.0722, s3 0.0815, s4 00.531, s5 0.596, s6 0.5059, s7 0.1626, s8 0.4131, s9 0.0308, s10 0.781, s11 0.0496, s12 0.0722, s13 0.0815, s14 0.531, s15 0.596, s16 0.5059, s17 0.1626, s18 0.4131, s19 0.0308, s20 0.781. The result of the training process is still far from the target value. In the epoch 100 and 1000 still equally have not fulfilled its value, this is because the process is still a little so the network has not really recognize the pattern. To obtain an optimal value should be done several times iterations on trained data and a large epoch value will affect the results.

Table 6. Results of training.

Epoch	s1	s2	s3	s4	s5	s6	s7	s8	s9	s10
100	0.0133	0.1403	0.0597	0.013 5	0.0042	0.0607	0.0589	0.5339	0.265	0.1983
1000	0.0496	0.0722	0.0815	0.531	0.596	0.5059	0.1626	0.4131	0.0308	0.781
Target	0	1	1	1	0	0	1	1	1	0

Table 7. Advanced training results.

Epoch	s11	s12	s13	s14	s15	s16	s17	s18	s19	s20
100	0.0133	0.1403	0.0597	0.0135	0.0042	0.0607	0.0589	0.5339	0.265	0.1983
1000	0.0496	0.0722	0.0815	0.531	0.596	0.5059	0.1626	0.4131	0.0308	0.781
Target	1	1	1	1	1	0	0	1	1	1

The next stage is the testing process using 10 sample data, with log sig activation parameters; target error 0.1; learning rate 0.1, then the results obtained in table 8 and time calculation in table 9. BPNN test results have accuracy on each subject sample as follows: s1 80.21%, s2 50.45%, s3 88.23%, s4 81.23%, s5 78.12% , s6 77.34%, s7 79.89%, s8 80.23%, s9 81.09%, s10 78.34%. Nguyen Widrow has the following results: s1 86.98%, s250.78%, s3 49.8%, s4 80.23%, s5 43.90%, s6 88.23%, s7 76.12%, s8 79.89%, s9 92.78%, s10 87.45%. Test results using BPNN have better accuracy with 90% pattern recognition, while Nguyen Widrow recognizes 70% pattern. In table 9 is the computational time of the test results of BPNN and Nguyen Widrow. IBRA obtained the following results s1 0.08 second, s2 0.1934 second, s3 0.0469 second, s4 0.0428 second, s5 0.0296 second, s6 0.1344 second, s7 0.238 second, s8 0.135 second, s9 0.019 second, s10 0.0999 second. Nguyen Widrow obtained the following results: s1 0.1106 second, s4 0.0164 second, s3 0.0667 second, s6 0.0993 second, s7 01034 second, s8 0.0411 second, s9 0.1074 second, s10 0.0567 second. The computational time BPNN has an average iteration time of 0.1019 seconds while the Nguyen Widrow has an average iteration time of about 0.0624 seconds faster.

Table 8. Test results.

	s1	s2	s3	s4	s5	s6	s7	s8	s9	s10
x1	1	1	0	0	0	1	1	0	1	0
x2	1	1	0	1	1	1	1	1	0	0
x3	1	0	0	1	0	1	0	1	1	0
x4	1	0	1	1	0	0	0	1	1	0
x5	1	0	0	1	0	0	1	1	1	0
x6	1	0	0	1	1	0	1	0	0	0
x7	1	1	1	1	1	0	1	0	1	1
x8	1	0	1	0	0	1	1	0	1	1
Target	1	0	0	1	0	1	1	1	1	0
Nguyen widrow (%)	86.98	50.78	49.8	80.23	43.90	88.23	76.12	79.89	92.78	87.45
BPNN (%)	80.21	50.45	88.23	81.23	78.12	77.34	79.89	80.23	81.09	78.34

Table 9. Computation time.

Patient	x1	x2	x3	x4	x5	x6	x7	x8	Nguyen widrow	BPNN
s1	1	1	1	1	1	1	1	1	0.1106	0.08
s2	1	1	0	0	0	0	1	0	0.0164	0.1934
s3	0	0	0	1	0	0	1	1	0.0667	0.0469
s4	0	1	1	1	1	1	1	0	0.0059	0.0428
s5	0	1	0	0	0	1	1	0	0.0149	0.0296
s6	1	1	1	0	0	0	0	1	0.0993	0.1344
s7	1	1	0	0	1	1	1	1	0.1034	0.238
s8	0	1	1	1	1	0	0	0	0.0411	0.135
s9	1	0	1	1	1	0	1	1	0.1074	0.019
s10	0	0	0	0	0	0	1	1	0.0567	0.0999
			Time (seconds)						0.0624	0.1019

4. Conclusions

This study aims to optimize the network computing time BPNN training by applying the method of Nguyen Widrow on the diagnosis of dental and mouth disease Angina Ludwig with dataset obtained from the hospital. From the experimental results it is known that the average iteration value of BPNN using Nguyen Widrow is much faster which is about 0.0624 dni 0.1019 seconds (without Nguyen Widrow). In contrast, for pattern recognition needs, it was found that back propagation without Nguyen Widrow was much better with an accuracy of 90% (only 70% with Nguyen Widrow).

Reference

[1] Nyoman Kusuma Wardana, Alfa Antariksa, Nazrul Effendy, 2008, Perancangan Sistem Pakar Untuk Diagnosa Penyakit Mulut Dan Gigi Menggunakan Bahasa Pemrograman Clips, Seminar Nasional Aplikasi Teknologi Informasi 2008 (SNATI 2008), 57-60

[2] Mahaputri AR, 2013, Angina Ludwig Pada Pasien Laki-laki Dewasa Muda Karena Infeksi Odontogen, Medula, Volume 1, Nomor 5, Oktober, 2013

[3] Okan Erkaymaz , Mahmut Ozer , Matjaž Perc, 2017, Performance of small-world feedforward neural networks forthe diagnosis of diabetes, Applied Mathematics and Computation, 22–28

[4] D vally, CH V Sarma, 2015, Diagnosis Chest Diseases Using Neural Network and Genetic Hybrid Algorithm. Journal of Engineering Research and Applications, ISSN : 2248-9622, Vol. 5, Issue 1, 20- 26

[5] Mehdi Neshat, Azra Masoumi, Mina Rajabi and Hassan Jafari, 2014, Diagnosing Hepatitis Disease by Using Fuzzy Hopfield Neural Network, Annual Research & Review in Biology, 2709-2721

[6] Bakpo, F. S.1 and Kabari, L. G, 2009, Diagnosing Skin Diseases Using an Artificial Neural Network, Artificial Neural Networks - Methodological Advances and Biomedical Applications, 253-270, DOI: 10.1109/ICASTECH.2009.5409725

[7] X. Shao, "Based on Two Swarm Optimized Algorithm of Neural Network to Prediction the Switch's Traffic of Coal," 2011 International Symposium on Computer Science and Society, pp. 299–302, Jul. 2011

[8] Bennett, C. J., Stewart, R. a., & Lu, J. W. (2014). Forecasting low voltage distribution network demand profiles using a pattern recognition based expert system. Energy, 67, 200–212

[9] Dewi, Candra dan M. Muslikh, 2013, Perbandingan Akurasi Backpropagation Neural Network dan ANFIS untuk Memprediksi Cuaca. Journal of Scientic Modelling & Computation, Vol. 1, No. 1

[10] Pan, S., Iplikci, S., Warwick, K., & Aziz, T. Z. (2012). Parkinson's Disease tremor

IOP Conf. Series: Journal of Physics: Conf. Series **1007** (2018) 012050 doi:10.1088/1742-6596/1007/1/012050

classification – A comparison between Support Vector Machines and neural networks. Expert Systems with Applications, 39 (12), 10764–10771. doi:10.1016/j.eswa.2012.02.189

[11] B. Al-kazemi and C. K. Mohan, Training feedforward neural networks using multi-phase particle swarm optimization. ICONIP'02, Proceedings of the 9th International Conference on Neural Information Processing Orchid Country Club, Singapore, (2002), pp. 2615–2619

[12] C. Blum and K. Soch, Training feed-forward neural networks with ant colony optimization: An application to pattern classification, Fifth International Conference on Hybrid Intelligent Systems (HIS 2005), Rio de Janeiro, Brazil (2005), pp. 233–238

[13] Y. P. Liu, M. G. Wu, and J. X. Qian, evolving neural networks using the hybrid of ant colony optimization and BP algorithms. In Advances in Neural Networks-ISNN, Springer, Berlin Heidelberg (2006), pp. 714–722

[14] C. Ozturk and D. Karaboga, Hybrid artificial bee colony algorithm for neural network training, IEEE Congress on Evolutionary Computation (CEC) New Orleans, USA (2011), pp. 84–88

[15] Nasib Ridwan Manalu, Pilipus Tarigan, 2016, Perancangan Aplikasi Untuk Pengenalan Karakter Tulis Tangan Menggunakan Metode Nguyen Widrow, Pelita Informatika Budi Darma, Volume : XV, Nomor: 1, ISSN : 2301-9425, 37-40

[16] H. R. Maier and G. C. Dandy, "Neural network based modelling of environmental variables: a systematic approach," Mathematical and Computer Modelling, vol. 33, no. 6-7, pp. 669–682, 2001

[17] A Turnip and D. Soetraprawata, The Performance of EEG-P300 Classification Using Backpropagation Neural Network, Mechatronics Electrical Power and Vehicular Technology, 04 (2013), 81- 88

[18] Eldon R. Rene, M. Estefanía López, Jung Hoon Kim,and Hung Suck Park, Back Propagation Neural Network Model for Predicting the Performance of Immobilized Cell Biofilters Handling Gas-Phase Hydrogen Sulphide and Ammonia," BioMed Research International, volume 2013 (2013), Article ID 463401, 9 pages

Acknowledgments

This research support by Faculty of Technology and Computer Science, Universitas prima Indonesia.

IOP Conf. Series: Journal of Physics: Conf. Series **1007** (2018) 012051 doi:10.1088/1742-6596/1007/1/012051

Analysis of investment IT planning on logistic company using COBIT 5

Syafrial Fachri Pane, Rolly Maulana Awangga, Rd. Nuraini and Siti Fathonah

Department of Informatics Engineering, Politeknik Pos Indonesia, Bandung, Indonesia

E-mail: syafrial.fachri@poltekpos.ac.ud

Abstract. IT Investment planning to review goals business strategy development needs to support IT governance. Planning its investments will be assessed with a framework is issued by ISACA that COBIT 5 focus on domain EDM, APO, BAI, and MEA. Mapping from framework COBIT 5 with Val IT lead nine domain consisting of EDM01, EDM02, APO01, APO02, APO05, APO06, APO07, BAI01, and MEA01. Article search google using RACI diagram generating mapping organisation telecoms and authority structures for the review identify respondents from any domain. Under selected area be used to review counting process capability level, and for a discussion of development, priorities make recommendations shall perform in accordance process needs PT.XYZ logistics company. The translating process capability level determined by the company prior currently at on level 4. The capability level domain process in the EDM is level 4, while at domain APO, BAI, and MEA are on level 3 and gap on APO, BAI, and MEA. The company recommendation domain subscription APO, BAI, and MEA to increase the capabilities of the rate review process to achieve the targets that have can determine the company.

1. Introduction

Information Technology (IT) is a need for all companies and business process like hospital[1] and mining company[2]. The study conducted 92 percent of respondents support the activities of logistics companies with a real system [3]. Logistics Company PT. XYZ is one company that utilises the role of information technology to support all business processes run by the company to be active and efficient. Business Focus of PT.XYZ Logistics Company emphasises on creating the value of logistics business through the consolidation and comprehensive integration of multimodal freight forwarding services which include land, sea, air and warehousing services. The strategic is planning of PT.XYZ logistics company through the acquisition of the internal and external company by considering strategic risk (according to PT Pos Indonesia, 2013). Figure 1 is the structure of the company's business processes (PT Pos Indonesia, 2014).

IOP Conf. Series: Journal of Physics: Conf. Series **1007** (2018) 012051 doi:10.1088/1742-6596/1007/1/012051

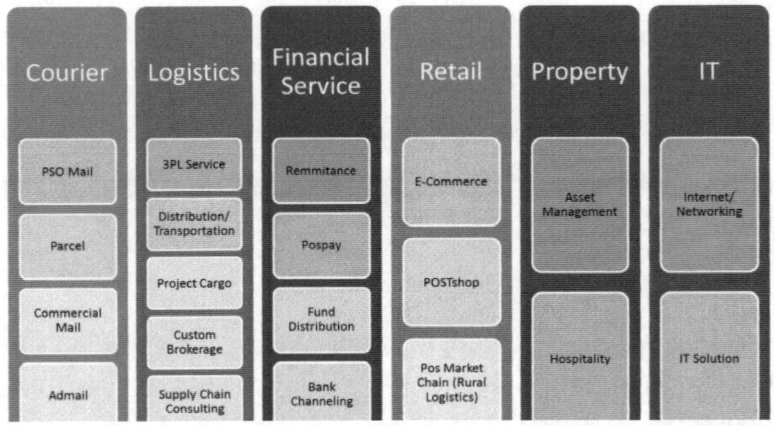

Figure 1. Company business structure.

To run the entire business focuses on the company requires significantresources, one of them is the one that focuses on planning IT investment by the organisation for information technology needs. The support the IT investment planning, a good IT governance must include in the company. This logistics company, PT. XYZ has already gotexcellent resource support either in financial or inadequate infrastructure. Planning the IT investment should be tailored to the strategic logistics of PT. XYZ. The following are the activities and financing plans for 2013 to 2016 used to perform the realisation phase of the transition phase of the IT investment planning process, as follows (PT Pos Indonesia, 2013).

Table 1 stated that the realisation of revenue has not been achieved and has not met the RKAP (according to PT Pos Indonesia, 2014) PT. XYZ Logistics Company created aplanned budget on RJP document (Long Term Plan) with an allocated budget from 400 billion to 1 Trillion. We can conclude that the budget was massive, and would need supervision and evaluation of all activities related to IT investment planning, whether or not it ran accordingly (by the company's business strategy), or vice versa. Kemper [4] defined oversight as determining what has been, that is, to evaluate work performance and, where appropriate, to apply corrective actions so that the outcome of the work is by the established plan.

Table 1. Revenue from business services activities.

Component	Year					Average	
	2009	2010	2011	2012	2013	Revenue	Growth
RJP	1.081.432	1.297.719	1.490.588	1.063.446	1.480.200	1.282.677	11,3%
RKAP	721.770	874.924	916.580	1.063.446	1.480.200	1.011.384	20.3%
REALIZATION	682.242	830.335	966.279	1.112.298	1.551.340	1.028.499	23,2%

The process of IT investment planning used the existing SOP (Standard Operational Procedure). Because there were several internal problems, such as the lack of evaluation and control of the management toward the IT Investment planning itself. A newspaper mentioned the result was not at level best. The Organization is one of the foundations to determine some existing policies and decisions that exist,and one of them implemented in the company activities of IT investment planning. Decision-making is the act of election of one or more possibilities[5], but this is hardly an election between right and wrong but is often the choice between the almost right and the wrong. Decision-making is the general responsibility of all executives, regardless of their functional area or management level. Managers are required, every day, to make decisions that create the future of their organisations as well as their futures[6]. Factors that often occurred in the field were that there was a mismatch process of IT investment planning resulting in ignorance and the absence of periodic evaluation. No wonder if the business focus did not work correctly, a largenumber of budgets was not well controlled, the decision made by business units was overlapping.Then the researchers can conclude there will be problems and adverse effects on the survival of the company for a while.

The researcher will look for solutions to the problems that have been described above by emphasising the approach of evaluation and control. One way of doing evaluation and control can be to use the COBIT framework 5. The focus of the framework is expected to answer and clarify the company's IT investment planning. COBIT 5 [7]. Incorporates COBIT 4.1, Val IT 2.0 and IT Risks as well as concepts from the Information Business Model for a detailed framework for effective business governance and IT governance. COBIT 5 [8] adopts the ISO 38500 view of the need for IT governance and IT management and uses the Evaluate, Direct and Monitor model of ISO 38500. Val IT and COBIT 5 are strictly related frameworks one another. Because of the function of the Val IT framework focuses on IT investments, researchers use them to serve as a reference in determining and mapping existing domain processes to COBIT 5.

Proper IT invasibility management should also have the ability to accommodate business unit involvement in every decision maker[9]. The IT Governance Institute[10] Organizations are increasingly recognising the importance of alignment between corporate business strategies and IT services. Revealed that one of the causes of failure in IT investment is the lack of business unit involvement in IT Investments[11]. Good IT investment is one of the cornerstones of IT governance[12]. IT management is needed to manage IT services in the company and synergise the IT services with every non-IT business unit (working unit) activity. Therefore, any improvements made to IT investment management will have an impact on corporate IT governance.This research is:

a) Analyze corporate information technology governance in planning IT investment to interact with corporate business strategy by mapping IT goals using the framework COBIT 5 focusing on EDM APO, BAI and MEA domains.
b) Analyze the planning IT investment in companies and other business units that match the task and responsibility by RACI mapping diagram using COBIT Framework 5 focusing on EDM, APO, BAI and MEA domains.
c) The Top provides suggestions and recommendations to the company from the results of analysis and assessment of current and next level of IT maturity with the achievement of capability level focus on EDM, APO,BAI and MEA domains.

2. Methods

The methodology used in this thesis research is by the COBIT 5 framework focusing on EDM, APO, BAI and MEA domains. This method used as a reference in the assessment of IT investment planning by mapping the company strategy, mapping authority and responsibility of the organisation and provides recommendations for improvement of the current planning by measuring the capability level in the company. In assessing IT investment planning, Base Practices (BP) and Work Product (WP) were used to generate a capability level that focused on EDM, APO, BAI and MEA domains.

2.1. RACI Chart

IT Governance Institute [13] RACI stands for Responsible, Accountable, Consulted, Informed. COBIT 4.1 explains that the RACI chart serves to show the role and responsibility of a function in the organisation ofspecific activity in the IT control objective. Roles and responsibilities are two things that are strictly related to the decision-making process. A decision can make by the parties who do have authority as decision makers. RACI is applied to every activity within the IT control objective to support the success of IT processes across all four domains. The purpose of giving these roles and responsibilities is to clarify the activity, as well as the means to determine the role and other functions of a particular activity. The RACI chart defines what and to whom to delegate, consisting of R (Responsible), A(Accountable), C(Consulted), I(Informed).

IOP Conf. Series: Journal of Physics: Conf. Series **1007** (2018) 012051 doi:10.1088/1742-6596/1007/1/012051

2.2. Capability Level

Capability Level = $(0 * y_0) + (1 * y_1) + ... + (5 * y_5)$ (1)
The equation 1 is formula calculation process for capability level. Description of the one formula is $Y_n(Y_0...Y_5)$ the number of processes at n. Z issome evaluated processes.

3. Experiment and Result

The result and discussion in doing the mapping and assessment of capability level along with its recommendation.

3.1. Mapping of the Val IT Framework with COBIT 5

a) Evaluate, Direct, and Monitor (EDM) consist of
   - EDM01 consists of EDM01.01, EDM01.02
   - EDM02 consists of EDM02.02, EDM02.03
b) Align, Plan, and Organize (APO) consists of
   - APO01 consists of APO01.02, APO01.07
   - APO02 consists of APO02.01
   - APO05 consists of APO05.01, APO05.02, APO05.03, APO05.04, APO05.05
   - APO06 consists of APO06.01, APO06.03
   - APO07 consists of APO07.01, APO07.05
c) Build, Acquire, and Implement (BAI) consists of
   - BAI01 consists of BAI01.02, BAI01.03, BAI01.04, BAI01.05, BAI01.06, BAI01.14
d) Monitor, Evaluate, and Assess (MEA) are from
   - MEA01 consists of MEA01.01, MEA01.02, MEA01.03, MEA01.04, MEA01.05

The mapping of the framework revised with IT VAL then it was continued by doing strategic mapping of the company's business to IT Related Goals. The strategic company is as follows:

a) PT. XYZ, Logistics Company as a Public Company, i.e. the company will sell some of its sharesthe public through Initial Public Offering. Through some considerations, the Board of Directors decides that the IPO conducted through a subsidiary (X1).
b) Repositioning from Postal Company to Trusted Postal Services. (X2)
c) Revitalizing the existing services coupled with two new business lines, namely retail business and property business (X3)
d) Developing the company fromasingle company to multi companies. (X4)

3.2. Mapping IT Related Goals

The process of mapping at this stage is concerned about looking at the relationship between strategic goals with Enterprise Goals that exist in COBIT 5. Then establish a connection between strategic targets with the Enterprise Goals are substantial, it does give a P or a mean Primary, and the relationship of the relationship between the real imperative with Enterprise Goals is not dominant. Results from Enterprise Goals mapping strengthened Financial Indicators, Customers, Internals, Learning& Growth from corporate strategic targets rather powerful on average (P).

3.3. Mapping of Questions and Assessment

The process the questions based on the criteria of the Process Assessment Model (PAM) of COBIT 5 will be designed The COBIT 5 serves to measure the capability level that has been determined by the company with the capability level contained in COBIT 5 and will adjust to the COBIT 5 domain used in this study. The results of the question and assessment mappings are enabled to support alignment and continuity as well as a clear linkage to all domains used in assessing the capability levels. Furthermore, RACI mapping diagram of the organisational structure, roles and responsibilities then matched with nine selected domains.

IOP Conf. Series: Journal of Physics: Conf. Series **1007** (2018) 012051 doi:10.1088/1742-6596/1007/1/012051

3.4. RACI Mapping Diagrams

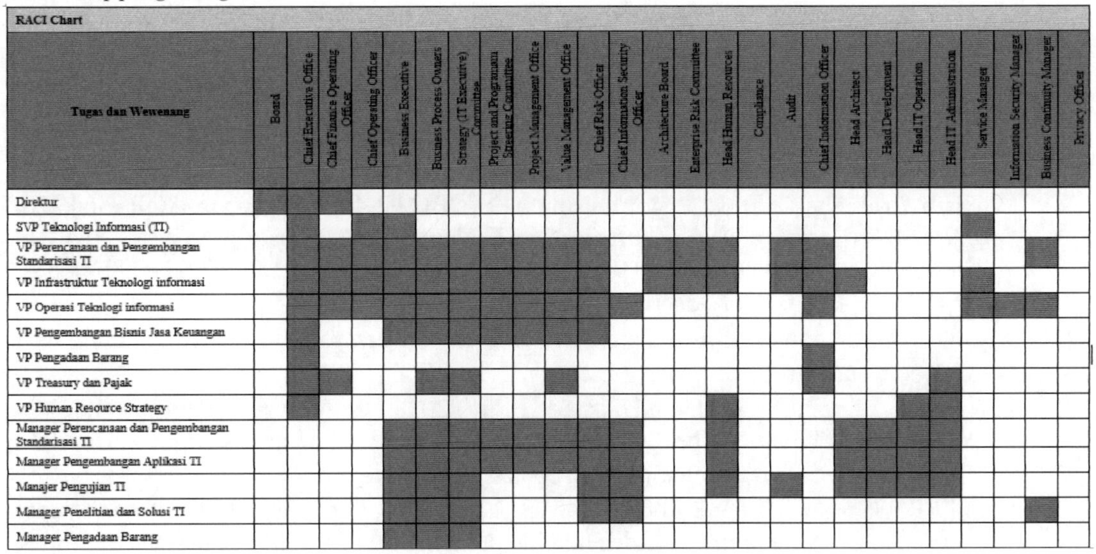

Figure 2. Company business structure.

Then the results of RACI mapping Diagram in figure 2 is also used to assist in the process of questionnaires and interviews, as it becomes more evident and reduces the errors in determining the questions that deserve to be given and gives the assessment of being effective and efficient.

3.5. Capacity Level Capitulation

The stages of this recapitulation stem from the assessment obtained from the questionnaire tabulation data that already has validity and reliability, and focuses on the tested domain then mapped so that it has the value of each question and generates the best practices level and work products as follows infigure 3.

No	Proses COBIT 5	Capability Level Existing	Capability Level Target	Capability Gap
1.	EDM01 – Ensure Governance Framework Setting and Maintenance	4	4	0
2.	EDM02 – Ensure Benefit Delivery	4	4	0
3.	APO01 – Manage the IT Management Framework	3	4	1
4.	AP002 – Manage Strategy	3	4	1
5.	APO05 – Manage Portfolio	3	4	1
6.	APO06 – Manage Budget and Cost	3	4	1
7.	APO07 – Manage Human Resource	3	4	1
8.	BAI01 – Manage Programmes and Projects	3	4	1
9.	MEA01 – Monitor, Evaluate and Assess Performance and Conformance	3	4	1

Figure 3. Rekapitulasi capability gap.

4. Conclusions

Results are IT-related mapping Target dominant to P (Primary). In the RACI diagram results in a close relationship between the organisational structure with the EDM, APO, BAI and MEA domains. Value of Capability Level domain EDM is level 4, level APO 3, Level 3 BAI, MEA level 3 and GAP 1. improvement recommendation that must expose 101 activity from best practice and work product. The results of best practices and work products recommendations should be made periodically to minimise GAP. This study added the business case for IT investment planning more complete.

References

[1] Awangga,R.M., Fathonah, N. S. and Hasanudin, T. I. 2017., Colenak: GPS tracking model for post-stroke rehabilitation program using AES-CBC URL encryption and QR-Code 2017 2nd International conferences on Information Technology, Information Systems and Electrical Engineering (ICITISEE) (IEEE) pp. 255–260, ISBN:978-1-5386-0658-2,URL http://ieeexplore.ieee.org/document/8285506

[2] Awangga,R.M, Pane, S.F, Tunnisa, K. and Suwardi,I.S., 2018., TELKOMNIKA (Telecommunication Computing Electronics and Control)

[3] Pane, S. F., Awangga R.M. and Azhari, B. R. 2018., K Means Clustering and Meanshift Analysis for Grouping the Data of Coal Term in Puslitbang tekMIRA TELKOMNIKA (Telecommunication Computing Electronics and Control)

[4] Mulki,J.P., Jaramillo, F. and Locander,W.B 2006., Effects of ethical climate and supervisory trust on salesperson's job attitudes and intentions to quit Journal of Personal Selling & Sales Management 26 19–26

[5] Drucker,P.F., 1995., Managing the non-profit organisation: Practices and principles Managing the non-profit organisation: Practices and principles (Taylor & Francis)

[6] Miles,R.E, Snow,C.C., Meyer,A.D. and Coleman, H. J. 1978., Organizational strategy, structure, and process Academy of management review 3 546–562

[7] ISACA, 2012., COBIT 5: A Business Framework for the Governance and Management of Enterprise IT (ISACA)

[8] Van, Grembergen. W., and De, Haes. S. 2012., A research journey into enterprise governance of IT, business/IT alignment and value creation, Business Strategy and Applications in Enterprise IT Governance 1–13

[9] Kerzner, H. 2004., Advanced project management: Best practices for implementation (John Wiley & Sons)

[10] Institute, I. G., 2007., COBIT Mapping: Mapping of TOGAF 8. 1 with COBIT 4. 0 (ISACA)

[11] Luftman, J., Kempaiah, R., and Nash, E., 2006., Key issues for IT executives 2005 MIS Quarterly Executive

[12] Pattberg, P., Widerberg, O., Isailovic, M., and Dias, Guerra, F. 2014., Mapping and measuring fragmentation in global governance architectures: a framework for analysis

[13] McSherry, R., and Pearce, P.201., Clinical governance: a guideto implementation for healthcare professionals (John Wiley & Sons)

IOP Conf. Series: Journal of Physics: Conf. Series **1007** (2018) 012052 doi:10.1088/1742-6596/1007/1/012052

Development and implementation of web based infrastructure for problem management at UNPRI

Rico WijayaDewantoro[*], Sumita Wardani, Rudy, Batara Surya Perdana Girsang, Abdi Dharma

Computer Science Departement, Faculty of Technology and Computer Science, Universitas Prima Indonesia, Medan, Indonesia

Email : *ricowijayadewantoro@gmail.com

Abstract. Information technology drastically affects human way of thinking. It has entered every part of human life and also became one of the most significant contributors to make human life more manageable. Reporting a problem of facilities and infrastructure in Universitas Prima Indonesia was done manually where the complainant have to meet the responsible person directly and describe how the problem looks like. Then, the responsible person only solve the problem but have no good documentation on it like Five Ws and How. Moreover, the other issue is to avoid a person who is mischievous for giving false reports. In this paper, we applied a set of procedures called Universitas Prima Indonesia Problem Management System (UNPRI-PMS) which also integrated with academic information system. Implemetation of UNPRI-PMS affects all of the problems about facilities and infrastructure at Universitas Prima Indonesia can be solved more efficient, structured, and accurate.

1. Introduction

Universitas Prima Indonesia is one of the private universities in Medan, North Sumatera with complete infrastructures and also located in four different places [1]. Reporting and handling an infrastructure problem manually using maintenance form has its own limitation. The poor service delivery due to paper-based is the main problem as required an extensive time to review the collected data [2]. It could because of mistaken problem handling, problem recorded unsystematically, and difficult to recognize an information in form because the data is written by human [3]. Management of an operational in people, materials, infrastructure and technology is named goods and services [4]. In order to make maintenance management effectively, there needs to be an integrated maintenance management process from the reporting stage to the maintenance execution for infrastructure [2]. The maintenance management processes started from complaints by staffs that brought them to the office counter to report defects. Then the reports are received by maintenance management Unit for identification and are recorded in database for immediate action and reference. Lastly, the field coordinator instructs electrical technician and mechanical technician for the maintenance execution of facility [5].

In this research, we applied a set of procedures called Universitas Prima Indonesia Problem Management System (UNPRI-PMS) which also integrated with academic information system. With the integration, students or staffs are able to report any issues about facilities and infrastructure. All of the progression will be documented well and also prevented from false reports. Implementation of

IOP Conf. Series: Journal of Physics: Conf. Series **1007** (2018) 012052 doi:10.1088/1742-6596/1007/1/012052

UNPRI-PMS affects all of the problems about facilities and infrastructure at Universitas Prima Indonesia can be solved more efficient, structures, and accurate.

2. Method

2.1 State Transition Diagram
Unified Modelling Language (UML) is a diagram that describe an object class with transition from one state to another state and visualize the design of a system (Hendri, 2007:90). State Transition Diagram is created for dynamic objects, so in each class could have more than one state diagram[6,7]. The transition from one state to another one can be discovered by the construct of the 'machine' and this merely effectuation that the system can be in a finite quantity of assorted states. Whatever the system and where we get a various output, it depends on what has happened ahead, this is a finite state system. The exemplary can be as elaborate or as abstract as it need to be is one of the asset of the state transition proficiency. Where a portion of the system is more than all important a greater depth of appendage can be modeled [13].

2.2 Web Based Applications
Web-based applications have high reliability, usability, more secured, takes a shorter time to market and shorter product life cycles, and also require continuous maintenance [8,11]. Many frameworks for web application development have been introduced and require a lot of resources, includes developers and tools [9,11]. In small software companies, they involved 10 to 50 employees generally for working on a project [10,11].

2.3 Integrated Information System Based on Web Services
Web Service technologies is considered to be the best solution of the problem of software complexity and reusability. Generally, web services are modular applications or functions, which are independent and self-describing, that can be discovered and called across the Internet or an enterprise intranet. Web services are based on open standards such as WSDL (Web Services Definition Language) for description, SOAP (Simple Object Access Protocol) for communication and UDDI (Universal Description, Discovery, and Integration) for register and discover services. Service choreography and orchestration is the next stage in developing and extending web services paradigm and addresses the need of composing several services in some business logic in order to achieve more complex and meaningful processes. The most common technologies for services composition appear to be BPEL and XPDL[12].

2.4 Workflow
To achieve a goal, workflow is needed to automate sets of procedure which had been initiate to completion [15,16]. Workflow Management Coalition (WfMC) stated that automation is a business process neither whole nor partial of documents, information, or task which is flow from one to another based on the initial behaviour. There are some benefits that can be gained from implementation of workflow management:
1. Reduced operating cost
2. Improved productivity
3. Faster processing times

By using the concept of the workflow management system, all of the initial procedures can be translated into a workflow form[14].

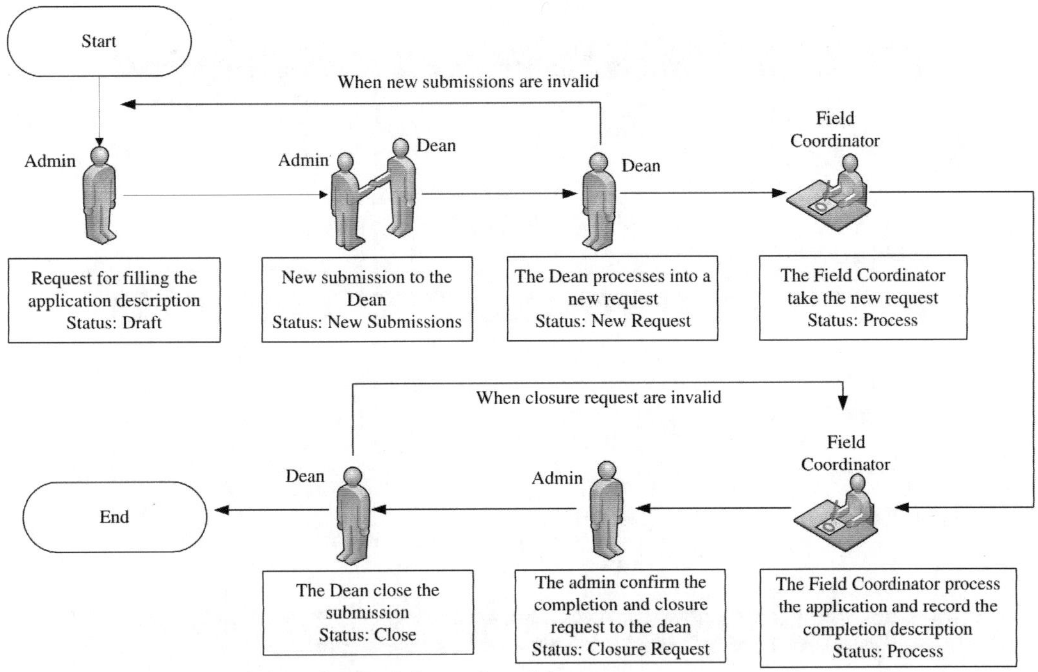

Figure 1. Workflow of problem management system.

3. Design

For a submission of problem, firstly, a staff have to write a description and save it as draft. Secondly, the staff update the status of a problem from draft to a new submission. In order to update the status, the staff have to tell the system that who the submission will be known (in this case, the problem have to be known by dean). Once the problem confirmed and validated by the dean, the status have to update from new submission to new request. If the problem is declined, it will be returned to the staff who submit that problem, so the status will be back to draft. Nevertheless, the problem will be shown in field coordinator page and to be responsible to follow up the problem. The field coordinator have to update the status from new submission to in progression. In any update of the problem, the field coordinator have to tell the staff who submit that problem by the system. Once the problem solved, the staff have to update the status from in progression to closing request. Lastly, the dean will recheck the progression of the problem and also update the status from closing request to closed once the solution confirmed. If the solution declined, the problem will be return to the field coordinator to resolved.

Figure 2. Login session.

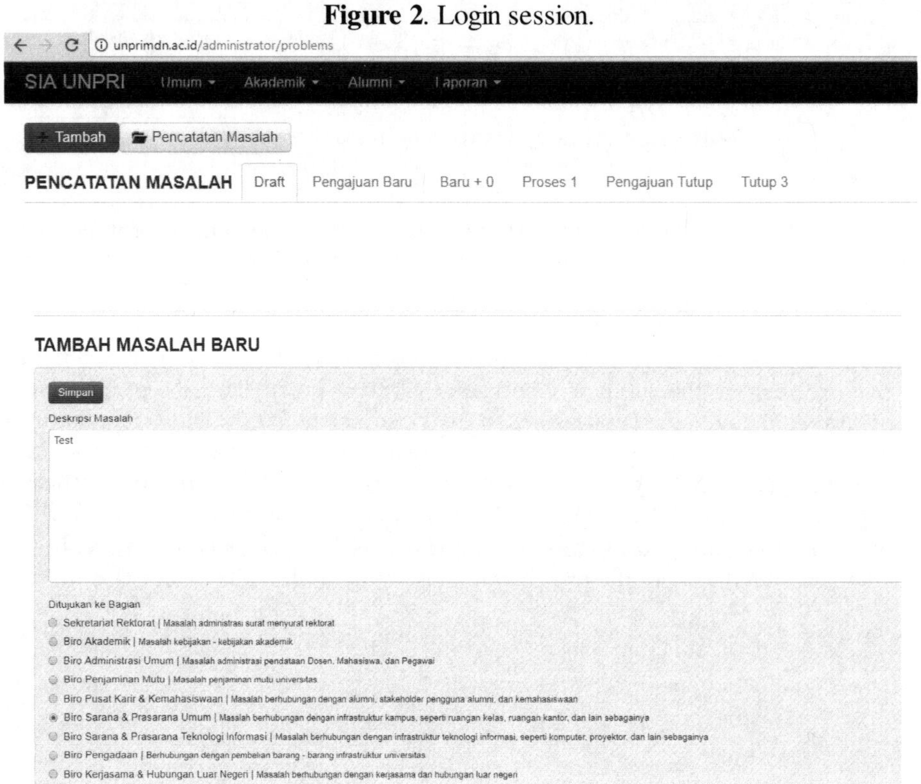

Figure 3. Request for filling the application description. status: draft.

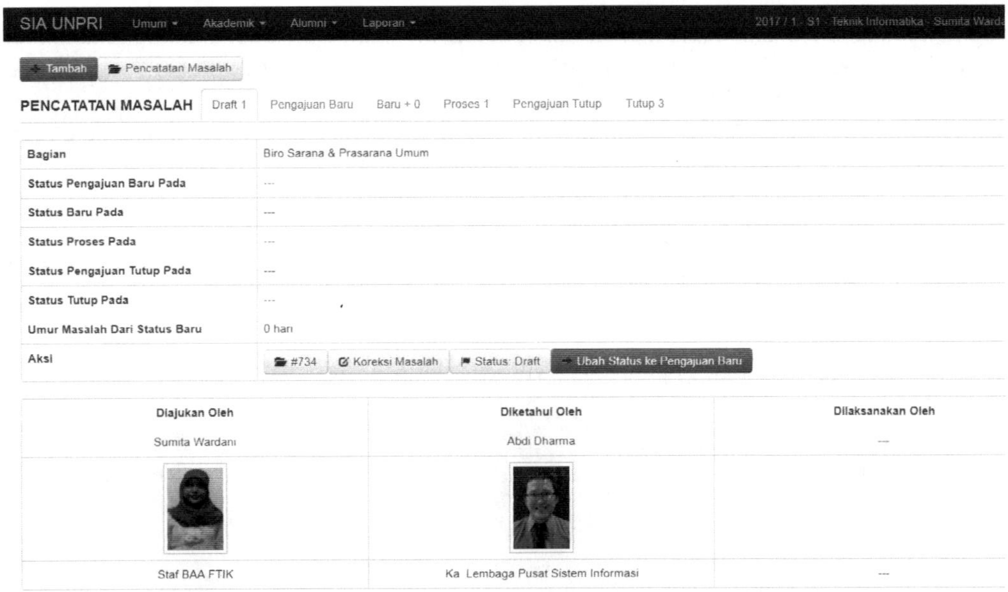

Figure 4. New submission to the dean. status: new submissions.

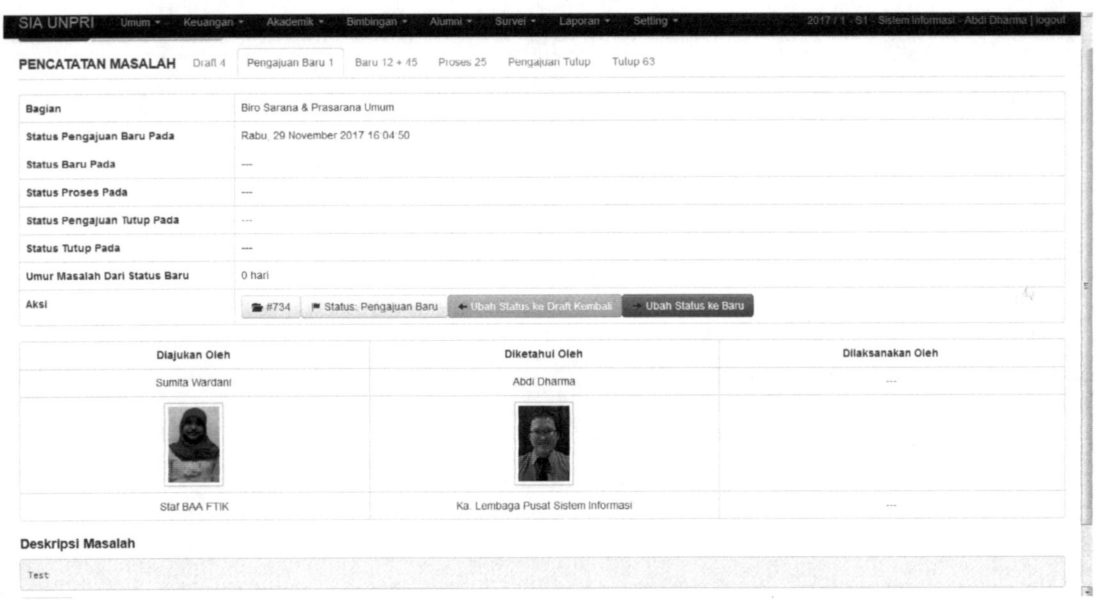

Figure 5. The dean processes into a new request. status: new request.

IOP Conf. Series: Journal of Physics: Conf. Series **1007** (2018) 012052 doi:10.1088/1742-6596/1007/1/012052

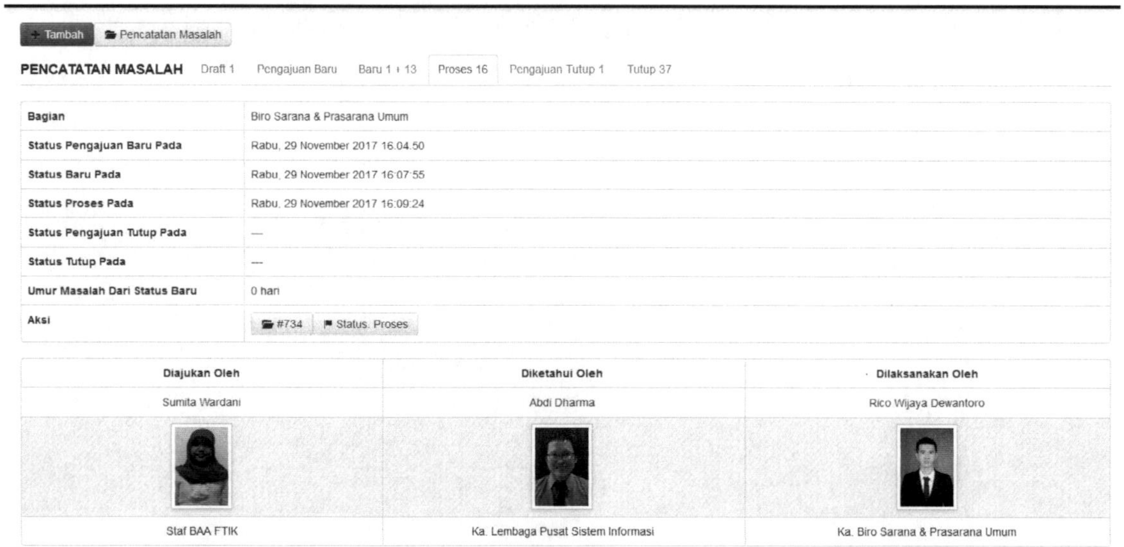

Figure 6. The field coordinator take the new request status: process.

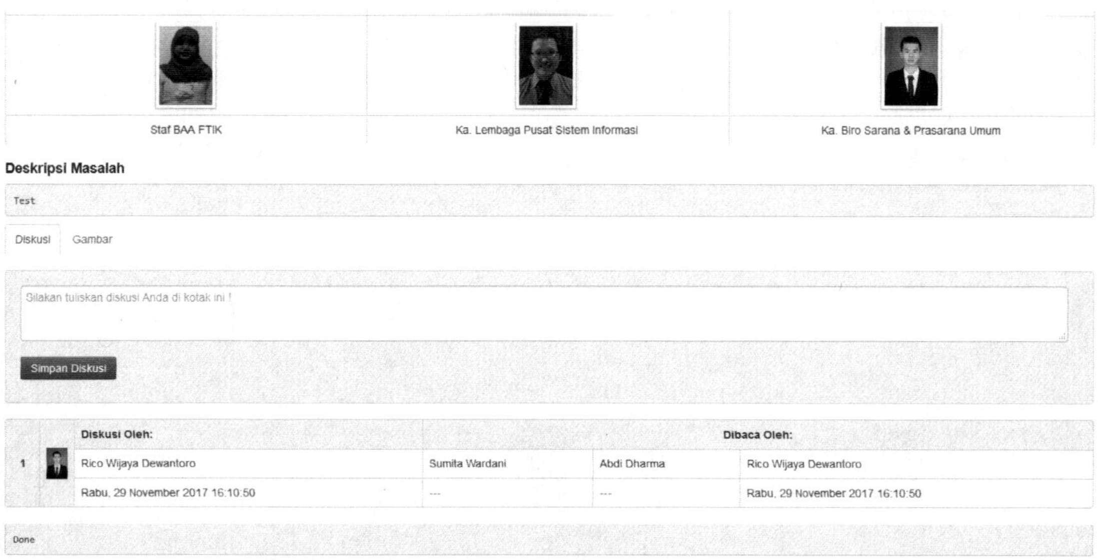

Figure 7. The Field Coordinator process the application and record the completion description. status: process.

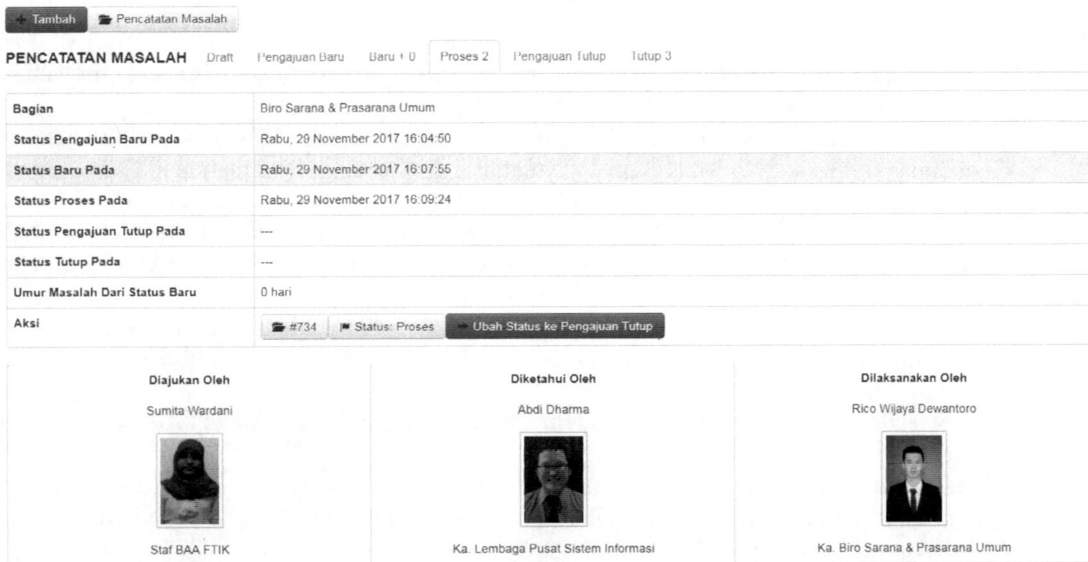

Figure 8. The admin confirm the completion and closure request to the dean. status: closure request..

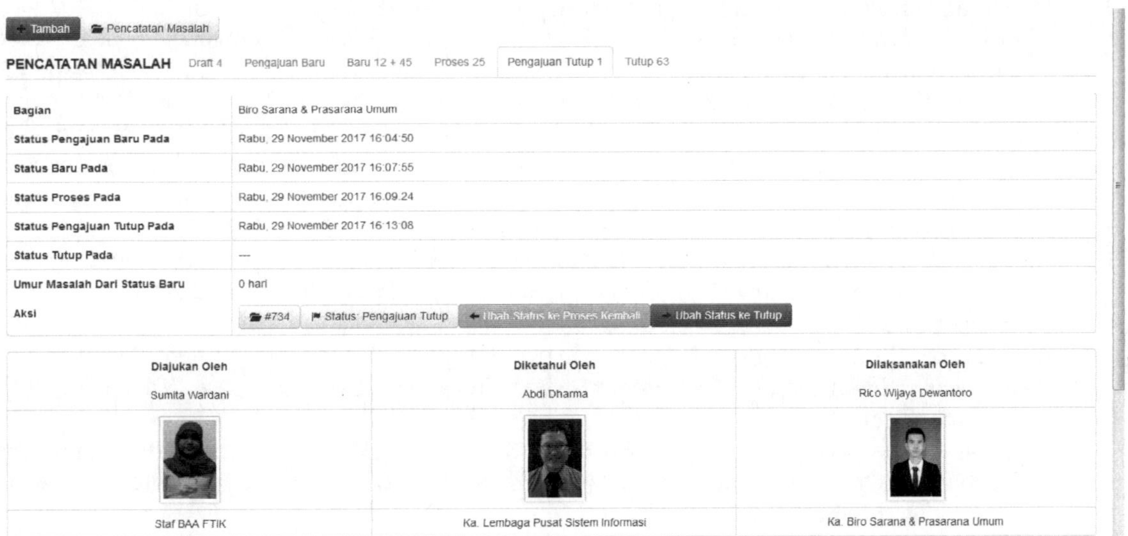

Figure 9. The Dean close the submission. status: close.

4. Result and Discussion

UNPRI-PMS application is designed to fulfill the infrastructure's services in Universitas Prima Indonesia. The application is an online system, so it is easy to access neither anytime nor anywhere with only internet access. Staff can log in to the system with a link http://unprimdn.ac.id/administrator/login by a browser. Implementation of UNPRI-PMS in Universitas Prima Indonesia is a proof that technology able to improve quality of services. In conclusion, with the implementation of UNPRI-PMS affects all of the problems about facilities and infrastructure at Universitas Prima Indonesia can be solved more efficient, structured, and accurate.

Reference

[1] A. Dharma, Mardi Turnip, Siti Aisyah, 2016, Infrastruktur Sistem Informasi Murah Berbasis Cloud Dengan Saas Di Universitas Prima Indonesia, Prosiding Seminar Nasional Inovasi dan Teknologi Informasi (SNITI-3), ISSN: 2548-4540

[2] Z. A. Ismail, Narimah Kasim, 2012, Improving Maintenance Management Practices on Conventional Method at Malaysian Polytechnic, Proceedings International Conference of Technology Management, Business and Entrepreneurship 2012 (ICTMBE2012)

[3] Lazim M. H. M., Samad N. A. A. (2011). Masalah Melaporkan Kerosakan Komputer di Jabatan Teknologi Maklumat dan Komunikasi, Politeknik Tuanku Syed Sirajuddin. Politeknik Tuanku Syed Sirajuddin Digest (PTSS Digest)

[4] N. E Myeda, Syahrul Nizam Kamaruzzaman, Michael Pitt, 2011, Measuring the performance of office buildings maintenance management in Malaysia, Journal of Facilities Management Vol. 9 No. 3, 1472-5967

[5] Hassan S. (2010). Developing Online Help Desk for Politeknik Tuanku Sultanah Bahiyah. Master's Thesis.Universiti Utara Malaysia

[6] Y. Pradifta, Terttiaavini, Suryati, 2015, Sistem Infomasi Seleksi Berkas Lamaran Calon Guru Secara Online Pada Sma Lti Indo Global Mandiri Palembang, JURNAL INFORMATIKA GLOBAL VOLUME 6 No.1, ISSN ONLINE : 2477-3786

[7] Rodríguez, Daniel, Rachel Harrison, and Manoranjan Satpathy. "A generic model and tool support for assessing and improving Web processes." In Software Metrics, 2002. Proceedings. Eighth IEEE Symposium on, pp. 141-151. IEEE, 200

[8] Ahmed E. Hassan and Richard C. Holt "Migrating Web Frameworks Using Water Transformations", Proceedings of the 27th Annual International Computer Software and Applications Conference (COMPSAC"03) 0730-3157/03 2003 IEEE

[9] Laporte, Claude Y., A. Renault, J. M. Desharnais, N. Habra, M. Abou El Fattah, and J. C. Bamba. "Initiating software process improvement in small enterprises: Experiment with micro-evaluation framework." In SWDC-REK, International Conference on Software Development, University of Iceland, Reykjavik, Iceland, pp. 153-163. 2005

[10] N. Uikey, Ugrasen Suman, 2015, A Lifecycle Model for Web-based Application Development: Incorporating Agile and Plan-driven Methodology, International Journal of Computer Applications (0975 – 8887) Volume 117 – No. 19

[11] K. Trichkov, Elisaveta Trichkova, Elena Ivanova, Integrated Information System Based on Web Services, International Conference on Computer Systems and Technologies - CompSysTech'07

[12] D. Jagli, Sunita Mahajan, Akanksha Gupta, Sheetal Waghmar, 2015, A New Innovation Technique Of State Transition Testing Used For Dbt, Informatics Engineering, an International Journal (IEIJ), Vol.3, No

[13] H. Wiranata, Workflow Berbasis Notifikasi Untuk Platform Elearning Guna Mendukung Pembelajaran, Explore – Jurnal Sistem Informasi dan Telematika ISSN 2087-2062

[14] Talaway, I. P. "Adaptive Workflow Management System Untuk Menangani Perubahan Pada Struktur Organisasi". Universitas Indonesia, Fakultas Ilmu Komputer, Jakarta (2004)

[15] H. Rizkyanto, Sholiq, Hanim Maria Astuti, 2012, Pembuatan Perangkat Lunak Untuk Workflow Pengelolaan Surat Dinas Bagian Surat Keluar di Pemerintah Kabupaten Buton Utara, JURNAL TEKNIK ITS Vol. 1,ISSN: 2301-9271

Acknowledgments

This research was supported by the Computer Science Departement, Faculty of Technology and Computer Science, Universitas Prima Indonesia (UNPRI) Medan.

Biomass analysis at palm oil factory as an electric power plant

Yusniati*, Luthfi Parinduri, Oris Krianto Sulaiman

Faculty of Engineering, Universitas Islam Sumatera Utara, Indonesia

E-mail: *yusniati@ft.uisu.ac.id, luthfip@yahoo.co.id, oris.ks@ft.uisu.ac.id

Abstract. Biomassa found in palm oil mill industryis a by-product such as palm shell, fiber, empty fruit bunches and pome. The material can be used as an alternative fuel for fossil fuel. On PTPN IVpalm oil millDolokSinumbah with a capacity of 30 tons tbs/hour of palm fruit fiber and palm shells has been utilized as boiler fuel to produce steam to supplyboilers power plant. With this utilization, the use of generators that using fossil fuel can be reduced, this would provide added value for the company. From the analysis, the fiber and shell materials were sufficient to supply 18 tons/hoursteam for the boiler. Shell material even excess as much as 441,5 tons per month. By utilizing the 2 types of biomass that is available alone, the electricity needs of the factory of 734 Kwh can be met. While other materials such as empty bunches and pome can be utilized to increase the added value and profitability for the palm oil mill.

1. Introduction

Biomass is an organic material made from plants and animals. Biomass contains stored energy derived from the sun. Plants absorb solar energy in a process called photosynthesis. Chemical energy in plants will be passed on to animals and people who eat them[2]. Indonesia has great potential to utilize biomass from palm oil by-products as an alternative fuel source. Indonesia's palm oil is one of the fastest growing commodities. Increasing the area of oil palm plantations has encouraged the growth of processing industries, including palm oil mills that produce crued palm oil (CPO).Palm oil mill is an industry loaded with processing residues. Palm oil mill only produces 25-30% of main products in the form of 20-23% CPO and 5-7% of palm kernel. While the remaining 70-75% is waste that can be classified into three groups namely liquid waste, solid waste, and gaswaste. The amount of solid waste produced by the palm oil processing mill ranges from 40 - 41% of every ton of processed palm.[3]Palm oil mill waste is actually a component of pollutants, but it can be usedas a source of electrical energy.

2. Palm Oil Mill Electricity Needs

The processing of palm oil into CPO is through several stages that require the consumption of electrical energy. The greater the production capacity, the complexity of processes and automation, the consumption of electrical energy in need is higher. The general parameters of power consumption in the palm oil factory are 17-19 Kwh/ton TBS. Ideally palm oil mills are able to independently meet their energy needs. Fiber waste and palm shells can be used for boiler fuel as a steam generator used for turbine power generation, as well as a steam source for boiling and processing.

The energy source installed at the palm oil mills with a capacity of 30 tons/ hour is 2 (two) generators with power 400/500(Kw/KVA), 1 (one) 200 kW genset and 2 (two) steam turbine

IOP Conf. Series: Journal of Physics: Conf. Series **1007** (2018) 012053 doi:10.1088/1742-6596/1007/1/012053

generators each with a power of 800 kW and 742 Kw/816 Hp that can operate alternately or together.[1] A 200 kW power generating set is operated to supply domestic and lighting the needs when the plant is in an inactive state and the turbine is not working yet. [5]Generator with 2 x 400 kW capacity is operated for ignition and the factory's first process until the factory produces fiber and shell for boiler fuel and the boiler capable of producing steam with capacity expected to drive steam turbine to produce electrical energy continuously.

Turbine can operate normally if steam pressure is around 18-21 bar. If the boiler's working pressure shows a downward trend of up to 15 bar then the turbine can not be burdened for the factory process and then there will be trip, so to keep the process does not stop abruptly, the engine room operator immediately activate the 400 kw generator to be in sync with turbine. [4] If this situation often occurs the consequence is the increase in operational costs due to the use of diesel fuel and the increase of bollieropratordifficultybecause he must immediately shoveling fuel into the boiler furnace to increase combustion heat and increase the steam pressure that should be enough in the supply of fuel feeding conveyor.

3. Electrical Energy Consumption

To know the characteristics and usage of the electrical load, can be read with measuring device that mounted at engine room panel in the form of kW-meter and amperemeter. While the electrical energy used is measured through the kWh-meter contained in each of the plant's panels. Loads will fluctuate and adjust the power requirements of the engine or electricity used by each unit. The use of electric power for processing is more dominant by 77.62%. Domestic loads came in second place at 16.75%. While other expenses such as head office, mill office, Workshop, and street lighting have small values ranging from 0.5 to 3%. So the use for this load does not have a big effect on the power that is borne against the plant [6].

The domestic electrical load is large enough to contribute to the usage of electrical power. The use of electrical power from this domestic load is borne by the palm oil processing indusry so that the calculation of the consumption of electrical energy against the Kwh/ton TBS will also be affected.

In actual conditions for domestic loads, the average electricity usage is recorded in an average at 17.30-21.00. This happens because the time is a break time and most people tend to use electricity to turn on the house lights, watch television or other devices that require electricity. As for the processing in the factory operating conditions remain stable. The difference in the electrical power at the factory is used for lighting loads. Safety on domestic panels is used to meet the domestic load. When one power distribution line is done to the office and housing, the domestic auto panel should not be turned off.

4. Research Metodology

This research was done like this flow chart below:

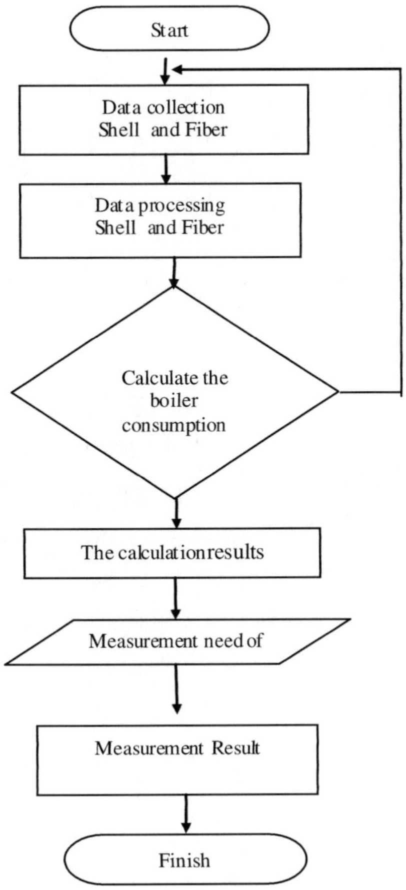

Figure 1. Flow chart of research path.

5. Biomass Utilization

The amount of biomass that is by-product of palm oil mill can be seen through Table 2 below:

Table 1. Electricity consumption of palm oil mill 30 Ton TBS/Hour. **

No.	STATION	INSTALLED		OPERATING		Df (%) *)
		(Kw)	(A)	(A)	(Kw)	
1.	Reception & Sterilizer	147	279	175	92	63
2.	Threshing	149	283	88	46	31
3.	Pressing	240	456	200	105	44
4.	Clarification	171	325	30	16	9
5.	Oil Storage	23	44	12	6	27
6.	Depericarcarper& Kernel	281	534	280	147	52
7.	Boiler Control	230	437	320	168	73
8.	WTP	193	367	63	33	17
9.	Boiler Demint	76	144	20	11	14
10.	Effluent Treatment	60	114	45	24	31
11.	Factory Lighting	75	142	50	26	35
12.	Domestic Lighting	50	95	40	21	42
13.	Street Lighting	53	100	55	29	54
	Total	**1748**			**734**	**42**

*) Df = Demad Factor

**) Source :Factory

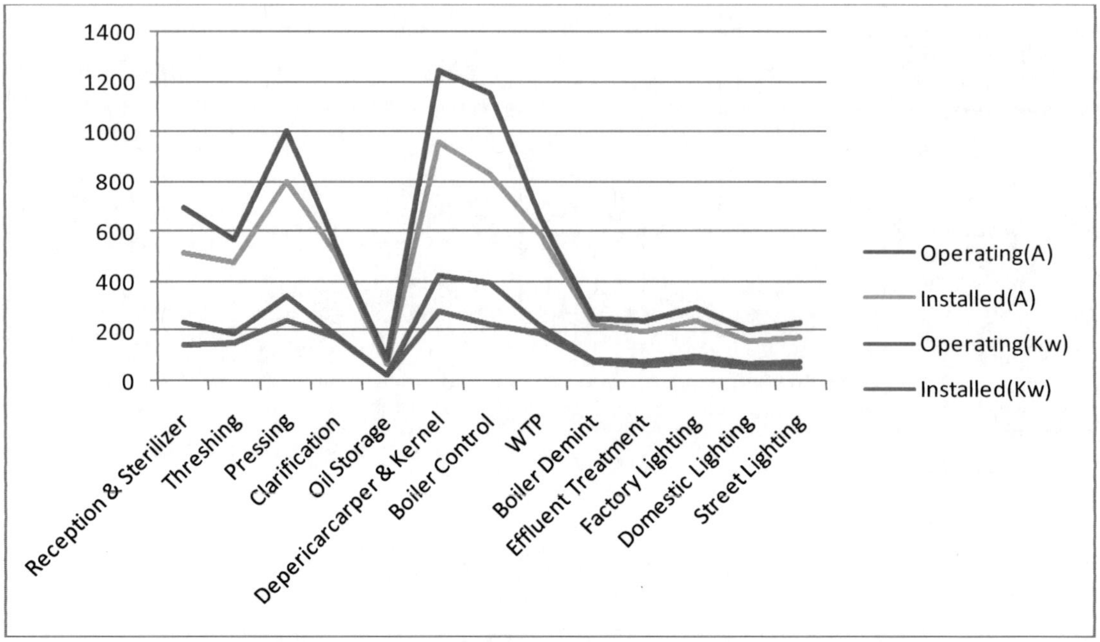

Figure 2. Electricity consumption of palm oil mill 30 Ton TBS/Hours.

Table 2. Biomass of palm oil mill.

No.	Biomass	Form	Sum*)	Calori (Kcal)**)
1.	JanjangKosong	Solid	22 – 23%	4492/kg
2.	Serat (Fiber)	Solid	12 – 14%	2637 – 4554 /kg
3.	Cangkang (Shell)	Solid	6- 8%	4105 – 4802/kg
4.	POME	Liquid	2 ton	4695 – 8569 /m3

*) Percentage of TBS processed
**) 1 Kcal = 4.187 Joule = 1,163 wh
***) Ref.No[3]

IOP Conf. Series: Journal of Physics: Conf. Series **1007** (2018) 012053 doi:10.1088/1742-6596/1007/1/012053

In Dolok Sinumbah palm oil mill this is fit the biomass boiler specification materials used as fuel is fiber and shell.[5]

Table 3. Boiler Specification used by palm oil mill.

No.	Description	Simbolic	Spesifikasi
1.	Steam Capacity	(Q)	20.000kg/h
2.	Steam Temperature	(tu)	280° C S.H
3.	Steam Peassure	(P)	20 kg/cm²
4.	Bait water temperature	(ta)	90^0 C
5.	η Kettle= 73%	η	73%
6.	Fuel	Fibers	75%
		Shells	25%

The usage of steam for palm oil processing with Triple Peak System on Sterilizer = 0.6 tonnes of steam/ton of TBS, so that Total vapor required = 30 tons of TBS x 0.6 tonnes of steam/ton TBS = 18 tonnes of steam/hour.

The amount of fuel available on the palm oil mill with a capacity of 30 tonnes of TBS/hour is generated /hour each: [8]

- Fiber = 14% x 30 ton = 4,2 ton
- Shell = 8% x 30 ton= 2,4 ton

Heating value calculation for fiber material (N.O):
Composition
Water = 39,8% x 4200 Kg = 1670 gg
NOS = 55,6% x 4200 kg = 2340 kg
OIL = 4,65% x 4200 kg= 190 kg

Heating Value
NOS = 3850 Kcal/kg
OIL = 8800 Kcal/kg
Heat Evaporation Water = 600 Kcal/kg
NO fiber = [(2340 x 3850) + (190 x 8800) – (600 x 1670)]/ 4200 = 2305 Kcal/kg

Heating value calculation for shell materials (N.O):
Composition
Water = 23,5% x 2400 kg = 564 kg
NOS = 75,9% x 2400 kg = 1821 kg
OIL = 0,6% x 2400 kg = 15 kg
Heating Value NOS = 4700 Kcal/kg
OIL = 8800 Kcal/kg
Heat Evaporation water = 600 Kcal/kg

NO shell = [(1821 x 4700) + (15 x 8800) – (600 x 564)]/ 2400 = 3480 Kcal/kg
Steam production of all fiber fuels is 4200 kg
ηKettle = [Q (**Δ**Entalphy)] / Gbb x NO
0,73 = [Q (**710,9 – 90,03**)] / 4200 x 2305
Q = [(4200 x 2305) x 0,73] / 620,87
Q = 11383 kg steam/hour

(note on P = 20 kg/cm2 tu 280^{0}C entalphy = 710,9 and tu 90^{0}C entalphy = 90.03)

Shell fuel is needed to supply 18 tons of steam/hour

η Kettle = [Q (**Δ**Entalphy)] / Gbb x NO
0,73 = [Q (**710,9 – 90,03**)] / Gbb x 3480
Gbb = [(18.000 – 11.383) x 620,87] / (0,73 x 3480)

IOP Conf. Series: Journal of Physics: Conf. Series **1007** (2018) 012053 doi:10.1088/1742-6596/1007/1/012053

Gbb = (4.108.297) / (2,540,4)
Gbb = 1.517 kg

To meet the steam requirement of 18000 kg/hour, the available crane fuels are still excessive or left as much as = 2,400 - 1,517 = 883 kg or as many as 441.5 tons/month (20 hours work/day, 25 days per month).

6. Conclusion

The by-product of the palm oil processing mill is a biomass that can be used as an alternative to fuel oil. In DolokSinumbah palm oil mill with a capacity of 30 tons of TBS/h has used fiber and shells to heat the boiler as a steam generator that used to drive the steam turbines. The electricity generated is used to meet all electricity needs of 734 Kwh.

From the analysis result it still have got excess of shell fuel equal to 883 kg/hour after all the requirement of steam is fulfilled. Thus, this palm oil mill can use the biomass that it produces with a safe enough availability for all its electricity needs, besides there are still potential biomass that can still be utilized such as empty bunches and POME which is liquid waste. Other biomass ingredients can be used to gain added value to the company while increasing profitability. This article is expected to inspire and contribute to all existing palm oil factories in Indonesia so as to suppress the use of national energy in the use of fossil energy that is increasingly scarce, expensive and environmentally unfriendly.

References

[1] IrwanIftadi, ,*KelistrikanIndustri*, CetakanPertama, GrahaIlmu, Yogyakarta. 2015.
[2] Sam Sum Ting, Dr;,*BiomassUtilization: Challenges and Future outlook*, KuliahUmum – JurusanTeknik Kimia – InstitutTeknologi Medan, 2016.
[3] Syukry Othman,*Overview of Palm Kernel Shell*, SelesaKreatif Resources, KuliahUmum – JurusanTeknik Kimia –InstitutTeknologi Medan, 2016.
[4] Tri Watiningsih, dkk.,*PembangkitTenagaListrik*, CetakanPertama, GrahaIlmu, Yogyakarta, 2014.
[5] *Data – data InstalasiPabrik Unit Usaha DolokSinumbah*, PT. Perekebunan Nusantara–IV (Persero), Medan. 2014.
[6] https://ivanemmoy.wordpress.com/2013/11/29/kelistrikan-pabrik-kelapa-sawit/.
[7] Setyardjo, Joko. "*Boiler Uap*" Edisikedua, PT.PradnyaParamita, Jakarta 1989.
[8] BimaDewantara,"*TinjauanTeknikPerubahanKerja Steam Dram di Boiler akibat Blow downpada PLTU unit 3 dan 4 (Studikasus di PT PJB UP Gersik)* ", InstitutTeknologiSepuluh November, Surabaya, 2010.
[9] Abdul Kadir, *"Energi"*Edisikedua, SumberDaya, Inovasi, TenagaListrikdanPotensiEkonomi, 1995.
[10] Karyono SA, MT, Ir, *"Boiler Uap Dan SistemTenagaUap"*PusdiklatMigas 2002.
[11] Abdullah, K. *Biomass Energy Potentials and Utilization in Indonesia.* Laboratory ofEnergy and Agricultural Electrification, Department of Agricultural Engineering, IPB and Indonesian Renewable Energy Society (IRES). Institute Pertanian Bogor, Indonesia ,2004.
[12] Loebis, B. & Tobing, P.L. *Potensi pemanfaatan limbah kelapa sawit*. Buletin Perkebunan. 0:49-56.1989.
[13] Ma, A.N., Choo, Y.M. & Cheah, K.Y. Development of Renewable Energy in Malaysia. Malaysian Palm Oil Board (MPOB). 2003.
[14] Menon, N. R, Rahman, Z. A & Bakar, N. A. Empty Fruit Bunches Evaluation: Mulch in Plantation vs. Fuel for Electricity Generation. Oil Palm Industry Economic Journal, Vol. 3, No. 2, p. 15-20. Malaysian Palm Oil Board. 2003.
[15] Davia, *Biomassa Energi*, data book, Edition 2, Oak Ridge, Tenesse, US 2009.
[16] Mujeebu and Abdullah, Biomassa Based Cogeneration and Trigeneration for effective Heat Recovery and Waste Management, 2009.

IOP Conf. Series: Journal of Physics: Conf. Series **1007** (2018) 012054 doi:10.1088/1742-6596/1007/1/012054

Constant voltage and constant current control implementation for electric vehicles (evs) wireless charger

MarojahanTampubolon, LaskarPamungkas, Yao Ching Hsieh, Huang Jen Chiu

Department of Electronic and Computer EngineeringNational Taiwan University of Science and TechnologyTaipei, Taiwan (ROC)

Email: marojahantampubolon@gmail.com

Abstract. This paper presents the implementation of Constant Voltage (CV) and Constant Current (CC) control for a wireless charger system. A battery charging system needs these control modes to ensure the safety of the battery and the effectiveness of the charging system. Here, the wireless charger system does not employ any post-regulator stage to control the output voltage and output current of the charger. But, it uses a variable frequency control incorporated with a conventional PI control. As a result, the size and the weight of the system are reduced. This paper discusses the brief review of the SS-WPT, control strategy and implementation of the CV and CC control. Experimental hardware with 2kW output power has been performed and tested. The results show that the proposed CV and CC control method works well with the system.

1. Introduction

The concept of Wireless Power Transfer(WPT) had been introduced by Nikola Tesla more than one hundred years ago[1]. Nowadays, we can find various applications of the WPT from a small power such as biomedical applications[2]to a largepower such as Electric Vehicles (EVs) charging system[3-5]. In term of power transfer application, there are two types of WPT technologies. Firstly, inductive coupling based WPT which is commonly called Inductive Power Transfer (IPT). Secondly, capacitive coupling based WPT which is called Capacitive Power Transfer (CPT). WPT system technologies contribute to safety and flexibility due to the cordless system. It is also more durable to the environment condition compared to a non-contactless system because it is not affected by dirt, dust, and chemical [6, 7]. A typical system of wireless charging for EVs, as shown in Fig. 1, consists of an AC-DC stage, inverter, compensator, coils, rectifier, and battery. The AC-DC stage serves either as the power factor corrector or a pre-regulator stage for the WPT system. It should be noted, in some cases, a post-regulator is added to control the output voltage and the output current of the WPT.However, in this paper, instead of adding a post regulator stage, a variable frequency control is employed to regulate the output voltage and current. Since the WPT is used to charge the battery of the EVs, a Constant

IOP Conf. Series: Journal of Physics: Conf. Series **1007** (2018) 012054 doi:10.1088/1742-6596/1007/1/012054

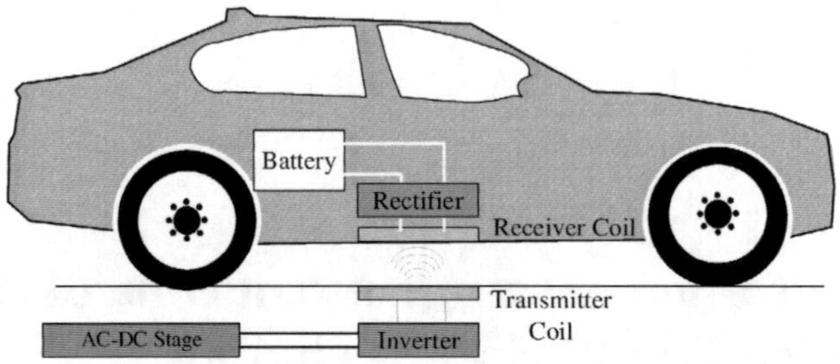

Figure 1. Typical inductive power transfer system for EVs.

Current (CC) and a Constant Voltage (CV) charging mode have to be applied to ensure the safety of the battery. This paper proposes the method to implement the CC and CV control for WPT. In this paper, we will discuss the brief fundamental theory of WPT, CC/CV control method, its realization, and the experimental results.

2. Characteristic of SS-WPT

IPT technologies employee the magnetic field to transfer the power from transmitter coil to the receiver coil. Since the transmitter and receiver coils separated with a relatively large gap, the coupling coefficient of the coupled coils is very low (0.1-0.3). At this condition, a direct power transferring is ineffective and inefficient.Many papers have introduced many types of the compensator to solve this problem. The basic types have been discussed in [8]. Series-series compensator is one of the basic compensator types that consists a capacitor connected in series with the coils in both sides of the WPT as shown in Fig. 2. Inverter stage converts the DC voltage ($V_{in(DC)}$) to a square wave AC voltage ($V_{in(AC)}$) with a magnitude gain $2\sqrt{2}/\pi$. The reversed process occurs in the rectifier stage, at which the magnitude gain of the output voltage (V_o) to the output AC voltage ($V_{o(AC)}$) is $\pi/2\sqrt{2}$. For this reason, the total magnitude of the SS WPT is determined by the compensator circuit as expressed in (1). A detailedanalysisofSS WPThas been discussed in [9].In (1) we can see the relationship of the voltage gain with the hardware parameters of compensator that includes the mutual inductance of the coil (M), self-inductance of transmitter side (L_1) and receiver side (L_2), compensator capacitors (C_1 and C_2), equivalent AC load ($R_{L(AC)}$) and angular frequency ($\omega=2\pi f$). The output power (P_o) of the WPT is linear with the equivalent AC resistance of the system due to the relationship $P=V^2R^{-1}$. The plot of various output powers and mutual inductances are presented in Fig.3. These plots based on these parameter: L_1 = 432μH,L_2 = 442μH,andC_1 = C_2=10 nF. The value of M correspondent with the air gap distance between the transmitter and receiver coils such that M=89.2 μH is a representation of fully aligned coil with 200mm air gap.

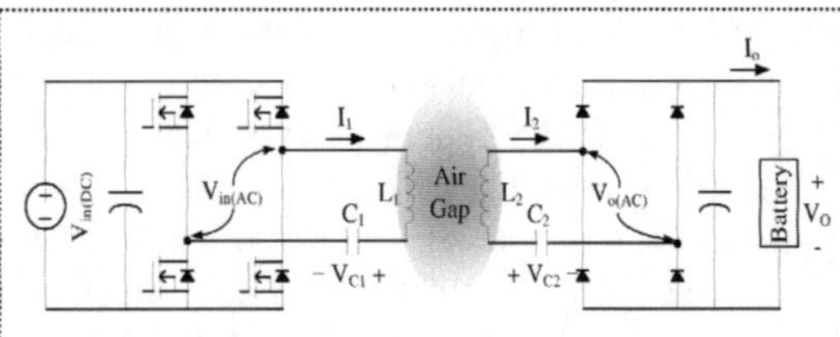

Figure 2. Topology of a Series-series compensated WPT.

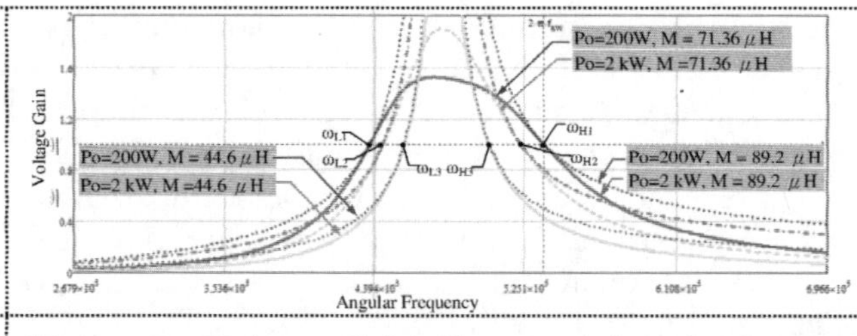

Figure 3. Voltage Gain Characteristic of SS WPT against load and mutual inductance.

IOP Conf. Series: Journal of Physics: Conf. Series **1007** (2018) 012054 doi:10.1088/1742-6596/1007/1/012054

$$G_v = \left|\frac{V_o}{V_{in(\mathrm{DC})}}\right| = j\omega M R_{L(AC)} \left(Z_s \left(j\omega L_1 + \frac{1}{j\omega C_1} \right) + \omega^2 M^2 \right)^{-1} \tag{1}$$

$$Z_s = j\omega L_2 + \left(j\omega C_2 \right)^{-1} + R_{L(AC)} \tag{2}$$

The characteristic of the SS-WPT can be seen in Fig.3. It's clear that the voltage gain of the system depends on the operating frequency, mutual inductance, and load. However, there are conditions where the voltage gain of the system does not change against the variation of loads under the same mutual inductance. These marked frequencies with ω_{L1}, ω_{H1}, ω_{L2}, ω_{H2}, ω_{L3}, and ω_{H3} are the independent load frequencies of the system under various mutual inductances. There are two independent load frequencies for each value of the mutual inductance. The first is on the left-hand side of the curve, and the second is on the right-hand side of the curve. We can see that if the mutual inductance decreases the first independent switching frequency (with L subscript) is decreased, and in contrary whenever the mutual inductance increases the second independent switching frequency (with H subscript) is increased.Selecting frequency at the independent load frequency is preferable because it provides a better efficiency of the system and reducing the effect of the variable load. However, because the voltage is controlled by using the variable frequency, there should be some consideration to applying this type of control. It will be discussed in the next section.

3. Control strategy of SS-WPT with CV and CC controller

3.1. Frequency selection

Practically, the mutual inductance of the coil is not constant at a fixed value. It always changes due to the misalignment and the air gap variation of the coupled coils. It results in a different independent load frequency of the system as shown in Fig 4. Because there are two independent load frequencies for each mutual inductance condition, the designers have to select the operating frequency region of the system. The frequencies ω_{L1}, ω_{L2}, and ω_{L3} are in the capacitive region of the system in which the realization of the Zero Current Switching (ZCS) is achievable. On the other side, ω_{H1}, ω_{H2}, and ω_{H3} are in the inductive region of the system. In this region, the Zero Voltage Switching (ZVS) can be achieved. If the ZVS is preferred rather than the ZCS, the frequency selection must be in the inductive region and vice versa. In this paper, the selected frequency range is in the inductive region. The range of the switching frequency should be selected carefully. Selecting frequency range too far from the resonant frequency generates more reactive power that results in lower efficiency due to the large conduction loss of the system.Fig 4 shows two examples of the frequency range selection. The Range I include the capacitive and the inductive region in one control mode. It cancause control failure due to the inconsistent behaviour of the voltage gain in the different region. Therefore, the frequency range should be limited to only one region for one control mode. For this case, selecting Range II is better compared to selecting Range I. The limitation of the minimum and maximum mutual inductance should be defined firstly to be able to select a more precise range of frequency.This mutual inductance limitation is related to the allowable misalignment and allowable air gap of the system. In this paper, the nominal resonant frequency of the system is selected to be 85 kHz, and the frequency range is 83 kHz to 91 kHz.

3.2. CV and CC controller

Fig.5 shows the typical charging steps of a Li-Ion battery [10]. Assuming the battery was initially at the low-level state of charge at which the battery voltage is lower than V_{min}. In this state, the Equivalent Series Resistance (ESR) of the Li-Ion battery is large enough. Therefore the constant charging current is set to a lower

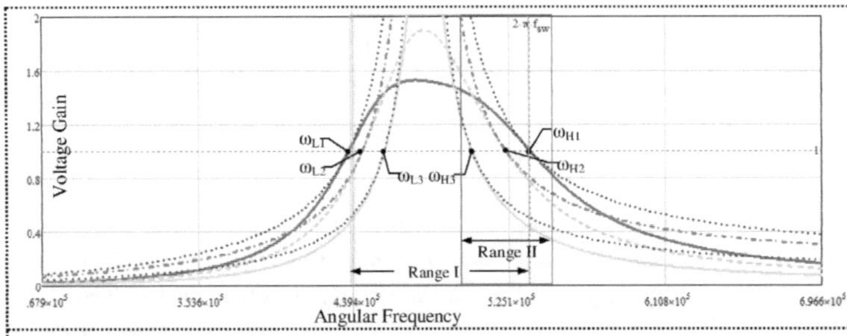

Figure 4. SS-WPT switching frequency selection consideration.

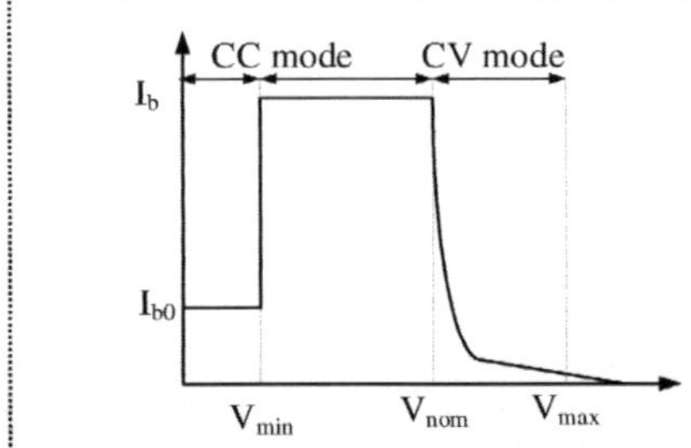

Figure 5. Three charging steps of Li-Ion Battery.

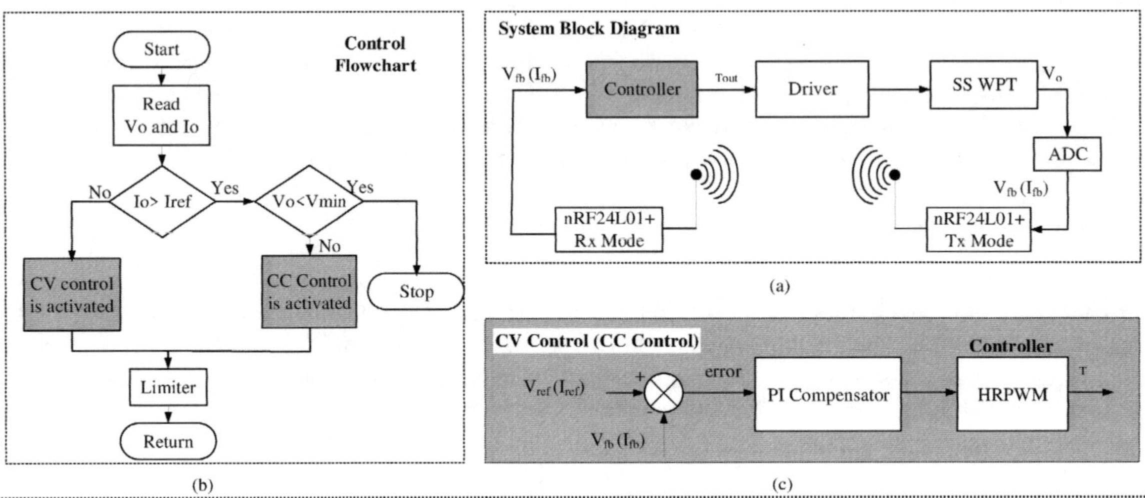

Figure 6. Control block of SS-WPT. (a) System block diagram. (b) Flowchart of CV and CC controller. (c) CV and CC control block.

value I_{bo}. Once the battery voltage reaches V_{min}, a larger charging current (I_b) can be applied until the voltage reaches V_{nom}. The CV mode is activated when the voltage of the battery reaches V_{nom} at which the battery is about to fully recharged. In the CV region, the voltage is almost constant, but the current drops to the smallest value until it stop when the voltage reaches V_{max}.

$$PI_{out}(n) = K_p.e(n) + i(n); \quad i(n) = k_i.e(n) + i(n-1) \tag{3}$$

4. Implementation and Experimental Results

A prototype of SS-WPT with a hardware specification of P_o=3kW has been developed. Both DC input voltage $V_{in\ (DC)}$ and the output DC voltage (V_o) are 400V. The MOSFET type and wireless communication module are FCH072N60F and nRF24L0+ single chip 2.4GHz transceiver respectively. The system has been tested under 2kW output power with CV and CC control. The CC mode is configured to be activated at 5A output current.The SS WPT system has been tested by using an electronic load configured in constant resistance mode. The testing result under the 2kW output power condition is presented in Fig. 7. At this condition, the efficiency of the DC-DC stage achieved 97.9%.The CV and CC mode output voltage and current as shown in Fig. 8 shows that the CV mode is active when the current smaller than the setting current otherwise is CC mode. The transition from CV to CC mode and from CC to CV could perform well. A small spike occurs at the beginning of the CV to CC transition because of the abrupt changing of output current and the controller need a little time to control.

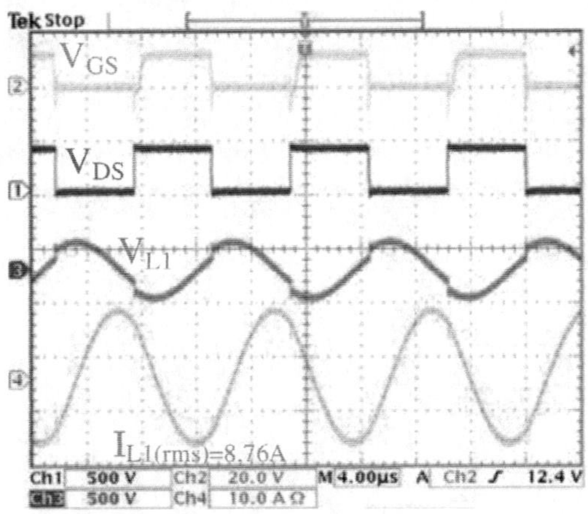

Figure 7. Key-waveforms of SS-WPT.

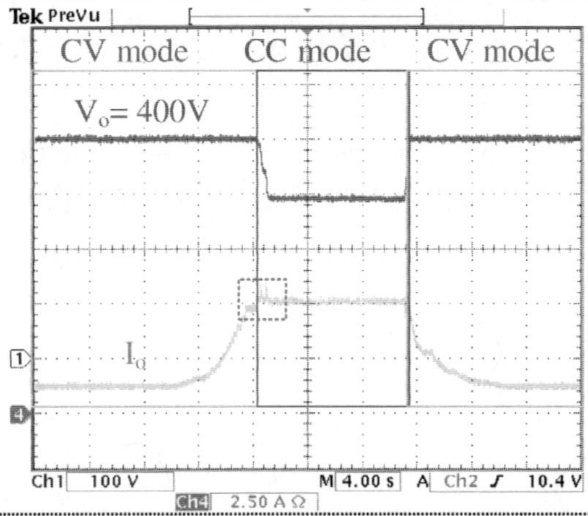

Figure 8. Output voltage and output current at CV and CC operation.

5. Conclusion

A brief analysis and theory of the SS WPT and the implementation of the CV and CC control has been discussed in the previous sections. A hardware validation also has been successfully performed. We can conclude that the proposed technique can be used to implement the controller for wireless charging. Moreover, the hardware efficiency of the DC-DC stage could achieve 97.9% at full load condition with air gap 200mm.

6. Acknowledgment

The authors would like to acknowledge the financial support of the Ministry of Science and Technology of Taiwan through grant number MOST-106-2218-E-011 -011 -MY2.

References

[1] N. Tesla, "Apparatus for transmitting electrical energy," USA, Dec, 1914.

[2] S.-n. Suzuki, M. Ishihara, and Y. Kobayashi, "The Improvement of the Noninvasive Power-Supply System Using Magnetic Coupling for Medical Implants," *IEEE Transactions on Magnetics,* vol. 47, no. 10, pp. 2811-2814, Oct. 2011.

[3] W. Li, H. Zhao, S. Li, J. Deng, T. Kan, and C. C. Mi, "Integrated {LCC} Compensation Topology for Wireless Charger in Electric and Plug-in Electric Vehicles," *IEEE Transactions on Industrial Electronics,* vol. 62, no. 7, pp. 4215-4225, 2015.

[4] U. K. Madawala and D. J. Thrimawithana, "A Bidirectional Inductive Power Interface for Electric Vehicles in V2G Systems," *IEEE Transactions on Industrial Electronics,* vol. 58, no. 10, pp. 4789-4796, 2011.

[5] S. Li and C. C. Mi, "Wireless Power Transfer for Electric Vehicle Applications," *IEEE Journal of Emerging and Selected Topics in Power Electronics,* vol. 3, no. 1, pp. 4-17, 2015.

[6] J. T. Boys, G. A. Covic, and A. W. Green, "Stability and control of inductively coupled power transfer systems," *IEE Proceedings - Electric Power Applications,* vol. 147, no. 1, pp. 37-43, 2000.

[7] G. A. Covic and J. T. Boys, "Modern Trends in Inductive Power Transfer for Transportation Applications," *IEEE Journal of Emerging and Selected Topics in Power Electronics,* vol. 1, no. 1, pp. 28-41, 2013.

IOP Conf. Series: Journal of Physics: Conf. Series **1007** (2018) 012054 doi:10.1088/1742-6596/1007/1/012054

[8] W. Chwei-Sen, O. H. Stielau, and G. A. Covic, "Design considerations for a contactless electric vehicle battery charger," *IEEE Transactions on Industrial Electronics,* vol. 52, no. 5, pp. 1308-1314, 2005.

[9] Y. C. Hsieh, Z. R. Lin, M. C. Chen, H. C. Hsieh, Y. C. Liu, and H. J. Chiu, "High-Efficiency Wireless Power Transfer System for Electric Vehicle Applications," *IEEE Transactions on Circuits and Systems II: Express Briefs,* vol. PP, no. 99, pp. 1-1, 2016.

[10] J.-J. Chen, F.-C. Yang, C.-C. Lai, Y.-S. Hwang, and R.-G. Lee, "A High-Efficiency Multimode Li–Ion Battery Charger With Variable Current Source and Controlling Previous-Stage Supply Voltage," *IEEE TRANSACTIONS ON INDUSTRIAL ELECTRONICS,* vol. 56, no. 7, pp. 2469-2478, JULY 2009.

IOP Conf. Series: Journal of Physics: Conf. Series **1007** (2018) 012055 doi:10.1088/1742-6596/1007/1/012055

A low cost indoor localization system for mobile robot experimental setup

S Adinandra*, A Syarif

Department of Electrical Engineering, Faculty of Industrial Technology, Universitas Islam Indonesia, Indonesia

E-mail: *s.adinandra@uii.ac.id

Abstract. Indoor localization becomes one of the most important part in mobile robot system. One fundamental requirement is to provide an easy-to-use and practical localization system for real-time experiments. In this paper we propose a combination of a recent open source virtual reality (VR) tools, a simple MATLAB code and a low cost USB webcam as an indoor mobile robot localization system. Using the VR tools as a server and MATLAB as a client, the proposed solution can cover up to 1.6 [m] × 3.2 [m] with the measurement position accuracy up to 1.2 [cm]. The system is insensitive to light, easy to move and can be quickly set up. A series of successful real-time experiments with three different mobile robot types has been conducted.

1. Introduction

Locomotion and positioning are two important parameters in mobile robot systems. Typical locomotion for mobile robot are differential drive, synchronous drive, tricycle drive, car type, tank type and walking type, see Figure 1 for examples. Each robot has its own potential and uniqueness. For example, the differential drive and car type are well known for their simplicities and functionalities. Both are widely used in application despite the lack of sideways movements. On the other hand, the synchronous drive walking type and the tank type are more flexible to environment changes. However, these types are typically designed to move slow to cover the overall working space and different terrains. (1)

In order to locate themselves in the working area, robots are equipped with localization systems. In mobile robot systems, localization is divided into two methods: indoor and outdoor localizations. For outdoor localization, global positioning system (GPS) is the main option that currently takes over almost the entire outdoor system. This phenomenon occurs due to the increasing accuracy of GPS combined with the easiness to find one as well as the low cost of the system. The DARPA challenge is one famous example of where mobile robots (car) autonomously move using GPS as their main navigation system, see. http://archive.darpa.mil/grandchallenge/.

However, GPS cannot be used indoor. Positioning of mobile robots is done by means of sensors attached to the robots or environment. The first method is sometimes called as relative indoor positioning system. A set of sensor is attached the robot or group of robots so that position and orientation of a robot relative to other robots or environment are known. Example are the use of ultrasound sensor, infrared, or on board camera. The advantage of relative method is robustness to environment changes. Since sensors are embedded in robot system, changing environment can be easily compensated by the robot. However, it requires a non-trivial algorithm to proses the environment information so that the position of the robot is obtained. Example of this method can be found in (2), (3), (4), (5), (6), (7), (8), and (9).

The second method, where sensors are attached to the environment, are sometimes called absolute localization. The "absolute" can be interpreted as the absolute position of the robot obtained within the environment. Since the sensors are attached to the environment, the position of the robot is always in the correct dimension of the environment. Typical solution is to attached camera system that covers the environment. The absolute method is not robust to environment changes but does not requires a nontrivial algorithm since regular image processing method can be used. Examples can be found in (10), (11), (12), and (13).

(a) Differential drive (b) Car type (c) Tank type

Figure 1 . Example of mobile robot locomotion.

The relative indoor localization is suitable for a mature mobile robot system, when a highly accurate localization with non-trivial algorithm. However, price can be another issue. For testing a simple test and experiment of a new locomotion algorithm, the absolute localization is preferable due to simplicity in installation. A moderate accuracy can be easily obtained, less robust to environment changes. Yet, the overhead camera used to capture the robot position typically is a camera that has a wide opening angle and insensitive to light. Thus, it is usually an expensive camera.

This paper proposes an easy-to-use and low cost solution to the absolute indoor localization system. Easy-to-use is interpreted as the quickness to set up the camera system. The low cost means the camera must be a webcam with cost less than 150 USD. The main contributions of this paper are as follow: *i)* an easy to set-open source and low cost indoor localization system, *ii)* a customable client-server for easy scaling up the system in terms of number of robots. This paper provides the low cost version to the solution used in (11), (14), (15), and (16).

The rest of this paper is organized as follow. In preliminaries section, the concept of ARToolkit and how the open source software can be customized are presented. In system design section, the detailed of proposed solution is explained. In experimental results and analysis, the performance of the proposed solution is given. Finally, at the end of this paper conclusions and recommendation are given.

2. Preliminaries

The overhead camera localization system typically consists of two main components: a camera and an image processing system. The image processing system is a system that can read a specific marker. In this work, we modify an open source virtual reality kit that emerges in popularity. The software is chosen because of its low computing power needs. In this section a quick overview of the software architecture and detailed explanation of the overhead camera indoor localization system is given.

2.1. AR Toolkit: an open source virtual reality software

The emerging popularity of Virtual Reality (VR) in gaming, supported by many smartphone, opens up a wide variety of open source VE programming tools. One of them is the ARToolkit (www.artoolkit.org). AR Toolkit is an open source platform that enable the user to create different type of virtual reality application. The website state the following:
ARToolKit 6 is a fast and modern open source tracking and recognition SDK which enables computers to see and understand more in the environment around them. It is being built from the ground up using modern computer vision techniques, up to the minute coding standards and new technologies developed in-house at DAQRI. ARToolKit 6 is being released under a Free and Open Source License that will allow the AR community to use the software in commercial products as well as for research, education and hobbyist development.

The AR Toolkit concept and architecture are shown in Figure 2. ARToolkit read custom marker so that the position, orientation (poses) and property of the marker can be read and processed by the software. At the user side, depending on the customized code, a specific virtual view will pop up. The openness of ARToolkit enable user to make any virtual environment, e.g. boxes, dolls, animals, vehicles, and others. As depicted in Figure 2, the "Unity native pluggin" tackle the computer vision task and provide two important data from the marker reading, i.e. the poses and unique property of individual marker. Poses data is important to determine marker's poses, i.e. virtual object poses. The property of marker is then visualized as different object according user's design. This concept is typical for an image processing system. However, ARToolkit is able to display a 3D virtual object while keeping the computation needs low. This feature is very important since authors have done test using other software that result in higher computation needs.

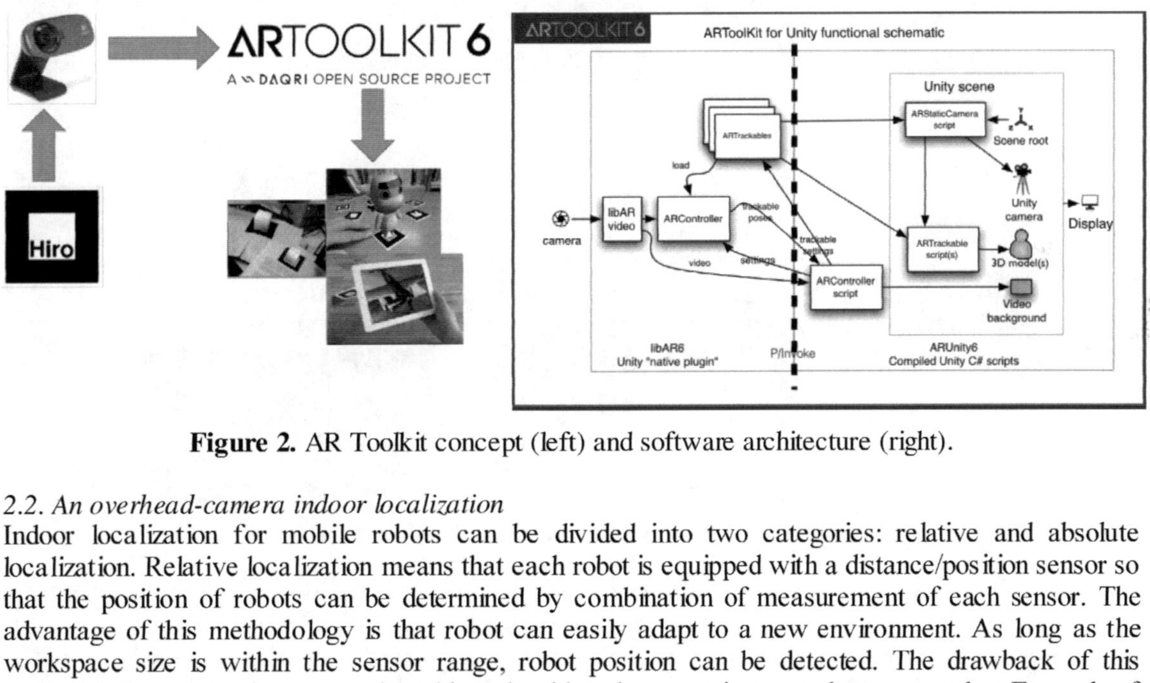

Figure 2. AR Toolkit concept (left) and software architecture (right).

2.2. An overhead-camera indoor localization

Indoor localization for mobile robots can be divided into two categories: relative and absolute localization. Relative localization means that each robot is equipped with a distance/position sensor so that the position of robots can be determined by combination of measurement of each sensor. The advantage of this methodology is that robot can easily adapt to a new environment. As long as the workspace size is within the sensor range, robot position can be detected. The drawback of this method is the need to have a good tracking algorithm that sometimes can be too complex. Example of this method can be found in (3), (17), and (18).

The absolute localization is typically built using a camera system. One or more cameras are attached on top of a certain working space. Using a specific set of markers, the position and orientation of the robot can be tracked. Examples of this setup can be found in (14) and (19), see Figure 3. The advantage of this method is the simplicity in tracking the marker. Many easy-to-use and open source vision system are available. The main drawback of this system is less flexibility when the size of the area changes. However, for testing and fast prototyping of a new algorithm this method is preferable due to its easiness in installation.

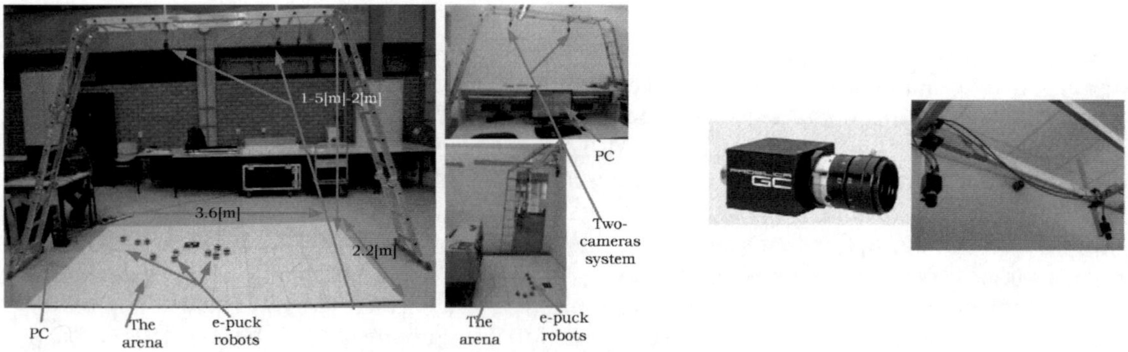

Figure 3. An overhead camera indoor localization system for mobile robot experimental setup (14).

3. System design

3.1. Proposed solution

In this work we modify the ARToolkit so that the reading of the marker can be accessed using other software. The proposed solution is depicted in Figure 4.

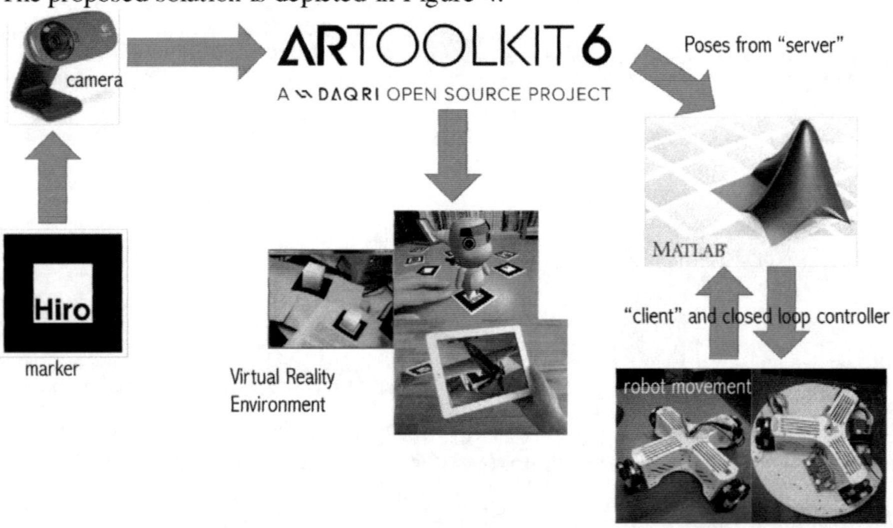

Figure 4. The proposed indoor localization system.

Although the proposed solution looks trivial, the reality says otherwise. ARToolkit is an open source software that can be customized using only the ARToolkit SDK. This situation is not suitable to our purposes since we practically only need the position and orientation of the markers that will be put on top of the robot. Thus, we create **a client-server method** using socket programming to access the marker position and orientation, where the ARToolkit act as server and MATLAB as the client. The choice of using MATLAB is simply for fast prototyping and does not count for the "low cost" term. C or Java can also be used as a client. Figure 4 also shows that once the localization is set, a closed loop control mechanism can be easily implemented. Thus, testing, experimenting and evaluating with new algorithms become quick and easy.

3.2. AR toolkit modification: client-server

Zoom in to the detail, the proposed solution works like the scheme shown in Figure 5. From the "Unity scene" block, the resulting scene root is ported into UDP communication. In this way the position and orientation of the robot can be read in MATLAB (or other client). The relative position of the marker is calibrated to the real size of the working space inside the server. The UDP communication gives an easy way to customize data frame segmentation. The segmentation is needed to identify different marker ID's position and orientation.

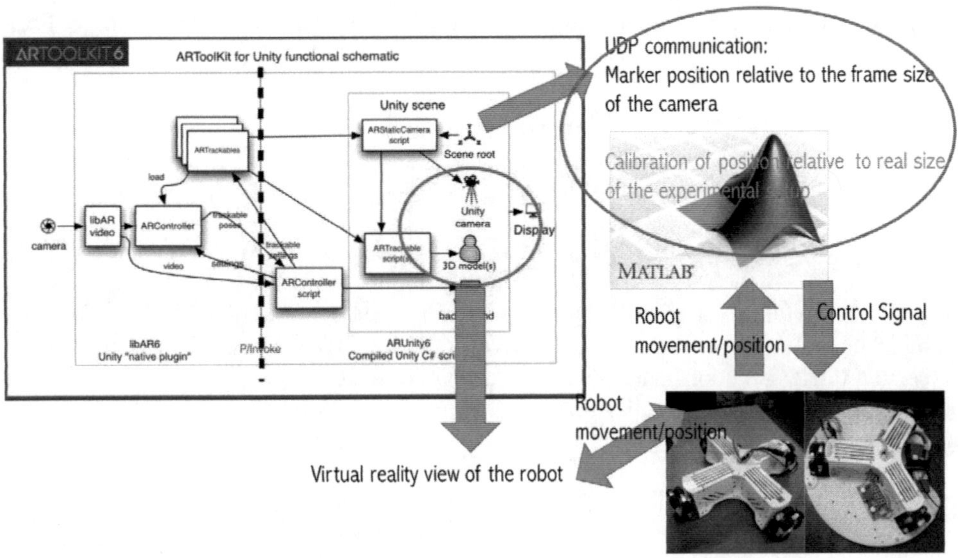

Figure 5. The detailed client-server proposed solution.

MATLAB, as the client, once receive information from the ARToolkit via UDP communication then computed the necessary control algorithm to minimize the difference between target and real-time robot movement. The ARToolkit continuously read the marker on top of the robot so that real-time movements of the robots can be tracked. In parallel, the nice 3D virtual reality still can be viewed. Furthermore, the 3D look can also be easily customized according to user specification.

3.3. System integration

The first step in integration, is to calibrate the relative frame position with respect to the real size of the working space. Since the goal is to have a quick and easy implementation, the calibration is done using the simplest way possible. Once the server and client is set, we need to do the following.

Suppose that the working space of the robot is as depicted in Figure 6. Put a marker to a position that later will be referred as original position, i.e. (0,0), in Figure 6 is shown by the "o" marker. Then from that point try to move the marker in two directions that later will be referred as "*x*" dan "*y*" direction/coordinates. While moving find the furthest distances/poinst from the "o" point that still can be tracked by ARToolkit, in Figure is denoted by "xx" and "yy". At each "o", "xx", and "yy" point, record the value read by the software. Once recorded, use the simple equation

$$y = mx + c \qquad (1)$$

to get the correct position of the marker with respect to real working space. Equation (1) need to be computed twice, each for x and y direction. The resulting calibration can be embedded either at server or client side. Once the calibration is completed, the client is ready for testing, experimenting or evaluating any control algorithm or mobile robot movements.

It is to be noted that in this work, for performance evaluation, the proposed solution is tested with omnidirectional mobile robot equipped with control algorithm presented in (11).

4. Experimental results and analysis

4.1. Calibration results

We test the proposed solution in a working space with the size of 1.6 [m] × 3.2 [m], see Figure 6 (right). We use a webcam that cost less than 100 USD, positioned at the middle of the working space 1.5 [m] above the ground. Using the procedure described in Section 3.4, the resulting calibration for the test area is as follow:

$$x_{real} = 0.49x_{read} - 6.223 \tag{2}$$

$$y_{real} = 0.501y_{read}\text{-}7.314 \tag{3}$$

where x_{real} and y_{real} represent the real position of the robot in the working space and x_{read} and y_{read} represent the reading (in pixels) by the overhead camera and ARToolkit. To get an easy measurement verification, the working space is covered with a layer of banner marked with boxes that indicates a certain Cartesian coordinate position, see Figure 7. It is to be noted that the system can be directly used at other place with similar size as long as the camera is positioned 1.5 [m] above the ground and at the center of gravity of the area.

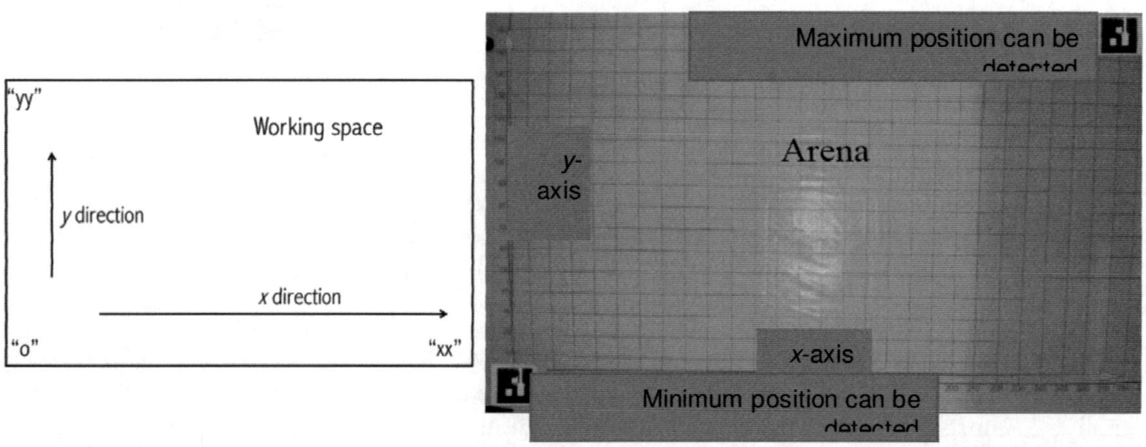

Figure 6. Example of working area and how to calibrate the resulting measurement from ARToolkit.

4.2. Static experiments: test the accuracy of the measurement

The static experiments are mainly to check the consistency of the tracking of robot position with static objects. The experiment scenario is simple. Three markers/robots are placed on the area. We modify the position and orientation of the robots. For each modification we record the position captured by the server-client as well as the manual measurement of center of the robots on the area. Some of the results are shown in Figure 7. The standard deviation of the difference between the client reading and manual measurements of each markers are in the range of 0.8 [cm] to 1.2 [cm] both for *x* and *y direction*, which indicate the accuracy of the proposed system. The accuracy cannot go lower because of the error in measuring the position of the robots, i.e. actual robot position is measured by manually computed the center of the robot at the area, see Figure 7.

4.3. Dynamic experiments: test of trajectory tracking control problem

For the dynamic experiments, two scenarios are considered. For the first scenario, similar scenarios to the static experiments are considered. However, in the dynamic case, the robots are moved manually and automatically (without controller) so that the real-time measurement are recording data of moving robots/markers. All robots are of holonomic type, i.e. three omni wheels and four omni wheels type.

The resulting experiments are shown in Figure 8 and Figure 9. As illustrated in Figure 8, the recording for multiple object is done perfectly with similar accuracy to the static measurement. The left hand side of Figure 8 shown the direction of the movement of the robots, while the right hand side shows recorded data. In addition, in Figure 9 the robots are given orders to make a circle movement via a certain set of individual wheel speed. Since it is not controlled, individually robot executes the circle movement differently that results in random movement. Similar to the result in Figure 8, the random movement can be captured real-time with the sampling time for each measurement given in bottom-right of Figure 9. These two results show that the proposed solution can handle different type of movements and speeds of the robots.

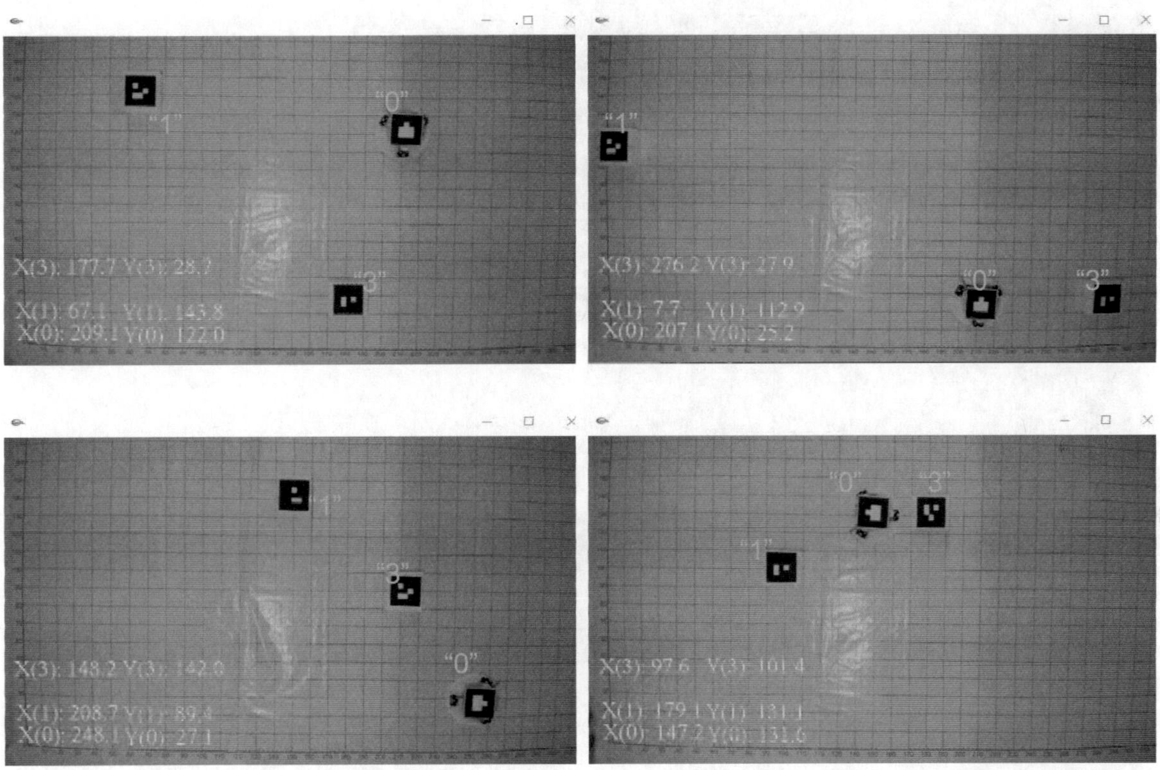

Figure 7. Static experimental results. The area is covered with a banner marked by some Cartesian coordinates. The measurement is at the real robot is computed at the center of individual robot/marker. The “0”, “1”, and “3” indicates the marker IDs.

The second dynamic experiments also the last experiment is to test a trajectory tracking controller using the proposed solution. This is similar to the circular test, only now the robot’s movements are controlled. Without losing generality, primarily based on the first dynamic test, the test is done only with one robots. The performance of a group robots should follow the single robot results.

For the sake of clarity, similar reference trajectory as in (7) is considered, i.e. the robot must follow an-eight shape like geometry. The reference is timed for each sampling time. Due the limited working

IOP Conf. Series: Journal of Physics: Conf. Series **1007** (2018) 012055 doi:10.1088/1742-6596/1007/1/012055

space only movement in quadrant one is considered. Similar control algorithm as in (7) is implemented for a three wheels omnidirectional robot. The result is depicted in Figure 5.

As can be observed in Figure 9, the recorded movement are smooth and keep the accuracy as the previous two experiments. The robot can track the given reference (indicated by the solid black line). Thanks to the control algorithm proposed in (7), the robot movement is coinciding with the reference. The proposed indoor localization is able to record the mobile robot movement well.

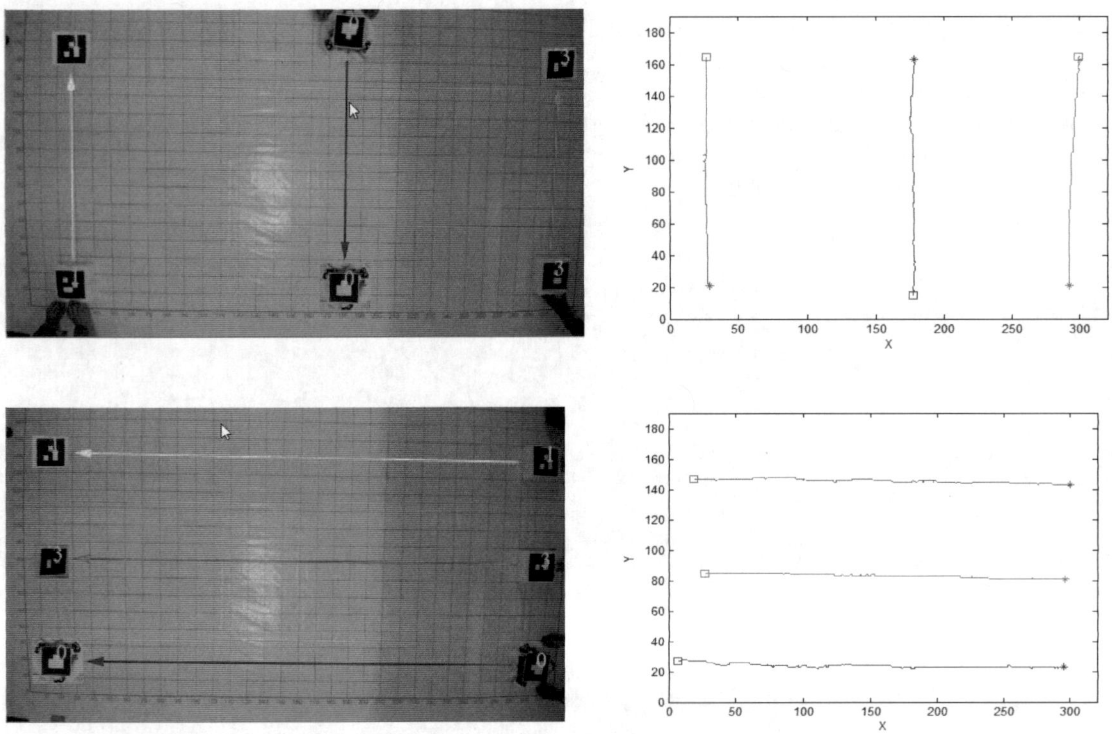

Figure 8. The resulting of dynamic experiments (1): manually move the robots.

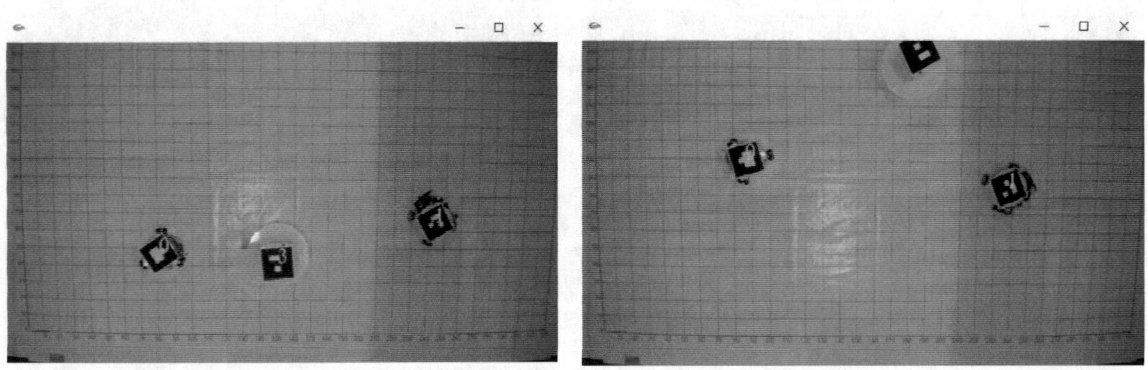

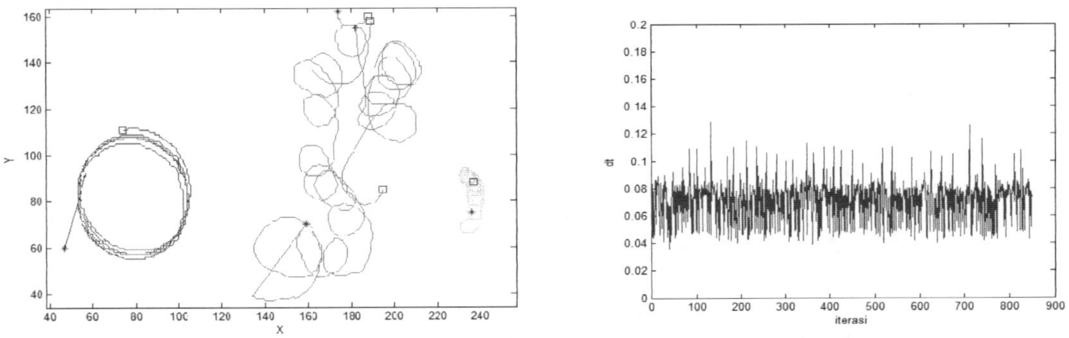

Figure 9. The resulting of dynamic experiments (2): uncontrolled circular task; top-left: initial position, top-right: end position, bottom-left": recorded movement, bottom-right: sampling time.

Although the real-time measurement is influenced by the performance of the control algorithm, the results suggest that the proposed solution fulfil the purpose to have an easy to implement indoor localization system. The effect of different control gain can be observed easily, as well the performance of other control algorithm with restriction on integer reading at the client side.

Compare to other method, e.g. (11), (14) or (20), the system provide less accurate position measurement. However, the result in Figure 10 suggests that for fast prototyping and testing a control algorithm, the proposed solution serves the purpose of having an easy-to-set and low cost indoor localization well.

5. Conclusions

In this work we proposed an easy to use and low cost indoor localization system using a combination of a cheap webcam and a customized ARToolkit-MATLAB client-server image processing system. The system can track several mobile robots/marker in real time with the accuracy of 0.8 [cm] to 1.2 [cm] both for x and y direction respectively. The system has been validated using three mobile robot system as well as a closed loop trajectory tracking control system. The proposed solution is suitable for rapid prototyping a control algorithm or analysing mobile robot characteristic.

For future work the calibration process will be done automatically. It means that once the system start it first will calibrate the relative camera measurement with respect to the real working space size without the help of any operator. Furthermore, another communication protocol can be revisited so that a better sampling time can be achieved.

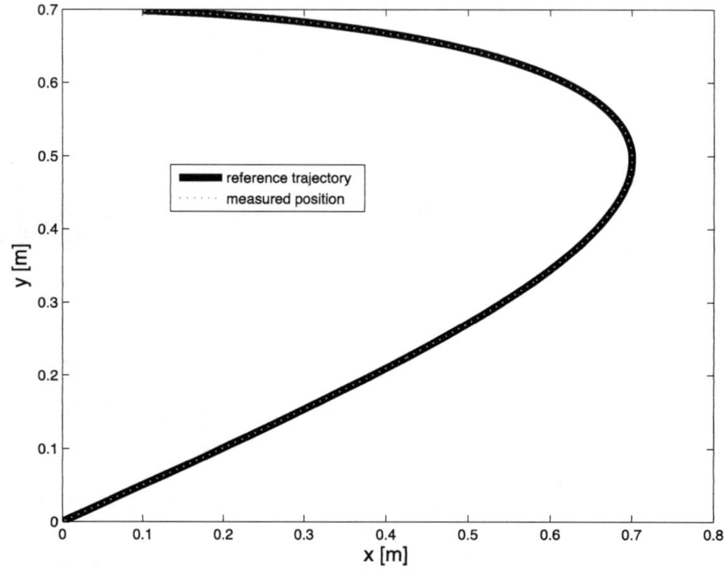

Figure 10. The resulting movement of the trajectory tracking control problem test.

References

[1] Siciliano B and Khatib O 2008 *Handbook of Robotics* (Springer-Verlag Berlin)

[2] Di Giampaolo E, Martinelli F 2014 Mobile Robot Localization Using the Phase of Passive UHF RFID Signals *IEEE Transactioan on Industrial Electronics* vol 61(1) pp 365–76.

[3] Borenstein J, Evererr H, Feng L, and Wehe D 1997 Mobile Robot Positioning: Sensors and Techniques. *J Robot Syst.* vol 14(4) pp 231–49.

[4] Kim S and Kim B 2013 Dynamic Ultrasonic Hybrid Localization System for Indoor Mobile Robots *IEEE Transactioan on Industrial Electronics.* vol 60(10) pp 4562–73.

[5] Mi J and Takahashi Y 2016 Design of an HF-Band RFID System with Multiple Readers and Passive Tags for Indoor Mobile Robot Self-Localization *Sensors* vol 16

[6] Cheng H, Chen H, and Liu Y 2015 Topological Indoor Localization and Navigation for Autonomous Mobile Robot. *IEEE Trans Autom. Sci. Eng.* vol 12(2) pp 729–38.

[7] Chen K, Mi Y, Shen Y, Hong Y, Chen A, and Lu M. 2017 SparseLoc: Indoor Localization Using Sparse Representation *IEEE Access* vol 5 pp 20171–82

[8] Kang W and Han Y. 2016 SmartPDR: Smartphone-Based Pedestrian Dead Reckoning for Indoor Localization *IEEE Sens. J.* vol.15(5) pp 2906–16.

[9] Zayets A and Steinbach E 2017 Robust WiFi-based indoor localization using multipath component analysis *Proceedings of International Conference on Indoor Positioning and Indoor Navigation (IPIN)* pp 1–8

[10] Kostic D, Adinandra S, Caarls J, Van De Wouw N, and Nijmeijer H 2009 Collision-free tracking control of unicycle mobile robots *Proceedings of the IEEE Conference on Decision and Control*

[11] Adinandra S and Ratnawati DA 2016 A Practical Coordinated Trajectory Tracking for A Group of Mixed Wheeled Mobile Robots with Communication Delays *TELKOMNIKA (Telecommunication Comput Electron Control* vo. 14(4) pp 1397–407.

[12] Winterhalter W, Fleckenstein F, Steder B, Spinello L, and Bugard W 2015 Accurate Indoor Localization for RGB-D Smartphones and Tablets given 2D Floor Plans *ProceedingIEEE/RSJ International Conference on Intelligent Robots and System* pp 3138–43.

[13] BniLam N, Ergeerts G, Subotic D, Steckel J, and Weyn M 2017 Adaptive probabilistic model

using angle of arrival estimation for IoT indoor localization *Proceedings International Conference on Indoor Positioning and Indoor Navigation (IPIN)* pp 1–7
[14] Kostic D, Adinandra S, Caarls J, Van De Wouw N, and Nijmeijer H 2010 Saturated control of time-varying formations and trajectory tracking for unicycle multi-agent systems *Proceedings of the IEEE Conference on Decision and Control*
[15] Garcia De Marina H, Cao M, and ayawardhana B 2015 Controlling rigid formations of mobile agents under inconsistent measurements *IEEE Tran.s Robotics and Autonomous Systems* vol 31(1) pp 31–9.
[16] Yang Q, Fang H, Cao M, and Chen J 2016 Distributed trajectory tracking control for multiple nonholonomic mobile robots. *IFAC-PapersOnLine* vol 49(4)pp 31–6
[17] Yi S-Y and Choi B-W 2004 Autonomous Navigation of Indoor Mobile Robot Using Global Ultrasonic System. *Mobile Robots: Perception and Navigation* pp 383–93
[18] Li I-H, M.C C, Wang W-Y, Su S-F, and Lai T-W 2014 Mobile Robot Self-Localization System Using Single Webcam Distance Measurement Technology in Indoor Environments. *Sensors* vol 14 pp 2089–109.
[19] Adinandra S, Caarls J, Kostic D, Verriet J, and Nijmeijer H 2014 Flexible transportation in warehouses *Automation in Warehouse Development* (Springer-Verlag Berlin)
[20] Sadowska A, van den Broek T, Huijbers H, Kostic D, Van De Wouw N, and Nijmeijer H 2011 A virtual structure approach to formation control of unicycle mobile robots using mutual coupling *International Journal of Control* vol 84(11) pp 1886–902

Acknowledgments
The authors would like to thank Department of Electrical Engineering, Faculty of Industrial Technology, Universitas Islam Indonesia for providing financial supports of this work.

IOP Conf. Series: Journal of Physics: Conf. Series **1007** (2018) 012056 doi:10.1088/1742-6596/1007/1/012056

Passivity-Based Control for Two-Wheeled Robot Stabilization

Nur Uddin[1,2], **Teguh Aryo Nugroho**[2], **and Wahyu Agung Pramudito**[2]
[1]Dept. of Informatics, Universitas Pembangunan Jaya, Tangerang Selatan, Banten, Indonesia.
[2]Dept. of Electrical Engineering, Universitas Pertamina, Jakarta Selatan, Jakarta, Indonesia.

E-mail: `nur.uddin@upj.ac.id`

Abstract. A passivity-based control system design for two-wheeled robot (TWR) stabilization is presented. A TWR is a statically-unstable non-linear system. A control system is applied to actively stabilize the TWR. Passivity-based control method is applied to design the control system. The design results in a state feedback control law that makes the TWR closed loop system globally asymptotically stable (GAS). The GAS is proven mathematically. The TWR stabilization is demonstrated through computer simulation. The simulation results show that the designed control system is able to stabilize the TWR.

1. Introduction

A two-wheeled robot (TWR) is a kind of mobile robots. The TWR is basically structured by a robot body and two wheels. The wheels are supporting the body. The two wheels support makes the robot to have high maneuverability which becomes the advantage of TWR compared to other mobile robots, for examples: three-wheels robot, four-wheeled robot, and legged robot. The high maneuverability makes TWR to be potential for many applications, for examples: transportation, surveillance, logistic, exploration, rescue, and leisure. Segway and hover board are two commercial product example of the TWR.

On another hand, the two wheels support provides a problem to the TWR. The two wheels support makes the TWR is statically unstable, where the robot is not able to stand by self. A stabilization is required to make the TWR to be stable such that it can stand by self. An active stabilization systems is commonly applied to stabilize the TWR. The active stabilization system is a feedback control system which utilizes deviation angle between actual position and vertical position to generate control torque for stabilizing the TWR. The active stabilization system consists of a controller, an actuator, and sensors. The actuator is to generate a control torque for the stabilization. The control torque is calculated by a controller base on a control law and states feedback. The states feedback is obtained trough states measurement using sensors.

Control law for the TWR stabilization can be designed using the available control theories. For model-based control design, dynamics model of the TWR is required. The dynamics model is a mathematics equation system which represents the TWR dynamics. Dynamic model of the TWR can be obtained through system modeling. The system modeling applies physical laws to derive dynamics equation of a plant. System modeling of a TWR can be done by applying the Newton's laws of motion as presented in [1, 2]. TWR modeling by applying the Euler-Lagrange method is presented in [3, 4]. The TWR system modelings result in a non-linear dynamics system.

Control system design of a non-linear system is a challenge work. It has more complexity than control system design of a linear system. Control system design using non-linear control theories is not straight forward as using linear system control theories. Simplifying the complexity is commonly done through approximating the non-linear system by a linear system. This approximation is known linearization. Through the linearization, linear control theories can be applied in control system design of a non-linear system but with a limitation [5].

Linear quadratic regulator (LQR) is one of the most popular linear control theories. Studies of applying LQR for TWR stabilization have been presented [6–8]. A comparison of TWR stabilization using PID control and LQR control has been presented in [6]. The study results show that both PID and LQR are able to stabilize the TWR, but the LQR controller has better control performance than the PID controller. LQR is a full states feedback control such that measurements of all states are required. However, not all of the states are measurable because limitation of the sensors availability. Observer provides a solution for the unmeasured states. The observer can be used to estimate the unmeasured states. Study on applying observer for states estimation and applying the estimated states in the states feedback control of TWR stabization has been presented in [8]. Linear control method is applicable for stabilizing a TWR through linearization. However, the linearization is only valid for a limited region around an equilibrium point. Therefore, a control system for TWR stabilization which is design through linearization is only able to stabilize the TWR for a limited operating area and the TWR closed-loop system is locally asymptotically stable.

The TWR is stabilized for the whole operating area if the closed loop system globally asymptotically stable (GAS). The GAS is achieved by satisfying the Lyapunov's stability theorem [9]. The Lyapunov's stability theorem is fundamental theorem of non-linear control theories. Therefore, applying non-linear control theories for TWR stabilization will a GAS of the closed loop system. Several studies on applying non-linear control theories for TWR stabilization has been presented. TWR stabilization using backstepping control method has been presented in [2, 10, 11]. Backstepping control method offer a systematically procedure for obtaining a control law and stability proof simultaneously [12]. Accomplishing the backstepping procedure will result in a states feedback control law which guaranties the closed loop system to be GAS. However, the backstepping method may result in a complex states feedback control law which makes difficulty in real-time system implementation [13]. Another study on TWR stabilization using Lyapunov-based control method has been presented in [14]. All of the non-linear control system designs for TWR stabilization resulted in GAS such that the TWR closed-loop system is stabilized for the whole operating area.

Passivity-based control (PCB) is another non-linear control method. The PCB is basically to make the closed loop system passive. Passivity is one of the most physically interesting concepts in systems and control theory [15]. It is inspired by nature that a passive system is asymptotically stable. Passivity is related to stability property in an input output sense, where a system is stable if a bounded input energy is supplied to the system, it results in a bounded output energy [16].

This paper is presenting a study of non-linear control system design for TWR stabilization using passivity-based control. The study is concerned on longitudinal mode only. Presentation of the paper is organized as follows. Section I provides introduction, motivation, and related works of this study. Model of TWR is described in Section II. Passivity-based control design for stabilizing the TWR is described in Section III. Performance of the TWR stabilization is evaluated through computer simulations and is presented in Section IV. Finally, Section V concludes the work.

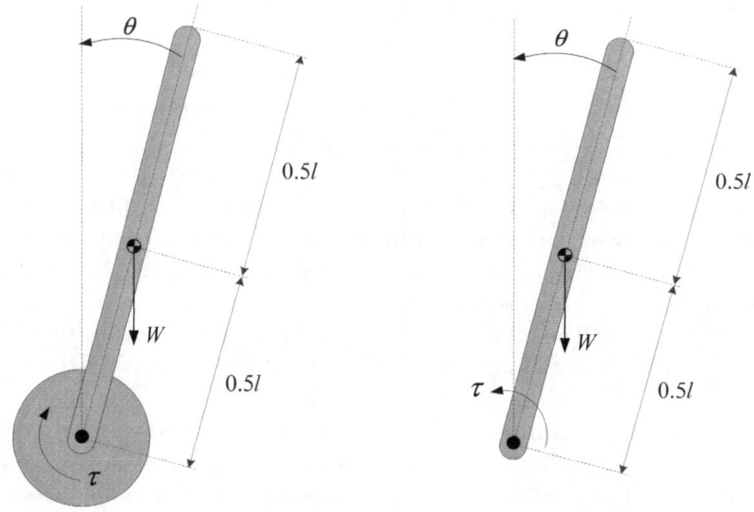

Figure 1. (a) Two-wheeled robot (TWR) model. (b) Free-body diagram of the TWR model.

2. Two Wheels Robot Model

A two-wheeled robot (TWR) is basically consisted of a robot body and two wheels. The wheels support the robot body. A TWR model is shown in Figure 1a. The robot body is represented by a linkage where the center of mass is assumed at the middle linkage. It is assumed that both wheels are connected by a shaft and driven by a motor. When a torque disturbance is given to the TWR, it makes the TWR to pitch or rotate such that robot position deviates from the vertical axis. Figure 1a shows a TWR at the pitching position with pitching angle θ. At the pitching position, weight of the robot gives a moment to the robot and the robot is to fall down. It is a reason why the robot is statically unstable. In order to keep the robot stable, a torque is required to counter the torque disturbance and the moment due to the weight. This required torque is called the control torque. When the motor is operating, it provides torque to rotate the wheels. Friction of the wheel and the floor results in reaction torque to the body. The reaction torque can be applied as the control torque for stabilizing the robot. Therefore, the motor has two functions: driving the wheels and an actuator for robot stabilization.

Figure 1b shows free body diagram of the robot. There are two working moments on the robot body, i.e.: moment due to the body weight and the reaction torque. The working moments on the robot body are evaluated at the wheel axis. Applying the Newton's second law results in the following dynamic equations:

$$\Sigma M = I\ddot{\theta} \tag{1}$$

$$\tau - \frac{1}{2}mgl\sin\theta = I\ddot{\theta} \tag{2}$$

where M is moment, I is inertia of the robot body, θ is the pitching angle which is the robot body deflection with respect to vertical axis, m is the mass of robot body, g is the gravity acceleration, l is the length of the robot body, $\ddot{\theta}$ is the pitching acceleration of the robot body, and τ is the control torque. Defining $I_r = \frac{1}{2}mgl$ and rearranging (2) result in:

$$I\ddot{\theta} + I_r\sin\theta = \tau. \tag{3}$$

IOP Conf. Series: Journal of Physics: Conf. Series **1007** (2018) 012056 doi:10.1088/1742-6596/1007/1/012056

3. Passivity-Based Control Design

Passivity based control is purposed to make a closed loop system to be passive. Theory of the passivity-based control is given as follows [9]:

Theorem 1 *If a system*

$$\dot{x} = f(x,u) \tag{4}$$

$$y = h(x,u) \tag{5}$$

is passive with a radial unbounded positive definite storage function and zero-state observable, then the origin $x = 0$ can be globally stabilized by $u = -\phi(y)$, where ϕ is any locally Lipschitz function such that $\phi(0) = 0$ and $y^T\phi(y) > 0$ for all $y \neq 0$.

In order to stabilize the TWR, defined position error of the robot:

$$e = \theta - \theta_r \tag{6}$$

where e is the position error, θ is the actual position angle, and θ_r is the desired position angle. For the TWR stabilization, the desired position is the robot at vertical position ($\theta_r = 0$). Time derivatives of the position error are:

$$\dot{e} = \dot{\theta} \quad \text{and} \quad \ddot{e} = \ddot{\theta}. \tag{7}$$

Substituting (6) and (7) into (3) results in error dynamics of the TWR system and is given as follows:

$$I\ddot{e} + I_r \sin e = \tau. \tag{8}$$

Define states:

$$x_1 = e, \quad x_2 = \dot{e} \tag{9}$$

such that (8) can be expressed in a states equation as follows:

$$\begin{aligned} \dot{x}_1 &= x_2 \\ \dot{x}_2 &= -\tfrac{I_r}{I}\sin x_1 + \tfrac{1}{I}\tau. \end{aligned} \tag{10}$$

TWR stabilization is purposed to make $e = 0$ and $\dot{e} = 0$ as time goes to infinity, i.e. $x_1(\infty) = 0$ and $x_2(\infty) = 0$. It is an explicit statement that the (10) should be GAS. For achieving the GAS, define a radially positive definite function:

$$V = \frac{1}{2}x_1^2 + \frac{1}{2}Ix_2^2. \tag{11}$$

Time derivative of V is given as follows:

$$\begin{aligned} \dot{V} &= x_1\dot{x}_1 + Ix_2\dot{x}_2 \\ &= x_1x_2 + Ix_2\left(-\frac{I_r}{I}\sin x_1 + \frac{1}{I}\tau\right) \\ &= x_1x_2 + x_2\left(-I_r\sin x_1 + \tau\right) \\ &= x_2\left(x_1 - I_r\sin x_1 + \tau\right). \end{aligned} \tag{12}$$

Let choosing $\tau = I_r \sin x_1 - x_1 + v$, substituting it into (12) result in:

$$\begin{aligned} \dot{x}_1 &= x_2 \\ \dot{x}_2 &= \tfrac{1}{I}\left(-x_1 + v\right), \end{aligned} \tag{13}$$

Table 1. Simulation Parameters

Parameter	Symbol	Value	Unit
Mass of the rod	m_r	0.1	kg
Length of rod	l	0.2	m
Inertia of the rod	I_r	13×10^{-4}	$kg.m^2$
Control gain	λ	2	

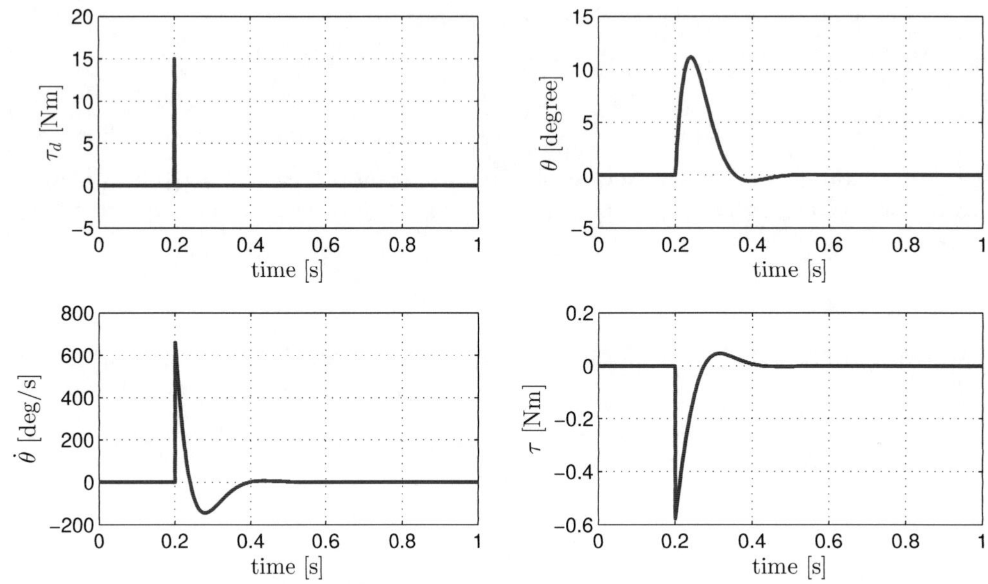

Figure 2. TWR stabilization when the robot is disturbed by an impulse torque 15 Nm.

and substituting into (12) results in:

$$\dot{V} = x_2 v. \tag{14}$$

By defining v as the input and $y = x_2$ as the output, the system (13) is passive with V in (11) as the storage function.

Giving a condition $v = 0$ and $y = 0$ for (13) implies $x_2 = 0$ and $x_1 = 0$. It shows that (13) is zero-state observable. Therefore, (13) is globally asymptotically stabilized by $v = -\phi(x_2)$ with any function ϕ such that $\phi(0) = 0$ and $y^T\phi(y) > 0$ for all $y \neq 0$. For this case, let choosing $v = cx_2$ with $c < 0$.

4. Simulation

Stabilizing the TWR using the designed control system is evaluated through computer simulation. Parameter values for this simulation is given in Table 1. It is assumed that all system states are measured and noise are neglected. In the simulation, the robot position is initially at $\theta = 0°$. At $t = 0.2$ second, the robot is disturbed by an impulse torque 15 Nm. Figure 2 shows the simulation result. The torque disturbance makes the robot to pitch about 11°. As the control system is active, the controller give command signal to the motor to generate control torque such that the robot position return to $\theta = 0°$. This result shows that the robot closed loop system is globally asymptotically stable. The design controller results in fast response of stabilization such that the robot return origin in about 0.4 seconds.

5. Conclusion

A control system design for two-wheeled robot stabilization using passivity-based control has been presented. Applying the method results in a state feedback control law which makes the closed loop system of the robot globally asymptotically stable. The design results in a fast TWR stabilization system which is able to stabilize the robot in about 0.4 seconds.

6. Future Works

This work is a part of research project on autonomous robot. This work will be continued by implementing the designed controller in a real TWR.

References

[1] Li J, Gao X, Huang Q, Du Q and Duan X 2007 Mechanical design and dynamic modeling of a two-wheeled inverted pendulum mobile robot *Automation and Logistics, 2007 IEEE International Conference on* (IEEE) pp 1614–1619

[2] Nomura T, Kitsuka Y, Suemitsu H and Matsuo T 2009 Adaptive backstepping control for a two-wheeled autonomous robot *ICCAS-SICE, 2009* (IEEE) pp 4687–4692

[3] Akesson J, Blomdell A and Braun R 2006 Design and control of yaipan inverted pendulum on two wheels robot *Computer Aided Control System Design, 2006 IEEE International Conference on Control Applications, 2006 IEEE International Symposium on Intelligent Control, 2006 IEEE* (IEEE) pp 2178–2183

[4] Kim Y, Lee S H and Kim D H 2011 Dynamic equations of a wheeled inverted pendulum with changing its center of gravity *Control, Automation and Systems (ICCAS), 2011 11th International Conference on* (IEEE) pp 853–854

[5] Slotine J J E, Li W *et al.* 1991 *Applied nonlinear control* (Prentice hall Englewood Cliffs, NJ)

[6] Nasir A, Ahmad M and Ismail R R 2010 *World Academy of Science, Engineering and Technology* **70** 227–232

[7] Asali M O, Hadary F and Sanjaya B W 2017 *International Journal of Electrical and Computer Engineering (IJECE)* **7** 2008–2017

[8] Uddin N, Nugroho T A and Pramudito W A 2017 Stabilizing two-wheeled robot using linear quadratic regulator and states estimation *Proc. of the 2nd International conferences on Information Technology, Information Systems and Electrical Engineering (ICITISEE)* (Yogyakarta, Indonesia)

[9] Khalil H K 2002 *Noninear Systems* 3rd ed (Prentice-Hall, New Jersey)

[10] Thao N G M, Nghia D H and Phuc N H 2010 A pid backstepping controller for two-wheeled self-balancing robot *Strategic Technology (IFOST), 2010 International Forum on* (IEEE) pp 76–81

[11] Tsai C C and Ju S Y 2010 Trajectory tracking and regulation of a self-balancing two-wheeled robot: A backstepping sliding-mode control approach *SICE Annual Conference 2010, Proceedings of* (IEEE) pp 2411–2418

[12] Krstic M, Kanellakopoulos I and Kokotovic P V 1995 *Nonlinear and adaptive control design* (Wiley)

[13] Uddin N and Gravdahl J T 2011 Active compressor surge control using piston actuation *Proc. of the ASME Dynamics System and Control Conference* (Virginia)

[14] Uddin N 2017 Lyapunov-based control system design of two-wheeled robot *2017 International Conference on Computer, Control, Informatics and its Applications (IC3INA)* (Jakarta, Indonesia)

[15] Hatanaka T, Chopra N and Spong M W 2015 Passivity-based control of robots: Historical perspective and contemporary issues *Decision and Control (CDC), 2015 IEEE 54th Annual Conference on* (IEEE) pp 2450–2452

[16] Ortega R, Perez J A L, Nicklasson P J and Sira-Ramirez H J 2013 *Passivity-based control of Euler-Lagrange systems: mechanical, electrical and electromechanical applications* (Springer Science & Business Media)

Acknowledgment

The author acknowledges financial support from Universitas Pertamina through internal research grant for a research project on Autonomous Robot System (ARS) for the year of 2017.

IOP Conf. Series: Journal of Physics: Conf. Series **1007** (2018) 012057 doi:10.1088/1742-6596/1007/1/012057

Design and Development of the Trash Spliter with Three Different Sensors

Despaleri Perangin Angin[1*], Hendrik Siagian[1], Eka Dodi Suryanto[1], Rahayu Sashanti[1]and Marcopolo[1]

Faculty of Technology and Computer Sciences, Universitas Prima Indonesia, Indonesia

Email: *despaleriperanginangin@yahoo.com

Abstract. Trash has become a major problem in everyday life. Until now there is no right method to handle it. This paper discusses a method of development of the trash splitter with three different sensors. There are three sensors are used infrared, metal, and light sensors. The results obtained are more effective with the results obtained show the devices have similar accuracy garbage sorting is a metal (98%), organic (26.67%), paper (32%), and plastics (58%). The accuracy of the mixed waste sorting is a metal (94.67%), organic (28%), paper (12%), and plastics (41.3%).

1. Introduction

In human life, trash is a problem that has not been handled properly. There are so many processes in human activities that generate trash so the number continues to increase every time [1-2]. In Indonesia, the classification of waste that is often used organic trash, and inorganic trash. Meanwhile, when viewed from the source, the existing trash in urban areas can be divided into several kinds. The types of urban trash are trash from residences, commercial areas trash, industrial trash, litter public places, and medical trash. [3] In his everyday life, every human being produces a number of garbage in the solid form of 1-3 kg. Untreated trash can cause problems and should be overcomed. The amount of waste can be overcome by recycling.

To recycle, trash must be sorted first. The sorting process is useful for separating trash by type. To distinguish the type of trash, often the sorting process is done manually by using human power. Humans will sort the type of trash as with a predetermined category. That way, the waste can be reprocessed into useful goods and have economic value. But along with the development of the age and the growing rate of increasing trash, the sorting of waste by manual method becomes not optimal. Lack of human resources in the process of sorting waste and the amount of waste that must be sorted, making a lot of trash that should be recycled ends directly in the final disposal.

Therefore, the problem of the excess amount of trash cannot be resolved by applying the manual sorting method. To overcome the problem, has been investigated tool to shrink the type of trash. Sorting trash can be done by destroying trash prior [4-5]. The method of sorting trash by destroying trash first, there is another method that is by utilizing the sensor to distinguish the type of trash [6]. The sorting of large amounts of trash is easier to use by using sensors.

2. Methodology

Sensors which can be used include the infrared sensor, metal sensor, and light sensor. The sensor consists of a transducer with or without a signal amplifier and processor formed in one sensing system. In control system environment and robotics, the sensor gives the similarity that resembles the eyes, hearing, nose, tongue which will then be processed by the controller as his brain [7-9]. So with a visible light sensor application using a particular method, the trash can be detected and classified according to their kind. More effective and faster sorting can be achieved. The sorting results are expected to be more accurate and faster than the conventional method of manual sorting using human labor [10].

As in the process of designing and testing tools, required components, and other supporting tools. Components and support tools will be divided into two parts, namely hardware components (hardware) having the form of software components (software) which consists of a program or initialize. In the design of automatic waste sorting tool, used several components that act as sensors (inputs) and actuators (output). In addition to sensors and actuators, also used as a microcontroller device that will run the given command. The computer is required as a liaison media between hardware and software. specification of tools, components, and supporting devices are shown in table 1.

Table 1. Supporting tool system.

No	Name	Specification	Amount
1	Laptop	Dell inspiron	1 piece
2	Microcontroller	Arduino Uno	1 piece
3	Program	Arduino IDE	1 piece
4	Infrared Sensor	FC-51 (DC 5V)	3 pieces
5	Metal Sensors	SN04 (DC 5V)	1 piece
6	Light Sensor	LDR (DC 5V)	1 piece
7	Servo	SG-90 (DC 5V)	4 pieces
8	Conch Fan	Brushless DC 5V	1 piece
9	Laser Module	DC 5V	1 piece
10	Battery	Power Bank 5V	2 pieces

As a guide for designing automated waste sorting tools, it is necessary to formulate the necessary system workflows first.The system workflow will be explained using the block diagram.The block diagram of automatic waste sorting tool design can be seen in figure 1.

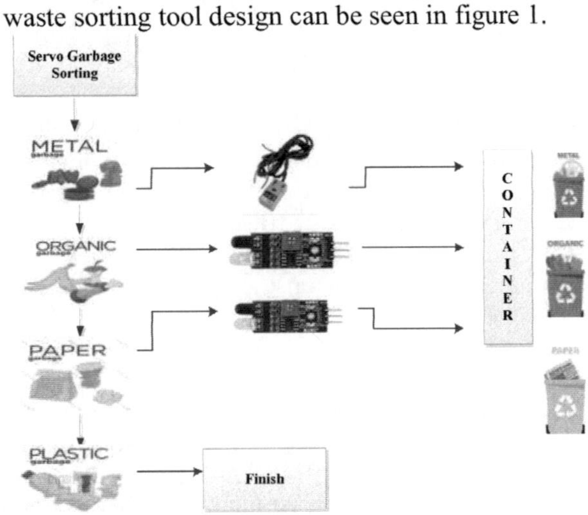

Figure 1. System block diagram.

IOP Conf. Series: Journal of Physics: Conf. Series **1007** (2018) 012057 doi:10.1088/1742-6596/1007/1/012057

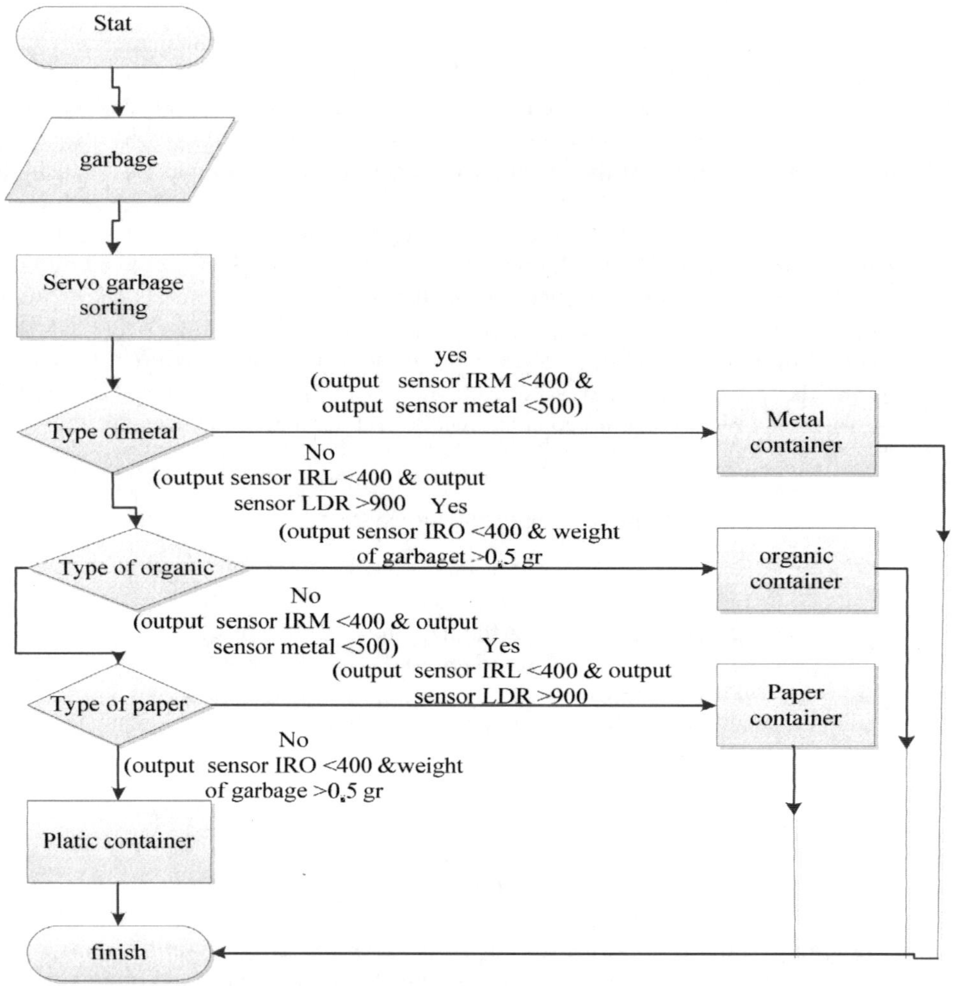

Figure 2. Flowchart.

The software used is Arduino IDE. Arduino IDE (Integrated Development Environment) is software used to program the microcontroller Arduino [11-13]. The tools will be designed based on system requirements. The design of automatic garbage sorting tool will consist of: (1) The design of an object detecting the presence of garbage, (2) design of sorting metal, (3) Design of sorting organic, (4) Design paper sorting, (5) The design of the plastic sorting.

Software design includes initialization program on the microcontroller, sensor initialization, and initialization of additional support tools. Detect the presence of waste serves to indicate that there is a waste to be processed by the tool. Consists of infrared and servo distance sensors. The metal divider will serve as a metallic waste scavenger. Consists of metal and servo sensors. Organic sorting will serve as organic waste sorting by using the difference method of waste weight. Organic waste tends to be heavier than plastic waste and paper waste. Consists of infrared sensors, servo, and fan pusher. Paper sorting will serve as paper waste sorter by utilizing light reflection difference method by paper. The paper tends to reflect the light on it. It consists of an infrared sensor, light sensor, servo, and laser. Plastic sorting will function as a plastic waste sorter by utilizing the difference reflection method oflight. Plastics tend to pass on the lights that hit them. It consists of an infrared sensor, light sensor, servo, and laser. The plastic dividing part is the same place in the paper sorting section.

The automated trash sorting tool uses multiple sensors to distinguish the type of trash.Using sensor will affect the outcome of a trash sorting. Placement of the sensor that precisely matches the characteristics of waste which will be detected so that the waste can be sorted. Metallic sensors used to detect objects made from the best metal. The method is used the placement of metal sensors at the beginning of the tool gives good results. The metal sensor is used a roko sn04 with a type of inductive metal sensor. Some of the sensors used can be seen in figure 2.

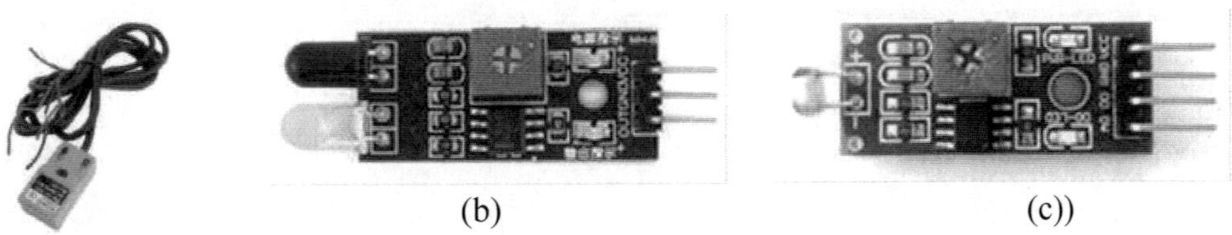

(b) (c))

Figure 3. (a) Sensor roko Sn04. (b) Infrared sensor FC-51.(c)Sensor light LDR.

Infrared sensors can be used to detect the presence of organic waste. But the method of detecting organic waste by way of differences in the weight of organic waste with waste paper and plastic waste is less precise. Often the system fails to sort out organic waste. The infrared sensor used is FC-51. Sensor FC-51 can be seen in figure 2.b[5]. Infrared sensors have been able to detect the presence of objects. However, the use of light sensors still is less than optimal. Differences in paper size and improper paper position affect the light sensor detection results. The method of sorting paper with plastic using the light difference is still less than optimal.

Infrared sensors have been able to detect the presence of objects. The use of light sensors that are synchronized with infrared sensors can provide better detection results than paper waste results. The light sensor used is a sensor module LDR (light dependent resistor). LDR will be very resistant when not exposed to light, otherwise, the value will be very low LDR resistance when exposed to very bright light [14-15].

3. Results and Discussions

Testing the tool is done by entering the garbage object directly. The experiment was carried out with 2 types of garbage, example similar waste, and mixed waste. Each experiment was conducted 15 times to get the average score. After doing the design and testing tool will get the test results in the form of quantitative data. The results of the form design tools can be seen in figure 3.

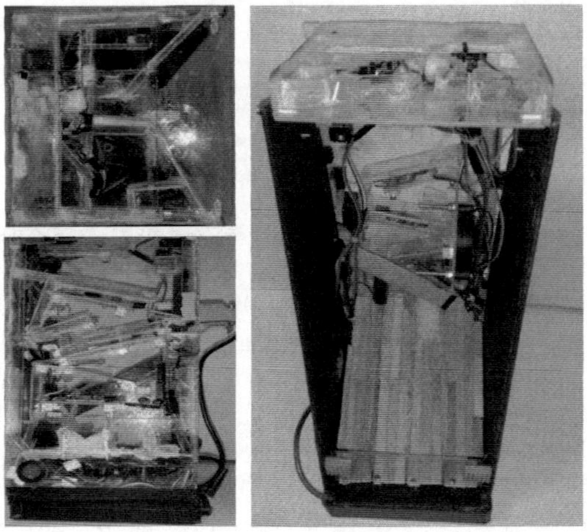

IOP Conf. Series: Journal of Physics: Conf. Series **1007** (2018) 012057 doi:10.1088/1742-6596/1007/1/012057

Figure 4. Results design.

The results of testing tools indicate the sensors used have been able to distinguish types of waste metal, organic, paper, and plastic. The test results of the tool can be seen in each of the test results tables below. table 2 shows the test results for similar waste. table 3 shows the test results for mixed waste. table 4 shows the average results for both types of testing.

Table 2. Resul1ts of similar waste testing.

Type	Conclusion					Accuracy
	Paper container	Plastic container	Metal container	Organic container	Failed (stuck)	
Metal	0	0	147	0	3	98%
Organic	26	56	1	40	27	26,67 %
Paper	48	51	0	25	26	32 %
Plastic	34	87	0	11	18	58 %
Total					12.3 %	53,67 %

Table 3. Mixed garbage test result.

Type	Conclusion					Accuracy
	Paper container	Plastic container	Metal container	Organic container	Failed (stuck)	
Metal	2	4	142	2	0	94,67%
Organic	28	40	22	42	18	28 %
Paper	18	55	28	39	10	12 %
Plastic	34	62	24	20	10	41,3 %
		Total			6.3 %	44 %

Table 4. Mixed garbage test result.

Type			Conclusion			Accuracy
	Paper container	Plastic container	Metal container	Organic container	Failed (stuck)	
Metal	2	4	289	2	3	97,3 %
Organic	54	96	23	82	45	27,3
Paper	66	106	28	64	36	22
Plastic	68	149	24	31	28	49,67 %
		Total			9,3 %	48,8 %

4. Conclusion

From the test results, it can be seen that the choose types of waste metal are the most accurate sorting, followed by sorting plastic waste, organic waste, and paper waste. The test results showed similar waste separation is a metal (98%), organic (26.67%), paper (32%), and plastics (58%). The accuracy of the mixed waste sorting is a metal (94.67%), organic (28%), paper (12%), and plastics (41.3%). Still, there is an error in the form of garbage stuck tool and fault detection by the sensor.

References

[1] Law No. 18 of 2008*Waste Management,*Jakarta: 2008
[2] Antoni R, Sarwoko M and Sunarya U,*Analysis and Implementation of Sensor System in Automatic trash Can with Fuzzy Method Based Microcontroller*, Faculty Electrical Engineering of Universitas Telkom
[3] E, Damahuri, Dan T.Padmi,*Waste Management,*Institut Teknologi Bandung 2010
[4] F. Restu,Engineering and Crushing Waste Picker Automatic With simple controls control systems in the Internal Scale State Polytechnic Batam <Integration Journal ISSN: 2085 - 3858, 2013
[5] Nurhadi M I, Siregar S and Hendrarini N, Mechanical Design System sold Waste, Faculty of Applied Sciences, ISSN : 2442-5826, 2015
[6] Russel MH, MH Chowdhury, MS Uddin, A. Newaz, And MM Talukder,Development Of Automatic Smart Waste Sorter Machine, International Conference on Mechanical, Industrial and Materials Engineering 2013 (ICMIME2013) 1 to 3 November, 2013, RUET, Rajshahi Bangladesh, Paper Id: AM - 012, 2013
[7] Septriyaningrum I A, Nugrahadi, D T and Ridwan Ichsan, Design and Development Of Prototype, Journal of Computer Science , Vol 3 ISSN: 2406-7857, 2016
[8] Setiawan D, Syahputra T and Iqbal M, Design of opening tools and cover of automatic garbage truck based on microcontroller
[9] Derry Tri Saputra, Application of Ultrasonic Sensor HC-SR04 On Valve Control System For Water Supplier With Arduino Uno Controlled RFID Control, State Polytechnic of Sriwijaya Palembang, 2015
[10] Mahmuddul Hasan Russell, Mehdi Hasan Chowdhury, Md.Shekh Naim Uddin, Ashif Newaz, And Md.Mehdi Masud Talukder,*Development Of Automatic Smart Waste Sorter Machine,*International Conference on Mechanical, Industrial and Materials Engineering 2013 (ICMIME2013) November 1-3, 2013, RUET, Rajshahi, Bangladesh.Paper Id: AM-012,2013
[11] E.Mulyana & R. Kharismanan, Design of Early Warning Equipment Flood Hazard with Arduino Uno R3 Microcontroller, Citec Journal ISSN: 2354-5771, 2014
[12] GEraldFElix, AndJiangDening,*Enchanced Bank Vault (Strong Room) Security System Design,*International Journal of Science and Research (IJSR) ISSN (Online): 2319-7064, 2015
[13] Elasya Y, Notosudjono D and Wismiama E, Application Of Ultrasonic Sensor Based On ATMga328 Microcontroller For Designing Smoking Place, Electrical Engineering Study Program, Faculty of Engineering, Pakuan University
[14] Eduardus Tuluk,Implementation Tools Pest Repellent Bird In Rice Field Area Using Ultrasonic Wave Based Microcontroller ATmega168,vol.vii Number 21 November 2012 -Journal of Information Technology ISSN: 1907-2430, 2012
[15] Z. Richard.Embedded Systems Handbook, Taylor and Francis Group.2006

Acknowledgments

The research is supported by thematic programs through the faculty of technology and computer science, Universitas Prima Indonesia, Indonesia.

IOP Conf. Series: Journal of Physics: Conf. Series **1007** (2018) 012058 doi:10.1088/1742-6596/1007/1/012058

Design of automatic mobile trolley using ultrasonic sensors

Eka Dodi Suryanto*, **Hendrik Siagian, Despaleri Perangin-Angin, Rahayu Sashanti, Suthes Yogen**

Faculty of Technology and Computer Science, Universitas Prima Indonesia. Indonesia

E-mail: *ekadodisuryanto@gmail.com

Abstract. An automatic mobile trolley was a prototype of wheel robot that serves as a trolley or shopping cart. This paper proposed an automatic mobile trolley using ultrasonic sensors. It can follow human movement automatically. It did not need to be encouraged or withdrawn. It would make an easier shopping for people as customers. The trolley controlled by a microcontroller module unit. It can stop , turn right, turn left, forward and backward. It can follow wherever they go, during they were in range. Based on the test results, the trolley succeeded to move forward by 80%, move backward 80%, turn left, 70%, turn right 70%, and stop 80%.

1. Introduction

Shopping mall is a place where most people from all walks of life will get their daily needs ranging from food product, apparels, toiletries; gardening tools electrical appliances, and others. The numbers of small and large shopping malls keep on increasing over the years throughout the world due to the demand of the public. Thus, the level of advancement of shopping mall system and infrastructure also varies. There are still plenty of spaces for improvement in terms of providing quality shopping experience to the consumers. Consumers often face problems and inconvenience when shopping [1].

Nowadays, in mall for purchasing variety of items it requires trolley. Every time customer has to pull the trolley from rack to rack for collecting items [2]. It needs a lot of power to be done when shopping. But, trolleys in supermarkets were needed. A lot of goods in the shopping cart would be to push or pull, so the customers will limit the activity of the hand. When focused on pushing the trolley, buyers will often miss a lot of goods sold in supermarkets and only buy the important goods only, and of course this could harm the supermarket company [2-4].

The technology represents to retailers opportunity to reduce costs and to improve services, allowing to attend clients quickly, precisely and supplying personalised services [5]. All around the world, robots are entering the shopping centers, attesting that robotic technology is gaining ground. The rationale for this choice lies in the fact that retail shops need to offer trendy yet low-cost substitutes to e-commerce, reducing the operating costs of personnel management [6].

In order to give the best services, it had been need a trolley that can follow automatically human movement. The direction could be detected using transducers, sensors or other detector. For example, the detector may be sensitive to weight, heat, sound, dielectric constant, and so on [7,8]. Design was divided into several parts namely the design of hardware, hardware relationships, the initialization of sensors and auxiliaries, and software design. As various kinds of robots have been developed to

IOP Conf. Series: Journal of Physics: Conf. Series **1007** (2018) 012058 doi:10.1088/1742-6596/1007/1/012058

replace human works. Ultrasonic range sensors have been recently used for range detection and obstacle recognition in robots because of their low price, high efficiency, and relatively simple structure. While ultrasonic waves have better directivity than low-frequency waves in general, they are attenuated more rapidly. Ultrasonic sensor uses sound waves for its ranging and the speed of sound is influenced by a number of environmental parameters [7-9].

2. Sensor implementation

Detection and tracking of objects in the side-near-field have attracted much attention for the development of advanced driver assistance systems. Ultrasonic sensors applied the principle of 'time of flight' (TOF) to measure distance, which computed the travel time of ultrasonic echo reflected from the target [10]. Therefore, the performance of ultrasonic sensors highly depends on the reflective characteristics (e.g., shape, surface material) of the target surface [11].

Thus, an accurate and reliable model of ultrasonic sensors is of critical importance to the design of detection and tracking algorithms. In the literature, the models can be divided into two types based on their modeling principles: physical models and empirical models. The physical models are derived from the fundamentals of sensor working process, which simulated the errors based on the mechanisms of generation. For example, Kuc and Siegel designed a three-level model including geometric distance calculation, attenuation from reflection and transmission, and object recognition based on threshold methods [12-14].

In this design, we used sensors to control the distance and direction of user movement and location. Ultrasonic sensors were used as a detector. The ultrasonic sensor was also can be used as a directional detection [10-12]. Ultrasonic sensors are often used in automation tasks to measure distance, position changes, level measurement, such as presence detectors or in special applications. Some types allow you to adjust the sensitivity using a potentiometer or digitally. In the case of sensors that can be connected via the communication interface to the PC, it is possible to set detailed parameters of all the sensor's operating range and measured distances [15-17]

Three sensors were used to detect a person that walked in front of it. We used ultrasonic sensors type HCSRF04. The sensors were installed in the front side of the trolley. Design of the sensors can be seen on figure 1. We assumed that the user will walk in front of the trolley, so that the sensors will detect the user and follow them wherever they go. Sensors detected the user based on the distance. So, we had to set the distance that sensors can detect first.

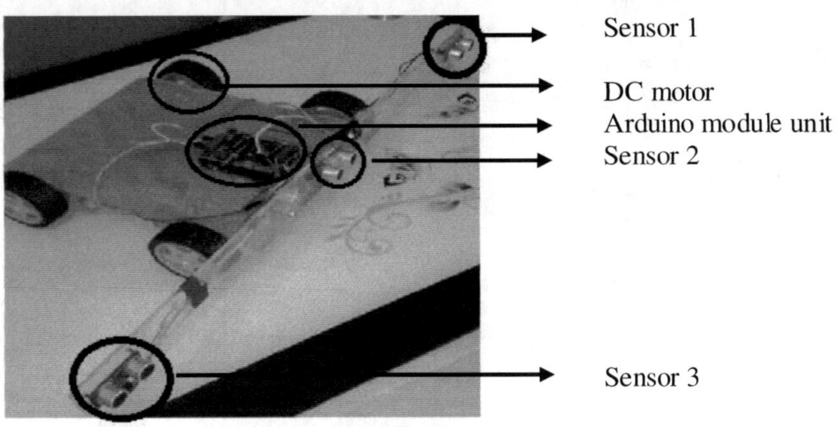

Figure 1. Design of the trolley.

IOP Conf. Series: Journal of Physics: Conf. Series **1007** (2018) 012058 doi:10.1088/1742-6596/1007/1/012058

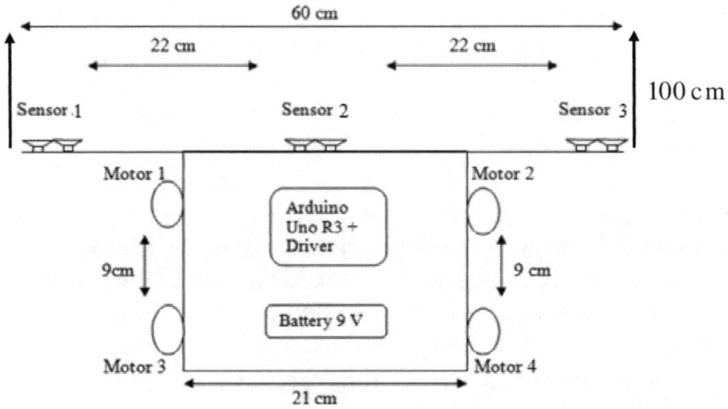

Figure 2. Sensors and dc motors position.

We used 4 DC motor as the driver. Motors will run when the sensors got the right distance and detected somebody that walked in front of it. It will drive the trolley to move forward, backward, turn left, turn right and stop. In the design of the component connection *adafruit* motor shield mounted directly on *Arduino* (plugging) above it, and other power source components such as sensors and motors also comes from *adafruit* shield motor, so that the 9V-DC power from the outside is given to adafruit shield motor. We used Arduino as the controller. Arduino must be programmed first in order to perform its function. *Arduino* programming used the IDE (Integrated Development Environment) program. In order to the prototype could work properly, ultrasonic sensors were used to detect the user. It must be set the value and distance measure first. Error in setting the value and distance measure would cause error detection. The program was designed base on the flowchart in Figure 3.

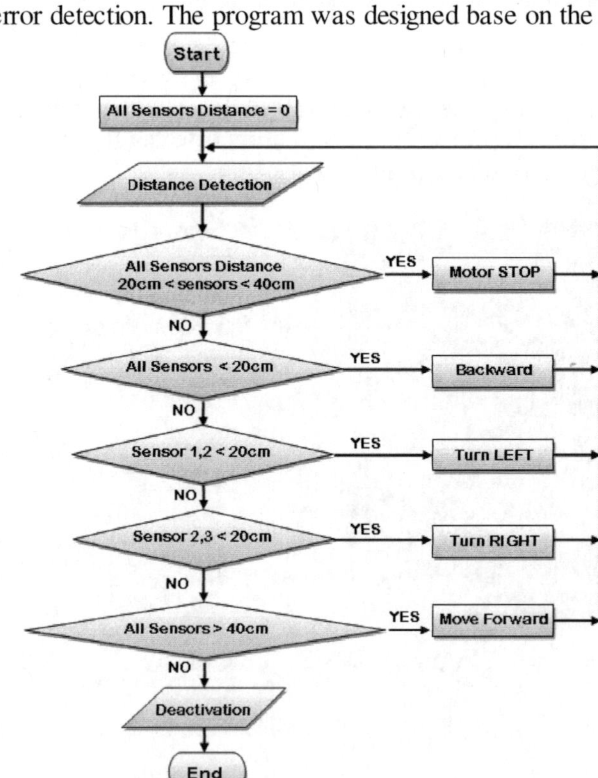

Figure 3. Flowchart of trolley program.

The distance measure must be set on the sensor to detect the customer in front of it. When the value on ultrasonic sensor was 0 cm or not detecting would make the automatic trolley turned on at the beginning. When value on sensor <= 20cm; sensor 2 < 20 cm and sensor 3 < 20 cm, the trolley would move straight. When the value on the sensor 1 > 20 cm and < 40 cm, sensor 2 > 20 cm or < 40 cm and Sensor 3 = 0 cm, the trolley would move to the right. When the value on sensor1 = 0 cm, sensor 2 > 20 cm or < 40 cm, sensor 3 > 20 cm or < 40 cm, the trolley would move to the left. When value on sensor 1 and Sensor 3 = 0 cm, sensor 2 > 20 cm or < 40 cm, the trolley would stop (at rest but still detect). When value on Sensor 1 and Sensor 3 = 0 cm, sensor 2 > 40 cm, the trolley would move forward. When the value on sensor 1 > 40 cm, Sensor 3 = 0 cm and sensor 2 > 40 cm, the trolley would move forward. When value on sensor 1 = 0 cm, Sensor 3 > 40 cm and sensor 2 > 40 cm, the trolley would move forward.

3. System Design

The program was designed so that the trolley can move forward based on the distance data from the sensor results. The program was built for 5 types of processes, that were forward, backward, and stop, left, and right. The program logic for forward, backward, right backward, and backward left using desbin logic, whereas to stop using if logic. The logic designed on the trolley was to make forward movement, change direction to the right, change direction left and backwards. Forward was when the trolley was over a distance of 40 cm, stops when the distance was more than 30 cm and smaller than 39 cm, and backward when the distance was less than 20 cm. The trolley could go forward, backward, stop, right backward, and backward left when it met the condition above. The variables in this program were US1, US2, US3, Jmin, bit1backward, bit2backward, bit3backward. *Jmin* was the minimum distance, *bitbackward* was the data of the form 1 and 0 (binary) obtained from the measurement sensor. If the distance is smaller or equal to Jmin, then *bitbackward* = 1; if the distance is greater or equal to 41, then *bitbackward* = 0.

The working principle of the ultrasonic sensor was transmit or sent a beam of ultrasonic waves, and then measured the time required until the arrival of the reflection of the object [18-20]. Arduino was used as the controller. Arduino was an open source electronic platform on hardware and software that was flexible and easy to use in creating interactive objects or environments [21]. Trolley prototype used the base material of mica plastic (acrylic), and used dc motors as its actuator. We could see the simple block diagram of automatic trolley working principle in figure 4.

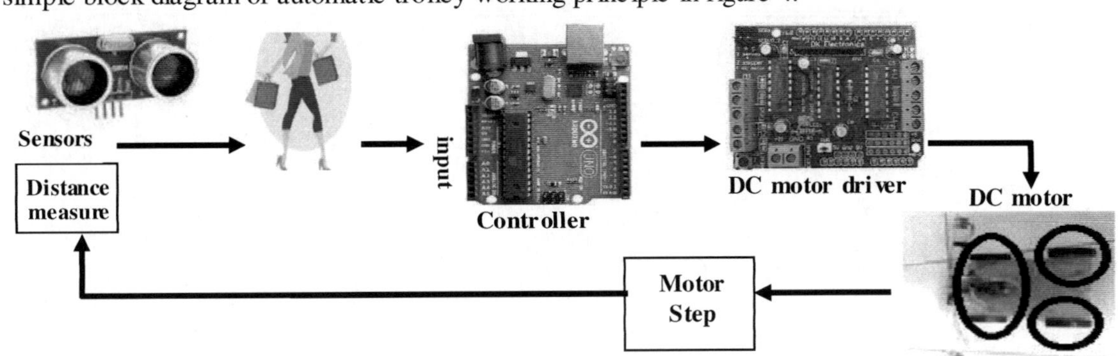

Figure 4. Block diagram of automatic trolley.

By prototyping this method, developers and customers could interact with each other during the system creation process. Frequently, a customer simply defines in general what it wanted without specifying what output was required, what processing and data were needed. On the other hand,

IOP Conf. Series: Journal of Physics: Conf. Series **1007** (2018) 012058 doi:10.1088/1742-6596/1007/1/012058

developers were less concerned about the efficiency of algorithms, operating system capabilities and interfaces that connect human and computer. It took good cooperation between them, so that developers would know exactly what the customer want.

4. Result and Discussions

4.1 Ultrasonic Sensor Test

This test was performed to find out whether the ping sensor had succeeded to detect the distance. Testing was done by adding newping library on Arduino idea software, first sensor was declared as US1, second sensor was declared as US2, third sensor was declared as US3. The measurement results were obtained in centimetre. The test was done to obtain the distance value. It was done by closed and kept the position of the object in front of the sensor, to know the sensitivity when given a barrier object. A few test was done for the same distance.

Testing for sensor sensitivity level was done by comparing the sensor measurement results with the actual distance of the object using a ruler. The results could be seen in Table 1.

Table 1. Comparison between ruler and sensor.

NO.	MEASUREMENT USING RULER (CM)	MEASUREMENT USING SENSOR (CM)
1.	7	6
2.	5	4
3.	20	20
4.	30	30
5.	40	40

4.2 Trolley Test

Overall, the trolley testing on the initial condition when the trolley was activated, the trolley would measure the distance to the object in front of the middle, front left and front right trolley by using ultrasonic sensors located on the center of front right and left trolley body. And then, the distance would be displayed on the computer monitor serial program through Arduino module unit. The trolley test results were shown in Table 2.

Table 2. Trolley Test Result.

Number	Logical	Trolley Design	Result (%)
1	Forward	All motor forward	80 %
2	Backward	All motor backward	80 %
3	Turn Left	Motor 1 and 3 move backward, motor 2 and 4 move forward	70 %
4	Turn Right	Motor 2 and 4 move backward, motor 1 and 3 move forward	70 %
5	Stop	All motor stop	80 %

5. Conclusion

The trolley design could follow the human movement in front of it. It was controlled by the distance and follows the user. Arduino module unit could control the trolley Arduino can control the Trolly to move forward, backward, turn right, turn left and stop. Based on the test results, the trolley succeeded to move forward by 80%, move back 80%, turn left, 70%, turn right 70%, and stop 80%. Trolly can follow human movement but cannot increase speed and distinguish who was its owner. The optimal measurement range of sensor measurement was obtained at a distance of 20 cm upwards to the end of the distance (in this case limited to 100 cm).

References

[1] Kalyani Dawkhar, Shraddha Dhomase, Samruddhi Mahabaleshwarkar. 2015. Electronic Shopping Cart For Effective Shopping based on RFID. International Journal Of Innovative Research In Electrical, Electronics, Instrumentation And Control Engineering. *Vol. 3.*

[2] J.S.Awati, S.B.Awati. 2012. Smart Trolley in Mega Mall. International Journal of Emerging Technology and Advanced Engineering. Vol. 2. Pp 474 – 477.

[3] Ankush Yewatkar, Faiz Inamdar, Raj Singh, Ayushya, Amol Bandal. 2016. Smart Cart with Automatic Billing, Product Information, Product Recommendation Using RFID & Zigbee with Anti-Theft. 7th International Conference on Communication, Computing and Virtualization 2016. PP 793 – 800

[4] Aboli Hanwate 1, Poonam Thakare. 2012. SMART TROLLEY USING RFID. International Journal of Research In Science & Engineering. PP 417 – 420

[5] Ankit A. Agarwal, Saurabh K. Sultania, Gourav Jaiswal, P. Jain. 2011. RFID Based Automatic Shopping Cart. IISTE. VOL. 1. PP 39 – 44.

[6] Francis, J., Drolia, U., Mankodiya, K., Martins, R., Gandhi, R., & Narasimhan, P. (2013, October). MetaBot: Automated and dynamically schedulable robotic behaviors in retail environments. In Robotic and sensors environments (ROSE), 2013 IEEE international symposium on (pp. 148e153).

[7] Nauriana, 2009. *"Design Of Wheel Robot Avoidance Obstruction,".*

[8] N. U. Azizah, 2014. *"Design Prototype Of Distance Detection With Ping Sensor Based OnArduino,"* Islam NegeriSyarifHidayatullah University, Jakarta,

[9] Kirtan Gopal Panda, Deepak Agrawal, Arcade Nshimiyimana, Ashraf Hossain. 2016. Effects of environment on accuracy of ultrasonic sensor operates in millimetre range. Perspectives in Science 8. pp 574 – 576.

[10] S. J. Sokop, 2016. *"Trainer Peripheral Interface Based On Arduino Uno Microcontroller"* UNSRAT, Manado.

[11] Shengbo Eben Li, Guofa Li, Jiaying Yu, Chang Liu, Bo Cheng, Jianqiang Wang, Keqiang Li. 2018. Kalman Filter-Based Tracking Of Moving Objects Using Linear Ultrasonic Sensor Array For Road Vehicles. Mechanical Systems and Signal Processing 98. PP 173–189.

[12] C. Xu, P. Zhang, H. Wang, Y. Li, C. Lv. 2015. Ultrasonic echo waveshape features extraction based on QPSO-matching pursuit for online wear debris discrimination, Mech. Syst. Signal Process. 60–61 pp 301–315.

[13] R. Kuc, M.W. Siegel 1987. Physically based simulation model for acoustic sensor robot navigation, IEEE Trans. Pattern Anal. Mach. Intell. PAMI-9 (766–778).

[14] J. Majchrzak, M. Michalski, G. Wiczyn´ ski. 2009. Distance estimation with a longrange ultrasonic sensor system, IEEE Sensor. 9 (7) 767–773.

[15] Adafruit Motor Shield, 2015. Datasheet Library, Adafruit Industries.

[16] R.Zurawski. 2006. *Embedded Systems Handbook,* Taylor and Francis Group.

[17] L. Koval, J. Vaňuš, P. Bilík. 2016. Distance Measuring by Ultrasonic Sensor. International Federation of Automatic Control. PP 153 – 158.

[18] E. Tuluk, 2012. *"Implementation Of Bird Pest Repellent In Rice Fields Using Ultrasonic Waves Based On Microcontrollers Atmega168,"* Respati University, Yogyakarta.

[19] J. Fraden, 2004. *Handbook of Modern Sensors Physics, Design andApplications.* California : Springer, ISBN 0-387-00750-4.

[20] J. Fraden, 2010. *Handbook of Modern Sensors,*California : Springer, ISBN 978-1-4419-6465-6.

IOP Conf. Series: Journal of Physics: Conf. Series **1007** (2018) 012058 doi:10.1088/1742-6596/1007/1/012058

[21] M. N. Meizani. 2015. "*Prototype Of Electronic Glasses For The Blind Based On Microcontroller Using Ultrasonic Sensor,*" Tanjungpura University.

Acknowledgment
This research was supported by all members in Faculty of technology and computer science, Universitas Prima Indonesia, Indonesia.

IOP Conf. Series: Journal of Physics: Conf. Series **1007** (2018) 012059 doi:10.1088/1742-6596/1007/1/012059

Brain Mapping of Low and High Implusivity based P300 Signals

Arjon Turnip[1], DwiEsti K.[1], Taufik Hidayat[1], and Teddy Hidayat[2]

[1]Technical Implementation Unit for Instrumentation Development
Indonesian Institute of Sciences, Bandung, Indonesia
[2]Faculty of Medicine, UniversitasPadjadjaran, Bandung, Indonesia

Email: *jujhin@gmail.com, arjon.turnip@lipi.go.id

Abstract. Impulsiveness is defined as action without good planning and with little consideration the consequences. Impulsive actions are typically poorly conceived, prematurely expressed, or inappropriate to the undesirable situation such as abuse of drugs. Instead of taking treatment for an addiction subject, it is better take prevention. In this paper, an implusivity detection based EEG-P300 potential is proposed. Twenty four subjects consist of three groups (addiction, methadone, and control) are involved in the experiment. Five different pictures (one picture related drug is used as a target) were randomly flashed to the subjects. The subject is asked to comfortly sit in a chair and to silently count the appearance number of the target. The high amplitude of the P300 component with shortest latency and dominant brain activity are indicated by high implusive group.

1. Introduction

Impulsiveness is a drive based on consciously or unconsciouslydesire. Acting impulsively is an action based on the impulse to express a desire [1, 2]. Acting impulsively also means acting without thinking about the action first, a behavior that wants to get immediate feedback from the environment, or impatient behavior to delay its desire.The cause of a person behaving impulsively is a case related to attention and active behavior, so there is no definite cause. According to experts, there are many factors that encourage a person to behave impulsively, such as the nature that tends to hyperactive, temperamental, family environment influences, gender, and characteristics of parents also play a role.

Impulsivity haswidely contributed to a psychopathology, including bipolar disorder[3], attention deficit hyperactivity disorder [4], borderline personality disorder [5],pathological behaviors associated with Parkinson's disease [6], and substance addictions [7]. Impulsivity is a multidimensional concept which can be represented into at least three distinct forms: trait impulsivity, impulsive action, and impulsive choice [1, 8]. Behavioral tendencies or personality traits, such as impulsivity and sensation-seeking, have been associated with various substance-use disorders relating to alcohol [9], tobacco [10], opiates [11], and cocaine [12]. The relationships between behavioral impulsivity and substance-related attentional biases were significantly stronger than the relationships between self-reported impulsivity and substance-related attentional biases [13]. Measuring of the impulsivity behavioral will also be critical for understanding of the relationships betweenimpulsivity and drug abuse. This findings will help thedevelopment method for prevention and treatment of the drug abuse.

IOP Conf. Series: Journal of Physics: Conf. Series **1007** (2018) 012059 doi:10.1088/1742-6596/1007/1/012059

Electroencephalography (EEG) equipment is now becoming more accessible in the general market, allowing a variety of research related to brain activity easier and wider. The brain-computer interface(BCI) is a method of communication between the user through brain signaling activity with the external device without any help of muscle contraction [14]. The brain signal activity used in BCI can be measured using invasive or noninvasive techniques. The EEG equipment is one of the noninvasive techniques.When many neurons which produce a small electrical voltage become active in the brain, their combination can be detected on the scalp surface with the help of electrode sensors or EEG. These very small electrical signals are amplified and recorded for the latter known as brain waves (represent brain activity that occurs in different areas of the brain). The BCI-based technology developers strive to make BCI more user-friendly, real-time with high accuracy, suitable for voluntary general users such as clinical patients and those with disabilities. Also it must be noted that the current use of BCI is not only used by patients but also by healthy users with various applications [15-25]. Thus, the researcherwill do the best to provide an improvement until the tool can be easily found in the market and is expected to operate well with just one channel. In the end anyone who developed the BCI should be really useful for those who need it. In this paper,ananalysis of recorded EEG signals and brain mappingfor early detection of implusivity is developed.All involved subjects received total brain map analysis. This study demonstrated that there are increased brain disturbances in drug abusers and dependents which can be identified and measured using the EEG tools.

2. Experiment Method

Acquisition is the process of taking data from each subject with the preparation and recording stages. The recording process is conducted using Mitsar EEG-202 equipment which consists of 31 channels.The 10 channels used in the experiment are Fp1, Fp2, F7, F3, Fz, F4, F8, P3, Pz, and P4. The channels of P3, Pz, and P4 are located in the central part of the brain (parietal lobe) which responsible for the subject feels of stimulation. The others channels are used as a drug-related channels (located in the frontal lobe that acts on the control center of the feeling while the drugs attack). In the process of brain signals recording the given the stimulus in the form of flashed images (similar drugs that expected to produce higher impulse)is used. In the experiment things to note are the installation of EEG sensors, tool calibration, and the selection of channels to be used.In the experiment, 16 male rehabilitaion subjects (8 subjects with high impulsivity and 8 subjects with low impulsive) were hired from Hasan Sadikin General Hospital (HSGH) of Bandung-Indonesia.Before the experiment, all subject were interviewed by the medical doctor according to their personal data and time period information of the drug abuse. Then the subject completed the questionnaires. Also, urine tests were performed on all subjects. During the experimental period it has been determined that no subjects are taking prescription. The experiment was supported by the Ethical clearance from RSHS Committee to make sure that the conducted experiment was based on the clinical ethic. Moreover, all subject were provided by written informed consent and rewarded with IDR 150.000 for transport compensation.

Before recording begins, subjects are given an explanation of the recording steps and some rules that the subject must perform. During the recording process, subjects are required to do little movement because the movement of the subject can be read by the instrument that will signal the noise and greatly affect the results of the recording. Signal recording using software compatible with mitsar is WinEEG. The visual stimulus already in video form is activated simultaneously with the WinEEG program, so the subject can respond to the stimulus and the recorded EEG signal. Estimated recording time of each session is 2 seconds. Each subject must pass through 10 sessions already available, so each subject will perform the recording process for 20 seconds.Participants (i.e., has been informed about the entire experiment procedure) were requested to contribute in an experiment of the EEG signals record of the brain activities. The experiemnt consist of three session which are before, ten minutes and one hours after methadone consumption, respectively. The EEG technologist will attach several flat metal discs (electrodes) to different places on the head, using a sticky paste to keep contact between the electrodes and the scalp. During the experiment, the electrodes impedance was retained under 5 KΩ such that a high-quality of the EEG data is obtained.After prerparation, each subject were sitting relaxed on a chair while concentrate to the flashed stimuli on the monitor. They were asked to count how many time the drug picture was appear. The sequence appearance of the flashed picture (shown in Fig. 1) were randomly displayed at the separate but sequentialexperiment. In this paper, we are only considering the channels of Fp1, Fp2, F7, F3, Fz, F4, F8, P3, Pz and P4. The recorded EEG signals were amplified and preprocessed with embended softwares on WinEEGsystem with sampling rate about 500 Hz. The experiment process is shown in Fig. 2. The stimuli was run on a different computer with the EEG machine was attached to record and save the data.

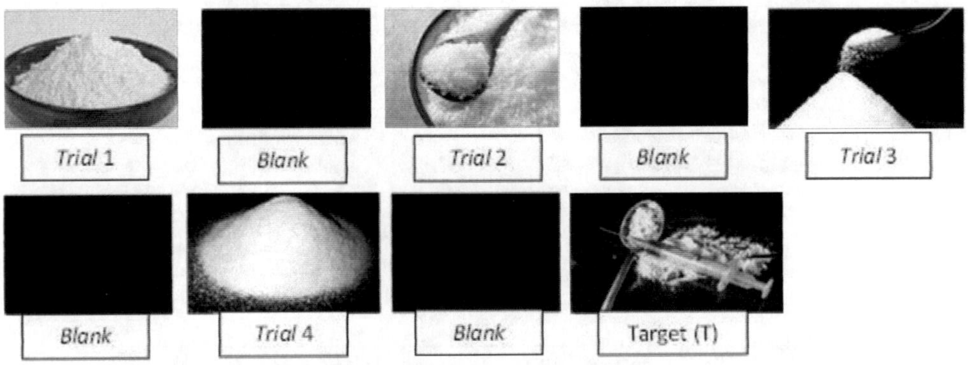

Figure 1. Flashed similar stimuli with (e) as a target [24].

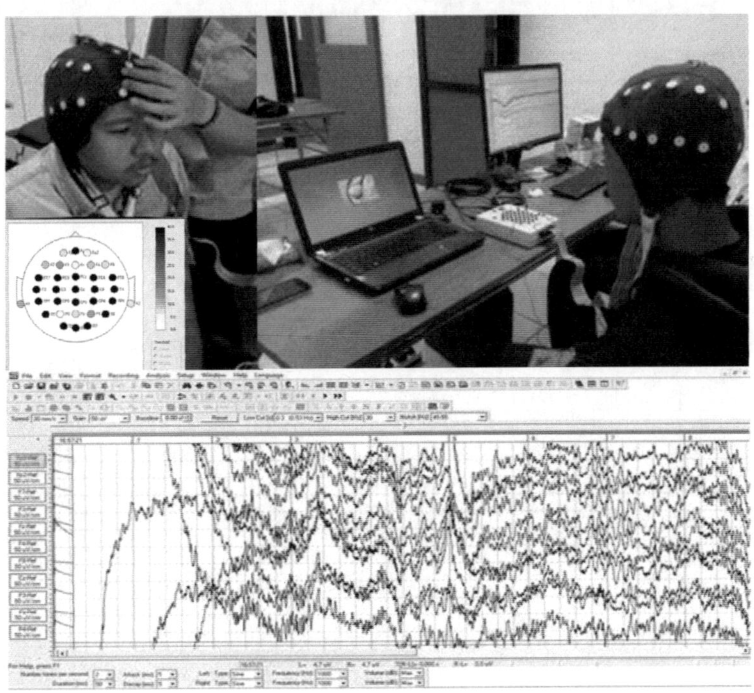

Figure 2. Experiment setup.

3. Results and Discussions

Raw data is the result of recording EEG signals that have not experienced signal processing or still contain noise and artifact. It is characterized by an irregular waveform signal. Raw data of EEG signals for subject A3 and B3 can be seen in Fig. 3(a) and (b) to indicate the low and high implusive group of subjects. The raw data of EEG signal is filtered using bandpass filter with frequency range

IOP Conf. Series: Journal of Physics: Conf. Series **1007** (2018) 012059 doi:10.1088/1742-6596/1007/1/012059

0.5 Hz to 30 Hz because in EEG signal research, the frequency with information is in the range of 0.5 Hz - 30 Hz. So the signal that is outside the frequency will be categorized as an artifact or noise. The bandpass filter results in the EEG signal shown in Fig. 4(a) and (b) which low and high implusive subjects. Figs. 5 – 7 are the extracted signals using wavelet method. Each target for each channels are average from Fig. 5 as shown Fig. 6. Each target and nontarget trial in the extracted signals in Fig. 5 are average in as shown Fig. 7. The average method is is used to shows the gerenal form of extracted P300 component.

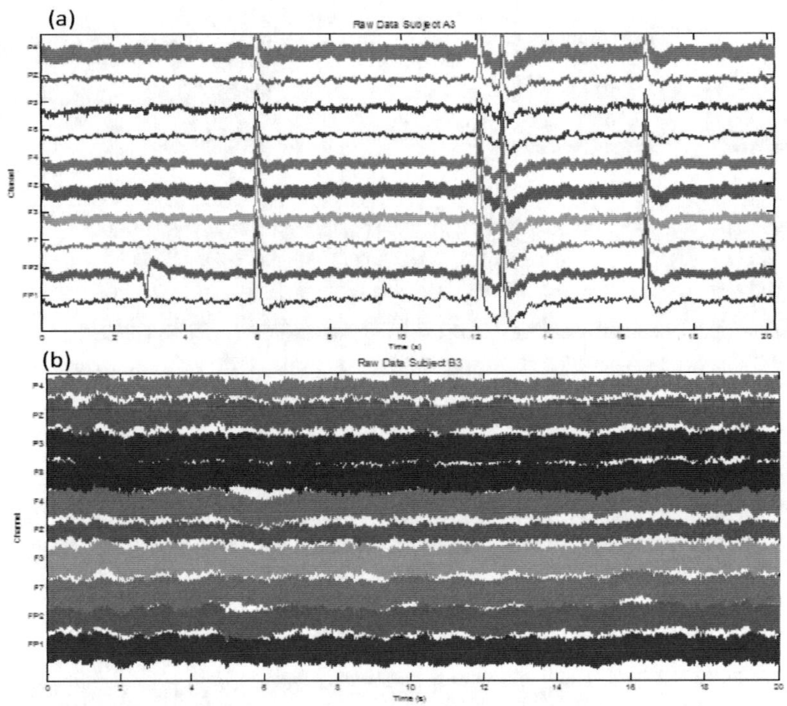

Figure 3. Raw data of (a) low (b) high implusive subjects.

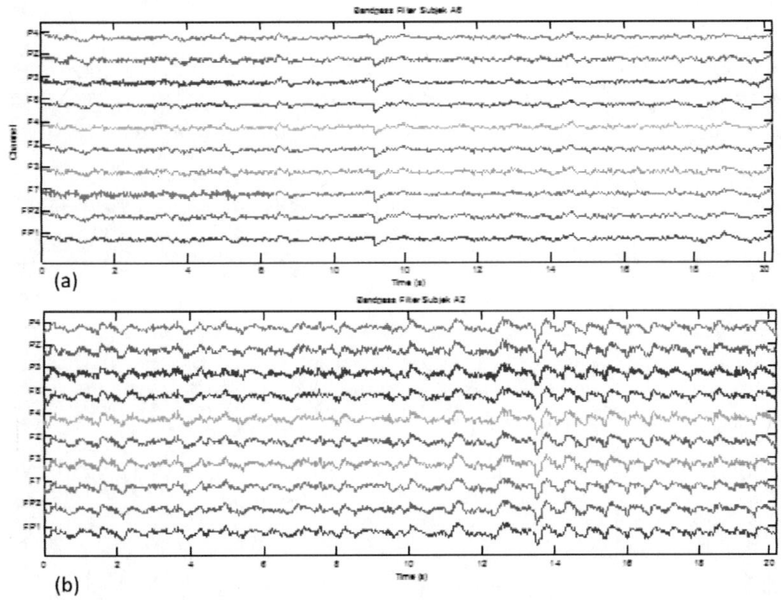

Figure 4. Filtered signals (a) low and (b) high implusive signals.

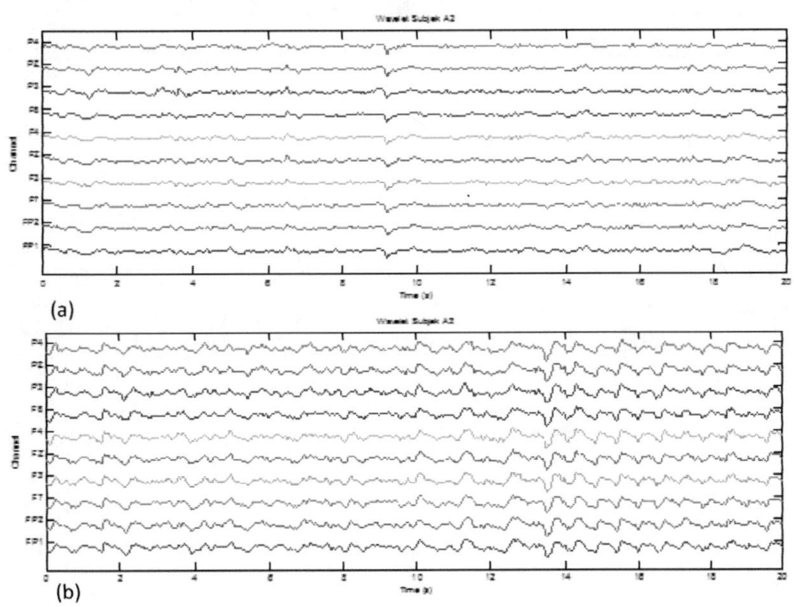

Figure 5. Extracted signal using Wavelet (a) low and (b) high implusive.

IOP Conf. Series: Journal of Physics: Conf. Series **1007** (2018) 012059 doi:10.1088/1742-6596/1007/1/012059

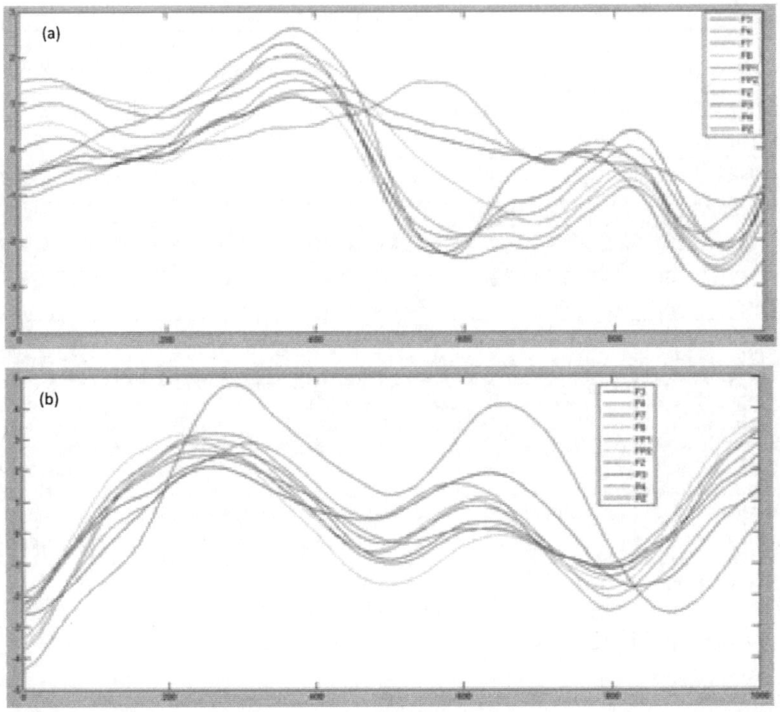

Figure 6. P300 component of each selected channels: (a) low (b) high implusive subjects.

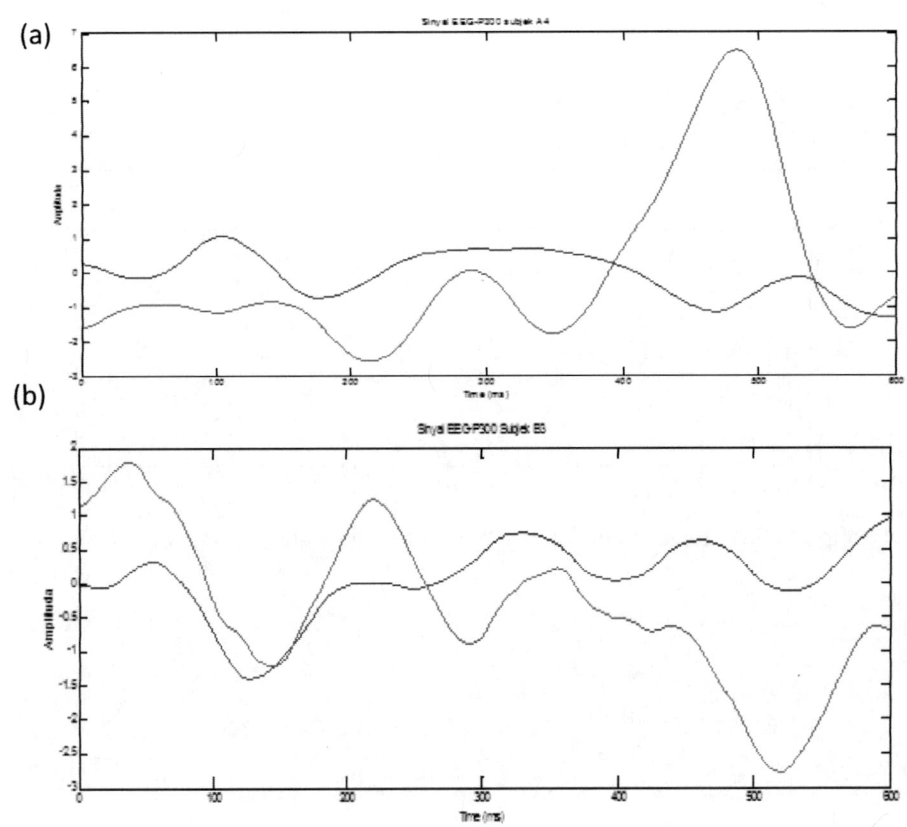

Figure 7. The average P300 component of the target and nontarget: (a) low and (b) high implusive subjects.

The maximum amplitude (Fig. 7(a)) of about 6and 1.5 micro volts for the target of low and high implusive, respectively. The latency is about 600 ms and 250 ms for the target of low and high implusive, respectively. The higher amplitude is obtained for the low implusive but longer latency by mean the responds to the given stimulus is slower compare with the high implusive subject. The high amplitude of the target may triggered by the high desire of the subject. Latency also looks significantly different for low and high implusive subject. This difference can be conclude that the subject is impulsively express their desire. Moreover, subjects tend to be more focused and especially more responsive to stimuli inputs. The difference between these two features seems to be easier to describe the identification of the implusiveness.

Fig. 8 shows brain activity maps (left is low implusive and right is high implusive) which obtained from the P300 signal from two out of the 16experimenter subjects of channels of Fp1, Fp2, F7, F3, Fz, F4, F8, P3, Pz and P4, respectively. Red color indicates more dominant activity of the brain than other areas. Since te experiment is based on the visual stimuli, the higher brain activity is provided in the optical area for most of the channels. For low implusive, the higer brain activity is obtained from channels F8 and Pz whilst for high implusive is obtained from channels F3, Fz, F8 and Pz.

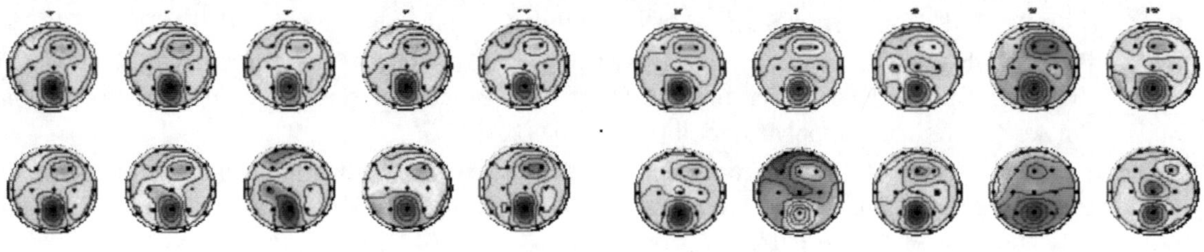

Figure 8. Brain Mapping of (left) low and (right) high implusive subject: channels of Fp1, Fp2, F7, F3, Fz, F4, F8, P3, Pz and P4.

4. Conclusions

In this paper, an implusivity detection based EEG-P300 potential is proposed. Sixteen subjects consist of two groups (low and high implusive) are involved in the experiment. The subject is asked to comfortly sit in a chair and to silently count the appearance number of the target. Based on the extracted P300 component and the mapping of the brain activity, the low and high implusive subject is successfully identified. The high amplitude of the P300 component with shortest latency and dominant brain activity are indicated by high implusive group.

References

[1] Ersche KD, Turton AJ, Pradhan S, et al. Drug addiction endophenotypes: impulsive versus sensation-seeking personality traits. Biol Psychiatry. 2010; 68(8):770–3

[2] Moeller FG, Barratt ES, Dougherty DM, et al. Psychiatric aspects of impulsivity. Am J Psychiatry. 2001; 158(11):1783–93

[3] Winstanley CA. The utility of rat models of impulsivity in developing pharmacotherapies for impulse control disorders. Br J Pharmacol. 2011; 164(4):1301–21

[4] Swann AC. Impulsivity in mania. Curr Psychiatry Rep. 2009; 11(6):481–7

[5] Winstanley CA, Eagle DM, Robbins TW. Behavioral models of impulsivity in relation to ADHD: translation between clinical and preclinical studies. Clin Psychol Rev. 2006; 26(4):379–95

[6] Bornovalova MA, Lejuez CW, Daughters SB, et al. Impulsivity as a common process across borderline personality and substance use disorders. Clin Psychol Rev. 2005; 25(6):790–812

[7] Housden CR, O'Sullivan SS, Joyce EM, et al. Intact reward learning but elevated delay

discounting in Parkinson's disease patients with impulsive-compulsive spectrum behaviors. Neuropsychopharmacology. 2010; 35(11):2155–64
[8] Evenden J. Impulsivity: a discussion of clinical and experimental findings. J Psychopharmacol. 1999; 13(2):180–92
[9] Petry NM. Delay discounting of money and alcohol in actively using alcoholics, currently abstinent alcoholics, and controls. Psychopharmacology (Berl). 2001; 154(3):243–50
[10] Kassel JD, Shiffman S, Gnys M, et al. Psychosocial and personality differences in chippers and regular smokers. Addict Behav. 1994; 19(5):565–75
[11] Madden GJ, Petry NM, Badger GJ, Bickel WK. Impulsive and self-control choices in opioiddependent patients and non-drug-using control participants: drug and monetary rewards. Exp Clin Psychopharmacol. 1997; 5(3):256–62
[12] Garavan H, Hester R. The role of cognitive control in cocaine dependence. Neuropsychol Rev. 2007; 17(3):337–45
[13] Coskunpinar A, Cyders MA. Impulsivity and substance-related attentional bias: A meta-analytic review. Drug Alcohol Depend. 2013; 133(1):1–14
[14] J. R. Wolpaw et al., Brain-computer interface technology: A review of the first international meeting, IEEE Trans. Rehabil. Eng., Vol. 8(2), pp. 164-173, 2000b
[15] A. Turnip, D. E. Kusumandari, M. F. Amri, A. I. Simbolon, M. A. Suhendra, S. IsKandar, F. F. Wirakusumah. "Detection of Drug Effects on Brain Activity using EEG-P300 with Similar Stimuli", IOP Conference Series: Materials Science and Engineering, 2017
[16] S. Ray, M. Haney, C. Hanson, B. Biswal, S. J. Hanson, "Modeling Causal Relationship Between Brain Regions Within the Drug-Cue Processing Network in Chronic Cocaine Smokers," Neuropsychopharmacology, 2015
[17] X. Liang, Y. He, B. J. Salmeron, H. Gu, E.A. Stein, Y. Yang, "Interactions between the Salience and Default-Mode Networks Are Disrupted in Cocaine Addiction. J Neurosci," Vol. 35(21), pp. 8081–8090, 2015
[18] A. Turnip, K.-S. Hong, and M.-Y. Jeong, "Real-time feature extraction of P300 component using adaptive nonlinear principal component analysis", BioMedical Engineering OnLine, vol. 10, 2011
[19] A. Turnip and K. S. Hong, "Classifying mental activities from EEG-P300 signals using adaptive neural network", Int. J. Innov. Comp. Inf. Control, vol. 8(7), 2012
[20] A. Turnip, S. S. Hutagalung, J. Pardede, and, D. Soetraprawata, "P300 detection using multilayer neural networks based adaptive feature extraction method", International Journal of Brain and Cognitive Sciences, vol. 2, no. 5, pp. 63-75, 2013
[21] A. Turnip and M. Siahaan, "Adaptive Principal Component Analysis based Recursive Least Squares for Artifact Removal of EEG Signals", Advanced Science Letters, vol. 20, no.10-12, pp. 2034-2037(4), October 2014
[22] A. Turnip and D. E. Kusumandari, "Improvement of BCI performance through nonlinear independent component analisis extraction", Journal of Computer, vol. 9, no. 3, pp. 688-695, March 2014
[23] A. Turnip, Haryadi, D. Soetraprawata, and D. E. Kusumandari, A "Comparison of Extraction Techniques for the rapid EEG-P300 Signals", Advanced Science Letters, vol. 20, no. 1, pp. 80-85(6), January, 2014
[24] A. Turnip, A. I. Simbolon, M. F. Amri, R. H. Setiadi, and E. Mulyana, "Backpropagation Neural Networks Training for EEG-SSVEP Classification of Emotion Recognition ", Internetworking Indonesian Journal, vol. 9, no.1, pp. 53-57, 2017
[25] A. Turnip, M. F. Amri, M. A. Suhendra, and D. . Kusumandari, "Lie Detection Based EEG-P300 Signal Classified by ANFIS Method" Journal of Telecommunication, Electronic and Computer Engineering, vol. 9, no. 1-5, pp. 107-110, 2017

[26] Y. Hu, B. J. Salmeron, H. Gu, E. A.Stein, Y. Yang Y., "Impaired functional connectivity within and between frontostriatal circuits and its association with compulsive drug use and trait impulsivity in cocaine addiction," JAMA psychiatry, Vol. 72(6), pp.584–592, 2015
[27] K. M. Wisner, E. H. Patzelt, K. O. Lim, A. W. MacDonald, "3rd An intrinsic connectivity network approach to insula-derived dysfunctions among cocaine users," Am J Drug Alcohol Abuse, Vol. 39(6), pp. 403–413, 2013
[28] C. Lebel, T. Warner, J. Colby, et al., "White matter microstructure abnormalities and executive function in adolescents with prenatal cocaine exposure," Psychiatry Res., Vol. 213(2), pp. 161–168, 2013
[29] A. B. Konova, S. J. Moeller, D. Tomasi, R. Z. Goldstein, "Effects of chronic and acute stimulants on brain functional connectivity hubs," Brain Res. 2015
[30] M. R. Mitchell, I.M. Balodis, E.E. Devito, et al., "A preliminary investigation of Stroop-related intrinsic connectivity in cocaine dependence: associations with treatment outcomes," Am J Drug Alcohol Abuse, Vol. 39(6), pp. 392–402, 2013
[31] D. B. Spronk,J. H. van Wel, J. G. Ramaekers, and R. J. Verkes, "Characterizing the cognitive effects of cocaine: A comprehensive review," Neurosci Biobehav Rev., Vol. 37(8), pp. 1838–1859, 2013
[32] S. Potvin, K. Stavro, E. Rizkallah, J. Pelletier, "Cocaine and cognition: A systematic quantitative review," Journal of addiction medicine, Vol.8(5), pp. 368–376, 2014

Acknowledgment

This research was supported by the Technical Implementation Unit for Instrumentation Development, Indonesian Institute of Sciences, Faculty of Medicine (Padjajaran University) and, Insentif Riset Sistem Inovasi Nasional (INSINAS), Kementerian Riset, Teknologi, dan Pendidikan Tinggi (Ristekdikti), Indonesia.

IOP Conf. Series: Journal of Physics: Conf. Series **1007** (2018) 012060 doi:10.1088/1742-6596/1007/1/012060

Brain Mapping of drug addiction in witdrawal condition based P300 Signals

Arjon Turnip[1*], Dwi Esti Kusumandari[1], and Teddy Hidayat[2]

[1]Technical Implementation Unit for Instrumentation Development
Indonesian Institute of Sciences, Bandung, Indonesia
[2]Faculty of Medicine, Universitas Padjadjaran, Bandung, Indonesia

Email: *jujhin@gmail.com, arjon.turnip@lipi.go.id

Abstract. Drug abuse for a long time will slowly cause changes in brain structure and performance. These changes tend to occur in the front of the brain which is directly interfere the concentration and the decision-making process. In this study an experiment involving 10 drug users was performed. The process of recording data with EEG system is conducted during craving condition and 1 hour after taking methadone. From brain mapping results obtained that brain activity tend to occur in the upper layer of the brain during craving conditions and tend to be in the midle layer of the brain after one hour of taking methadone.

1. Introduction

The current sense of health becomes wider and more dynamic. This means that a person's health can not only be measured from the physical, mental, and social aspects, but also can be measured from his productivity. For someone who has not or does not work anymore, productive can be interpreted socially. Lately, drug abuse is something we're used to seeing around us. They do not know what negative impact they get after abusing the drug, which is in their minds is only a moment's pleasure. Increased drug abuse in Indonesia with social and economic impacts is increasingly alarming. Although in medicine most of the class of nakotics, psychotropic, and other adictivesubstance beneficial for the treatment, but if in wrong use will result very harm to individual and society especially young generation.No one has ever been trained and prepared to be a parent, and it is not easy to be a parent especially to be an ideal parent. It is difficult to find early evidence in a drug user, usually a user will be detected after reaching the addiction level. Moreover, until now there is no tool that can be used to detect early abuse of drugs.

The influence of drugs on the brainworks: drugs can change the atmosphere of feelings, the ways of thinking, an awareness and behavior of users. That is why narcotics are called psychoactive substances. There are several kinds of effects of drugs on the brain, such as inhibiting the brain work called depresansia (decrease awareness for drowsiness) such as opioid group (opiate, morphine, heroin, pethidine) and sedatives (sedativa and hypnotics). The drugs also affects the brain which responsible for the life feelings called the limbus systemas the center of pleasure in the brain. The drugs can also stimulate the brain works so-called stimulants that cause the user in fresh and enthusiasm, more confidence, and become familiar with others. However, this can cause the user can not sleep, anxiety, heart pounding in faster andincreases of blood pressure. Examples are amphetamines, ecstasy, methamphetamine, cocaine, and nicotine contained in tobacco. Everything related to the drugs are directly connected to the brain activities. Therefore, by measuring the brain activity with electroencephalography (EEG) equipment, the information of the drugs ubuser can be identified.

EEG equipment is now becoming more accessible in the general market, allowing a variety of research related to brain activity easier and wider. The aquired and processed of the brain signals which communicated with external devices is called brain-computer interface (BCI). The brain signal activity used in BCI can be measured using invasive or noninvasive techniques. The BCI based EEG equipment is one of the noninvasive techniques and it is not only used by patients but also by healthy users with various applications [1-7].Over the last few decades, EEG activities have been widely used to study the brain cognitive dysfunction and neurobiological alterations among drug addicts [8]. In addition to EEG spectral analysis to examine the effects of drug abuser on brain functionality (Davydov and Polunina, 2004; Franken et al., 2004; Polunina and Davydov, 2006; Fingelkurts et al., 2009) [9-12], brain event-related potential (ERP) components were evaluated as a reliable approach to study cognitive abilities related to information processing, selective attention, and memory updating of addicts [8]. Among various ERP components, measuring the amplitude and latency of mismatch negativity, P300, and P600 components in standard condition has attracted special attention in ERP evaluation among addicts [2,13,14]. Additionally, examining cognitive responses to drug-related stimuli showed promising results in providing better understanding of brain functional alterations during withdrawal and abstinence periods associated with addiction traits such as craving and anhedonia [15-16]. For this purpose, various types of ERP paradigms have been designed to probe subjects' evoked cognitive responses using deviant, affective, and drug-related stimuli [8]. P300 is associated with selective attention, memory renewal, motivation, stimulus significance, activation of inhibitory processes [17-19] and has been investigated among opioid dependents [15, 16, 20-24].

In this paper, an application of signal processing before the brain mapping is proposed.Brain mapping has opened important perspectives for the neurophysiological evaluation of drug abuser, for the discrimination of drug effects on the brain and for the study of the relationship between the brain and behavior. All involved subjects received total brain map analysis. This study demonstrated that there are increased brain disturbances in substance abusers and dependents which can be identified and measured using the EEG tools. These disturbances are not found in as great a number of non-drug abusers with similar psychiatric problems. The drug abusing subjects had the greatest number of total brain abnormalities. Moreover, increasing abnormalities were observed in the more severe drug abuser or dependents compared to the normal controls.

2. Experiment Method

An electroencephalogram is a test that measures and records the electrical activity of the subject brain. Special sensors are attached to their head and hooked by wires to a computer. The computer records the brain's electrical activity on the screen.In the experiment, 10 male rehabilitaion subjects were hired from Hasan Sadikin General Hospital (HSGH) of Bandung-Indonesia. Before the experiment, all subject were interviewed by the medical doctor according to their personal data and time period information of the drug abuse. Then the subject completed the questionnaires. Also, urine tests were performed on all subjects. During the experimental period it has been determined that no subjects are taking prescription. The experiment was supported by the Ethical clearance from RSHS Committee to make sure that the conducted experiment was based on the clinical ethic. Moreover, all subject were provided by written informed consent and rewarded with IDR 150.000 for transport compensation.

Participants (i.e., has been informed about the entire experiment procedure) were requested to contribute in an experiment of the EEG signals record of the brain activities. The experiemnt consist of three session which are before, ten minutes and one hours after methadone consumption, respectively. The EEG technologist will attach several flat metal discs (electrodes) to different places on the head, using a sticky paste to keep contact between the electrodes and the scalp. During the experiment, the electrodes impedance was retained under 5 KΩ such that a high-quality of the EEG

IOP Conf. Series: Journal of Physics: Conf. Series **1007** (2018) 012060 doi:10.1088/1742-6596/1007/1/012060

data is obtained.After prerparation, each subject were sitting relaxed on a chair while concentrate to the flashed stimuli on the monitor. They were asked to count how many time the drug picture was appear. The sequence appearance of the flashed picture (shown in Fig. 1) were randomly displayed at the separate but sequentialexperiment. In this paper, we are only considering the channels of Fp1, Fp2, F7, F3, Fz, F4, F8, P3, and P4. The recorded EEG signals were amplified and preprocessed with embended softwares on WinEEGsystem with sampling rate about 500 Hz. The experiment process is shown in Fig. 2. The stimuli was run on a different computer with the EEG machine was attached to record and save the data.

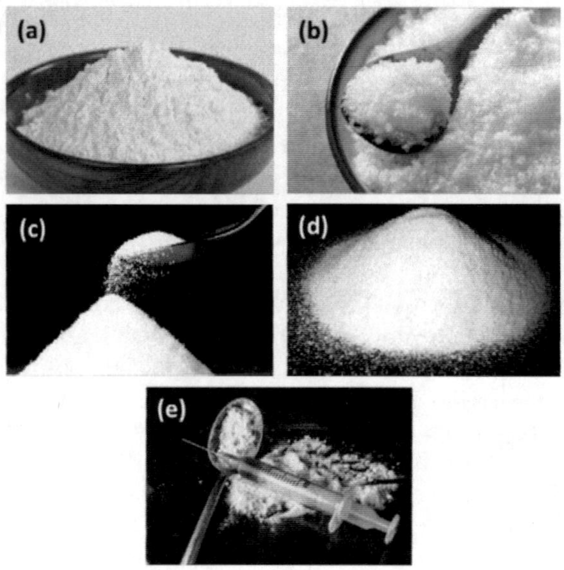

Figure 1. Flashed similar stimuli with (e) as a target [11].

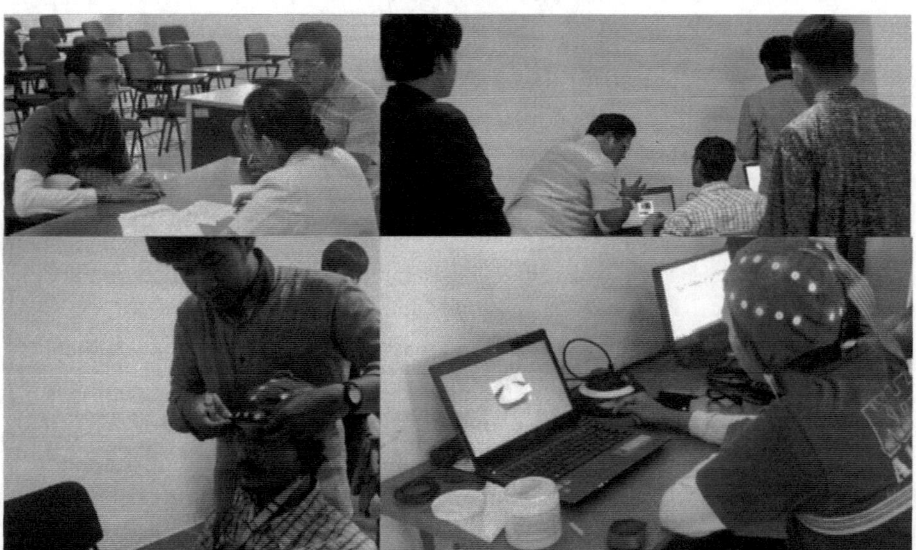

Figure 2. Experiment setup.

3. Results and Discussions

Raw data is the result of recording EEG signals that have not experienced signal processing or still contain noise and artifact. It is characterized by an irregular waveform signal. Raw data of EEG signals for subject 6 can be seen in Fig. 3. The raw data of EEG signal is filtered using bandpass filter with frequency range 0.5 Hz to 30 Hz because in EEG signal research, the frequency with information is in the range of 0.5 Hz - 30 Hz. So the signal that is outside the frequency will be categorized as an artifact or noise. The bandpass filter results in the EEG signal shown in Fig. 4. Figs. 5-7 show the extracted P300 signals using wavelet method which are Fig. 6 and Fig. 7 indicated the average P300 signals from Fig. 5.

To analyze the P300 features extracted during the paradigm experiment, data were fitted into a repeated model while tasks and features (amplitude/latency) were chosen as within-subjects factors. Results showed that the P300 amplitude of each subject before was lower compared with after taking methadone, while their latency was longer. Results also indicated that task type had a significant effect on between target and nontarget differences of P300 amplitude, and the after treatment analysis confirmed the higher P300 amplitude compared with nontarget.

The maximum amplitude of about 4 micro volts when craving condition tends to be higher when compared with the amplitude after one hour of drinking methadone which is about 1.25 micro volts. The high amplitude of the target is also followed by high amplitude of the nontarget, which means that the brain activity is much higher overall and tended to not focus. The differences can be seen in the post-drinking methadone (Fig. 7(b)) where the target amplitude is significantly higher than nontarget. The high amplitude of the target are triggered by the flashed image of the drug, while the high nontarget is suspected to be triggered by the craving condition. Latency also looks significantly different before and after taking methadone ie each about 390 ms and 240 ms. This difference can be interpreted as a change in the condition of the subject body including brain as a whole that is subject conditions become more comfortable after taking the drug. Moreover, subjects tend to be more focused and especially more responsive to stimuli inputs. The difference between these two features seems to be sufficient to describe the effect of drugs on brain activity.

Fig. 8 shows brain activity maps which obtained from the P300 signal from two of the 10 methadone subjects when before taking the drug or methadone (craving condition) and after one hour of taking the drug. Red color indicates more dominant the brain activity than other areas. The first that can be seen from Fig. 8 is that in the craving condition the brain activity tends to spread rather than after taking the drug (more active in the frontal lobe area). The spread of brain activity occurs in subject 1 which is the second line of the condition after taking the drug. These area is an occipital lobe and the increase in activity is thought to be due to visual effects of the stimuli. For both subjects, changes in brain activity in the frontal lobe appear to change from the outer layer to the deeper layers. this pattern of change is most clearly seen in subject 1 i.e., the third line craving condition and the second line after taking the drug.

Similar to the P300 signal occurs in the brain map of the beta wave before and after taking the drug, the concentration of brain activity occurs in the second row for each subject. The brain activity at 10 minutes after taking the drug is not significantly change compared with craving conditions. It can be concluded that the drug is very decisive to the dynamics changes of the brain activity. The Reasons why the results may not be helpful include moving during experiment, taking some medicines, drinking coffee, soda, tea, or other foods that contain caffeine before the test, being unconscious from severe drug poisoning. The EEG records changes in the brain waves that may not be in just one area of the brain. A problem affecting the entire brain such as drug intoxication or metabolic disorders that change the chemical balance in the body, including the brain may cause these kinds of changes.

IOP Conf. Series: Journal of Physics: Conf. Series **1007** (2018) 012060 doi:10.1088/1742-6596/1007/1/012060

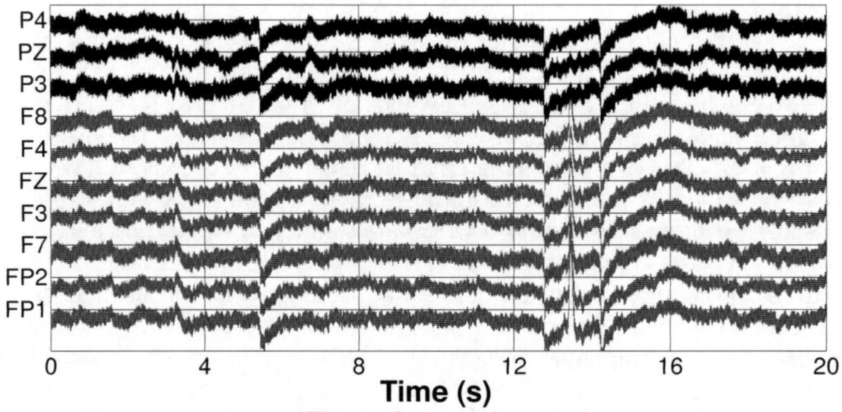

Figure 3. Raw data.

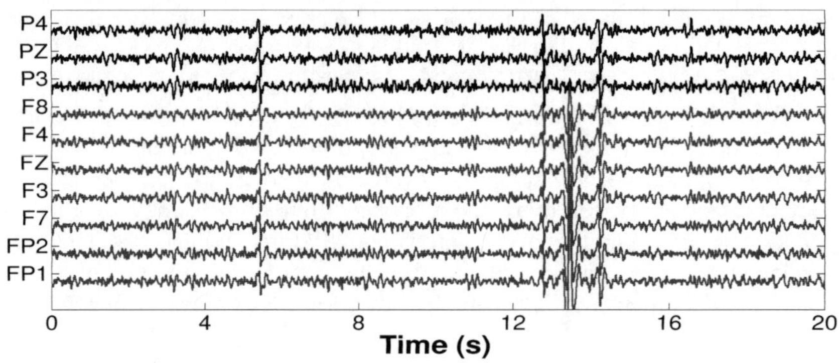

Figure 4. Filtered signals using Bandpass Filter method.

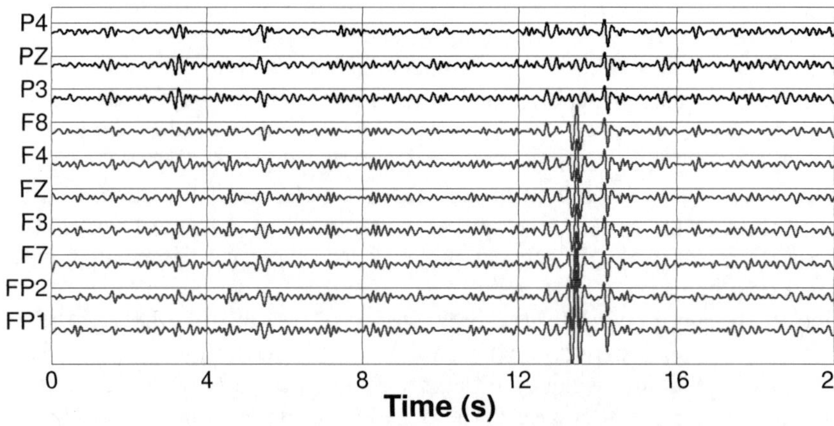

Figure 5. The extracted signals using wavelet method.

IOP Conf. Series: Journal of Physics: Conf. Series **1007** (2018) 012060 doi:10.1088/1742-6596/1007/1/012060

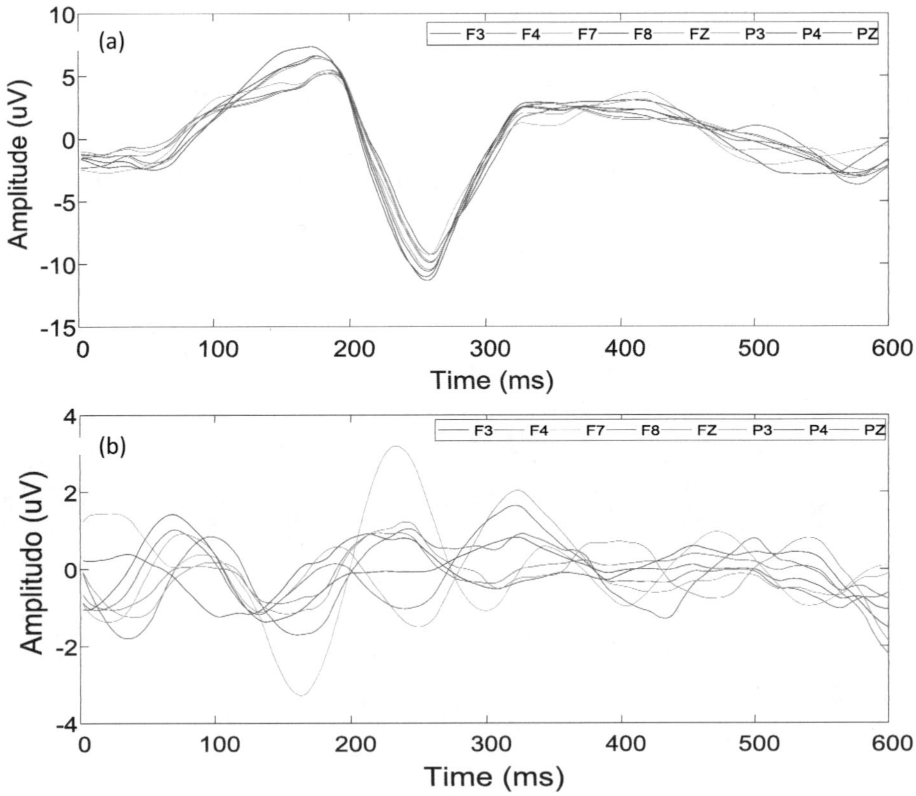

Figure 6. Extracted P300 signals (target) for each considered channels: (a) before and (b) aftertaking methadone.

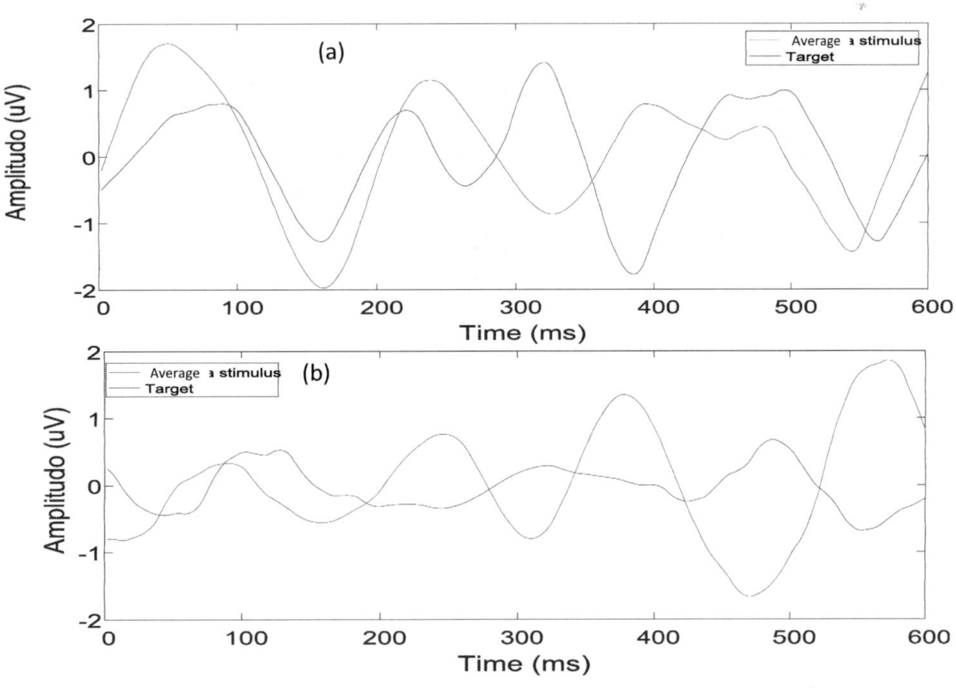

IOP Conf. Series: Journal of Physics: Conf. Series **1007** (2018) 012060 doi:10.1088/1742-6596/1007/1/012060

Figure 7. The Average Extracted P300 signals(target and nontarget): (a) before and (b) after taking methadone.

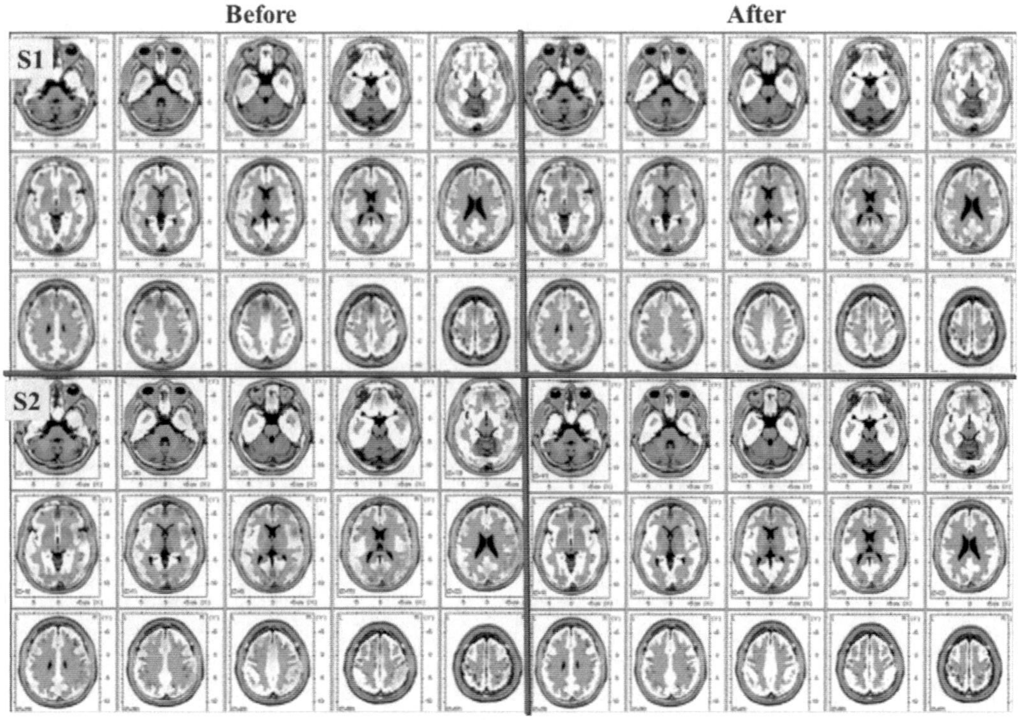

Figure 8. 3D Brain Mapping of the brain activities of two subjects before and after taking methadone.

4. Conclusions

The comparison effect of the drug abuse to the brain activities before (craving condition) and after taking the drug or methadone is provided. It is obtained that the maximum average amplitudes is highly higher in craving condition than after taking the drug. The abnormality of the brain activity in the craving condition is highly related with the drug abuse. The drug abusing subjects had the greatest number of total brain abnormalities. Moreover, increasing abnormalities were observed in the more severe drug abuser or dependents compared to the normal controls.

References

[1] J. R. Wolpaw et al., Brain-computer interface technology: A review of the first international meeting, IEEE Trans. Rehabil. Eng., Vol. 8(2), pp. 164-173, 2000b.

[2] A. Turnip, D. E. Kusumandari, M. F. Amri, A. I. Simbolon, M. A. Suhendra, S. IsKandar, F. F. Wirakusumah. "Detection of Drug Effects on Brain Activity using EEG-P300 with Similar Stimuli", IOP Conference Series: Materials Science and Engineering, 2017.

[3] S. Ray, M. Haney, C. Hanson, B. Biswal, S. J. Hanson, "Modeling Causal Relationship Between Brain Regions Within the Drug-Cue Processing Network in Chronic Cocaine Smokers," Neuropsychopharmacology, 2015.

[4] X. Liang, Y. He, B. J. Salmeron, H. Gu, E.A. Stein, Y. Yang, "Interactions between the Salience and Default-Mode Networks Are Disrupted in Cocaine Addiction. J Neurosci," Vol. 35(21), pp. 8081–8090, 2015.

[5] A. Turnip and M. Siahaan, "Adaptive Principal Component Analysis based Recursive Least Squares for Artifact Removal of EEG Signals", Advanced Science Letters, vol. 20, no.10-12, pp. 2034-2037(4), October 2014.

[6] A. Turnip and D. E. Kusumandari, "Improvement of BCI performance through nonlinear independent component analisis extraction", Journal of Computer, vol. 9, no. 3, pp. 688-695, March 2014.

[7] A. Turnip, A. I. Simbolon, M. F. Amri, R. H. Setiadi, and E. Mulyana, "Backpropagation Neural Networks Training for EEG-SSVEP Classification of Emotion Recognition ", Internetworking Indonesian Journal, vol. 9, no.1, pp. 53-57, 2017.

[8] Motlagh F, Ibrahim F, Menke JM, Rashid R, Seghatoleslam T, Habil H.2016. Neuroelectrophysiological approaches in heroin addictionresearch: a review of literatures. J Neurosci Res 94:297–309.

[9] Davydov DM, Polunina AG. 2004. Heroin abusers' performance on the Tower of London test relates to the baseline EEG alpha2 mean frequency shifts. Prog Neuropsychopharmacol Biol Psychiatry 28:1143–1152.

[10] Franken IHA, Stam CJ, Hendriks VM, van den Brink W. 2004. Electroencephalographic power and coherence analyses suggest altered brain function in abstinent male heroin-dependent patients. Neuropsychobiology 49:105–110.

[11] Polunina AG, Davydov DM. 2006. EEG correlates of Wechsler adult intelligence scale. Int J Neurosci 116:1231–1248.

[12] Fingelkurts AA, Fingelkurts AA, Kivisaari R, Autti T, Borisov S, Puuskari V, Jokela O, Kahkonen S. 2009. Methadone restores local and remote EEG functional connectivity in opioid-dependent patients. Int J Neurosci 119:1469–1493.

[13] A. Turnip, K.-S. Hong, and M.-Y. Jeong, "Real-time feature extraction of P300 component using adaptive nonlinear principal component analysis", BioMedical Engineering OnLine, vol. 10, 2011.

[14] A. Turnip and K. S. Hong, "Classifying mental activities from EEG-P300 signals using adaptive neural network", Int. J. Innov. Comp. Inf. Control, vol. 8(7), 2012.

[15] Lubman DI, Yucel M, Kettle JWL, Scaffidi A, MacKenzie T, Simmons JG, Allen NB. 2009. Responsiveness to drug cues and natural rewards in opiate addiction associations with later heroin use. Arch Gen Psychiatry 66:205–213.

[16] Jiang YP, Liu H, Xu P, Wang Y, Lu GH. 2011. Effect of electroacupuncture intervention on cognition attention bias in heroin addiction abstinence-a dot-probe-based event-related potential study. Chin J Integr Med. 17:267–271.

[17] A. Turnip, S. S. Hutagalung, J. Pardede, and, D. Soetraprawata, "P300 detection using multilayer neural networks based adaptive feature extraction method", International Journal of Brain and Cognitive Sciences, vol. 2, no. 5, pp. 63-75, 2013.

[18] A. Turnip, Haryadi, D. Soetraprawata, and D. E. Kusumandari, A "Comparison of Extraction Techniques for the rapid EEG-P300 Signals", Advanced Science Letters, vol. 20, no. 1, pp. 80-85(6), January, 2014.

[19] A. Turnip, M. F. Amri, M. A. Suhendra, and D. . Kusumandari, "Lie Detection Based EEG-P300 Signal Classified by ANFIS Method" Journal of Telecommunication, Electronic and Computer Engineering, vol. 9, no. 1-5, pp. 107-110, 2017.

[20] Bauer LO. 2001. CNS recovery from cocaine, cocaine and alcohol, or opioid dependence: a P300 study. Clin Neurophysiol 112:1508–1515.

[21] Marques-Teixeira JE, Barbosa MFS. 2005. Emotional states and informational brain processing in drug addicts free of drugs: an ERPs study. Int J Psychiatry Clin Pract 9:213–220.

[22] Singh SM, Basu D, Kohli A, Prabhakar S. 2009. Auditory P300 eventrelated potentials and neurocognitive functions in opioid dependent men and their brothers. Am J Addict 18:198–205.

[23] Ceballos NA, Bauer LO, Houston RJ. 2009. Recent EEG and ERP findings in substance abusers. Clin EEG Neurosci 40:122–128.

[24] Duncan CC, Barry RJ, Connolly JF, Fischer C, Michie PT, N€a€at€anen R,Polich J, Reinvang I, Van Petten C. 2009. Event-related potentials inclinical research: guidelines for eliciting,

recording, and quantifying mismatchnegativity, P300, and N400. Clin Neurophysiol 120:1883–1908.

Acknowledgment

This research was supported by the Technical Implementation Unit for Instrumentation Development, Indonesian Institute of Sciences, Faculty of Medicine (Padjajaran University) and, Insentif Riset Sistem Inovasi Nasional (INSINAS), Kementerian Riset, Teknologi, dan Pendidikan Tinggi (Ristekdikti), Indonesia.

IOP Conf. Series: Journal of Physics: Conf. Series **1007** (2018) 012061 doi:10.1088/1742-6596/1007/1/012061

Convenient synthesis of Mn-doped Zn (O,S) nanoparticle photocatalyst for 4-nitrophenol reduction

Noto Susanto Gultom[1], Hairus Abdullah[1,2], Dong-Hau Kuo[1]*

[1]Department of Materials Science and Engineering, National Taiwan University of Science and Technology, No.43, Sec. 4, Keelung Road, Taipei 10607, Taiwan

[2]Department of Industrial Technology, Universitas Prima Indonesia, Medan, Indonesia

E-mail: *dhkuo@mail.ntust.edu.tw

Abstract. The conversion of 4-nitrophenol as a toxic and waste pollutant to 4-aminophenol as a non-toxic and useful compound by photocatalytic reduction is highly important. In this work, the solid-solution concept by doping was involved to synthesis earth-abundant and green material of Mn-doped Zn(O,S). Zn(O,S) with different Mn doping contents was easily synthesized at low temperature 90°C for 4-NP reduction without using the reducing agent of $NaBH_4$. The Mn-doped Zn(O,S) catalyst exhibited the enhancements in optical and electrochemical properties compared to un-doped Zn(O,S). It was found that 10% Mn-doped Zn(O,S) had the best properties and it could totally reduce 4-NP after 2h photoreactions under low UV illumination. The hydrogen ion was proposed to involve the 4-NP reduction to 4-AP, which is hydrogen ion and electron replaced the oxygen in amino (NO_2) group of 4-NP to form the nitro (NH_2) group. We also proposed the incorporation of Mn in Zn site in the Zn(O,S) host lattice could make the oxygen surface bonding weak for easily forming the oxygen vacancy. The more oxygen vacancy for more hydrogen ion would be generated to consume for 4-NP reduction.

1. Introduction

The development of industry such as pesticides, dyes, drugs, etc has improved the quality of human life on one side. In the other side, it also contributes the pollution to the environment. One of the waste from those industries especially pesticides industry is 4-nitrophenol (4-NP). The 4-NP is well known as nitroaromatic compounds with high toxicity level and hazardous waste [1]. Therefore, the US government through environmental protection agency (EPA) lists the 4-NP as one of the toxic pollutant. By considering these negative effects, the conversion of 4-NP to another compound with less toxic and useful is absolutely needed. One of the ways to convert the 4-NP is a photocatalytic reduction with 4-aminophenol (4-AP) as a product. 4-AP known as non-toxic chemical is largely used for pharmaceuticals, dyestuffs, paracetamol, and antipyretic drugs in industries [2, 3].

Several kinds of semiconductor photocatalysts have been widely used for the 4-NP reduction. However, most of them used the noble metal such as silver, gold, and platinum to improve the catalytic activity [4-7]. In large-scale application, the utilization of noble metal is inefficient due to the high cost in production. In addition, some of the noble metal like silver or gold is not so stable. Sometimes after being re-used for several times, the silver or gold nanoparticle aggregated and lowered the catalytic activity. To overcome this issue, the researchers introduce the stabilizing reagent. However, it makes the synthesis process a little complicated, inefficient and expensive. Therefore,

IOP Conf. Series: Journal of Physics: Conf. Series **1007** (2018) 012061 doi:10.1088/1742-6596/1007/1/012061

designing a photocatalyst with material's criteria of earth abundance in precursors, simple processing, no noble metals, high activity, and good stability is highly important. To specifically point out, the previous reports had generally used the sodium borohydride ($NaBH_4$) as a reducing agent in facilitating the conversion of the 4-NP to 4-AP. They proposed $NaBH_4$ as a hydrogen source to accelerate the reaction. Although the $NaBH_4$ is a cheap chemical, the excess of its chemical might cause environmental pollution.

In the previous report, we had synthesized the Zn(O,S) solid solution by combining the ZnO and ZnS for hydrogen evolution reaction [8]. In this work, we tried to optimize the application of Zn(O,S) by doping concept for 4-NP reduction. Doping concept has been widely applied in many applications in order to obtain the desired properties from a material. The better properties such as electronic and optical properties could be achieved by introducing the small amount of dopant so-called impurity into the host lattice. Manganese is one of the proper candidate dopant from transition metal group. In literatures, manganese had been shown as dopant to enhance the catalytic performance. For example, manganese doped hematite (α-Fe_2O_3) nanorods could enhance the efficiency of photoelectrochemical water splitting [9]. Abduhadir also had reported that Mn-doped WO_3 had the higher catalytic activity compared to un-doped WO_3 for methylene blue degradation [10]. Another report by Faouzi *et al.* showed that Mn-doped ZnO increased the surface area, decreased electron-hole recombination, and finally improved the catalytic activity for methylene orange degradation[11]. Based on the above considerations, we choose the manganese as a dopant to improve the performance of Zn(O,S). By combining the doping and solid solution concepts, the Mn-doped Zn(O,S) photocatalyst was systematically synthesized at low temperature with different Mn contents. Since our catalyst could produce hydrogen, $NaBH_4$ was not been used in this work. The hydrogen ion from the photoreaction Mn-doped Zn(O,S) was proposed for 4-NP reduction and the possible mechanism was elucidated in this paper.

2. Experimental section

The experimental procedures were similar to our previous work [12], except $MnCl_2$ at 0, 2.5, 5, 10, and 20 % base on the amount of zinc acetate hydrate was added for each Mn-doped Zn(O,S). Briefly, 20 mmol of zinc acetate hydrate, 2 mmol of manganese chloride, and 10 mmol of thioacetamide were dissolved in 800 mL deionized water. This solution was stirred at 400 rpm and heated at 90C for 4 h. After washing process the precipitation was dried by rotary evaporator. As-prepared catalyst was characterized by XRD, SEM, UV-VIS spectrometer, PL spectroscopy, HPLC and GC.

3. Results and discussion

3.1 X-ray diffraction Analysis (XRD)

The X-ray diffraction patterns of Mn-doped Zn(O,S) with different Mn contents were evaluated, shown in figure 1. Although the ionic radius of Mn (r= 80 pm) was larger than Zn (r= 74 pm), the X-ray pattern did not show any shift before and after being doped with Mn, which can be attributed to the little amount of Mn in the Zn(O,S) host lattice [13]. As confirmed by analyses of energy dispersive spectra (EDS), the actual amounts of Mn were 0.53, 0.89, 1.98, and 2.45 for the Mn-doped Zn(O,S) with the Mn precursors at 2.5, 5, 10, and 20%, respectively. The similar behavior was also found in the previous studies[14]. However, the XRD intensity of the catalyst after being doped was lower than un-doped, indicating that Mn was successfully doped to the Zn site in the Zn(O,S) host lattice to disturb the lattice flatness. All the peaks located between those of ZnO (JCPDS#65-2880) and ZnS (JCPDS#05-0566) showed a cubic crystal structure, similar to our previous work [8]. The peaks at 29.36°, 48.85°, and 58.38° can be assigned as (111), (220), and (311) planes, respectively. There were no other peaks for secondary phases like manganese oxide or manganese sulfide, which revealed that our catalyst was pure Mn-doped Zn(O,S).

3.2 Morphology and microstructure

IOP Conf. Series: Journal of Physics: Conf. Series **1007** (2018) 012061 doi:10.1088/1742-6596/1007/1/012061

The morphology of Mn-doped Zn(O,S) was observed by FE-SEM. Figure 2a shows the representative image of Mn-doped Zn(O,S). The morphology of aggregate nanoparticle was clearly observed with some small and big particles. The particle size is actually tiny, about 2.5 nm as confirmed by Scherrer calculation. Due to aggregation of some small particles, the particle size of Mn-doped Zn(O,S) had variation from 10 to 200 nm. There is no big difference of morphology before and after doping Mn even we increased the Mn content to 20%. Figures 2b-e show the elemental mapping for each element in Mn-doped Zn(O,S). The dot patterns explained the existence of zinc, manganese, oxygen, and

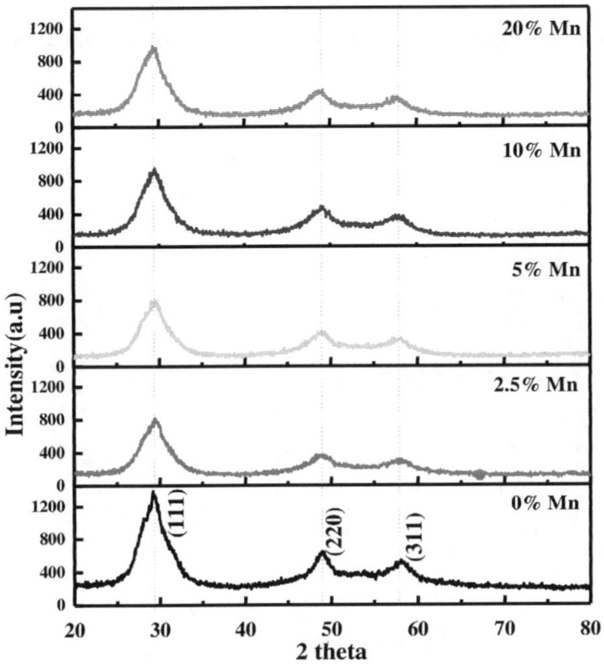

Figure 1. XRD patterns of Mn-doped Zn(O,S)

Figure 2. (a) Representative of FE-SEM image of Mn-doped Zn(O,S) and elemental mapping images for (b) Zn, (c) Mn, (d) O, and (e) S in 10% Mn-doped Zn(O,S).

Sulfur as constituents in our catalyst. The manganese as a dopant is not only much less than zinc but also is homogeneously distributed over Mn-doped Zn(O,S) nanoparticle. This provides another evidence that Mn has been successfully doped in Zn(O,S) host lattice.

IOP Conf. Series: Journal of Physics: Conf. Series **1007** (2018) 012061 doi:10.1088/1742-6596/1007/1/012061

3.3 Optical properties

To study the optical properties of Mn-doped Zn(O,S), the DRS and PL spectra were performed. The small amount of Mn could significantly increase the absorbance at UV range (figure 3). The absorbance of 10% Mn-doped Zn(O,S) was 2-fold higher compared to un-doped Zn(O,S) and had the highest absorbance. Figure 3b exhibits PL emission of Mn-doped Zn(O,S) with different Mn contents. The emission intensity of catalyst before and after doping is not quite different. However, the lowest emission intensity was found for 10% Mn-doped Zn(O,S). This indicates that it has the lowest charges recombination or the fastest charge separation. To further study the optical properties of Mn- doped Zn(O,S) the Tauc plot was performed to calculate the band gap of catalysts. The band gap values of Mn-doped Zn(O,S) were 3.49, 3.62, 3.62, 3.59, and 3.61 eV for using precursors at 0, 2.5, 5, 10 and 20% Mn, respectively. The substitution of Mn in the Zn(O,S) host lattice only enhanced the absorbance and did not significantly change the band gap energy.

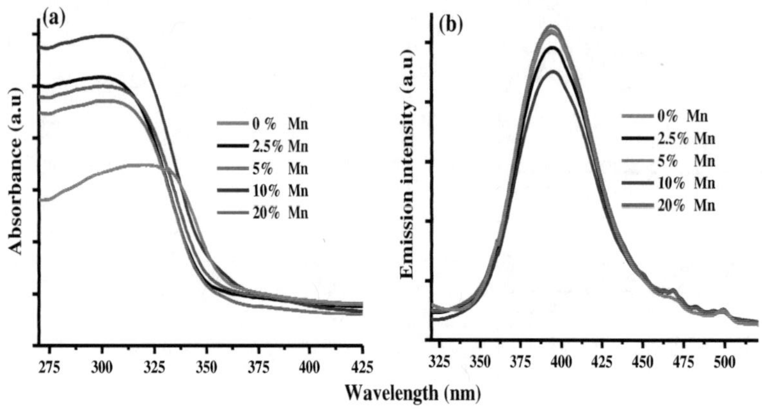

Figure 3. Diffuse reflectance spectra of Mn-doped ZnOS with different Mn contents.

3.4 Electrochemical impedance spectroscopy (EIS) analysis

The EIS analysis is a capable characterization to evaluate the charge transfer at the interface between catalyst and electrode. In the Nyquist plot, the charge transfer ability is indicated by the diameter of the semicircle. The higher diameter indicates slower charge transfer and vice versa. The electrochemical impedance spectra of Mn-doped Zn(O,S) with different Mn contents are shown in figure 4. To determine the charge transfer resistance (R_{ct}) of Mn-doped Zn(O,S), the measurement results were fitted to Randle equation circuit model (inset in the figure 4) in Z view software. The values of charge transfer resistance were 44.2, 27.8, 19.6, 11.3, 31.5 kΩ for using 0, 2.5, 5, 10, and 20% Mn precursors in doped Zn(O,S), respectively. The charge transfer resistance constantly decreased with increasing the Mn content up to 10%. This result concurred with the PL data in figure 3b, in which the 10% Mn-doped Zn(O,S) showed the fastest charge separation.

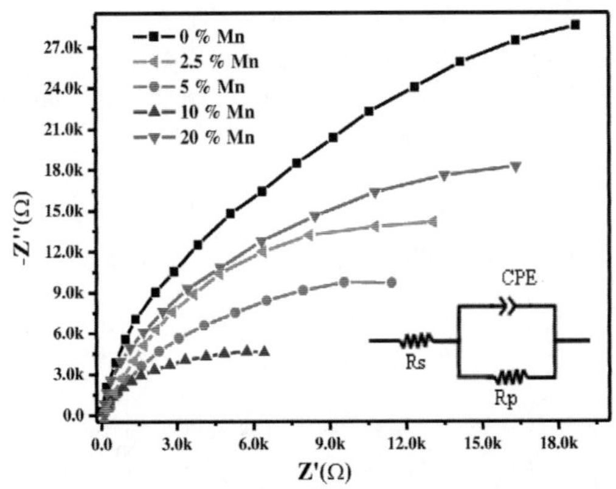

Figure 4. Electrochemical impedance spectra (EIS) of Mn-doped Zn(O,S).

3.5 Photocatalytic 4-NP reduction

Figure 5 shows the time-dependent UV-Vis absorbance spectra of 4-NP reduction at different conditions. It has been known that the peak absorbance of 4-AP and 4-NP located at 300 nm and 317 nm, respectively. In the presence of sodium borohydride the peak of 4-NP shifts to 400 nm due to the formation of 4-nitrophenolate [15]. Since we did not use the $NaBH_4$ in this work, the peak position for 4-NP was at 317 nm. Figure 4a shows the 4-NP reduction without using any catalyst. It can be seen clearly that the absorbance did not change even after 90 min photoreaction. It means there no 4-NP reduction happened in this condition. Next, we tested the performance of un-doped Zn(O,S) with the existence of 10 % ethanol, as shown in figure 5b. The absorbance slowly decreased and the peak shifted a little to the lower wavelength, indicating that the 4-NP reduction happened. However, the reduction was very slow. After introducing 10% Mn-doped Zn(O,S) catalyst without ethanol, the absorbance did not decrease furthermore, as shown in figure 4c and the reduction did not obviously happen. This can be explained as that more hydrogen ions for the significant 4-NP reduction are needed, while the reduction reaction to produce hydrogen ion was very low without using ethanol as sacrificial reagent, as shown in our previous reports [8, 12, 14]. The figure 5d shows the 4-NP reduction in the presence of 10% Mn-doped Zn(O,S) with 10% ethanol. The absorbance regularly decreased with increasing the photoreaction irradiation time and the peak was clearly shifted from 317 nm to 300 nm. Accordingly, at this condition, 4-NP was significantly reduced to 4-AP after 90 min reaction. To confirm the 4-NP reduction, further characterization was performed by using the HPLC.

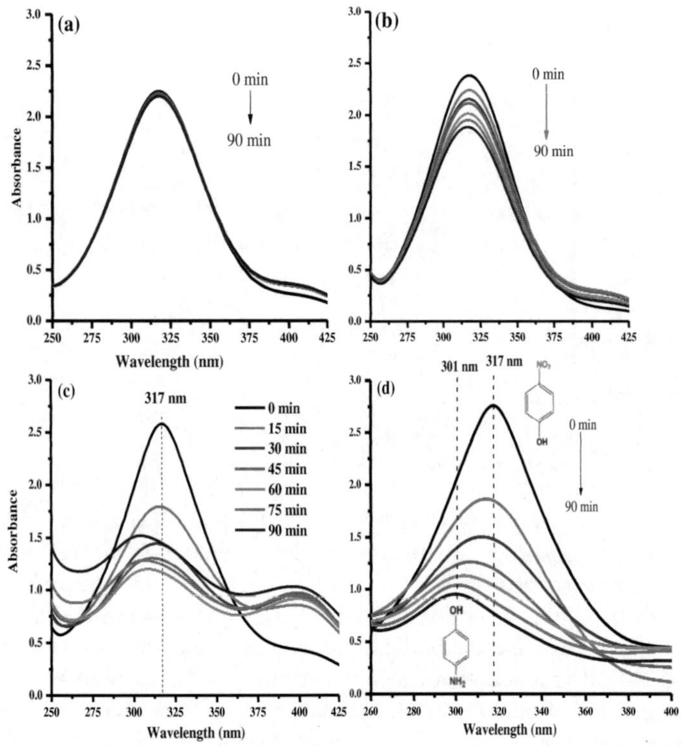

Figure 5. UV-Vis absorbance spectra of the 4-NP reduction. (a) without catalyst, (b) with Mn-free Zn(O,S) and 10% ethanol, (c) with 10% Mn-doped Zn(O,S) without ethanol, and (d) with 10% Mn-doped Zn(O,S) with10% ethanol.

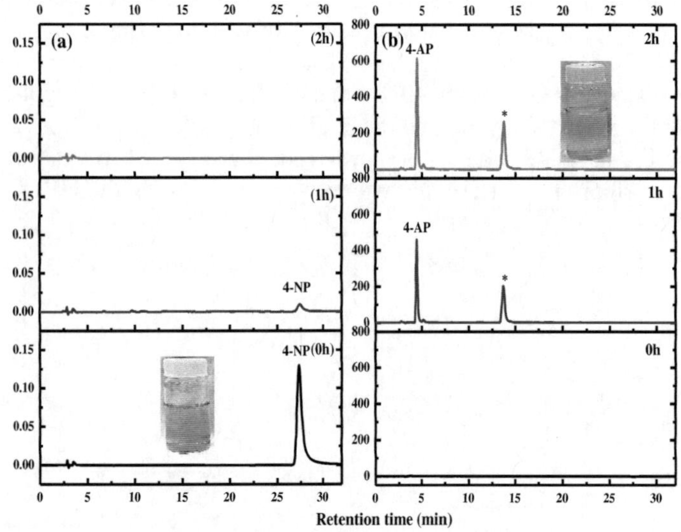

Figure 6. High performance liquid chromatography (HPLC) profiles achieved with (a) UV detector and (b) Fluorescence detector.

IOP Conf. Series: Journal of Physics: Conf. Series **1007** (2018) 012061 doi:10.1088/1742-6596/1007/1/012061

Figure 6 shows high performance liquid chromatography results for the 4-NP reduction with Mn-doped Zn(O,S) in 10% ethanol. Initially, we monitored the commercially available standard samples of 4-NP and 4-AP to identify their retention time. The retention time of 4-NP was found at 27.3 min by UV detector and for 4-AP at 4.3 min by the fluorescence detector. As soon as the photoreaction started, the 2 mL aliquot solution was taken at the time scale of 0h and was checked by the HPLC. It had the quite high intensity at 27.3 min (figure 6a) and there was no peak at 4.3 min (figure 6b), which means no formation of 4-AP. However, after 1 h photoreaction, the 4-NP intensity was much lower than 0h. The peak intensity of 4-NP dramatically decreased (figure 6a) with the concurrently rapid decrease in the 4-AP peak intensity (figure 6b). The disappearance of peak at 27.3 min after 2h proved that 4-NP was totally reduced to 4-AP. The photographs of colour change from yellowish to the clear solution (shown inset figure 6) also reveal the 4-NP reduction to 4-AP. The additional peak in figure 5b (signed as*) might be caused by the polymerization of 4-AP since it can easily polymerize even at the room temperature [16]. This HPLC results corroborate the reduction of 4-NP to 4-AP over the Mn-doped Zn(O,S) catalyst.

4. Kinetic mechanism

Figure 7 displays the schematic kinetic mechanism for hydrogen evolution-involved reaction and the 4-NP reduction to 4-AP. Once the catalyst is illuminated by the photo with energy larger than 3.6 eV, the photocarrier holes and electrons will be generated in the valence band and conduction band, respectively. Those electrons and holes need to migrate to the surface of the catalyst to participate the chemical reaction before their recombination together. The reaction for 4-NP reduction is initiated by the oxidation reaction of water with the formation of oxygen vacancy. Then following by the reduction of water via oxygen vacancy to form the hydrogen ion (H^+). We proposed hydrogen ion involved in the reduction of 4-NP to 4-AP which is hydrogen and electron would replace the oxygen in the nitro (NO_2) group to form the amino (NH_2) group. Therefore, the hydrogen evolution reaction in figure 8 with 30 ppm 4-NP was much lower compared to without 4-NP. Furthermore, we confronted the photocatalytic hydrogen evolution reaction result in figure 8 with HPLC result in figure 6. During the first hour photoreaction, almost no hydrogen gas evolution since most of the hydrogen ion was consumed to reduce the 4-NP. However, after the first-hour reaction the hydrogen evolution a little increased since the remain 4-NP was less (see figure 6a). Thereby, some of the hydrogen ions reduced the 4-NP to 4-AP and some others produced hydrogen gas. Finally, after the second-hour reaction, hydrogen evolution rate dramatically increased since the 4-NP has been totally reduced to 4-AP and all hydrogen ions now were utilized to produce the hydrogen gas. The better performance of the 10% Mn-doped Zn(O,S) is related to several issues such as the bonding strength, absorbance, charge recombination and charge transfer. As proved by PL result and EIS results, the 10% Mn doping showed the lowest charge recombination and the fastest charge transfer. Furthermore, the incorporation Mn is proposed to weaken the bonding and make the surface more active. Mn-doped Zn(O,S) with the best composition has significantly improved its photocatalytic activity.

IOP Conf. Series: Journal of Physics: Conf. Series **1007** (2018) 012061 doi:10.1088/1742-6596/1007/1/012061

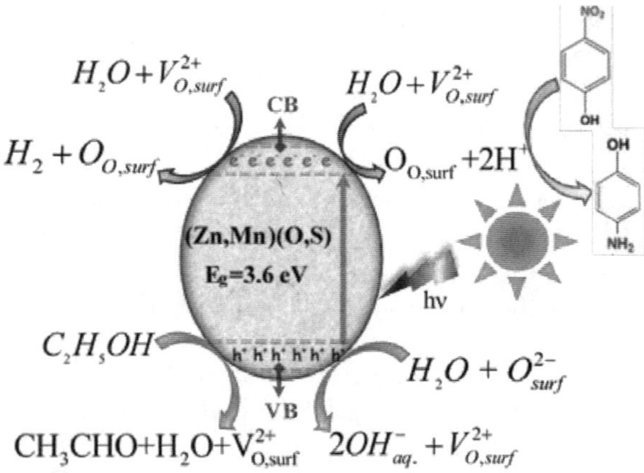

Figure 7. Schematic mechanism of the 4-NP reduction with Mn-doped Zn(O,S).

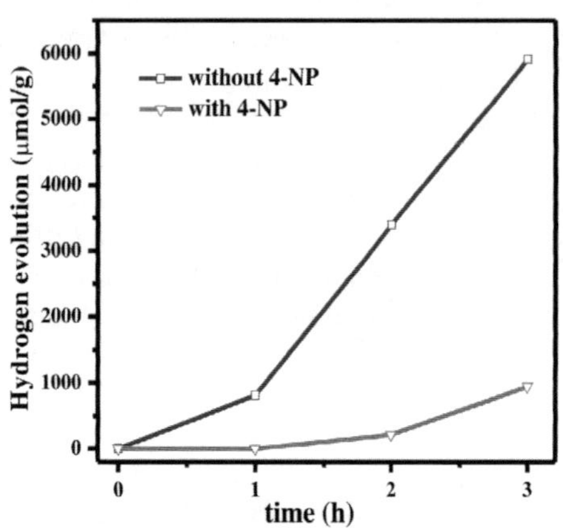

Figure 8. Photocatalytic hydrogen evolution with and without 30 ppm 4-NP.

5. Conclusions

The Mn-doped Zn(O,S) had been successfully synthesized by the convenient method at low temperature with different Mn contents. The incorporation of Mn in the Zn(O,S) host lattice not only increased the absorbance but also decreased the photocarrier electron-hole recombination as confirmed by DRS and PL spectra, respectively. Additionally, the Mn also could significantly decrease the charge transfer resistance. The 10 % Mn-doped to Zn(O,S) host lattice was found the best properties and it could totally reduce the 30 ppm 4–NP to 4-AP after 2-hour photoreaction without using the $NaBH_4$ as reducing agent. The hydrogen ion was proposed to involve the 4-NP reduction to 4-AP, which the hydrogen ion and electron replaced the oxygen in amino (NO_2) group to form the nitro (NH_2) group. The capability of Mn-doped Zn(O,S) to reduce the 4-NP was contributed by the synergetic effect of the high absorbance, low charge recombination, and fast charge transfer. The substitution of Mn in the side of Zn in the Zn(O,S) host lattice also weakened the oxygen bonding to easily forming the oxygen vacancy.

References

[1] L. Yang, S. Luo, Y. Li, Y. Xiao, Q. Kang, and Q. Cai, "High Efficient Photocatalytic Degradation of p-Nitrophenol on a Unique Cu2O/TiO2 p-n Heterojunction Network Catalyst," Enviromental science and technology, vol. 44, pp. 7641-7646, 2010.

[2] J.-J. Lv, A.-J. Wang, X. Ma, R.-Y. Xiang, J.-R. Chen, and J.-J. Feng, "One-pot synthesis of porous Pt-Au nanodendrites supported on reduced graphene oxide nanosheets toward catalytic reduction of 4-nitrophenol," Journal of Materials Chemistry A, vol. 3, pp. 290-296, 2015.

[3] S. R. Subashchandrabose, M. Megharaj, K. Venkateswarlu, and R. Naidu, "p-nitrophenol toxicity to and its removal by three select soil isolates of microalgae: The role of antioxidants," Environmental Toxicology and Chemistry, vol. 31, pp. 1980-1988, 2012.

[4] S. M. Ansar and C. L. Kitchens, "Impact of Gold Nanoparticle Stabilizing Ligands on the Colloidal Catalytic Reduction of 4-Nitrophenol," ACS Catalysis, vol. 6, pp. 5553-5560, 2016.

[5] S. Xiao, W. Xu, H. Ma, and X. Fang, "Size-tunable Ag nanoparticles immobilized in electrospun nanofibers: synthesis, characterization, and application for catalytic reduction of 4-nitrophenol," RSC Advances, vol. 2, pp. 319-327, 2012.

[6] Z. M. El-Bahy, "Preparation and characterization of Pt-promoted NiY and CoY catalysts employed for 4-nitrophenol reduction," Applied Catalysis A: General, vol. 468, pp. 175-183, 2013.

[7] F.-h. Lin and R.-a. Doong, "Highly efficient reduction of 4-nitrophenol by heterostructured gold-magnetite nanocatalysts," Applied Catalysis A: General, vol. 486, pp. 32-41, 2014/09/22/ 2014.

[8] H. Abdullah, D.-H. Kuo, and X. Chen, "High efficient noble metal free Zn(O,S) nanoparticles for hydrogen evolution," International Journal of Hydrogen Energy, vol. 42, pp. 5638-5648, 2016.

[9] Gurudaya, S. Y. Chiam, M. H. Kumar, P. S. Bassi, H. L. Seng, J. Barber, et al., "Improving the Efficiency of Hematite Nanorods for Photoelectrochemical Water Splitting by Doping with Manganese " ACS Appl. Mater. Interfaces vol. 6, p. 5852−5859, 2014.

[10] M. H. S. Abhudhahir and J. Kandasamy, "Photocatalytic effect of manganese doped WO3 and the effect of dopants on degradation of methylene blue," J Mater Sci: Mater Electron vol. 26, pp. 8307-8314, 2015.

[11] F. Achouri, S. Corbel, L. Balan, K. Mozet, E. Girot, G. Medjahdid, et al., "Porous Mn-doped ZnO nanoparticles for enhanced solar and visible light photocatalysis," Materials and Design, vol. 101, pp. 309-316, 2016-04-09 2016.

[12] H. Abdullah, N. S. Gultom, and D.-H. Kuo, "A simple one-pot synthesis of a Zn(O,S)/Ga2O3 nanocomposite photocatalyst for hydrogen production and 4-nitrophenol reduction," New Journal of Chemistry, vol. 41, pp. 12397-12406 2017.

[13] R. D. Shannon, "Revised Effective Ionic Radii and Systematic Studies of Interatomie Distances in Halides and Chaleogenides," Acta Crystallographica, vol. 32, pp. 751-767, 1976.

[14] N. S. Gultom, H. Abdullah, and D.-H. Kuo, "Enhanced photocatalytic hydrogen production of noble-metal free Ni-doped Zn(O,S) in ethanol solution," International Journal of Hydrogen Energy, vol. 42, pp. 25891-25902, 2017.

[15] O. Ahmed Zelekew and D.-H. Kuo, "A two-oxide nanodiode system made of double-layered p-type Ag2O@n-type TiO2 for rapid reduction of 4-nitrophenol," Physical Chemistry Chemical Physics, vol. 18, pp. 4405-4414, 2016.

[16] E. Tsuchida, M. Kaneko, and Y. Kurimura, "Oxidative polymerization of aromatic amines in aqueous solution of iron chelate," Die Makromolekulare Chemie, vol. 132, pp. 209-213, 1970.

Acknowledgments

This work was supported by the Ministry of Science and Technology of Taiwan under Grant numbers MOST 106-2633-E-011-002.

IOP Conf. Series: Journal of Physics: Conf. Series **1007** (2018) 012062 doi:10.1088/1742-6596/1007/1/012062

Photocatalytic antibacterial activity of copper-based nanoparticles under visible light illumination

Zong-Yan Wu[1], Hairus Abdullah[1,2], Dong-Hau Kuo[1*]

[1]Department of Materials Science and Engineering, National Taiwan University of Science and Technology, Taiwan
[2]Department of Industrial Engineering, Universitas Prima Indonesia, Indonesia

E-mail: *dhkuo@mail.ntust.edu.tw

Abstract. Copper oxide and sulfide nanoparticles after annealing treatment at 400 °Chave been characterized and tested for their bactericidal properties toward *Staphylococcus aureus* and *Escherichia coli* under the dark and LED light illuminated conditions. It was found that the nanoparticles with the formation of $CuS/Cu_2S/CuO$ nanoheterostructuresexhibited a great capability of killing *Staphylococcus aureus* and *Escherichia coli* with or without light illumination. The antibacterial activity of the nanoparticles was demonstrated and simply observed with colony counting method. A mechanism of the antibacterial behaviour had been proposed and elucidated in this work.

1. Introduction

Antibiotics have been mainly used in human life for several decades since 1940s. The utilization of antibiotics has been widely known in many applications such as curing for infectious diseases, combating cancer, compromising with immune system to protect human from cancer, preventing diseases, etc.[1]However, the more the antibiotics used, the more possibility the related bacteria will be resistant to it[2]. The mechanisms to protect bacteria from disruptive effect of antibiotics would be developed by altering the antibiotic target sites or metabolism pathway to reduce the drug accumulation from cell.[3, 4] For the illustration of the antibiotic resistance or inactivation of antibiotic mechanism, the beta-lactamase enzymes as product of bacteria can cleave the beta lactam ring, therefore the beta lactam antibiotics such as penicillin will be neutralized. *Salmonella aureus* adapted this mechanism or other resistance mechanism to dramatically reduce the antibacterial effects of large classes of traditional antibiotics. To overcome the inactivation of antibiotic mechanism, the works on improving antibiotics need to be constantly carried out. The pathogen resistance against antibiotic medication has appeared as a critical problem. To overcome this medication problem, nanoparticle application to kill bacteria via a mechanism to cause cell membrane rupture could be one of the great methods[5].

The main mechanism of nanoparticles (NPs) to kill bacteria in the dark andthe light illuminated conditions varies with different interactions between nanoparticle and bacteria membrane. Some proposed mechanisms are related to physical contact between nanoparticles and bacteria membrane, release of nanoparticle ions from its surfaces, and the enhanced specific surface area of nanoparticles as particle size decreases. The final result of the nanoparticle interaction causes a cell wall penetration or cell membrane damage. As claimed by many works, chemical interaction, particle shape, particle size, and zeta potential are the most relevant variables to affect antibacterial activities. The other advanced photochemical interaction between nanoparticles and bacteria cell membrane has been proved with our previous works and many other works.[1-3, 6, 7] Those works had proposed that the radical species which formed during photo reactions on catalyst surfaces played important roles to disrupt bacteria cell membrane. It has been proposed that active species such $OH^{\bullet}$, $O_2^{\bullet-}$, and H_2O_2, known as reactive oxygen species (ROS),[8] would be formed during the photocatalytic bactericide process. The produced ROS would permeate and induce the cell wall rupture as a further effect of cell membrane after an oxidation reaction.

To find new materials with low cost, durable, cost effective, and efficient antibacterial agents is considerably required. Many works[1-5, 9-11] revealed nanoparticle semiconductorswere great materials to kill bacteria via the photocatalytic mechanism. It is absolutely a different mechanism to inactivate bacteria compared to traditional one. The bactericidal property of nanoparticle semiconductor was regardless of the bacterialresistance mechanism as mentioned above.To the best of our knowledge, once the bacteria can be deactivated with photocatalytic mechanism, no resistance pathway can be developed by bacteria to protect themselves. Therefore, works on developing nanoparticle semiconductors as antimicrobial agents are considerably prospective for future medication. Li et al reported that Ag NPs and Ag^+ had strong antimicrobial properties against four bacteria strains, E. coli ATCC 8739, Pseudomonas aeruginosa ATCC 9027, Salmonella aureus ATCC 6538, and Salmonella epidermidis ATCC 12228.[12]However, high cost of silver has caused a drawback in large scale industrial application. The other attempts to minimize the utilization of silver without decreasing its property for antibacterial application had been carried out with other metal incorporation.[2, 13-15] The antibacterial application for NPs could be used in large scale, if the materials involved in their design were not precious metals. Therefore, to progress the utilization of nanoparticles in antibacterial work, materials with the characteristic of low cost, abundance in chemical, high efficiency, low toxicity, and facile powder processing are required. It was found copper sulfides and oxides were promising for further investigation to enhance their antibacterial behavior.[1, 14]

Based on our previous work[1] on SiO_2/Cu_2O for antibacterial application, this work simply focused on improving the antibacterial effect of copper oxide and sulfide NPs with annealing treatments at 400 °C for 3 h. The aim of this work is to indicate copper-based materials with low cost and toxicity are promising to be further improved for antibacterial application. The as-prepared CuO and commercially available CuS materials were tested toward *Staphylococcus aureus (S.aureus)*and *Escherichia coli (E.coli)*bacteria for respectively representing gram positive and negative strainswith initial bacterial concentration of 10^8 CFU/mL under the dark and the20 W LED light illuminated conditions. The bactericidal behaviors of CuO and CuS NPs were demonstrated and elucidated in this work.

2. Experimental procedure

2.1. Materials

All materials in this work were used as received without any further purification treatment.

2.2. Synthesis of copper-based nanoparticles

To synthesize CuO NPs, 4.8 g $Cu(NO_3)_2 \cdot 2.5\ H_2O$ was first dissolved in 300 mL distilled (DI) water by stirring for 30 min, followedby slowly adding 0.8 g NaOH (200 mL DI water) into the solution. The solution was stirred for 2 h at room temperature. The obtained precipitate was washed for three times with alcohol and dried in a rotary evaporator. To form copper oxide NPs, the as-prepared powder was annealed at 400 °C for 3 h in air. The other CuS nanoparticle powder was commercially obtained and annealed with the same condition.

2.3. Characterizations

The morphologies and microstructures of nanoparticles were examined by using field-emission scanning electron microscopy (FE-SEM, JSM 6500F, JEOL, Tokyo, Japan). Powder X-ray diffraction

(XRD) patterns of nanoparticles were recorded by Bruker D2-phaser diffractometer using Cu Kα radiation with a wavelength of 1.5418 Å.

2.4. Luria Bertani (LB) broth and agar preparation
LB broth and LB agar was prepared with 10 g tryptone, 5 g yeast extract, and 10 g NaCl in 1000 mL DI water under vigorous stirring for 1 h. During the preparation, the solution pH was maintained at 7–7.4. To remove all the possible contamination, the as-preparedLB broth solution was heated using autoclave for 20 min with the pressure and temperature were set to 15 psi and 121 °C, respectively. Finally, the LB agar was prepared with the sterilized LB solution by adding 1.5 wt% agar powder.After agar powder was totally dissolved, LB solution will change to solid medium at room temperature. The as-prepared LB broth and agar medium were kept at 4°C for future use.

2.5. Antibacterial experiment
The bactericidal effects of as-prepared CuO and CuS were carefully examined by a simple colony counting method. *S. aureus* and *E. coli* with the concentration of 10^8 cfu/mL was prepared and used in this work. To examine the bactericide effects, 3 mg of as-prepared nanoparticles was dispersed in 1 mL bacterial solution and illuminated with 20 watt LED lamp at a distance of 30 cm for 3 h or without light illumination under a constant shaking. The intensity of LED lamp is 12 W/m^2 as measured by photometer. An aliquot of 0.1 mL was taken hourly and spread uniformly on LB agar medium. The LB agar medium was then kept with face-down at 37 °C for 12 h to cultivate the bacteria. The amount of bacterial colonies grown on the agar medium were counted to simply determine the bactericidal abilities of as-prepared nanoparticles.

3. Results and discussion

3.1. X-ray diffraction (XRD) pattern analysis
All the as-prepared and commercially received nanoparticle powders were examined with XRD measurements to determine their phases before and after annealing process. Figure 1 shows the XRD patterns of CuO and CuS before and after annealing at 400 °C.

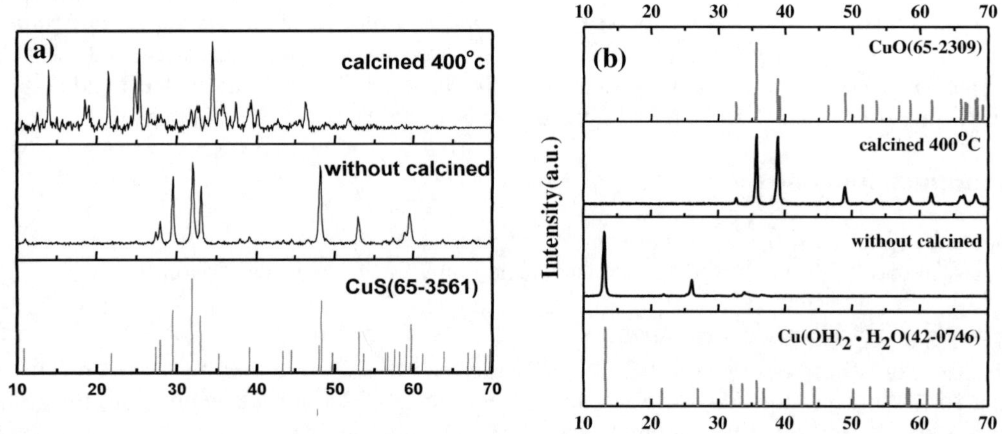

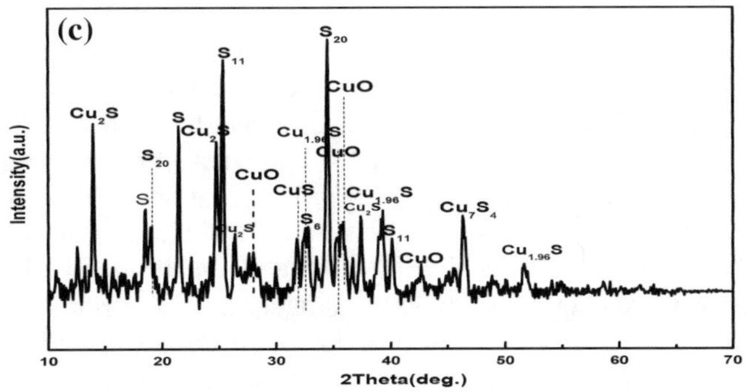

Figure 1. XRD patterns of (a) $Cu(OH)_2$ and (b) CuSbefore and after annealing at 400 °C and (c) phase identification of CuS after annealed at 400 °C.

As shown in Figure 1, $Cu(OH)_2$ phase (PDF#420746) instead of CuO was formed during the chemical synthesis, however CuO phase (PDF#65-2309) was shown up after annealing process. The XRD peaks of commercial CuS nanoparticles after annealing process showed many major phases of CuS, Cu_2S, and CuO which possibly formed nanoheterostructure particles. The peaks of sulfur in nanoparticle were also appeared due to the effect of 400 °C annealing that induced the liberation of sulfur from CuS NPs. Field emission scanning electron microscopy (FE-SEM) analysis was carried out to examine the morphology of the nanoparticles before and after annealing. Figure 2 shows the nanoparticles of CuO and CuS with the average particle sizes of 100–300 nm and 200–1000 nm, respectively.

Figure 2. FE-SEM images of (a,b) CuO and (c,d) CuS nanoparticles (a,c) before and (b,d) after annealing process.

IOP Conf. Series: Journal of Physics: Conf. Series **1007** (2018) 012062 doi:10.1088/1742-6596/1007/1/012062

All the copper-based nanoparticles were tested their antibactericidal properties toward *S. aureus* and *E. coli* with or without light illumination for 3 h, as shown in Figure 3 and 4. The amounts of bacteria colonies grown on agar medium were calculated and summarized in Table 1 and 2. Themore amount of bacteria grown on agar medium indicated the worse antibacterial property of the tested nanoparticles. The antibacterial testing results showed the best nanoparticle for simultaneously killing *S. aureus* and *E. coli* was CuS after annealed at 400 °C under LED light illumination. The great bactericide property was related to the nanoheterostuctures formation between Cu_2S, CuO, and CuS which formed during annealing process and induced the efficient photo generated electron and hole separation to proceed the formation of ROS which was considerably toxic to bacteria. The observable ability to kill bacteria in dark condition was mainly related to the dissolution of copper ions into solution or to the existing photo carriers which could proceed the photo reaction to oxidize cell membrane of bacteria[1].

Table 1. Amounts of *E. coli* bacteria cultivated on agar media for each hour after antibacterial experiments as shown in Figure 3.

Nanoparticles in different condition	*E. coli* amount (cfu)		
	1 h	**2 h**	**3 h**
$Cu(OH)_2$ - Dark	3,200	0	0
$Cu(OH)_2$ – LED Light	1320	0	0
CuS- Dark	$> 10^7$	$> 10^7$	$> 10^7$
CuS - LED Light	$> 10^7$	$> 10^7$	$> 10^7$
CuO (annealing at 400 °C) - Dark	23540	20910	19920
CuO (annealing at 400 °C) - LED Light	21,980	20,050	8,580
CuS (annealing at 400 °C) - Dark	3,280	70	0
CuS (annealing at 400 °C) - LED Light	20	0	0

IOP Conf. Series: Journal of Physics: Conf. Series **1007** (2018) 012062 doi:10.1088/1742-6596/1007/1/012062

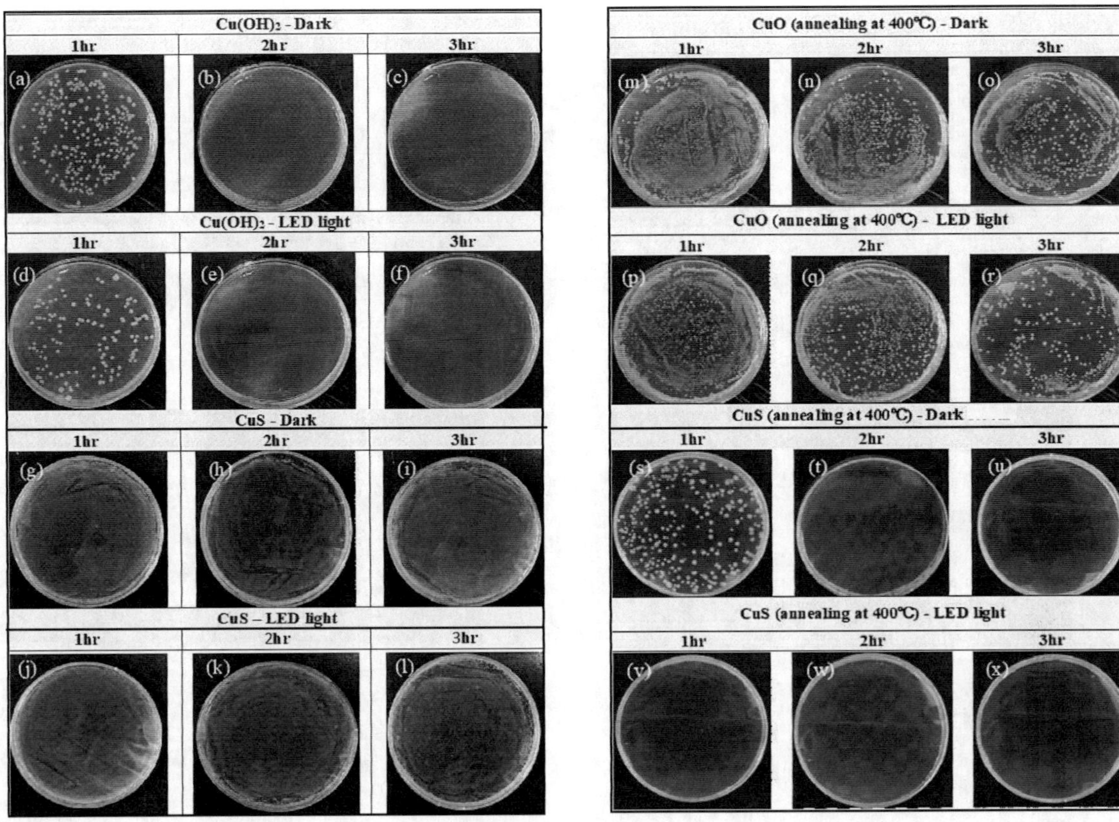

Figure 3. Agar media for *E. coli* cultivation after exposure to $Cu(OH)_2$, CuS, 400 °C-annealed CuO, and 400 °C-annealed CuS NPs for 1 to 3 h under the dark andLED illuminated conditions.

Table 2. Amounts of *S. Aureus* bacteria cultivated on agar media for each hour after antibacterial experiments as shown in Figure 4.

Nanoparticles in different condition	*S. aureus* amount (cfu)		
	1 h	2 h	3 h
$Cu(OH)_2$ - Dark	13,890	370	0
$Cu(OH)_2$ – LED Light	70	40	0
CuS - Dark	> 10 7	> 10 7	> 10 7
CuS - LED Light	> 10 7	> 10 7	> 10 7
CuO (annealing at 400 °C) - Dark	> 107	> 10 7	> 10 7
CuO (annealing at 400 °C) - LED Light	> 107	> 107	> 107
CuS (annealing at 400 °C) - Dark	23,560	70	0
CuS (annealing at 400 °C) - LED Light	20	0	0

IOP Conf. Series: Journal of Physics: Conf. Series **1007** (2018) 012062 doi:10.1088/1742-6596/1007/1/012062

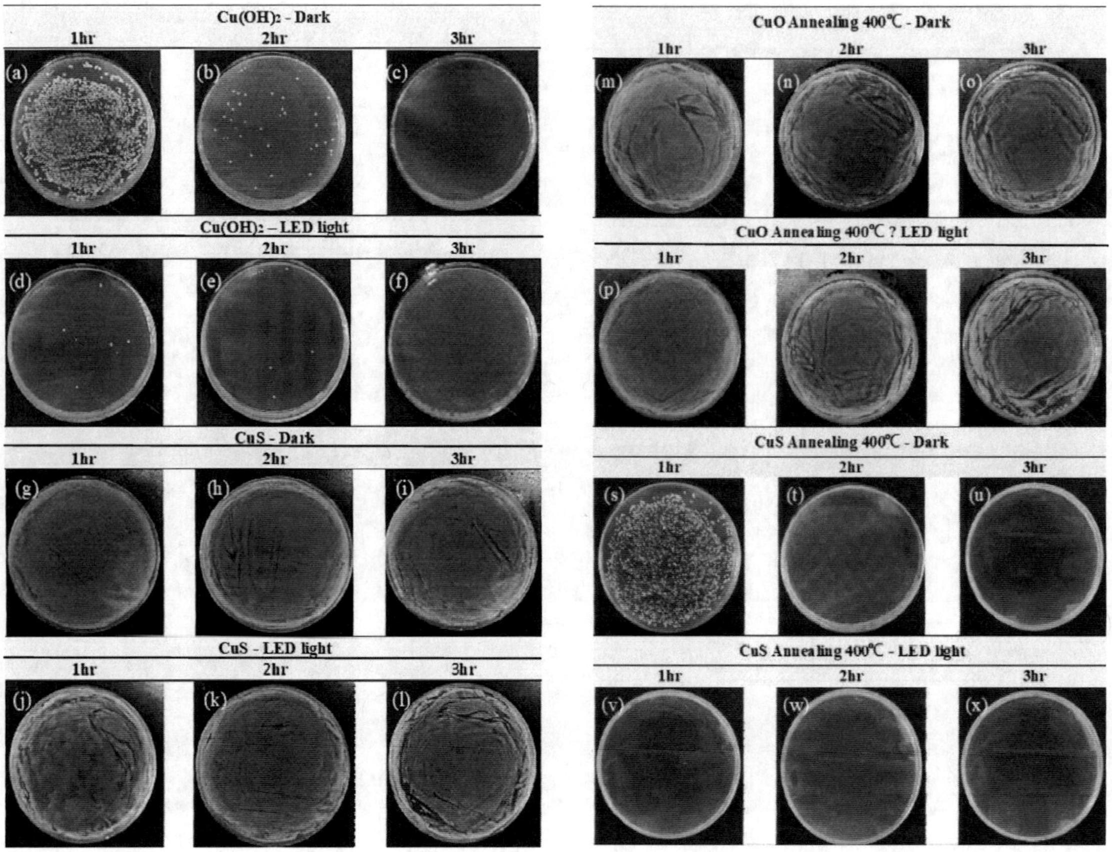

Figure 4. Agar media for *S. aureus* cultivation after exposure to $Cu(OH)_2$, CuS, 400 °C-annealed CuO, and 400 °C-annealed CuS NPs for 1 to 3 h under the dark and LED illuminated conditions.

4. Conclusions

The great photocatalytic bactericidal properties of copper-based NPs were found for the $CuS/Cu_2S/CuO$ nanoheterostructure which induced efficient photo carrier separation to further provide a photo reaction to form ROS leading to rupture cell membrane of bacteria. Copper-based NPs with the characteristics of low cost, abundance in chemical, high efficiency, low toxicity, and facile powder processing were found as prospective materials to combat bacterial infected diseases.

References

[1] Hou YX, Abdullah H, Kuo D H, Leu S J, Gultom N S and Su C H2018Compos. B Eng.133 166.

[2] Chuang K T, Abdullah H, Leu S J, Cheng K B, Kuo D H, Chen H C, Chien J H and Hu W T 2017J Photochem. Photobiol. A: Chem.337 151.

[3] Seil J T and Webster T J 2012Int.J. Nanomedicine7 2767.

[4] Shahidi S, Rashidian M and Dorranian D2017 Opt. Laser Technol. 99 145.

[5] Ioniţă M, Vlăsceanu G M, Watzlawek A A, Voicu S I, Burns J S and Iovu H 2017Compos. B Eng.121, 34.

[6] Malato S, Maldonado M I, Fernández-Ibáñez P, Oller I, Polo I and Sánchez-Moreno R 2016Mat. Sci. Semicond. Process. 42, 15.

[7] Scuderi V, Buccheri M A, Impellizzeri G, Di Mauro A, Rappazzo G, Bergum K, Svensson B G and Privitera V 2016Mat. Sci. Semicond. Process. 42 32.

[8] Abdullah H, Gultom N S and Kuo D H 2017New J. Chem.41 12397.

[9] Goryunova N A, Anshon A V, Karpovich I A, Leonov E I and Orlov V M1971 Phys. Status Solidi A2, K117.
[10] Mazhar ME, Faglia G, Baratto C, Comini E, Zappa D, Kumar R and Sberveglieri G 2014Procedia Eng.87 16.
[11] Wu C Y, Pan Z Q, Liu Z, Wang Y Y, Liang F X, Yu Y Q, Wang L and Luo L B 2017J . Nanopart. Res.19 35.
[12] Li W R, Sun T L, Zhou S L, Ma, Shi Q S, Xie X B and Huang X M 2017Int. Biodeterior.Biodegradation123 304.
[13] Guldiren D and Aydın S 2017Mat. Sci. Eng. C78 826.
[14] Yamamoto L M, Nunes J H B, Ribeiro M A, Ferreira A M d c, Lustri W R and Corbi P P 2017Polyhedron138 168.
[15] Fakhri A, Pourmand M, Khakpour R and Behrouz S 2015J.Photochem. Photobiol. B: Biol. 149 78.

Aknowledgement
This work was supported by the Ministry of Science and Technology of Taiwan under Grant numbers MOST 105-2218-E-011-013.

IOP Conf. Series: Journal of Physics: Conf. Series **1007** (2018) 012063 doi:10.1088/1742-6596/1007/1/012063

Design and Testing of UMM Vertical Ball Mill (UVBM) for producing Aluminium Powder

I S Aisyah[1],Wahyu Caesarendra[2] andAgus Suprihanto[2]

[1]Department of Mechanical Engineering, Faculty of Engineering, University of Muhammadiyah Malang, Malang, Indonesia
[2]Department of Mechanical Engineering, Faculty of Engineering, Diponegoro University, Semarang, Indonesia

E-mail: *siti@umm.ac.id

Abstract. UMM Vertical Ball Mill (UVBM) was intended to be the apparatus to produce metal powder with superior characteristic in production rate while retaining good quality of metal powder. The concept of design was adopting design theory of Phal and Beitz with emphasis on increasing of probability of success in engineering and economy aspects.Since it was designed as vertical ball mill, a new way to produce powder, then it need to be tested for the performance after manufactured. The test on UVBM was carried out by milling of aluminium chip for 5 (five) different milling time of 0.5 hours, 1 hour, 3 hours, 5 hours and 7 hours, and the powder product then be characterized for it morphology and size using Scanning Electron Microscope (SEM) and Sieve.The results of the study were the longer of the milling time, the finer of the powder. From the test results of SEM, the morphology of the powder with 5 variations of milling time were most of the powder in form of flake (flat), small round and angular (irregular). The distribution of powder size was best obtained on the variation of milling time 3 hours, 5 hours, and 7 hours with percentage of 200 mesh in size of 22.14 %, 64 % and 91.25 % respectively.

1. Introduction

Powder metallurgy is a process in transforming metal powder or its alloy with certain size and distribution into product without melting process [1]. Aluminium is categorized as light material and has good corrosion resistant, good conductor properties and other superior properties compared to other. Because of its popular application, it is one of the materials available in metal powder stock. Ball mill (BM) is one of the instruments / apparatus to produce metal powder. In terms of the capability, the powder may have micrometers or even nanometers in size. The mill consists of a vial (cylinder tube) to contain the materials and crusher balls. The mill then is rotated or vibrated in high frequency, depend on the necessity. It makes the materials which are trapped between ball crusher and vial wall will be deformed caused by crushing action between ball and vial wall. The deformation of material makes fragmentation of material structures to smaller structure [2].

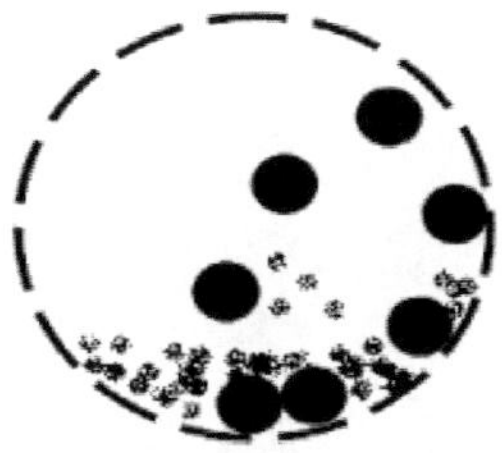

Figure 1. Ball crusher inside vial [2].

In industry, BM has important role in industry to produce powder both for big and small scale industry. To produce fine metal powder, it always need a crusher which suitable with its function and design. There are many type of ball mill to produce powder, one of them are Electric Discharge Assisted Mechanical Milling (EDAMM). This type of ball mill combined ball mill action and electric discharge to synthesize and process metal by promoting rapid transformation of freshly fracture particles and ions associated with the generated dusty plasma. Therefore, it works on the principle of interface transformation rather than diffusional transformation. In was proven that ball mill was able to produce powder both SiC and HfCN [3] [4].

In this research, we designed and tested of indigenous design and less advance of UVBM to produce aluminium powder with further goal to produce micro or even nano in size. This effort is important because nano size particles have taken over the material for drugs to be more efficient; cellular phone and laptop which is smaller in size and more efficient; and the car which more eco-friendly, etc.

2. Material and Methods

2.1. Aluminium

Aluminium was f founded by Sir Humphrey Davy in 1809, it was first reduced from its ore by H.C. Oersted in 1825. Aluminium also found in alloyed metal to increase it mechanical properties by adding Cu, Mg, Si, Mn, Zn, Ni. The addition also makes the alloys have good corrosion resistant, wear resistant and low expansion coefficient [5]. There are many application of aluminium powder such as for rocket propellant, personal hygiene equipment, boosting reliability of computer, refine several exotic materials, as frame materials for car and aircraft to save weight. Its application includes the parts for structural and non-structural one and is the key for developments of aluminium alloys [6].

2.2. UVBM design and testing

A flow process of design and testing of UVBM is presented in in figure 2.

IOP Conf. Series: Journal of Physics: Conf. Series **1007** (2018) 012063 doi:10.1088/1742-6596/1007/1/012063

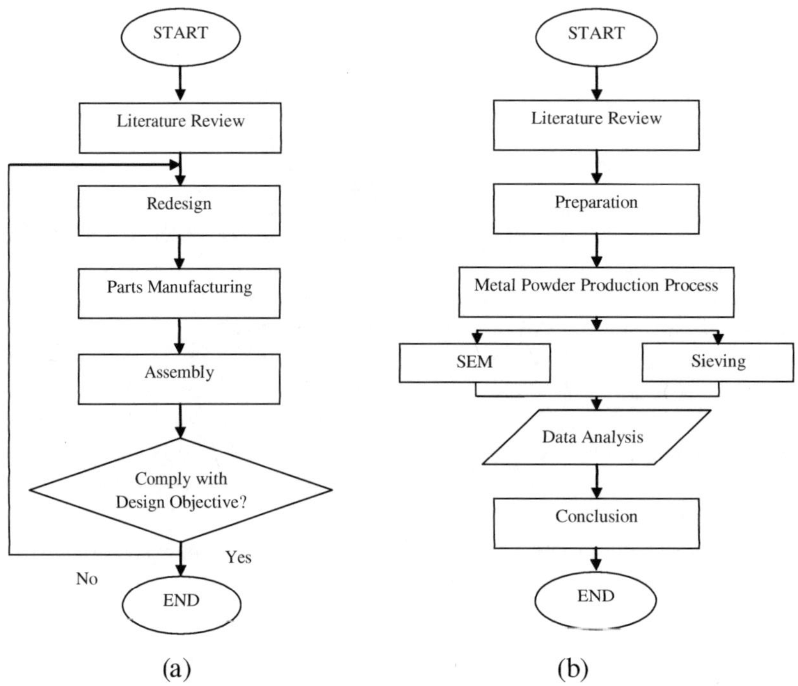

Figure 2. Flow chart of (a) design of UVBM and (b) testing of UVBM.

3. UVBM

3.1. Drawing of UVBM

The design process of UVBM was conducted by adopting Phal and Beitz design methods [7] and the steps were depicted in Figure 2.a. The result was depicted in the drawing as presented in Figure 3.

Figure 3. A 3D CAD drawing of assembled UVBM.

Figure 3 depicted the designed UVBM with the dimension as follows.

- Ball mill tube was manufactured from stainless steel with dimensions of 180 mm in height, 100 mm in diameter, and 10 mm in thickness.
- Ball crusher has diameter of 16 mm.

IOP Conf. Series: Journal of Physics: Conf. Series **1007** (2018) 012063 doi:10.1088/1742-6596/1007/1/012063

- 220 V electric motor (Aliens) with 1400 rpm, reduced to 60 rpm by a pulley drive.
- Pulleys was manufactured from cast iron with dimensions of 13 mm, 78 mm, and 300 mm
- V-belt was type A.
- Shaft was manufactured from cast iron with 12 mm in diameter.
- And frame was manufactured from steel box with dimension of 30 x 30 x 2 mm.

3.2. Testing of UVBM

After manufactured, a series of testing was performed on UVBM to examine the capability of ball mill to produce metal powder. As shown in Figure 2b, the test was conducted by milling aluminium chips for milling time of 0.5 hours, 1 hour, 3 hours, 5 hours, and 7 hours. The powder products then were examined by Scanning Electron Microscope (SEM) and sieving to reveal its morphology and its size. Data analysis then was conducted to make conclusion on its capability in producing metal powder in size distribution, morphology, and time production.

3.3. Characterization of powder

*3.3.1. Morphological analysis by SEM.*Figure 4 shows SEM microphotograph of grinded aluminium alloy (powder) by varying of milling times of 0.5 hours, 1 hour, 3 hours, 5 hours, and 7 hours. It was shown that most of the powder is in flake form and small fraction has circular, oval or irregular shape. In time variation of 0.5 hours, it can be seen that powder size is bigger and coarser than other time variation and mostly in flake shape. When the milling time getting longer, then the powder become finer. For milling time of 1 hour, the powder also became finer eventhough the shape still in flake. 3 hour of milling time variation made the powder became finer and other shape of powder, i.e. oval, circular and irregular, start to form. For last two time variations, i.e. 5 and 7 hours, the trend was also observed. More powder has finer size and the shape also varied in form of oval, circular and irregular shape. Also, it was observed that the thickness of the powder become thinner. The shapes for time variation of 5 hours were dominated by flake and irregular shape while for time variation of 7 hours was flake.

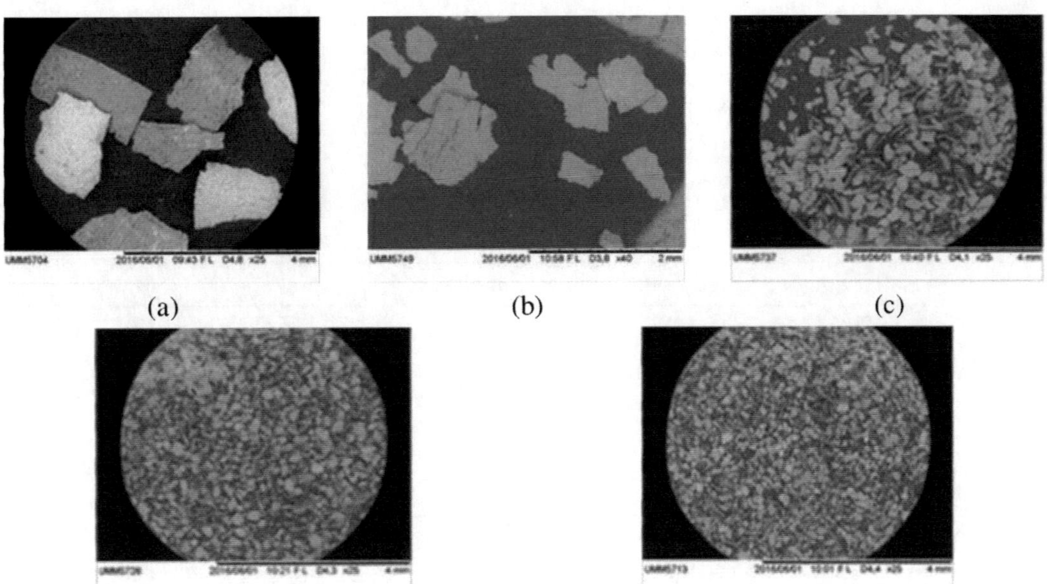
(a) (b) (c)
(d) (e)

Figure 4. SEM of aluminium powder with milling time variations of

(a) 0.5 hours;(b) 1 hour;(c) 3 hours;(d) 5 hours; and (e) 7 hours.

IOP Conf. Series: Journal of Physics: Conf. Series **1007** (2018) 012063 doi:10.1088/1742-6596/1007/1/012063

3.3.2. Powder size analysis by SEM.

Figure 5. SEM of aluminium powder for size analysis with milling time variations of

(a) 0.5 hours;(b) 1 hour;(c) 3 hours;(d) 5 hours; and (e) 7 hours.

Determination of powder size was conducted by SEM and sieving. Figure 5 show SEM microphotograph of aluminium powder with 5 milling time variations of 0.5 hours, 1 hour, 3 hours, 5 hours, and 7 hours. It can be seen that the size of powder was varied because of deformation effect caused by crushing force from crusher ball on aluminium. The tabulation of size of powder was given in Table 2 and the distribution was presented in Figure 6 above. From the table, the average size of powder from time variation of 0.5 hours, 1 hour, 3 hours, 5 hours, and 7 hours was 2028 μm, 753.6 μm, 259.6μm, 158.4μm, and 135.2 μm respectively.

Table 1. Dimension of powder for time varying of 0.5, 1, 3, 5, and 7 hours.

No.	Dimension of Powder (μm)				
	0.5 hour	1 hours	3 hours	5 hours	7 hours
1	2400	494	556	147	89.2
2	2040	694	480	205	256
3	1900	1200	58	298	179
4	1950	562	112	58	52.5
5	1850	818	92.3	84	99.3
sum	10140	3768	1298.3	792	676
mean	2028	753.6	259.6	158.4	135.2

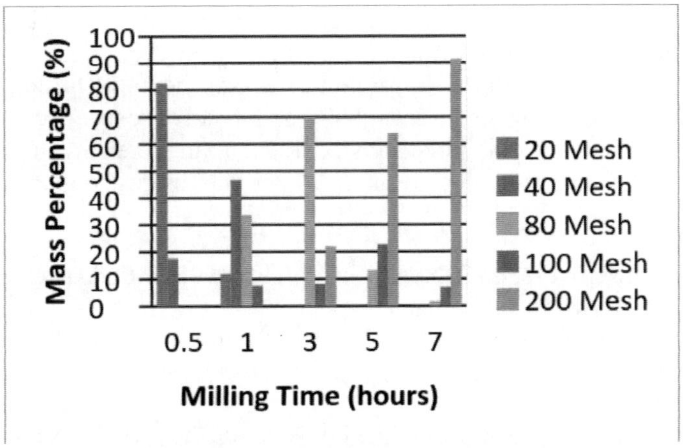

Figure 6. Distribution of powder size.

Graph in Figure 6 above shows mass percentage of metal powder for time varying of 0.5, 1, 3, 5, and 7 hours using 24 ball crusher of 16 mm. From the graph, it was shown that longer milling time gives finer size distribution. The centrifugal force act on rusher ball initiated by rolling ball mill's vial makes balls fell into the powder in the top curvature of ball mill and gave crushing effects. This crushing effect made chips deformed and broke into powder. The longer of milling time made crushing effect become obvious and finer powder was produced in larger number.

Another method of ball mill, i.e. horizontal one [8], produced finest powder at 500 rpm, followed by 700 rpm and 450 rpm. The crusher ball which gave finest one were 5 crusher balls with diameter of 42 mm, then 8 crusher balls with diameter of 27 mm, followed by 6 crusher balls with diameter of 42 mm then 6 crusher balls with diameter of 27 mm, and the last ones were 4 crusher balls with diameter of 42 mm and 7 crusher balls with diameter of 27 mm, the results from this results is better as depicted in Figure 7 below.The main differences in the process was for vertical ball mill gave stronger crusher effect compare to horizontal one so that finer powder produced.

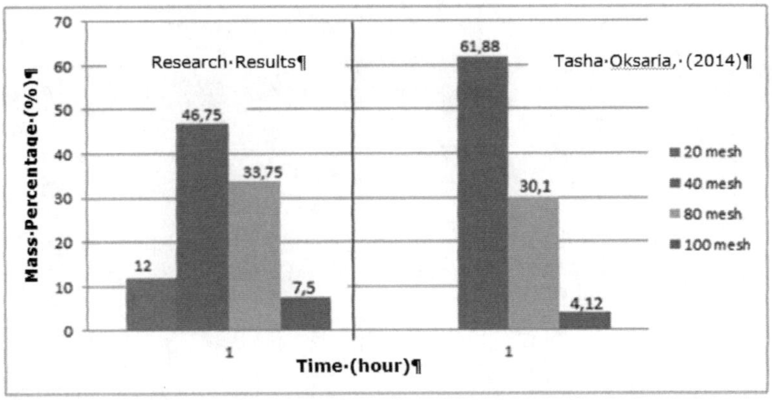

Figure 7. Comparison between our research and Oksaria's [8] one on percentage of sieving mass.

4. Conclusion

Regarding the results of the research, it can be concluded that longer the milling time will give finer powder. The morphology of the powder shows that the dominant shape was flake followed by circular and angular (irregular). Milling time of 7 hour produced finest powder with flake shape. Size distribution of 5 (five) different milling time shows different trends and mainly caused by deformation on aluminium chip caused by crusher balls.

IOP Conf. Series: Journal of Physics: Conf. Series **1007** (2018) 012063 doi:10.1088/1742-6596/1007/1/012063

The size distributions of powder product are as follows. On 5 (five) variations of milling time, i.e. 0.5 hours, 1 hour, 3 hours, 5 hours, and 7 hours the average size were 2028 μm, 753.6 μm, 259.6 μm, 158.4 μm, and 135.2 μm. The finest size distribution was on 7 hours of milling time, followed by 5 hours and 3 hours with mesh 200 were 91.25%, 64%, and 22.14% respectively.Compared to another type of ball mill, i.e. horizontal one, the trend for size distribution were similar with the longer time to mill and more crusher ball employed will give finer powder but our design give faster time.

References

[1] Suresh S., Mortensen A., and Needleman A., (1993). Fundamentals of Metal-matrix Composites, Butterworth-Heinemann
[2] Maurice, D., and Courtney, T. H., (1996). Milling dynamics: Part II Dynamic of SPEX mill and a One-dimensional Mill, Metallurgical and Materials Transactions A, Volume 27A, July, pp.1973-1979
[3] Aisyah I.S.,Wyszomirska M., Calka A., and Wexler, D. (2017). Nitrogenation of hafnium carbide powders in AC and DC plasma by Electrical Discharge Assisted Mechanical Milling. Journal of Alloys and Compounds. 715, pp. 192-198
[4] Aisyah I.S., Wyszomirska M., Calka A., and Wexler, D. (2016). Rapid transformation of hexagonal to cubic silicon carbide (sic) by electric discharge assisted mechanical milling. 36, pp. 1177-1186
[5] Surdia,T., and Saito,S., 1995, Pengetahuan Bahan Teknik, PT. Pradnya Paramita, Jakarta
[6] Gurganus, T.B., 1995, Aluminum Powder Application, Advanced Materials and Processes Journal, Volume 148, Issue 2
[7] Phal. G, and Beitz. W (1977), Engineering Design. Berlin, Heidelberg
[8] Tasha Oksaria.(2014). Pengaruh kecepatan putar drum dan jumlah bola penumbuk pada ball mill terhadap distribusi ukuran serbuk duralumin. Jurnal, Universitas Brawijaya

IOP Conf. Series: Journal of Physics: Conf. Series **1007** (2018) 012064 doi:10.1088/1742-6596/1007/1/012064

Effect of the Machined Surfaces of AISI 4337 Steel to Cutting Conditions on Dry Machining Lathe

Robbi Rahim[1], Suhardi Napid[2*], Abdurrozzaq Hasibuan[3], Siti Rahmah Sibuea[3] and Y Yusmartato[4]

[1]School of Computer and Communication Engineering, Universiti Malaysia Perlis, Malaysia
[2]Department of Mechanical Engineering, Universitas Islam Sumatera Utara, Medan, Indonesia
[3]Department of Industry Engineering, Universitas Islam Sumatera Utara, Medan, Indonesia
[3]Department of Electrical Engineering, Universitas Islam Sumatera Utara, Medan, Indonesia

E-mail: *suhardi.napid@uisu.ac.id

Abstract. The objective of the research is to obtain a cutting condition which has a good chance of realizing dry machining concept on AISI 4337 steel material by studying surface roughness, microstructure and hardness of machining surface. The data generated from the experiment were then processed and analyzed using the standard Taguchi method L9 (3^4) orthogonal array. Testing of dry and wet machining used surface test and microhardness test for each of 27 test specimens. The machining results of the experiments showed that average surface roughness (Raavg) was obtained at optimum cutting conditions when VB 0.1 mm, 0.3 mm and 0.6 mm respectively 1.467 µm, 2.133 µm and 2,800 µm for dry machining while which was carried out by wet machining the results obtained were 1,833 µm, 2,667 µm and 3,000 µm. It can be concluded that dry machining provides better surface quality of machinery results than wet machining. Therefore, dry machining is a good choice that may be realized in the manufacturing and automotive industries.

1. Introduction

Today's wet machining applications are still done in metal cutting industry for cutting steel [1]. This 16% cost is converted to the total cost of production from the automotive industry in America to 48 billion dollars, 1 billion mark in Germany and 78 billion yen in Japan [2]. Based on the above values, the recommended cutting fluid is eliminated then it can be imagined how much value can be saved so that the selling value of the product can be reduced. The impact of a fluid intake on wet machinery is not only a matter of cost but also on health and the environment. To solve the problem of this cutting fluid, the machining experts finally recommend dry machining [3]. Viewed from the ecological aspect that the green machining can also be referred to as dry machining.

Che Haroen and Ginting[4] reported that the tendency of dry machining is better than wet machining to lathe tool steel using a TIN-coated carbide tool. Surface integrity is affected by cutting depth (a), feeding (f), corner radius (r) and cutting fluids [5]. The cost advantages of dry machining include without cooling, no cooling pump, no filter and no sales of chips cleaners [6]. Without the use of cutting fluid in dry machining cutting will certainly lead to high cutting temperatures and greater frictional force occurring in the cutting area because temperature and friction will result in disruption to tool life and surface smoothness.

Canter [7] reports that without cutting fluid, excessive tool wear and worse surface packaging during machining, these two factors will increase the cost of fabrication and reduce productivity. If dry machining is implemented in AISI 4337 steel, then what may be encountered are:

a. AISI 4337 alloy steel is a ductile material, the absence of cutting fluid wiil cause friction and high heat.
b. Decrease in cutting speed and attachment of chip on surface will be obtained by chip / continuous chip with properties of workpiece AISI 4337 ductile.
c. Applications for dry machining will cause the machining surface hardness to be higher than wet machining. Dry machining performed on machining of several metal materials such as cast iron, carbon steel, titanium alloys

To overcome these problems it is necessary to change the machining method from wet machining to dry machining method, thereby reducing production costs and avoiding environmental pollution. From the above exposure it is deemed necessary to analyze the problem of surface roughness, surface hardness and surface microstructure machined as an ingredient for fabricating machine components. This study aims to obtain cutting conditions that have a good chance for the realization of dry machining concept on AISI 4337 steel material.

2. Material and Method

AISI 4337 steel cylindrical rod is machined using CNC machine (length = 200 mm and diameter 50 mm) using CNC machine. The standard condition of the workpiece with a ratio of length and diameter of more than 10 is not recommended on metal cutting [8]. In analyzing surface roughness and surface hardness was carried out with variations of VB wear, cutting variation and different tool geometric, see table 1 below. The result of machining lathe has 9 different cutting forms when performed on wet machining as well as dry machining.

By three variations of VB wear, ie 0.1 mm, 0.3 mm and 0.6 mm of 9 cutting forms, one optimum cutting method was chosen to obtain 3 optimum cutting forms for wet and dry machining. Thus can be compared wet and dry machinery results. Measurement of surface hardness of machining result used Micro hardness test equipment while surface roughness with surface test.

Table 1. Plan for Variations VB = 0.1 mm; 0.3 mm and VB = 0.6 mm with wet machining and dry machining.

	FACTORS			
Number of Experiments	V(m/min)	a (mm)	F(mm/r)	Gp(^{o})
HPB1;HPK1	200	1.0	0.15	6
HPB2;HPK2	200	1.5	0.20	12
HPB3;HPK3	200	2.0	0.25	18
HPB4;HPK4	250	1.0	0.20	18
HPB5;HPK5	250	1.5	0.25	6
HPB6;HPK6	250	2.0	0.15	12
HPB7;HPK7	300	1.0	0.25	12
HPB8;HPK8	300	1.5	0.15	18
HPB9;HPK9	300	2.0	0.20	6

Characteristics of the S / N ratio can be divided into three categories:

a. Nominal best characteristic:

$$S/N = 10\log \frac{\bar{y}}{s^2_y} \tag{1}$$

b. The smaller The Better:

$$S/N = -10\log \frac{1}{n}\left(\sum y^2\right) \tag{2}$$

c. The Larger The Better:

$$S/N = -10\log \frac{1}{n}\left(\sum \frac{1}{y^2}\right) \tag{3}$$

Data collected, processed and then analyzed using Taguchi method. The experimental design was carried out with the standard L 9 (3^4) orthogonal array. Standard L9 (3^4) orthogonal array has nine lines, three levels and four factors [9]

3. Results and Discussion

VB wear and tear connection with surface roughness at optimum cutting conditions for dry and wet machining with the meaning of a larger wear curve followed by increased roughness. The ratio of Ra values to dry and wet machining is not significant where dry machining has better curve characteristics because the Ra value is lower for the same cutting condition, see figure 1 below:

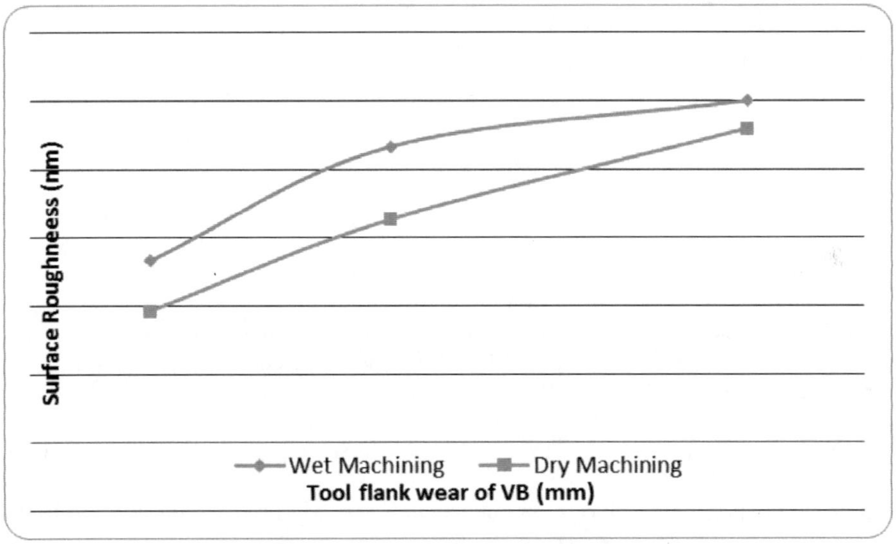

Figure 1.VB wear relationship with Surface roughness of HPK and HPB for optimum cutting conditions.

In other words, optimum dry machining cutting conditions (HPK8.0.6 dry machining = 2.8 μm) has a finer surface than wet machining (HPK8.0.6 dry machining = 3 μm)

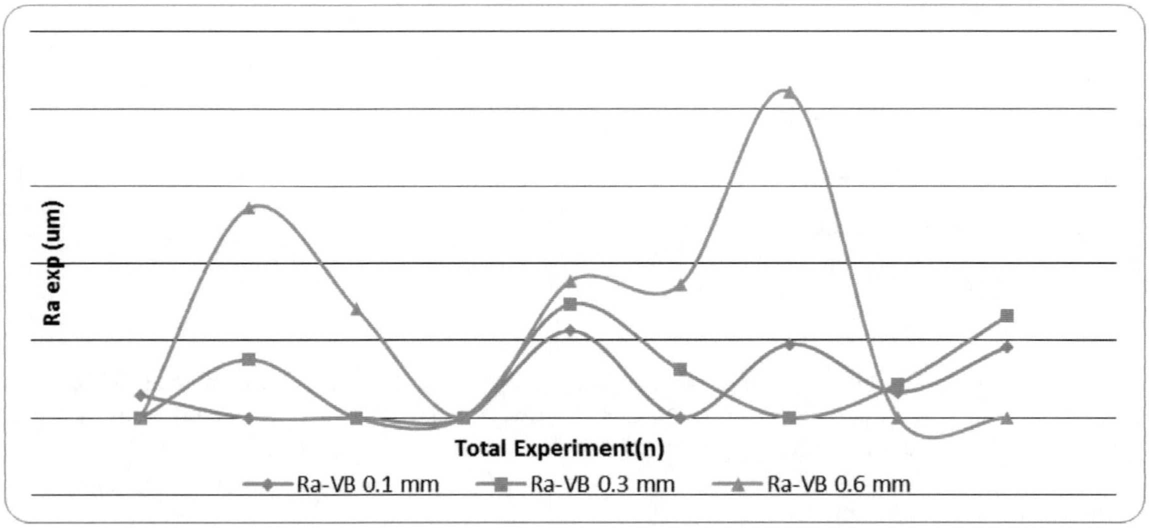

Figure 2. Relation of experimental amount with Ra exp on VB 0,1; 0.3 and 0.6 mm.

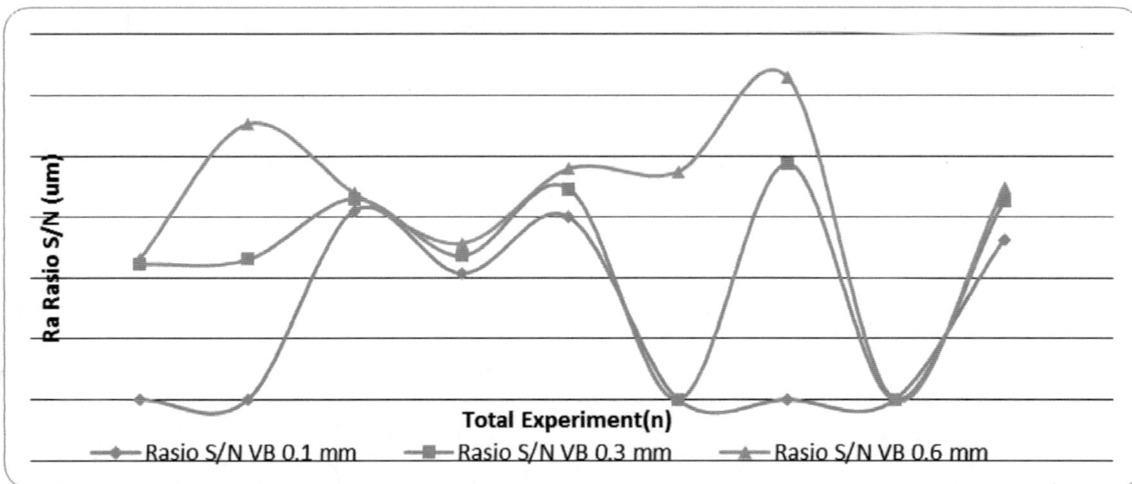

Figure 3. Relation of experimental amount with Ra Rasio S/N pada VB 0,1; 0,3 dan 0,6 mm.

Table 2. Two Ways for Surface Roughness, C1D3 Maximum Cutting Condition.

C_1					C_2				C_3			Total
D_1	0.322	1,112	1,159	2.593	1,319	1,634	1,738	4,691	1,501	1,734	1,895	5,130
D_2	0,697	0,973	1,872	3.542	0,954	1,152	2,264	4,370	1,350	1,936	2,647	5,933
D_3	0,425	0,657	0,894	**1.976**	1,037	1,182	1,286	3,505	1,556	1,625	1,698	4,879
Total				8,111				12,566				15,942

Table 3. ANOVA Pareto Analysis for Surface Roughness.

Added with Level Factor	Factors and Interactions				
	C x D	A	B	C	D
1	11,842	11,842	12,031	8,111	12,414
2	12,177	12,177	11,476	12,566	13,845
3	12,600	12,600	13,112	15,942	10,360
The sum of squares of the	0,865	0,865	4,408	92,568	15,269

difference (S) Contribution Rate (%)	8,140	8,163	70,154	5,509	7,628

For the geometry of tool 12° indicates a change of sub-surface structure which the heat on the work piece is high enough that the outer layer hardened by consequence the formation of a crater-like hole seen in figure 4 below.

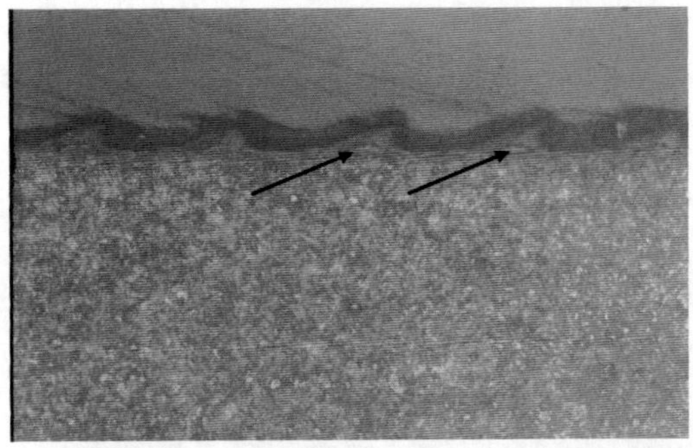

Figure 4. Micro Structure with HPK2,0.6.

With tool geometric 6° providing a softer outer surface layer, the phenomenon occurring in the outer surface layer is a white base structure enclosing a surface layer of ferrite like Figure 5 and Figure 6.

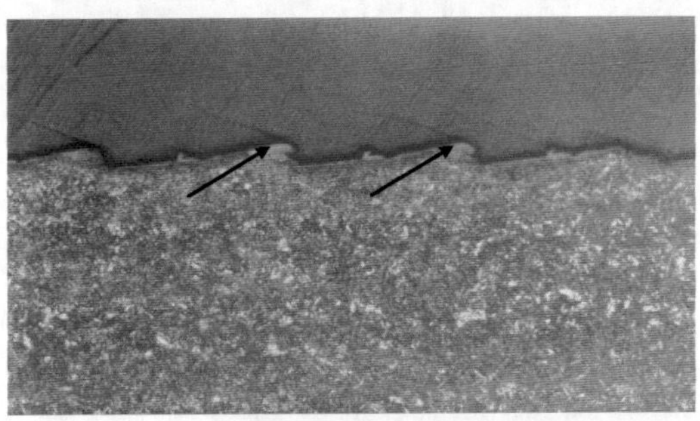

Figure 5. Micro structure with HPK5,0.6 for tool flank wear of VB in dry machining.

Figure 5 and figure 6 shows the wear of VB proportional to the surface hardness in which the above two graphs are the result of dry machining (HPK) and also the result wet machining (HPB) gives a difference in HV hardness. Wet machining hardness value of HV is lower due to low temperature factor on machining surfaces with the same VB wear. In the dry machining process the friction and heat that occurs will be very high where the tool will experience wear on the edge (VB) that is directly related to the workpiece surface. Flank wear will occur with variations of VB = 0.1 mm, 0.3 mm and 0.6 mm. Wear VB is higher then the workpiece surface will experience the friction and higher heat because end cutting edge of the tool with the consequence of the work surface of machining results will be harder. So the wear of VB greatly influences the hardness, the greater the wear of VB then the HV hardness increases. From the graph it can be seen that the hardness of HV is the dependent variable which is hardness as a function of VB wear. From the observations presented in Figures 6 and 7, fluctuations of Ra change in VB 0.1; 0.3 and 0.6 mm with 9 cutting conditions which in figure 6 the value of Ra change of the variation of VB data is processed through the S / N ratio equation as shown in Fig. 7. Figure 7 shows a change in Ra of a fluctuating S / N ratio. The average occurrence of a decrease is due to the change in the rate of small Ra while the rate of change Ra is large (from 1,467μm to 3.6μm) on HPK1 machining results. The change of Ra in VB (0.3-0.6) mm is relatively higher, this is due to the higher rate of Ra change as in HPK2 from 3,767 μm up to 13,567 μm and

HPK7 from 9.3 μm up to 21,067 μm. Of the three Ra changes to the VB wear variation, the resulting Ra value is low with 0.15 mm / r feed for the same so for the 9 cutting conditions with the same cutting depth resulted in Ra fluctuating as a result of significant influence of the feed in determining the value Ra.Based on Table 6 it turns out that C1D3 is the 8th cutting condition providing optimum cutting conditions in the form of surface roughness Ra, So with optimum cutting condition in C1D3 it can be said that Ra surface roughness (1.976 μm) is expected which according to dry machining criteria. From Figure 8 it is shown that wear VB = 0.6 mm has a mean S / N ratio higher than VB = 0.1 mm and 0.3 mm for each factor and the same interaction because the average Ra value or ratio value S / N is bigger.Based on Table 7 ANOVA analysis, there is a correlation between factor and interaction with contribution ratio as in Fig 9 where cutting condition B = (feeding f), A = (cutting speed V), C = (cut a depth), D = (tool geometric of GP) and interaction of CxD = (interaction of cutting depth and tool geometric) shows significant differences between B (feeding f) with A, CxD, D and C. The significant factors and interactions selected from the left side cumulatively contribute 70.154% B (feeding f) has a dominant effect in the machining process.

4. Conclusion

From the experiment can be taken several conclusions as follows:

a. Testing data Roughness and hardness of AISI 4337 alloy steel surface surfaces that machining surfaces that dry machining is better than wet machining where no significant difference is seen in Figures 5 and 12.
b. In the 3 optimum cutting conditions consisting of HPK1.0.1; HPK8.0.3 and HPK8.0.6 that the variation of VB HPK8.0.6 provides a more optimum cutting condition due to Ra surface roughness and hardness of HV is more ideal than wet machining through quality aspects of machining results and in accordance with ISO 3685 which is more cutting time 5 minutes.

References

[1] LiewWYH, Yuan, Ngoi BKA, Evaluation of Machining of Performance of STAVAX with PCBN
[2] Hattori M,Cost and Process information Modeling for Dry Machining, 2004
[3] SME, The Futue of Machining Technology and Porcess Machining Technology Association of The Society Manufacturing Engineering 2001
[4] Che Haroen, Ginting A, The Influences of tool wear and Tool Life on Surface Integrity During turning Tool Steel Using Uncoated Carbide, 2000
[5] Dawson and Thomas R.Kurfess, Tool life, Wear rates and Surfaces Quality In Hard Turning, journal 2004
[6] Bulloch H, Research & Technology Transfer Workgroup Dry Machining,2004
[7] Canter Neil M, The Possibilities and Eliminations of Dry Machining, 2003
[8] ISO 3685 , Tool Life Testing With Single Point Turning Tool, 1993
[9] N.Belavendram, Taguchi Techniques for Industrial Experimentation, Quality by Design, 1995

IOP Conf. Series: Journal of Physics: Conf. Series **1007** (2018) 012065 doi:10.1088/1742-6596/1007/1/012065

Comparison Between PCI and Box Girder in BridgesPrestressed Concrete Design

Cut Rahmawati[1*], Z Zainuddin[1], Syafridal Is[1] and Robbi Rahim[2]

[1]Department of Civil Engineering, Abulyatama University, Indonesia
[2]School of Computer and Communication Engineering, Universiti Malaysia Perlis, Malaysia

E-mail: *cutrahmawati@abulyatama.ac.id

Abstract. This research is done by comparing PCI and Box Girder types of prestressed concrete design. The method used is load balance. Previous studies have just discussed the differences in terms of effectiveness and economics. In this study, the researchers want to know the design process by comparing the working forces, the resulting moment, and the losses of the prestressed. As the case in this study, the researchers used the bridge with the span of 31 meters. The tendon pulling system was conducted with post-tensioning system. The analysis result showed that prestressed of the Girder box type sustained the greatest moment due to the combination of its own weight, additional dead load, lane load, and wind load of 44,029 kNm, while the biggest moment of PCI Girder was 7,556.75 KNm. The Girder beam box experiences greater moment and shear force than PCI Girder. This is the effect of the weight of its own Girderboxwaslarger than PCI Girder. The losses ofprestressed style of Girderboxand PCI Girder type were24.85%and 26.32%, respectively.Moreover, it showed that the type of Girder box is cheaper, easier, and more efficient than PCI Girder.

1. Introduction

The prestressed concrete is essentially a concrete in which the internal stresses of magnitude and the corresponding distribution are given so that the voltages caused by external loads are held up to a desired level. The strategies include the addition of compressive force to the structure to reduce or even to eliminate internal tensile strength and in this case,thecracks in the concrete can be removed. Externally prestressed concrete structure is broadly applied in the highway bridges, urban bridges, and railway bridges with the development of external prestress technology. In the design and construction process of externally prestressed concrete bridge, the prestress force is often determined according to the theory formula [1].Precast Concrete-I girder (PCI girder) and Box Girder are structural blocks that directly receive the traffic load after the slab, which then distribute the load to the columns and forwarded to the foundation.

Losses of prestress is a characteristic of all prestressed concrete members wherein the level of prestress force first applied to the member is reduced over time due to short- and long-term conditions[2].Steinberg showed that prestress losses are consistently calculated lower (by about 33 percent) when using deter-ministic or nominal parameters than when including statistical variability[3].Measured prestress losses included elastic shortening losses and time-dependent losses due to concrete shrinkage and creep. A coefficient to account for the relationship between the prestresslosses from the measured prestressing forces and the actual prestress losses from concrete compressive strains is proposed [4].Similar results were found for other prestress loss estimation methods and when using the box beam example [5].

The purpose of this plan is to know the difference of two prestressed concrete models in the construction world used. The observed differences will be focused on the loss force of presstresed and tendon trajectories. This research is planned to use bridge with the same middle span, which is 31 meter. Several previous studies have shown that the use of Girder boxes is more effective than PCI Girder. In this study, the researchers wanted to find out the differences in the design process, especially in the total estimation of the losses of prestress. In the calculation of the losses of prestressed, many factors are interrelated. Thus the researchers assess the estimation with lump sum loss more appropriately used.This research will be used for designers/planners in choosing bridge type. More importantly, the losses of prestress of PCI and Box Girder play a big role in the designer's ability.

The productivity and the efficiency of construction are measured not only by the requirements for the project, but also by the durability of the project [6].According to Kromel E. Hanna,GeorgeMorcous,and Maher K. Tadros (2009),Prestressed concrete Box Girders have several advantages [7]:

- The easy and speedy construction because of eliminatingconcrete forming and placing operations (forexample, the Arbor Rail Line Bridge in NebraskaCity, Neb., was erected and opened to traffic within 72 hrs)
- A shallow superstructure depth, which is often necessaryto maintain the required vertical clearance (forexample, an interstate bridge in Colorado has a spanto-depth ratio of 39)
- Low construction cost compared with I-girder bridgesand other competing systems
- Hollow portions inside the box girders that reduce theself-weight of the girders and provide space for gaslines, water pipes, telephone ducts, storm drains, andother utilities
- Improve bridge aesthetics because of the flat soffitand slender superstructure
- High torsional stiffness which is ideal for curvedbridgeconstruction

Vishal U. Misal , N. G. Gore , P. J. Salunke (2014) analyzed about the "Analysis and Design of Prestressed Concrete Girder".In theirstudy,which was observed by extracting result, they concluded that Box Girder is costlier than I girder. It also revealed that the losses is more in I girder as compared to Box Girder [8].

2. Materials and Methods

This research used bridge with the same middle span, which is 31 meter. The methods utilized in this study were by designing, analyzing and comparing the result from estimation of loss of prestress, bending moment and concrete cost.

a. Section Properties of Prestressed Concrete Beam

The prestressed concrete section used is a BoxGirderand PCI Girder with planned dimensions. Images and data of girder dimensions box are as follows:

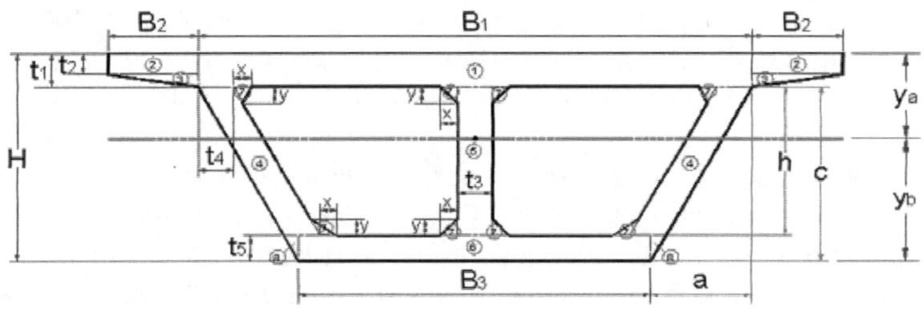

Code	Dimension (m)
B1	6.25
t1	0.35
B2	0.10
t2	0.25
H	1.6
t3	0.4
t4	0.4
B3	4.00
t5	0.25

(a). Type of Box Girder and Dimensions.

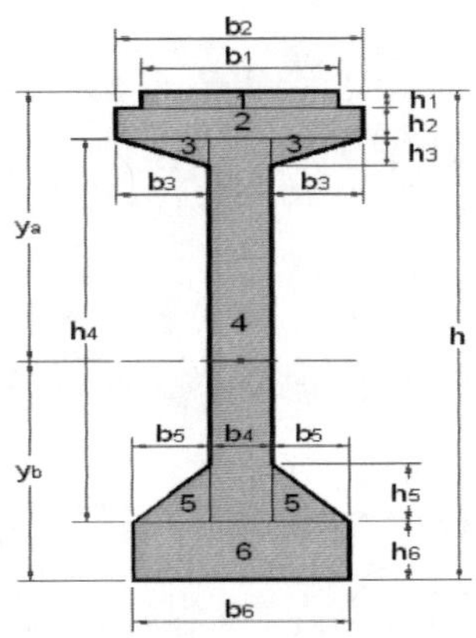

Code	Dimension (m)
b1	0.64
b2	0.8
b3	0.3
b4	0.2
b5	0.25
b6	0.7
h1	0.07
h2	0.13
h3	0.12
h4	1.15
h5	0.25
h6	0.25
h	1.60

(b). Type of PCI Girder and Dimensions.

Figure 1. The differences of prestressed concrete models and dimensions used.

From the results of planning, it was obtained that the girder beam area is (A) = 9.849 m^2 and the area of the Box Girder is 0.6523 m^2.

b. The planned Material Quality
The material used for Box Girder and slab of this bridge floor is the concrete with the quality of K-500 (f'c 50 Mpa). Prestressed steel tendons used in high quality steel types consist of wire, strands, or bar. For planning, it used Type strands which is the standard strands cable VSL (Vorspam System Losinger), which is uncoated 7 Wire Super Strands ASTM A-416 grade 270 with a cross-sectional area of 12.7 mm and ultimit tensile strength is 1860 Mpa[9].The tendon withdrawal system is implemented by post tensioning system. In this method, the concrete first printed with prepared holes (duct) or grooves for the placement of cable / strands. When the concrete is strong enough, then the cable/strands are pulled, the ends are buried, then the hole is grouted. The researchers used each respective reference for details in calculating losses according to AASHTO LRFD[10] and Standard Specifications [1]. The AASHTO LRFD method is commonly used in estimating deflection and cracking moment during the bridge design process, and the time step method has been considered a more accurate method when compared to measured losses [11].

3. Result and Discussion

a. The Position and Tendon Plotting
From the calculation results, it showed that the most number of tendons in the type of Girder Box is 21 tendons with 20 strands/tendon,whereas Girder PCI only requires 4 tendons with 19 strands / tendon. However, if it was seen from the shape and number of girder that must be provided for one bridge construction, then the Box Girder is more efficient in construction. In one segment of the bridge, it required two box girders, while for PCI Girder takes ten pieces.The Girder Box placement adds the aesthetics of the bridge especially for use as a fly over. Thus,in Indonesia, with the densely urban population and the increasing traffic,the use of girder box is felt very effective.

b. Moment, Shear Force and Combination of Load Beams
Based on the forces acting on the girder beam, the bridge will affect the maximum moment and shear force. The shear force and moment in the prestressed beam are the functions of the distance x which is measured along the longitudinal axis. One of step to find out is to draw a graph which shows how the Moment and Style Shear against X.

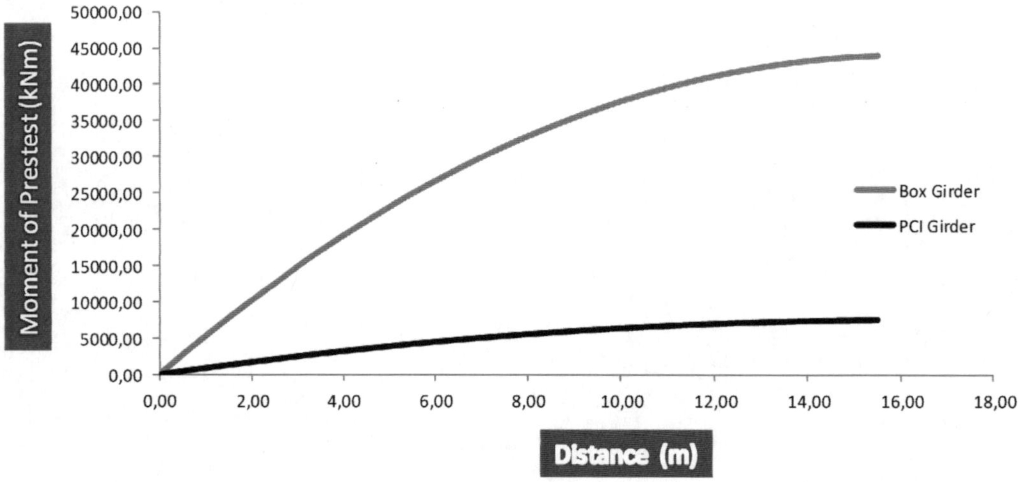

Figure 2. Bending moment diagram.

The picture above shows that the two types of prestressed concrete experienced the greatest moments due to a combination of their own weight, additional dead load, lane load and wind load. Box Girder presstresed experienced the greatest moment of 44,029 kNm, while in PCI Girder the biggest moment was 7,556.75 KNm.

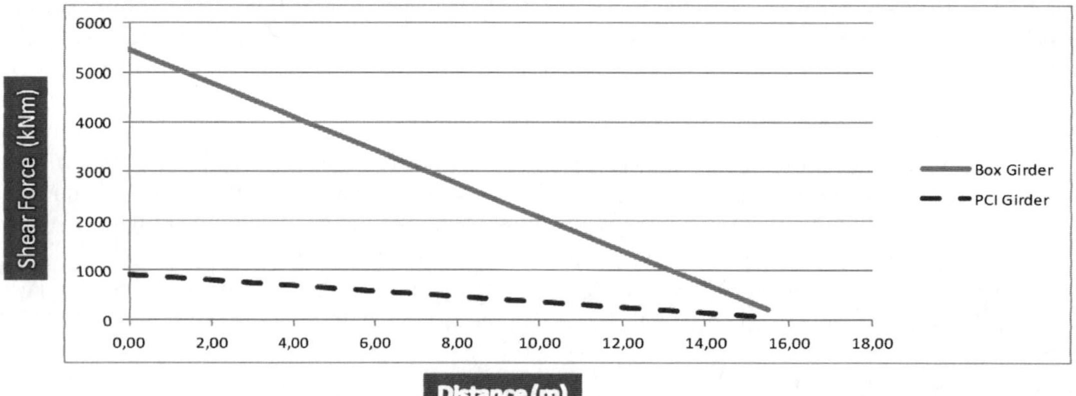

Figure 3. Shear force diagram.

The prestresed Girder box experienceds the largest shear force of 5.467.52 kNm, while on PCI Girder experienced the largest shear force of 913.53 KNm. The shear force was experienced by the Girder box is far above PCI Girder.

c. Lump Sum Estimates for Prestress Loss

The initial prestressed force which was applied to the concrete element undergoes a reduction process therefore it was determined the prestressed force stages such as loading, transfer to concrete, workload conditions to reach ultimate. The loss of the prestressed force in terms of the result of the overall calculation of the prestressed force was derived from the loss of prestressed force as in the following table.Theloss of prestressed in terms of: prestress due to elastic deformation of concrete, Relaxation of stress in steel, Shrinkage of concrete, Creep of concrete, Friction, Anchorage slip [12-16].

IOP Conf. Series: Journal of Physics: Conf. Series **1007** (2018) 012065 doi:10.1088/1742-6596/1007/1/012065

Table 1. Loss of each prestressed force.

Losses of Prestress	Type of PCI Girder		Type of Box Girder	
	Force (kPa)	The percentage of prestressed force	Force(kPa)	The percentage of prestressed force
Loss of stress due to friction	606.277	0.05	9,998.824	0.77
Loss due to anchorage slip	37,354.839	2.87	37,354.839	2.87
Loss due to deformation of concrete	82,793.364	6.36	25,685.178	1.97
Loss due to creep of concrete	20,351.012	1.56	82,192.310	6.31
Loss due to shinkage of concrete	92,930.588	7.14	92,930.589	7.14
Loss due to relaxation of stress in steel	108,588.755	8.34	75,447.197	5.79
Lumpsum of the loss of presstressed force		26.32		24.85

From the table above, it can be seen that the Box Girder was less experience the total losses of prestress force that is 24.85%, while PCI Girder is 26.32%. Relaxation in steel is the biggest cause of losses of prestress in PCI Girder by 8.34%. This is because the fixed voltage that occurs in PCI Girder is greater than that in Box Girder. BoxGirderexperienced the greatest losses of prestress due to the shrinkage of concrete. The concrete shrub on the prestress of concrete causes the shortening of the drawn wires (Stands) moreover it contributes to the loss of voltage. Concrete losses are influenced by cement type, aggregate and treatment methods. To reduce the prestress loss due to the shrinkage on prestressed concrete, it is advisable to use high quality concrete with low water/cement ratio. Gilbertson[11]recommended total loss of prestress form Box Girder about 15.2% and I Girder about 27.1%.

d. The Construction Cost

In the term of the construction, PCI Girder requires a lower cost than Girder Box. In this case, the cost of Box Girder construction is twice as much as the PCI Girder type. However, in its implementation, Box Girder is very effective if it is constructed in narrow areas and heavy traffic. This cause that Box Girder is more preferable by the contractor in working on long-span bridges.

4. Conclusions

Based on the results of the parametric study and the comparison of Box and PCI Girder Prestressed, several conclusions are drawn below:

a. The larger the fixed voltage occurs, the greater the losses of prestress due to the relaxation of the prestressing concrete. This occurs in the type of PCI Girder which lost the prestressed force due to relaxation in the larger steel that is 8.34%.
b. The shinkake on the prestressed concrete can cause the surface cracks so the proper concrete treatment is essential to prevent cracking. This is experienced by both types of prestressed concrete namely PCI and Box Girder. The percentage loss of the same prestressedforce is shown by both of these types of 7.14%.
c. The moment and the shear force that occurs are larger on the Box Girder than PCI Girder. This is as a result of its own weight from a large girder box material. Therefore, the number of tendons and the reinforcement of the Box Girder also becomes more numerous.

IOP Conf. Series: Journal of Physics: Conf. Series **1007** (2018) 012065 doi:10.1088/1742-6596/1007/1/012065

d. There are many other things that show the effectiveness of Box Girder used even though the required cost for construction is greater. From the side of Lump Sum Estimates for Prestress Loss, Box Girder shows better result.

References

[1] American Association of State Highway Transportation Officials. "Bridge Design Specifications, AASHTO". Washington, DC,USA, 4th edition, 2007
[2] E. P. Steinberg, "Probabilistic Assessment of Prestress Loss in Pretensioned Prestressed Concrete," *PCI Journal*, vol. 40, no. 6, pp. 76-85, 1995
[3] L.A.CaroJ.R, Martí-Vargas, P.Serna, "Prestress losses evaluation in prestressed concrete prismatic specimens," *Engineering Structures,* vol. 48, no. March 2013, pp. 704-715, 2013
[4] Gilbertson, Christopher G., "Proba-bilistic Assessment of Several Pre- stress Loss Methods," MSCE Thesis, Michigan Technological University, Houghton, MI, 2001
[5] Adamu, S., Hamid, R, A. (2012, May)."Lean Construction Techniques Implementation in Nigeria Construction Industry".*Canadian Journal on Environmental*, Construction and Civil Engineering, Vol. 3(4), pp.186-193
[6] Hanna, K. E., Morcous, G., &Tadros, M. K. (2009, Oct). "Transverse post-tensioning design and detailing of precast, prestressed concrete adjacentbox-girder bridges". *PCI Journal*, Vol.54(4), pp.160-174
[7] Vishal U. Misal , N. G. Gore , P.J. Salunke (2014, Jan). "Analysis and Design of Prestressed Concrete Girder". *International Journal of Inventive Engineering and Sciences (IJIES).* Vol. 2(2), pp.14-17
[8] ASTM Standard A 416/A416M – 06, "Standard Specification for Steel Strand, Uncoated Seven-Wire for Prestressed Concrete." ASTM International
[9] AASHTO, LRFD Bridge Design Specifications, Second Edition, American Association of State Highway and Transportation Officials, Washington, DC, 2000
[10] Gilbertson, CG, Ahlborn, TM, "A Probabilistic Comparison Of Prestress Loss Methods In Prestressed Concrete Beams," *PCI Journal*, vol. 49, no. 5, pp. 52-69, 2004
[11] Dr. D. Maity, IITG, "Lecture Note-17-Prestress Concrete" Internet:http://www.iitg.ernet.in/scifac/qip/public_html/cd_cell/chapters/dmaity_adv_struct_design/prestress%20concrete%20(17-23).pdf, [Oct.1, 2017]
[12] Lin, T. Y dan Burns, Ned. H. 1993. DesainStrukturBetonPrategang. Jakarta:Erlangga
[13] Nawy, Edward. G. 2001. BetonPrategangSuatuPendekatanMendasar. Jakarta:.Erlangga
[14] Raju, N. Krishna. (2007). Prestressed Concrete. (4th edition). [On-line]. Available: https://civiltechnocrats.files.wordpress.com/2013/11/prestressed-concrete-krishnaraju.pdf[July 12, 2016]
[15] D. J. Farquhar, "Cable Stayed Bridges," *ICE manual of bridge engineering*. pp. 357–381, 2008
[16] J. E. Donald, D. W. Kulp, and W. F. DeGrado, "Salt bridges: Geometrically specific, designable interactions," *Proteins Struct. Funct. Bioinforma.*, vol. 79, no. 3, pp. 898–915, 2011

Acknowledgments

Thanks to the prestressed concrete design team, consisting of the students and lecturers of Civil Engineering of Abulyatama University for their hard work.

IOP Conf. Series: Journal of Physics: Conf. Series **1007** (2018) 012066 doi:10.1088/1742-6596/1007/1/012066

Design of mechanical arm for an automatic sorting system of recyclable cans

Y Resti[1*], A S Mohruni[2], F Burlian[2], I Yani[2], and A Amran[1]

[1]Department of Mathematics, Universitas Sriwijaya
[2]Department of Mechanical Engineering, Universitas Sriwijaya

Email: *irsyadiyani@ft.unsri.ac.id

Abstract. The use of a mechanical arm for an automatic sorting system of used cans should be designed carefully. The right design will result in a high precision sorting rate and a short sorting time. The design includes first; design manipulator, second; determine link and joint specifications, and third; build mechanical systems and control systems. This study aims to design the mechanical arm as a hardware system for automatic cans sorting system. The material used for the manipulator is the aluminum plate. The manipulator is designed using 6 links and 6 join where the 6th link is the end effector and the 6th join is the gripper. As a driving motor used servo motor, while as a microcontroller used Arduino Uno which is connected with Matlab programming language. Based on testing, a mechanical arm designed for this recyclable canned recycling system has a precision sorting rate at 93%, where the average total time required for sorting is 10.82 seconds.

1. Introduction

Currently, many automatic systems have been built, especially automatic sorting system [1-7]. The automatic sorting system is designed by integrating software with hardware. The software can be an identification or recognition system, and most of these systems are built in the color of the object image being sorted [8-16], while the hardware is a system consisting of a set of objects sorting components (manipulators), a mechanical system, and a control system connected to a programming language.

The hardware in an automatic sorting system is designed by considering the objects to be sorted. The use of a mechanical arm for an automatic sorting system of used cans should be designed carefully. The right design will result in a high precision sorting rate and a short sorting time. The design includes first; design manipulator, second; determine link and joint specifications, and third; build mechanical systems and control systems. In the recycling industry of cans, automatic sorting system is used to automatically sort out the types of recyclable cans. The purpose of this paper is to design a mechanical arm as a hardware system for automatic sorting system of recyclable cans, where manipulators are made from aluminium plates, mechanical systems using servo motors as motors, and control systems using Arduino Uno as microcontrollers associated with Matlab programming language.

2. Methodology

The main component of a mechanical arm consists of three important parts, namely manipulators, mechanical systems, and control systems. In designing a mechanical arm for an automatic sorting system of recyclable cans, we use the following main steps: 1) design the manipulator, 2) determine the specification of links and joints, and 3) build the mechanical and the control system.

2.1 Design the manipulator

A manipulator consists of links and joints. The joint is used to connect each link. Links that act as components that move, lift, and push work objects are called end-effector. In this study, manipulators are designed using 6 links and 6 joints, where the 5th link is the end-effector and the 6th link as a gripper. The manipulator design begins by describing it in the form of a coordinate system at the initial position as shown in Figure 1, where X_j, Y_j, and $Z_j$are the local coordinates of each joint, and θ_j is the rotation angle of joint.

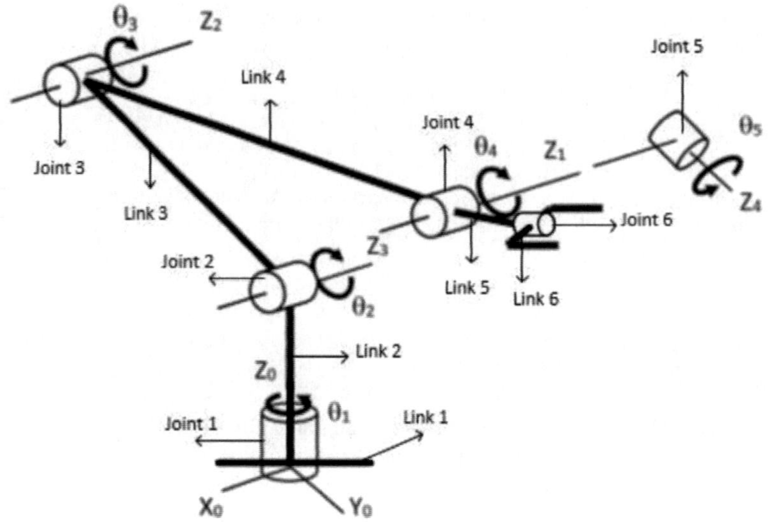

Figure 1. The coordinate system of manipulator design at the initial position.

To determine the rotation angle of the j -th joint to achieve a certain motion function used Denavit-Hartenberg (D-H) method. Let θ_j be the rotation angle of the j -th joint, $\alpha_j$is the rotation angle on the axis X_j, $d_j$is the translation on the axis Z_{j-1}, $a_j$is the translation on the axis X_{j-1}. Movement of the j -th joint from the $(j-1)$ -th joint with D- H method is defined as

$$^{j-1}A_j = \begin{bmatrix} Cos\theta_j & Cos\alpha_j Sin\theta_j & Sin\alpha_j Sin\theta_j a_j Cos\theta_j \\ Sin\theta_j & Cos\alpha_j Cos\theta_j & Sin\alpha_j Cos\theta_j & a_j Sin\theta_j \\ 0 & Sin\alpha_j & Cos\alpha_j & Cosd_j \\ 0 & 0 & 0 & 1 \end{bmatrix} \quad (2.1)$$

where the parameters a_j, α_j, and $d_j$are constant, while the parameter $\theta_j$changes when the j -th joint moves from the $(j-1)$ -th joint. The parameter $\theta_j$is obtained by creating a coordinate system of (2.1).

2.2 Determine the specification of links and joints

The material used for the link is acliric and stepper motor for each joint is used type M42SP-4NK. Specifications of the link and the joint consist of mass, force, torque and gear radius are calculated using (2.2) to (2.8). Suppose that for each the i -th link, p_i, l_i, t_i, and $\rho_i$are length, width, thickness, and density, the mass for the i -th link (m_i) are defined as,

$$m_i = p_i l_i t_i \rho_i \quad (2.2)$$

IOP Conf. Series: Journal of Physics: Conf. Series **1007** (2018) 012066 doi:10.1088/1742-6596/1007/1/012066

Suppose that for each the j -th joint, τ_j, N_j, and R_j, respectively are the torque, the number of gear, and the radius of gear. The force in the j -th joint (F_j) is defined as,

$$F_j = \frac{\tau_j}{R_j} \tag{2.3}$$

From (2.2) can be obtained torque, motion force, and gear radius for the i-th link defined as, respectively,

$$\tau_i = \frac{\tau_j N_i}{N_j} \tag{2.4}$$

$$F_j = F_i \tag{2.5}$$

$$R_i = \frac{\tau_i}{F_i} \tag{2.6}$$

where N_i is the number of gear for the i -th link. Furthermore, the energy of maximum movement for the i -th link (E_i) is defined as

$$E_i = F_i S_i = m_i g S_i \tag{2.7}$$

where g is the gravity of the earth and S_i is the maximum movement distance of the i-th link defined as,

$$S_i = \frac{\theta_i}{360^0} \text{x} 2\pi R_i \tag{2.8}$$

2.3. Build the mechanical and the control system

Mechanical system of the arm is built based on mechanical design and actuator system. The mechanical design is the basis for determining mechanical arm motility (flexibility), while the actuator system is a hardware system that moves the mechanical arm. The mechanical system is based on links and joints designed on the manipulator as shown in Figure 1, while the actuator system (driver) used is a stepper motor. For the control system on the mechanical arm is built by making software using Matlab software based on open loop control system (forward feed). The flow chart for the control system is given in Figure 2, while the open-loop control system is shown in Figure 3.

The feedforward control system can be expressed as a control system whose output is not reconsidered by the controller. The performance of the controller is not affected even though the mechanical arm may not have reached the desired target as per reference. These controls are suitable for mechanical arm operating systems that have actuators that operate based on logic feedback using step configuration in sequence, such as stepper motors. Stepper motor does not need to be fitted with sensors on its axis to know the final position. If in good condition and no more load problem, then stepper motor will rotate in accordance with controller command and reach a target position appropriately. If the actual motion has equal to the reference, then the controller input will be zero. This means that the controller no longer provides an actuation signal to the mechanical arm because the final target motion command has been obtained. The smaller the error is calculated, the smaller the controller steering signal to the mechanical arm until it reaches the steady state.

IOP Conf. Series: Journal of Physics: Conf. Series **1007** (2018) 012066 doi:10.1088/1742-6596/1007/1/012066

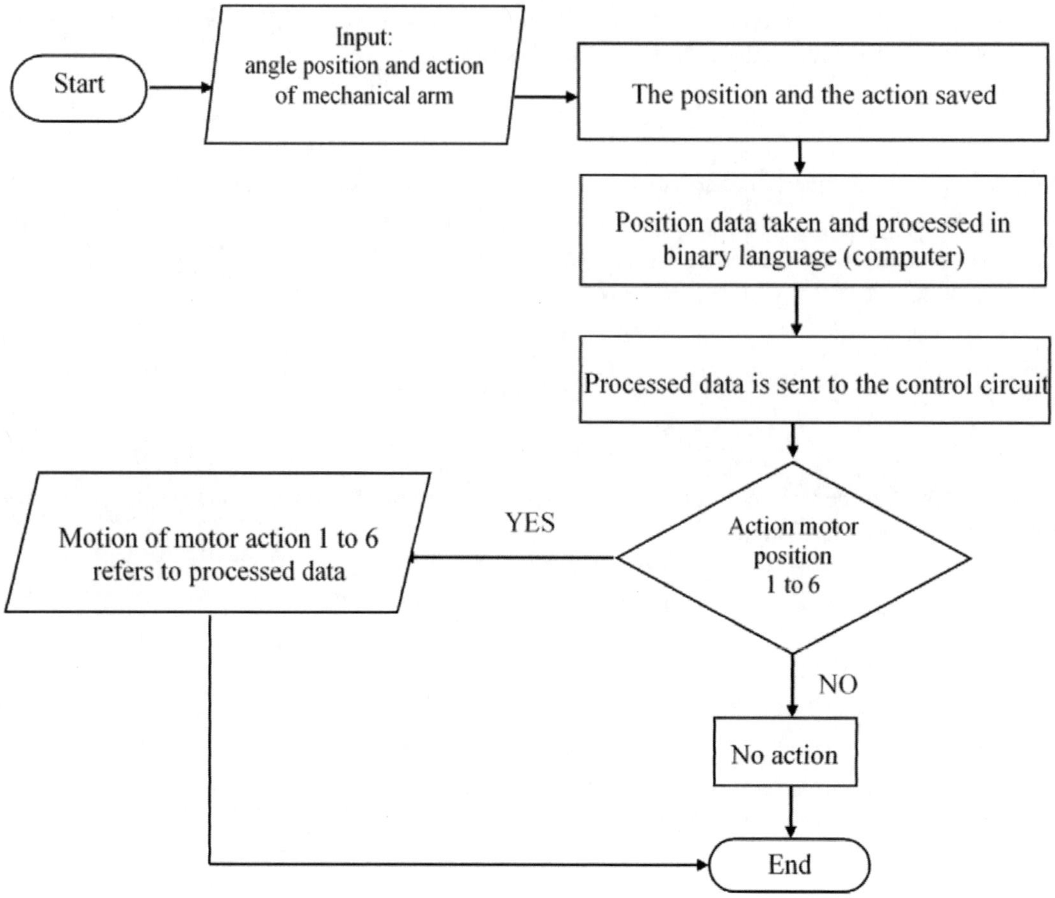

Figure 2. Flowchart for control system on the mechanical arm.

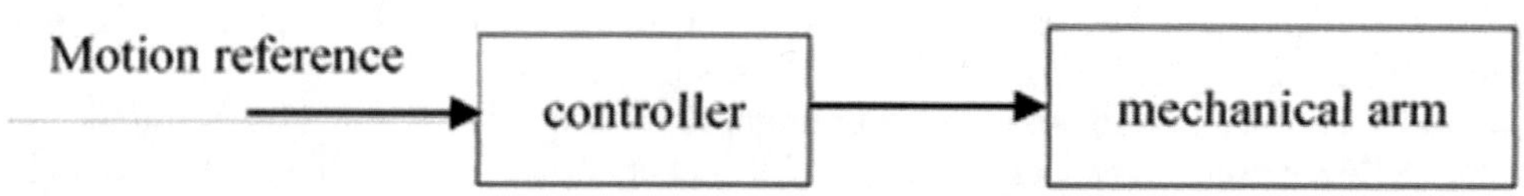

Figure 3. Open loop control system.

3. Result and Discussion

The link coordinate system of the mechanical arm (5 degrees of freedom) to determine the rotation angles of the joint is shown in Figure 4 with reference to Figure 1.

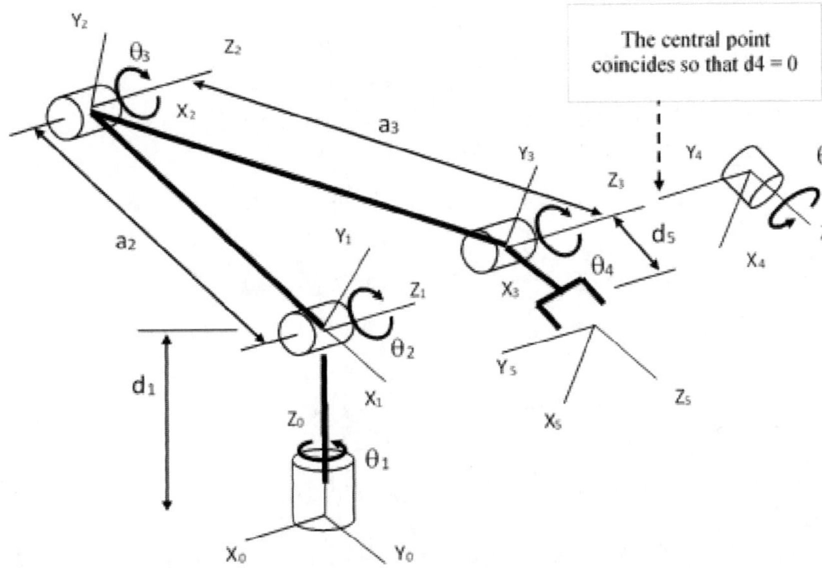

Figure 4. Link coordinates system of the mechanical arm with 5 degrees of freedom.

Based on Figure 4 and (2.1), the following 5 matrices are obtained and the parameters as in Table 1.

$$
{}^{0}A_1 = \begin{bmatrix} C_1 & 0 & -S_1 & 0 \\ S_1 & 0 & C_1 & 0 \\ 0 & -1 & 0 & d_1 \\ 0 & 0 & 0 & 1 \end{bmatrix} \quad
{}^{1}A_2 = \begin{bmatrix} C_2 & -S_2 & 0 & a_2C_2 \\ S_2 & C_2 & 0 & a_2S_2 \\ 0 & 0 & 1 & 0 \\ 0 & 0 & 0 & 1 \end{bmatrix} \quad
{}^{2}A_3 = \begin{bmatrix} C_3 & -S_3 & 0 & a_3C_3 \\ S_3 & C_3 & 0 & a_3S_3 \\ 0 & 0 & 1 & 0 \\ 0 & 0 & 0 & 1 \end{bmatrix}
$$

$$
{}^{3}A_4 = \begin{bmatrix} C_4 & 0 & S_4 & 0 \\ S_4 & 0 & -C_4 & 0 \\ 0 & 1 & 0 & 0 \\ 0 & 0 & 0 & 1 \end{bmatrix} \quad
{}^{4}A_5 = \begin{bmatrix} C_5 & -S_5 & 0 & 0 \\ S_5 & C_5 & 0 & 0 \\ 0 & 0 & 1 & d_5 \\ 0 & 0 & 0 & 1 \end{bmatrix} \tag{3.1}
$$

where $C_i = \cos\theta_i$, $S_i = \sin\theta_i$.

Table 1. Parameters of the link coordinate system.

The *j*-th joint	**Parameter of the link coordinate system**				
	θ_j	α_i	a_i	d_i	m_i
1	θ_1	-90^0	0	d_1	m_1
2	θ_2	0	a_2	0	m_2
3	θ_3	0	a_3	0	m_3
4	θ_4	90^0	0	0	m_4
5	θ_5	0	0	d_5	m_5
6	θ_6	0	0	0	m_6

For each of , Y, and Z axis, let n be the normal vector of mechanical arm, s be the sliding vector of mechanical arm, a be the approach vector of mechanical arm, and p be the position vector of mechanical arm. The kinematic equation of (3.1) is,

IOP Conf. Series: Journal of Physics: Conf. Series **1007** (2018) 012066 doi:10.1088/1742-6596/1007/1/012066

$$T = {}^{0}A_1\,{}^{1}A_2\,{}^{2}A_3\,{}^{3}A_4\,{}^{4}A_5 = \begin{bmatrix} n_x\, s_x & a_x p_x \\ n_y\, s_y & a_y\; p_y \\ n_z\, s_z & a_z p_z \\ 0\quad 0 & 0\quad 1 \end{bmatrix} \tag{3.2}$$

where,

$$\begin{aligned}
n_x &= C_1\big(C_2(C_3C_4C_5 - S_3S_4S_5) - S_2(S_3C_4C_5 + C_3S_4C_5)\big) - S_1S_5 \\
n_y &= S_1\big(C_2(C_3C_4C_5 - S_3S_4S_5) - S_2(S_3C_4C_5 + C_3S_4C_5)\big) - C_1S_5 \\
n_z &= -S_2(C_3C_4C_5 - S_3S_4S_5) - C_2(S_3C_4C_5 + C_3S_4C_5) \\
s_x &= -C_1\big(C_2(C_3C_4C_5 - S_3S_4S_5) - S_2(S_3C_4C_5 + C_3S_4C_5)\big) - S_1S_5 \\
s_y &= -S_1\big(C_2(C_3C_4C_5 - S_3S_4S_5) - S_2(S_3C_4C_5 + C_3S_4C_5)\big) + C_1C_5 \\
s_z &= S_2(C_3C_4C_5 - S_3S_4S_5) + C_2(S_3C_4C_5 + C_3S_4C_5) \quad (3.3) \\
a_x &= C_1\big(C_2(C_3C_4 + S_3C_4) - S_2(S_3S_4 - C_3C_4)\big) \\
a_y &= S_1\big(C_2(C_3S_4 + S_3C_4) - S_2(S_3S_4 - C_3C_4)\big) \\
a_z &= -S_2(C_3S_4 + S_3C_4) - C_2(S_3S_4 - C_3C_4) \\
p_x &= C_1\big(d_5\big(C_2(C_3C_4 + S_3C_4) - S_2(S_3S_4 - C_3C_4)\big) + a_3(C_2C_3 - S_2S_3) + a_2C_2\big) \\
p_y &= S_1\big(d_5\big(C_2(C_3S_4 + S_3C_4) - S_2(S_3S_4 - C_3C_4)\big) + a_3(C_2C_3 - S_2S_3) + a_2C_2\big) \\
p_z &= -d_5\big(C_2(S_3S_4 - C_3C_4) + S_2(C_3S_4 + S_3C_4)\big) - a_3(S_2C_3 + C_2S_3) - a_2S_2 + d_1
\end{aligned}$$

The mechanical system building process in this study is presented in Figure 5, while the control system is given in Figure 6. As input is a data of joint angle on a manipulator ruled by a gripper. This data is then processed and transmitted to the driver circuit through the parallel port communication system using Matlab software. Once the data is received by the driver circuit, the data will generate a signal (electric current) to run the stepper motor.

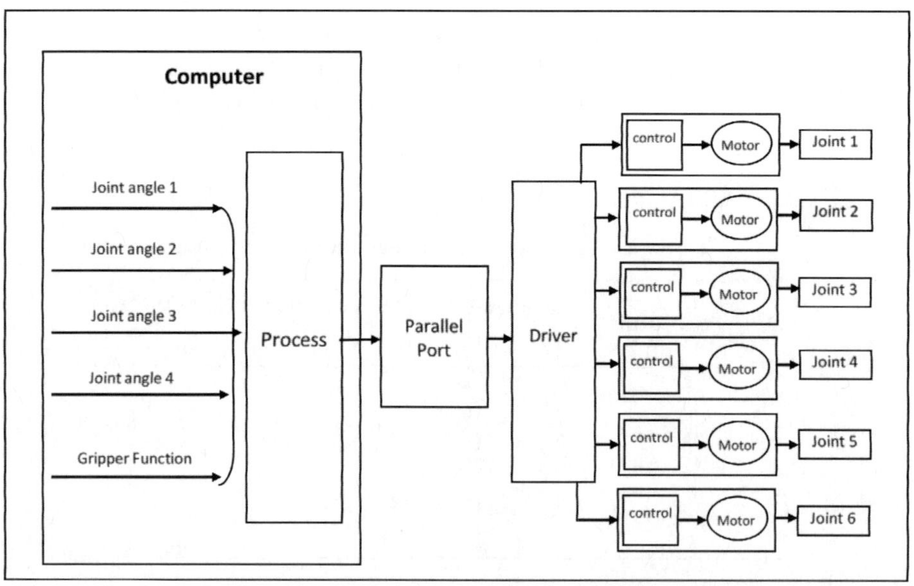

Figure 5. Mechanical system building process of the arm for sorting system.

IOP Conf. Series: Journal of Physics: Conf. Series **1007** (2018) 012066 doi:10.1088/1742-6596/1007/1/012066

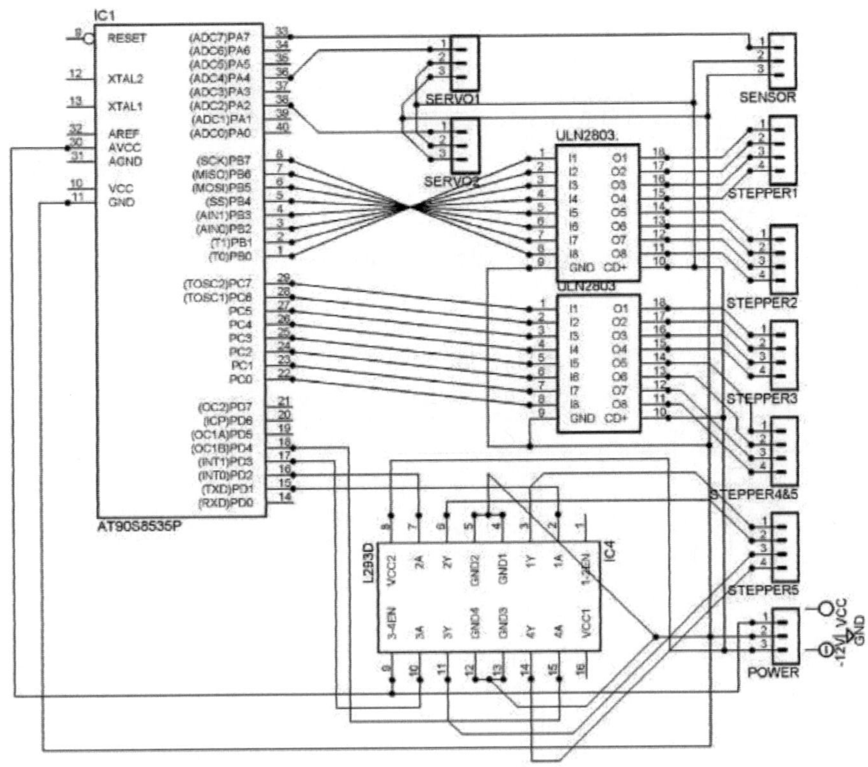

Figure 6. The mechanical arm control system.

Based on (2.2) - (2.8) and (3.1) - (3.3) obtained the mechanical arm design for sorting system of the recyclable cans with the specification as shown in Table 2.

Table 2. Specification of mechanical arm.

Model		Prototype of mechanical arm	
Structure		Vertical Articulated Mechanical	
Control Coordinate		5 Degree of Freedom (5-Dof) + Gripper	
Mechanical Mass		$\pm$1.4 Kg	
Moving Range	Base	$\pm 90^0$	524 Rotation
	Shoulder	0^0, $+120^0$	5280 Rotation
	Elbow	0^0, 150^0	900 Rotation
	Wrist	$\pm 75^0$	686 Rotation
	Wrist Rot.	$\pm 90^0$	90 Rotation
	Gripper	0^0, $+90^0$	108 Rotation
Weight carrying capacity		Max. 0.270 Kg	
Upper Distance		Max. 300 Mm	
Front Distance		Max. 220 Mm	
Gripper	Wide	40 Mm	
	Thickness	30 Mm	
Material		Fiberglass	
Power Capasity		$\leq$ 24 V	

IOP Conf. Series: Journal of Physics: Conf. Series **1007** (2018) 012066 doi:10.1088/1742-6596/1007/1/012066

The results of the motion testing of the total time required by the mechanical arm to sort the used cans are shown in Table 3, while the accuracy testing of the object sorted is presented in Table 4.

Table 3. The motion testing of the total time.

Testing	Total time (s)
1	11,63
2	11,41
3	10,38
4	10,25
5	10,44

Table 4. The accuracy testing of the object sorted.

Object	Testing															Success
	1	2	3	4	5	6	7	8	9	10	11	12	13	14	15	
1	√	√	√	√	√	√	√	√	√	√	√	√	√	√	√	15
2	√	√	√	√	√	√	√	√	√	√	√	√	-	√	√	14
3	√	√	√	√	√	√	√	√	√	√	√	√	√	√	√	15
4	√	√	√	-	-	√	√	√	√	√	-	√	√	√	-	11
5	√	√	√	√	√	-	√	-	√	√	√	√	√	√	√	13
6	√	√	√	√	√	-	√	√	-	√	√	√	√	√	√	13
7	√	√	√	√	√	√	√	√	√	√	√	√	√	√	√	15
8	√	√	√	√	√	√	√	√	√	√	√	√	√	√	√	15
9	√	√	√	√	√	√	√	-	√	√	√	√	√	√	√	14
10	√	√	√	√	√	√	√	√	√	√	√	√	√	√	√	15

From the test results, the average of the total time required by the mechanical arm to sort the used cans is 10.82 seconds, while the percentage accuracy rate is 93%. It is stated that mechanical arm as the recyclable cans separator can work with satisfactory results.

4. Conclusion

This design of mechanical arm for the automatic sorting system of recyclable cans has the accuracy rate in sort the used cans at 93 %, where the average of the total time required is 10.82 seconds. The rotation angle of joint which is the biggest moving range elbow, i.e., 150^0 to 900 rotations, while number of rotation, which is the most is shoulder, i.e., 5280 rotations. This mechanical arm has also the maximum weight carrying capacity at 0.270 Kg, the maximum upper and front distance respectively at 300 mm and 220 mm.

References

[1] Chang, MS 2012. Design of an Automated Sorting and Orienting Machine for Electronic Pins, Thesis, Massachusetts Institute of Technology

[2] Golmohammadi A, Bejaei F and Behfar H 2013. Design, Development and Evaluation of an Online Potato Sorting System Using Machine Vision, *International Journal of Agriculture and Crop Sciences*. www.ijagcs.com IJACS/2013/6-7/396-402

[3] Reddy, DVK 2014. Sorting of Objects Based on Colour by Pick and Place Robotic Arm and with Conveyor Belt Arrangement, *International Journal of Mechanical Engineering and Robotic Research, 3 (1) pp.67-74*

[4] Mouli CC, Jyothi P and Raju KN, 2014. Modeling and Implementation of Wireless Embedded Robot Arm for Object Sorting, *Journal of Electrical and Electronics Engineering, 9(4), pp.35-44*

[5] Poudarbani R, Ghassemzadeh HR, Seyedarabi H, Nahandi FZ and Vahed MM 2015. Study on an Automatic Sorting System for DateFruits, *Journal of the Saudi Society of Agricultural Sciences*, 14, pp. 83-90

[6] Myint KM, Htun ZMM and Tun HM 2016. Position Control Method For Pick And Place Robot Arm For Object Sorting System, International Journal of Scientific & Technology Research, 5(6), pp. 57-61

[7] Rudresh HG and Shubha PP 2017. Colour Sensor Based Object Sorting Robot, International Research Journal of Engineering and Technology, 4(8), pp. 504-507

[8] Khojastehnazhand MM, Omid M and Tabatabaeefar A 2010. Development of a Lemon Sorting System Based on Color and Size, *African Journal of Plant Science* 4(4), pp. 122-127, http://www.academicjournals.org/ajps

[9] Yung YK 2011. Color Sorting System with Robot Arm, Report. Universiti Teknikal Malaysia.

[10] Kale VR and Kurkani VK 2013. Object Sorting System Using Robotic Arm, International Journal of Advanced Research in Electrical, Electronics and Instrumentation Engineering, 2(7), pp. 3400-3407

[11] Amin U, Ahmad G, Liaqat N, Ahmed M and Zahoor S 2014. Detection & Distinction of Colors using Color Sorting Robotic Arm in a Pick & Place Mechanism, International Journal of Science and Research, 3(6), pp. 1164-1168

[12] Shen LJ and Hassan I 2014. Design and Development of Colour Sorting Robot, 2rd Eureka-Design and Development of Colour Sorting Robot.

[13] Wanve S, and Gawalwad BG 2015. Automatic color object sorting system, International Journal of Modern Trends in Engineering and Research, 2(7), pp.1423-1428

[14] Bhaskar PS, Dofe SS, Tripahi SS and Banode HR 2016. Colour Sorting System with Robotic Arm, International Journal for Innovative Research in Science & Technology, 2(10), pp. 267-272

[15] Nageswari K L, Sujana Mand Polaiah B 2016. Automatic Detection & Distinction of Colored objects using Color Sorting Robotic Arm for Industrial Application, *International Journal of Advanced Research in Management, Architecture, Technology and Engineering, pp. 328-331*

[16] Palok B and Shanmugam SA 2016. Design and Development of a 3 axes Pneumatic Robotic Arm, International Journal of Advanced Research in Electrical, Electronics and Instrumentation Engineering, 5(9), pp. 1-8

Acknowledgments
Authors wishing to acknowledge assistance or encouragement from colleagues, the special work of technical staff or financial support from organizations should do so in an unnumbered Acknowledgments section immediately following the last numbered section of the paper.

IOP Conf. Series: Journal of Physics: Conf. Series **1007** (2018) 012067 doi:10.1088/1742-6596/1007/1/012067

Identification of bearing failure using signal vibrations

Irsyadi Yani[1], Yulia Resti[2], Firmansyah Burlian[1]

[1]Department of Mechanical Engineering, Universitas Sriwijaya
[2]Department of Mathematics, Universitas Sriwijaya

E-mail: yulia_resti@mipa.unsri.ac.id

Abstract. Vibration analysis can be used to identify damage to mechanical systems such as journal bearings. Identification of failure can be done by observing the resulting vibration spectrum by measuring the vibration signal occurring in a mechanical system. Bearing is one of the engine elements commonly used in mechanical systems. The main purpose of this research is to monitor the bearing condition and to identify bearing failure on a mechanical system by observing the resulting vibration. Data collection techniques based on recordings of sound caused by the vibration of the mechanical system were used in this study, then created a database system based bearing failure due to vibration signal recording sounds on a mechanical system. The next step is to group the bearing damage by type based on the databases obtained. The results show the percentage of success in identifying bearing damage is 98 %.

1. Introduction

Bearing conditions in a mechanical system are crucial to the performance of a machine. There are a number of mechanisms that can cause damage to engine element bearings such as mechanical damage, crack damage, wear damage, lubricant deficiency, corrosion and plastic deformation, which in turn leads to damage to the engine system as a whole, other machine components. By knowing the type of damage occurring in a bearing appropriately, it will be able to reduce the use of new spare parts that require the manufacture of natural resources as raw materials and fuel is not small, reduce maintenance costs and or repair costs significantly, also can avoid the risk of accidents due to engine system damage.

The degree of bearing damage can be identified in several ways, such as observation of vibration signals, vibration signal recording, and sound signal recording. Over the past two decades, the identification of bearing damage has been the subject of extensive research. S. Simani, C. Fantuzzi [1] identified bearing damage based on vibration signal observation, while Trendafilova [2] identified bearing damage based on vibration signal recording. According to Kardille [3] that identify the type of damage to the fan-based recording sound signals from the induction motor, the advantages of a device identification system detection or engine-based recording vibration signal and voice signal recording is more easily measured, has a high level of accuracy and reliability.

Data retrieval techniques based on vibration signal recording can be seen in Trendafilova [1]. Trendafilova [1] identifies the type of damage to roller bearings, which is based on pattern recognition and analysis of key components of the measured vibration signal. The recorded signal has been processed before by applying the wavelet transform to extract high frequency areas. Four categories of signals considered, i.e. no signal indicating damage, deep race damage signals, outbreak damage signals, and roller element damage signals.

IOP Conf. Series: Journal of Physics: Conf. Series **1007** (2018) 012067 doi:10.1088/1742-6596/1007/1/012067

2. Methodology

This research will develop a bearing damage identification system based on noise, noise level removal by measuring the magnitude produced by sound signal from bearing rotation in real time. The stages in this study as a whole are given in Figure 1. The design of the software for the identification module is useful for matching sound (magnitude quantities) identified by reference to which the software has been designed in the database module. The results of the integration of both software modules are useful for the degree of damage to the bearing of an engine element. In the final stages of the study there are two targets: first, the intelligent computing system technology package that records the complete database, identifies the damage and matches it to the database and detects the level of damage in real-time.

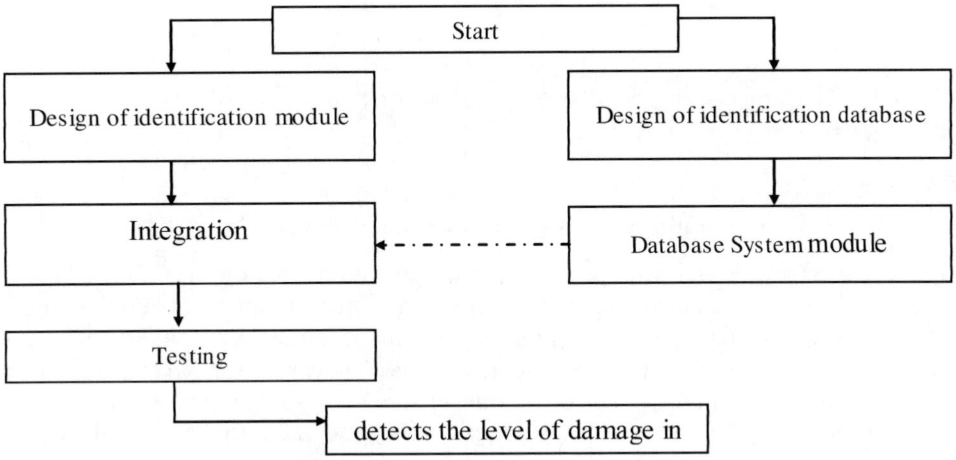

Figure 1. Identification system.

The data collection process has two stages, the first process of soaking bearing in H_2SO_4 3% with 4 variations period that is 5, 10, 15 and 20 days. The second process is the sound recording of the bearing vibration with the testing tool as the initial data for database creation, in this process will get the sound data bearing with wma format. Furthermore, the sound data is analyzed vibration to get the magnitude to be used as database module. The next step is to test for identification of bearing damage by recording bearing sounds randomly.

3. Results and Discussion

The result of sound magnitude analysis of the resulting vibration can be seen from the following figure.

IOP Conf. Series: Journal of Physics: Conf. Series **1007** (2018) 012067 doi:10.1088/1742-6596/1007/1/012067

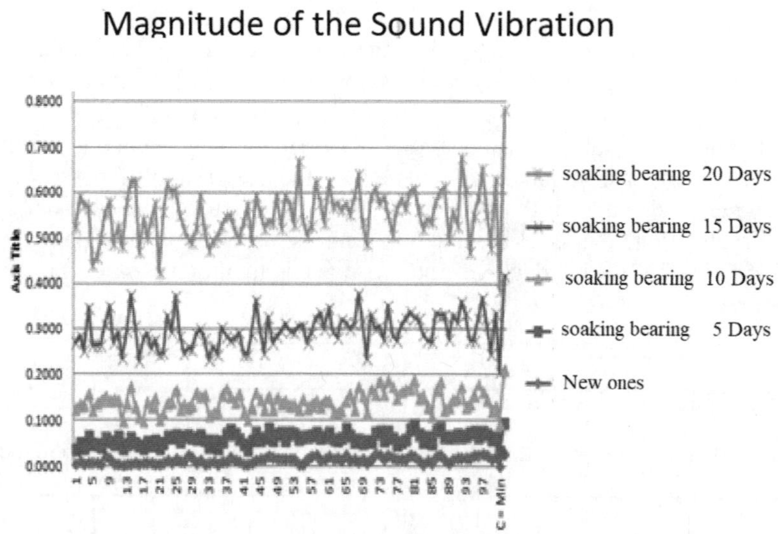

Figure 2. Magnitude of Sound Vibration.

The new bearing is as the initial size parameter with corrosion bearing $H_2 SO_4$ 3% as the test material is seen a significant comparison with other words a metal that has been exposed element $H_2 SO_4$ then corrosion propagation can occur significantly. From the analysis of the existing spectrum wave magnitude shapes, it can be seen that bearing variations have different magnetic spectrum images. This shows that the corrosion rate of the bearings has an influence on the wave magnitude. Values of Sound Magnitude (dB) Bearings can be seen in the following table.

Table 1. Table Values of Sound Magnitude (dB) Test Bearings.

Nomor	Baru	5 hari	10 hari	15 hari	20 hari
1	0.0066000	0.0360000	0.0920000	0.1880000	0.3191000
2	0.0086000	0.0343000	0.0958000	0.1171000	0.2204000
3	0.0104000	0.0382000	0.1152000	0.1490000	0.2247000
4	0.0272000	0.0424000	0.1051000	0.1260000	0.2555000
5	0.0053000	0.0542000	0.0972000	0.1440000	0.2448000
6	0.0106000	0.0437000	0.1149000	0.1569000	0.3085000
7	0.0086000	0.0573000	0.1010000	0.1423000	0.2386000
8	0.0250000	0.0428000	0.1012000	0.1249000	0.2166000
9	0.0191000	0.0373000	0.0788000	0.1788000	0.2325000
10	0.0077000	0.0538000	0.0934000	0.2035000	0.2292000
:	:	:	:	:	:
96	0.0193000	0.0464000	0.0777000	0.1898000	0.2828000
97	0.0227000	0.0462000	0.0877000	0.1360000	0.2747000
98	0.0289000	0.0421000	0.0729000	0.1502000	0.2956000
99	0.0278000	0.0425000	0.0631000	0.2032000	0.2693000
100	0.0175000	0.0441000	0.0651000	0.1657000	0.2632000
MIN	0.0053000	0.0314000	0.0600000	0.1158000	0.2106000
MAX	0.0289000	0.0595000	0.1152000	0.2066000	0.3634000

Table 2. Results of Bearing Identification Testing.

Number	Magnitudo (dB)	Status
1	0.0298000	Detected
2	0.0382000	Detected
3	0.0328000	Detected
4	0.0543000	Detected
5	0.0400000	Detected
:	...	...
:	...	...
80	0.0564000	Detected
81	0.0672000	Not detected
:	...	...
86	0.0550000	Detected
87	0.0597000	Not detected
88	0.0455000	Detected
:	...	...
:	...	...
100	0.0441000	Detected

From the experimental table, the results obtained from 100 times the test there are 98 data that was successfully detected in accordance with the database that was built while the undetectable voice data only reached 2 votes only. So the percentage of detection matches between voice data and train data is 98%.

4. Conclusion

From the previous description, it can be concluded that among others is.

1. Bearing failure occurring in this case is confirmed by several mutually supportive factors, namely the increase in the value of significant vibrations, the emergence of the natural frequency of bearing failure on the spectrum, and the characteristics of the spectrum, i.e. the appearance of sub-sync frequencies in the spectrum.
2. One of the natural frequencies detected is the natural frequency of the cage.

References

[1] S Simani, C Fantuzzi, RJ Patton , (2003) 'Model-Based Fault Diagnosis Techniques', Model-based Fault Diagnosis in Dynamic Systems Using Identification Techniques, p 19-60

[2] IrinaTrendafilova, (2010), 'An automated procedure for detection and identification of ball bearing damage using multivariate statistics and pattern recognition', Mechanical Systems and Signal Processing, p 1858-1869

[3] Kardile, B.A. (2012) 'Bearing Life Improvement of Centrifugal Blowers by Vibration Analysis', International Journal of Modern Engineering Research, Nov-Dec, p1

[4] Sayer, R. (1999) 'Vibration Analysis of Centrifugal Fans', Computational Systems, Inc

[5] Bleier, F. (1997) Fan Handbook, New York : McGraw-Hill

[6] Singleton, K. (2002) 'Overhung Fans Reliability Improvement Opportunities' , KSC Consulting LCC

[7] Scheffer, C. and Girdhar, P. (2004) Practical Machinery Vibration Analysis and Predictive Maintenance, Oxford:Newnes

[8] Thomson, W. and Dahleh, M. (1993) Theory of Vibration with Applications, 5th ed, Cheltenham: Prentice Hall

[9] Vibration Diagnostic Guide (2000) SKF Reliability Systems

IOP Conf. Series: Journal of Physics: Conf. Series **1007** (2018) 012067 doi:10.1088/1742-6596/1007/1/012067

[10] Peters, J., Beginning Vibration Analysis with Basic Fundamentals, Connection Technology Center, Inc.

[11] Robichaud, J.M. 'Reference Standards for Vibration Monitoring and Analysis' Bretech Engineering Ltd

[12] Cease, B.T.(2011), 'Machine Vibration Standards : Ok, Good, Better and Best' Cease Industrial Consulting

[13] Rukhanje, S., Pathan, S., & co. (2012) 'Condition Monitoring for Fault Diagnosis of Bearings' , Third Biennial National Conference on Nascent Technologies, February, pp1-2

[14] Bennett, R. (2006) 'Spherical Roller Bearings and Overhung Fan. A case history in condition monitoring, root cause analysis, and design.' , National CM Forum, August

[15] Upadhay, R.K., & co. (2013) 'Rolling Element Bearing Failure Analysis : A Case Study' , Case Studies in Engineering Failure Analysis, January, pp 15-17.

[16] Courtney, S. (2005) A Guide to Low Frequency Vibration Measurement, Mechanalysis (India) Limited

[17] Berry, J. (2000), Proven Method For Specifying Both Spectral Alarm Bands As Well As Narrowband Alarm Envelopes Using Today's Condition Monitoring Software Systems, 4th ed, Technical Associates of Charlotte, P.C

IOP Conf. Series: Journal of Physics: Conf. Series **1007** (2018) 012068 doi:10.1088/1742-6596/1007/1/012068

Implementation of Push Recovery Strategy Using Triple Linear Inverted Pendulum Model in "T-FloW" Humanoid Robot

R. Dimas Pristovani, D. Raden Sanggar, Pramadihanto. Dadet

EEPIS Robotics Research Center (ER2C), Politeknik Elektronika Negeri Surabaya (PENS), Indonesia

E-mail: *dimaspens@gmail.com, sanggar@pens.ac.id, dadet@pens.ac.id

Abstract. Push recovery is one of humanbehaviorwhich is a strategy to defend the body from anexternal force in any environment. This paper describes push recovery strategy which usesMIMO decoupled control system method. The dynamics system uses aquasi-dynamic system based on triple linear inverted pendulum model (TLIPM). The analysis of TLIPMuses zero moment point (ZMP) calculation from ZMP simplification in last research. By using this simplification of dynamics system, the control design can be simplified into 3 serial SISOwith known and uncertain disturbance models in each inverted pendulum. Each pendulum has different plan to damp the external force effect. In this experiment, PID controller (closed-loop)is used to arrange the damp characteristic.The experiment result shows thatwhen using push recovery control strategy (closed-loop control) is about 85.71% whilewithout using push recovery control strategy (open-loop control) it is about 28.57%.

1. Introduction

Push recovery challenge hasthe main goal, which withstands a strong push and keeps standing. The robot receives a push from an object that swings against its body. The robot receives one or more pushes and withstand successfully to fulfil the push recovery challenge [1]. The humanoid robot in this discussion is called "T-FLoW". In this discussion, we use aT-FLoW humanoid robot with 28 Degrees of Freedom (DoF) version. To complete the push recovery challenge, the MIMO decoupled control systemanalysisis usedto design the control system[2–3]. Another research about push recovery is also using similar action[11–21], but the difference in this discussion is the dynamics system approach and control system. With a lot of DoF, the dynamics model of T-FLoWisapproached by using Triple Linear Inverted Pendulum Model (TLIPM) [4–5].

With this approach, the dynamic system model of T-FLoW will be quasi-dynamics system model and itis analyzed by using thecalculation of zero moment point (ZMP) analysisas in the previous research[6–8]. By using this simplification of dynamics system, the control design can be simplified into 3 serial SISO with known and uncertain disturbance (mass and external force) in each inverted pendulum (IP).Each pendulum has different plan to damp the external force effect (uncertain disturbance). In this plan, PID controller is used to arrange the damping characteristic[9–10]. In the experiment result will explainingabout thecomparison between open-loop push recovery control strategy (without PID controller) and closed-loop (3 serial SISO) push recovery control strategy (with PID controller).

2. Originality

The differences with previous research about push recovery strategyarein the dynamics system approach and control system usage. As mentioned before, the dynamics system approach is using Triple Linear Inverted Pendulum Model (TLIPM). In previous research in reference [11-21], they are using Double Linear Inverted Pendulum Model (DLIPM) or SingleLinear Inverted Pendulum Model (SLIPM). The control system design is using MIMO decoupled control system analysis with PID controller. This proposed control system is used to smoothing the movement of each IP. In another research in reference [11-21], they are not using MIMO control analysis. The last difference is in the domain input for thecontrol system. In our control system is using cartesian space input for the control system and another research is using joint space input for their control system.

3. Mechanical Design (T-FLoW)

Mechanical design of T-FLoW humanoid robot is shownin figure 1. T-FLOW humanoid robot has 2 version of mechanical design which is using 28 Degree of Freedom (DoF) and using 38-DoF. But, in this discussion, 28-DoF version of T-FLoW humanoid robot will be used. The configuration of 28-DoF is 16-DoF in the left and right leg, 3-DoF in the waist, 6-DoF in the left and right arm, and 3-DoF in the head. The total height and width of T-FLoW humanoid robot are about 112.67cm and 27.54cm. The configuration of each joint and number of actuators are shown in table 1.

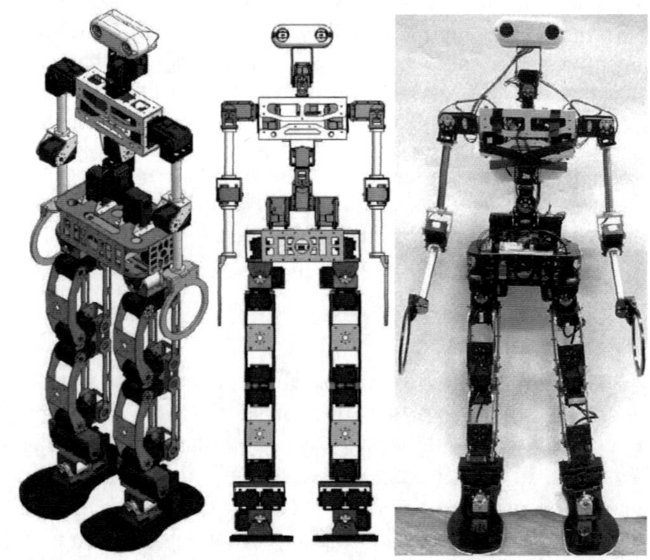

Figure 1. 28-DoF version mechanical design of T-FLoW humanoid robot.

Table 1. Degree of Freedom (DoF Configuration).

Parts	Degree of Freedom	Number of
Head	3	3
Right Arm	3	3
Left Arm	3	3
Waist	3	4
Right Leg	8	9

Left Leg	8	9
Total	**28**	**31**

The lower body is a left leg, right leg, and hip. T-FLoW humanoid robot using parallel link mechanical system called parallelogram mechanism in the left and right legs. The parallelogram mechanism is used to add more active torque in the legs. The upper body is a left arm, right arm, and waist. The upper body is using serial link mechanical system.

4. Quasi Dynamics System Model

In the last research of T-FLoW humanoid robot, ZMP classic calculation was used to obtain the dynamics characteristic. Where $P_{zmp,(x,y,z)} = [P_{zmp,x} \quad P_{zmp,y} \quad P_{zmp,z}]^T$ is position vector of ZMP from its origin position. The main equation tocalculate the position vector of ZMP for multi-body is shown in Equation 1as follows.

$$P_{zmp,(x,y)} = \frac{P_{CoM,(x,y)} m_{tot}\, g_z + P_{zmp,z}\dot{\mathcal{P}}_{(x,y)} - \dot{\mathcal{L}}_{(y,x)}}{m_{tot}\, g_z + \dot{\mathcal{P}}_z} \quad (1)$$

In which$P_{CoM,(x,y,z)}$ is the position vector of CoM.This position can be obtained from position calculation of each joint, which is used in the robot. $P_{zmp,(x,y,z)}$is the position vector of ZMP.This position is the dynamics characteristic of the system. $m_{tot}$is the mass total from the system. $\dot{\mathcal{L}}_{(x,y,x)}$ and $\dot{\mathcal{P}}_{(x,y,x)}$ is the angular acceleration (torque) and the linier acceleration (force). $g_z$isthe linier acceleration of gravitation in Z axis. When discussing control system, the dynamics system must be calculated in real time condition. Because of it, the dynamic system model of T-FLoW humanoid robot is approached by using Triple Linear Inverted Pendulum Model (TLIPM). The simplification design of dynamic system model based on TLIPMis shown in figure 2.

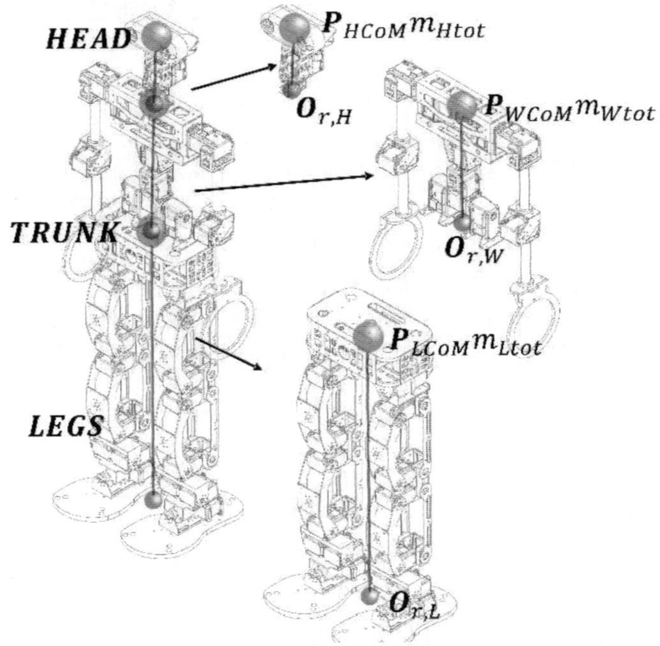

Figure 2. Part division of TLIPM in T-FLoW humanoid robot which is Head part (IP 1), Trunk part (IP 2), and

IOP Conf. Series: Journal of Physics: Conf. Series **1007** (2018) 012068 doi:10.1088/1742-6596/1007/1/012068

Legs part (IP 3).

The dynamics calculation of T-FLoW humanoid robot is divided into 3 parts which are Head part (IP 1), Trunk part (IP 2), and Legs part (IP 3). As mentioned before, the dynamics system is simplified from multi-body analysis into single body analysis based on the parts. Because of it, the position vector of ZMP for multi-body in Equation 1 from previous research is developed and simplified to obtain dynamics system based on Inverted Pendulum (IP) model.

$$\dot{\mathcal{P}}_{(x,y,z)}=\dot{\mathcal{P}}_{CoM,(x,y,z)}=m_{tot}\ddot{P}_{CoM,(x,y,z)} \text{ and } \dot{\mathcal{L}}_{(x,y,z)}=\dot{\mathcal{L}}_{CoM,(x,y,z)}=P_{CoM,(x,y,z)}m_{tot}\ddot{P}_{CoM,(x,y,z)} \tag{2}$$

$$P_{zmp,(x,y,z)} = [P_{zmp,x} \quad P_{zmp,y} \quad 0]^T \tag{3}$$

$$P_{CoM,(x,y,z)} = \sum_{i=1}^{n} m_i P_{i,(x,y,z)} \tag{4}$$

The explanation from Equation (2) and (3) above is included in Equation (1). In the IP models, the linear acceleration in Z axis is equal to zero ($\ddot{P}_{CoM,z} = 0$) and $g_{(x,y,z)} = [0 \quad 0 \quad -g_z]^T$(the gravitational acceleration vector).

$$P_{zmp,(x,y)} = \frac{\left(P_{CoM,(x,y)} m_{tot}\, g_z + P_{CoM,(x,y)} m_{tot} \ddot{P}_{CoM,z}\right)}{m_{tot}\, g_z + m_{tot}\ddot{P}_{CoM,z}} - \frac{\left(P_{CoM,z} m_{tot}\, g_{(x,y)} + P_{CoM,z} m_{tot} \ddot{P}_{CoM,(x,y)}\right)}{m_{tot}\, g_z + m_{tot}\ddot{P}_{CoM,z}} \tag{5}$$

$$P_{zmp,(x,y)} = \frac{\left(P_{CoM,(x,y)} m_{tot}\, g_z + 0\right)}{m_{tot}\, g_z + 0} - \frac{\left(0 + P_{CoM,z} m_{tot} \ddot{P}_{CoM,(x,y)}\right)}{m_{tot}\, g_z + 0} \tag{6}$$

$$P_{zmp,(x,y)} = \frac{P_{CoM,(x,y)}\cancel{m_{tot}\, g_z}}{\cancel{m_{tot}\, g_z}} - \frac{P_{CoM,z}\cancel{m_{tot}}\ddot{P}_{CoM,(x,y)}}{\cancel{m_{tot}}\, g_z} \tag{7}$$

$$P_{zmp,(x,y)} = P_{CoM,(x,y)} - \frac{P_{CoM,z}}{g_z}\ddot{P}_{CoM,(x,y)} \tag{8}$$

Equation (8) is the dynamics model of linear IP. This equation is used to calculate dynamics characteristic in each IP from dynamics system model in T-FLoW humanoid robot. Figure 3, figure 4, and figure 5 is transformation process from multi-body calculation into single body calculation.

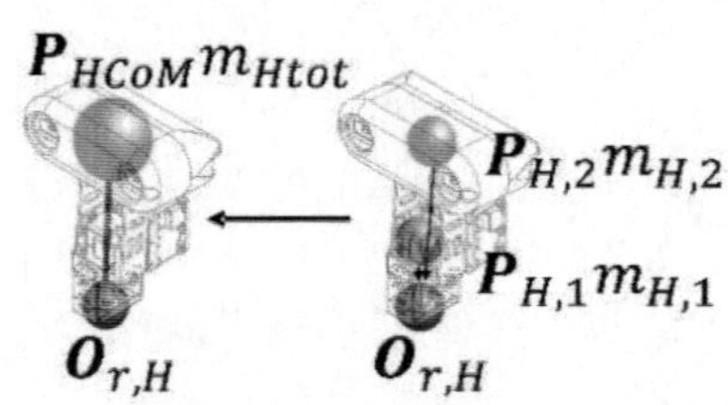

Figure 3. Mass and position distribution in inverted pendulum 1 (IP 1 – Head part), using 3 DoF.

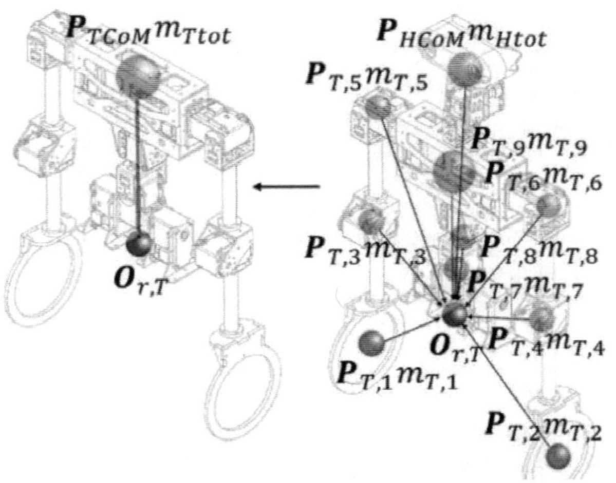

Figure 4. Mass and position distribution in inverted pendulum 2 (IP 2 – Trunk part), using 9 DoF.

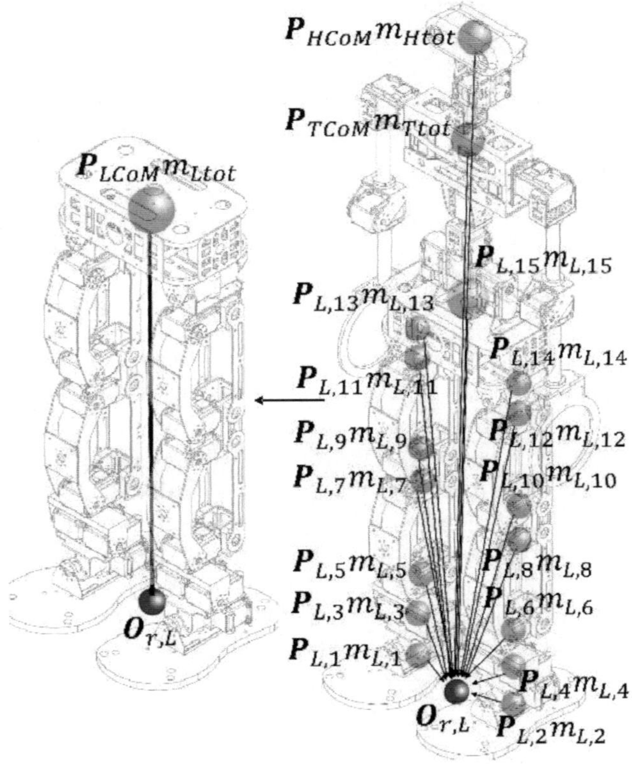

Figure 5. Mass and position distribution in inverted pendulum 3 (IP 3 – Legs part), using 16 DoF.

Based on figure 3, figure 4, and figure 5, the position vector of ZMP for each IP can be calculated by using the equationas follows.

$$P_{H,CoM,(x,y,z)} = \sum_{i=1}^{n=3} m_{H,i} P_{H,i,(x,y,z)} \tag{9}$$

$$P_{T,CoM,(x,y,z)} = \sum_{i=1}^{n=9} m_{T,i} P_{T,i,(x,y,z)} + P_{H,CoM,(x,y,z)} \tag{10}$$

$$P_{L,CoM,(x,y,z)} = \sum_{i=1}^{n=16} m_{L,i} P_{L,i,(x,y,z)} + P_{T,CoM,(x,y,z)} \tag{11}$$

$$P_{(H,T,L),zmp,(x,y)} = P_{(H,T,L),CoM,(x,y)} - \left({P_{(H,T,L),CoM,z}}/{g_z} \right) \ddot{P}_{(H,T,L),CoM,(x,y)} \tag{12}$$

Where H is the head part notation, T is the trunk partnotation, and L is the legs partnotation. $P_{H,CoM,(x,y,z)}$,$P_{T,CoM,(x,y,z)}$, and$P_{L,CoM,(x,y,z)}$ is the position vector of CoM in each IP.$m_{H,i}$, $m_{T,i}$, and$m_{L,i}$ is each jointmass in each IP.$P_{H,i,(x,y,z)}$, $P_{T,i,(x,y,z)}$, and$P_{L,i,(x,y,z)}$ is each joint position vector in each IP.This position vector can be obtained from forward kinematic (FK) analysis.$\ddot{P}_{H,CoM,(x,y)}$, $\ddot{P}_{T,CoM,(x,y)}$, and$\ddot{P}_{L,CoM,(x,y)}$ is linear acceleration of CoM in each IP, this linier acceleration can be obtained from acceleration sensor.$P_{H,zmp,(x,y)}$, $P_{T,zmp,(x,y)}$, and$P_{L,zmp,(x,y)}$is position vector of ZMPin each IP.

5. Control Design based on System Model

As mentioned in theintroduction, T-FLoW humanoid robot is doing a push recovery challenge. The push recovery strategy is how to maintain the robot's posture during an external force without moving the legs (step/walk).The Push recovery in this challenge hasthe main goal, which withstands a strong push and keeps standing.Therobotreceives a push from an object that swings against its body.

The push recovery strategy is adapting the human behavior. When human receives a small external force, they are only correcting the position of their body. The position correction is happening based on the force direction vector. When an external force has X direction (forward direction) and the external force is directly contacting to the upper body of the human, the human reaction is moving the upper body (trunk) into the same direction vector of external force and the legs are moving into the opposite direction vector of external force. This reaction is adapted by T-FLoW humanoid robot. As seen in figure 6, the T-FLoW humanoid robot is receiving an external force through the upper body (trunk). The upper body (trunk) of the robot is following the direction vector of external force. But, the legs and the head body is following the opposite direction vector of external force.

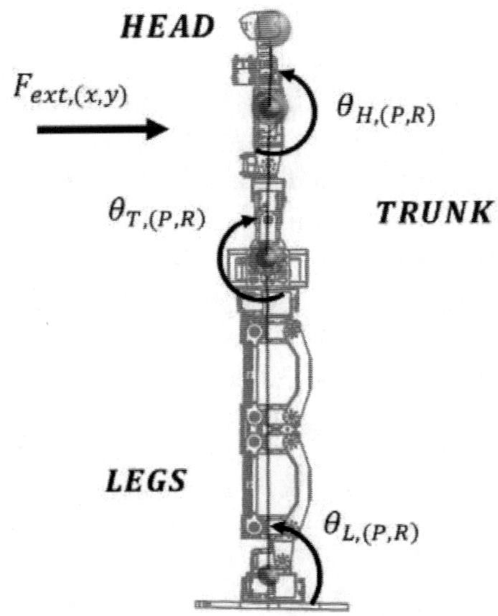

Figure 6. Push recovery strategy based on TLIPM in the T-FLoW humanoid robot.

The strategy to damp the external force is how to arrange the position vector of CoM in each IP to not be outside of support polygon (SP). For the first IP, the strategy is how to maintain the condition at the angular position reference which is the first IP is remaining in upright condition ($\theta_{H,P,Ref} = 0$). For second IP, the strategy is how to damp the external force disturbance because the external force is directly contacted with the second IP. In the second IP (Trunk), the arms are used to shift the position vector of CoM in the Trunk. For the third IP (Legs), the strategy is how to compensate the second IP during external force. The compensation is from direction of the angular position in second and third IP which is when second IP is moved into forward direction because the external force, so the third IP is moved into opposite direction (backward direction). This compensation is happened because the third IP is the most important part of the robot which is the main base of the robot body.

The MIMO decoupled control design based on system model is used in push recovery control strategy. As mention it before, the dynamic system of T-FLoW humanoid robot is using quasi-dynamic system model approach based on TLIPM.

With this approach, the MIMO decoupled control design is simplified into 3 serial SISO control design when implemented into the system. Basically, the MIMO control system example is described in Equation 13 and MIMO decoupled control system which is multi SISO control system example is described in the Equation 14.

$$\begin{bmatrix} P_1(t) \\ P_2(t) \\ \vdots \\ P_q(t) \end{bmatrix} = \begin{bmatrix} P_{11,zmp}(t) & P_{12,zmp}(t) & \cdots & P_{1p,zmp}(t) \\ P_{21,zmp}(t) & P_{22,zmp}(t) & \cdots & P_{2p,zmp}(t) \\ \vdots & \vdots & \ddots & \vdots \\ P_{q1,zmp}(t) & P_{q2,zmp}(t) & \cdots & P_{qp,zmp}(t) \end{bmatrix} \begin{bmatrix} U_1(t) \\ U_2(t) \\ \vdots \\ U_p(t) \end{bmatrix} \text{ or, } \theta(t) = P_{zmp}(t)U(t) \quad (13)$$

$$\begin{bmatrix} P_1(t) \\ P_2(t) \\ \vdots \\ P_q(t) \end{bmatrix} = \begin{bmatrix} P_{11,zmp}(t) & 0 & \cdots & 0 \\ 0 & P_{22,zmp}(t) & \cdots & 0 \\ \vdots & \vdots & \ddots & \vdots \\ 0 & 0 & \cdots & P_{qp,zmp}(t) \end{bmatrix} \begin{bmatrix} U_1(t) \\ U_2(t) \\ \vdots \\ U_p(t) \end{bmatrix} \text{ or, } \begin{bmatrix} P_1(t) \\ P_2(t) \\ \vdots \\ P_q(t) \end{bmatrix} = \begin{bmatrix} P_{11,zmp}(t) \\ P_{22,zmp}(t) \\ \vdots \\ P_{22,zmp}(t) \end{bmatrix} \begin{bmatrix} U_1(t) \\ U_2(t) \\ \vdots \\ U_p(t) \end{bmatrix} \quad (14)$$

Based on equation 12, The design of the control system (closed-loop) in the T-FLoW humanoid robot can be seen in figure 6 and the strategy to damping the external force is shown in figure 5. From figure 6, the total SISO control system (closed-loop) in this system is 6 part such as $U_{H,x}(t)$, $U_{H,y}(t)$, $U_{T,x}(t)$, $U_{T,y}(t)$, $U_{L,x}(t)$, and $U_{L,y}(t)$. But, these 6 parts of SISO control system is categorized into 3 IP parts which is first IP (Head) is $U_{H,x}(t)$ and $U_{H,y}(t)$, second IP (Trunk) is $U_{T,x}(t)$ and $U_{T,y}(t)$, and third IP (Legs) is $U_{L,x}(t)$ and $U_{L,y}(t)$.

The control system is work based on cartesian data input from dynamics system calculation. The input is position vector of ZMP ($P_{(H,T,L),zmp,(x,y)}$). The data input is compared with real position vector of each IP ($P_{(H,T,L),Real,(x,y)}$) and obtain the position vector error ($e_{(H,T,L),(x,y)}$). The position vector error ($e_{(H,T,L),(x,y)}$) is the input for PID controller ($U_{(H,T,L),(x,y)}$). The PID controller ($U_{(H,T,L),(x,y)}$) has output position vector for the system ($P^T_{(H,T,L),zmp,(x,y)}$), this position vector is used as a new data input in the T-FLoW humanoid robot. Because the data input is in the cartesian area (position vector), it must be transformed into joint area (angle joint) by using Inverse Kinematics (IK) of TLIPM. The Mechanical System (MS) is interferencewith uncertain disturbance (external force ($Q_{(H,T,L),(x,y)}$)) in each inverted pendulum (IP). To obtain the real-time condition of T-FLoW humanoid robot during external force (TLIPM), The Feedback System (FS) is used to measure it. The FS has several sensors to measure the data of T-FLoW humanoid robot during external force such as angular position (joint area). The feedback must be transformed into cartesian area by using Forward Kinematics (FK) of TLIPM and then this data is already to processed again. The control diagram of push recovery strategy is shown in figure 7 as follows.

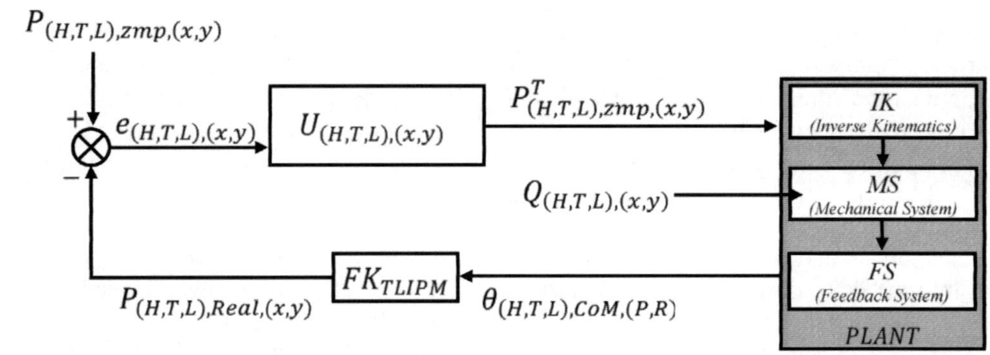

Figure 7. TLIPM with serial SISO control system design using PID controller in push recovery strategy.

The IK process is calculating the local transformation calculation in each IP.From position vector of ZMP in each IP, the angular position in each IP is obtained by using Equation 15.

$$\theta_{(H,T,L),zmp,(P,R)} = \tan^{-1}\left(P^T_{(H,T,L),zmp,(x,y)} \middle/ P_{(H,T,L),CoM,z}\right) \tag{15}$$

The FK process is similar to IK process, just calculating the local transformation calculation in each IP. From anangular position in each IP, the real position vector in each IP isobtained by using Equation 16 as follows.

$$P_{(H,T,L),Real,(x,y)} = P_{(H,T,L)CoM,z} \times \sin\theta_{(H,T,L),CoM,(P,R)} \tag{16}$$

Where, $\theta_{H,zmp,(P)}$, $\theta_{T,zmp,(P)}$, and $\theta_{L,zmp,(P)}$ is the angular ZMP position transformation result from position vector of ZMP in each IP with pitch rotation. $\theta_{H,zmp,(R)}$, $\theta_{T,zmp,(R)}$, and $\theta_{L,zmp,(R)}$ is the angular ZMP position transformation result from position vector of ZMP in each IP with roll rotation.$P^T_{(H,T,L),zmp,(x,y)}$isthe position vector from PID controller output.$P_{(H,T,L),Real,(x,y)}$is the position vector in real time condition. $\theta_{(H,T,L),CoM,(P,R)}$isthe angular position in real time condition from FS.

The general equation of PID controller for the 3 serial SISO push recovery strategy (closed-loop) is shown in Equation 17 and Equation 18 as follows.

$$e_{(H,T,L),(x,y)}(t) = \left(P_{(H,T,L),zmp,(x,y)}(t) - P_{(H,T,L),Real,(x,y)}(t)\right) \tag{17}$$

$$U_{(H,T,L),(x,y)}(t) = K_{p,(H,T,L),(x,y)}\left(e_{(H,T,L),(x,y)}(t)\right) + K_{i,(H,T,L),(x,y)}\int_0^{\tau}\left(e_{(H,T,L),(x,y)}(\tau)\right)d\tau + K_{d,(H,T,L),(x,y)}\frac{d\left(e_{(H,T,L),(x,y)}(t)\right)}{dt} \tag{18}$$

The value of $K_{p,(H,T,L),(x,y)}$, $K_{i,(H,T,L),(x,y)}$, and $K_{d,(H,T,L),(x,y)}$ in the PID controller will be tune by using Ziegler-Nichols tuning methods. This method is used as one of the way to tune the PID controller. This method is work based on step response requirement of the system (figure 8) [22]. The tuning formulas can be seen in table 2.

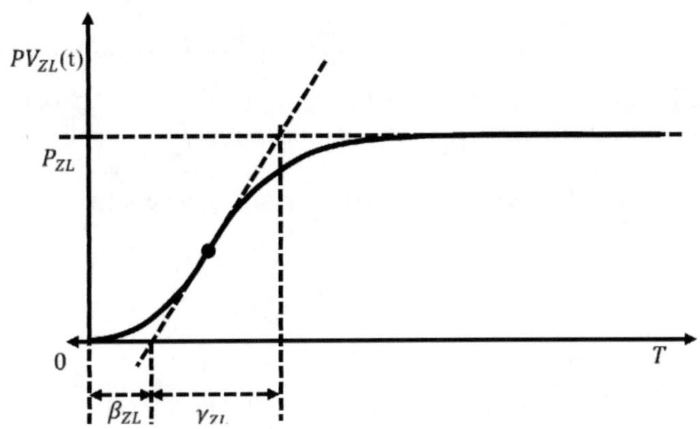

Figure 8. S-shaped step input response of the plants.

Table 2. Ziegler-Nichols tuning rule based on step response of plants.

Type of Controller	$K_{p(n)}$	$K_{i(n)}$	$K_{d(n)}$
P	γ_{ZL}/β_{ZL}	∞	0
PI	$0.9\gamma_{ZL}/\beta_{ZL}$	$\beta_{ZL}/0.3$	0
PID	$1.2\gamma_{ZL}/\beta_{ZL}$	$2\beta_{ZL}$	$0.5\beta_{ZL}$

6. Experiment Result

The experiment of push recovery in this section is using pendulum model to generate an external force. The robot is receiving the external force from the backward direction (X-axis).Before the experiment, the PID controller (closed-loop) is tuned by using Ziegler-Nichols tuning methods and the result of this tune can be seen in table 3. The pendulum (force generator) is drawn to certain angles so the pendulum will swing and causing an external force against the robot's body. The pendulum is swinging directly into trunk part of the robot body. Because in this experiment is only focus on the implementation of push recovery strategy during external force directly into the trunk with the only

IOP Conf. Series: Journal of Physics: Conf. Series **1007** (2018) 012068 doi:10.1088/1742-6596/1007/1/012068

forward direction (X-axis). The pendulum has weight about 2.33 Kg and length about 115 cm as table 4 as follows.

Table 3. Tuned Constanta of PID controller using Ziegler-Nichols tuning methods.

Part	$K_{p(Pitch)}$	$K_{i(Pitch)}$	$K_{d(Pitch)}$	$K_{p(Roll)}$	$K_{i(Roll)}$	$K_{d(Roll)}$
Head	2.4	0.2	0.05	1.5	0.16	0.04
Trunk	1.5	0.4	0. 1	1.8	0.2	0.05
Legs	1.92	0.5	0.125	2.0	0.3	0.075

Table 4. Specification of Pendulum (External Force Generator).

Part	Description
Length	115 cm
Distance	56 cm
Weight	2.23 Kg

By using this tuned constanta, the PID controller is ready touse in the push recovery control strategy. The experiment is comparing the open-loop push recovery control strategy and closed-loop (3 serial SISO) push recovery control strategy. The experiment is starting with a small certain angle in the pendulum to generate a small external force and the result can be seen in table 5.

Table 5. Implementation experiment result of push recovery strategy in T-FLoW humanoid robot.

Pendulum Angle (Force Generator) (Degree)	Without Control	With Control
5	Success	Success
10	Success	Success
15	Not Success	Success
20	Not Success	Success
25	Not Success	Success
30	Not Success	Success
35	Not Success	Not Success

From table5 can be seen the experiment result of push recovery strategy implementation comparison between open-loop control system and closed-loop control system. When the robot doing push recovery challenge using open-loop control system, the percentage of successful withstand is about 28.57% and robot falls is about 71.43%. When the robot doing push recovery challenge using closed-loop control system, the percentage of successful withstand is about 85.71% and robot falls is just about 14.29%. From the experiment also known the maximum value of external force that can be handled by push recovery strategy is about 30 degrees of pendulum angle. In figure 9 shows the push recovery strategy at a maximum angle about 30 degrees of pendulum angle and the comparison in each IP can be seen in figure 10, figure 11, and figure 12. From the position vector comparison shows when using closed-loop control system, the robot withstands and keeps standing (red line) and when using open-loop control system, the robot is fall (blue line).

IOP Conf. Series: Journal of Physics: Conf. Series **1007** (2018) 012068 doi:10.1088/1742-6596/1007/1/012068

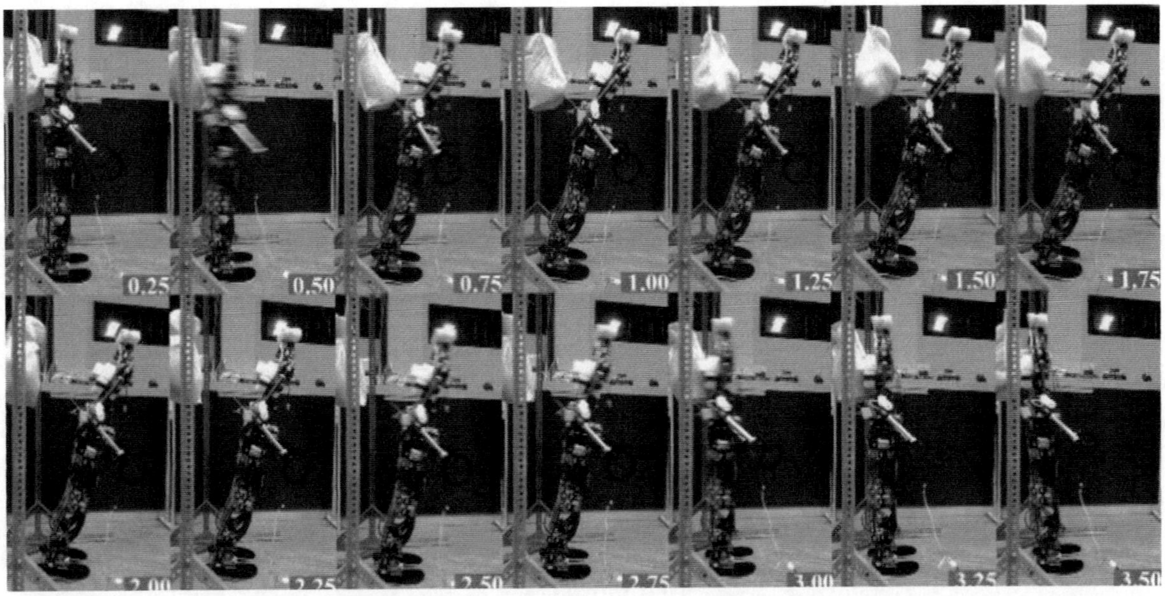

Figure 9. Implementation sample of push recovery strategy from table 5 during maximum external force (30 degrees of pendulum angle). In this figure shows, the robot controlsystemreached the critical condition because of theposition vector of ZMP almost outside SP.

In figure 10, figure 11, and figure 12, there are Fall area and Safe area. Fall area is area when the robot is impossible to maintain the balance. When the robot reaches this condition even in short time, the robot will fall. The safe area is area when therobot is still possible to balancing the body. When the robot reaches this condition, the robot has a chance to maintain the body balance to avoid fall condition.

The Inverted Pendulum 1 is shown in figure10. In data without PID controller (open-loop), the standard deviation (SD) of data is about 14.12 and the maximal minimal angle of data is about (46 ~ -1) degree. In data with PID controller (closed-loop), the SD of data is about 5.163 and the maximal minimal angle of data is about (29 ~ -18) degree

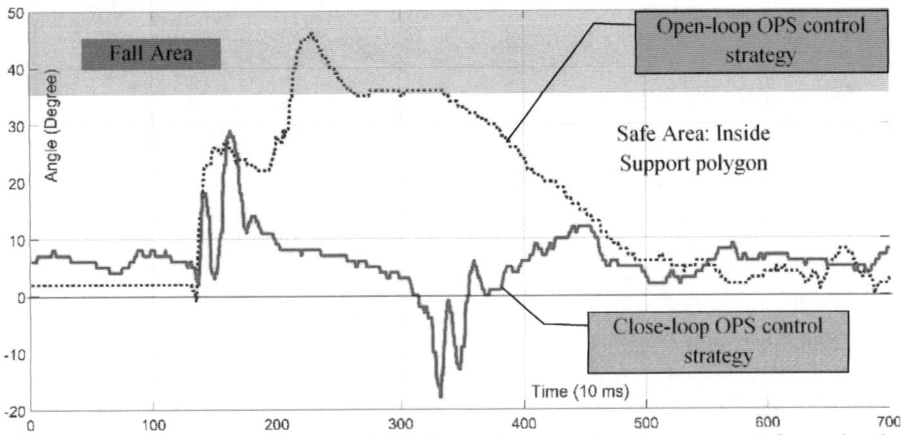

Figure 10. Position vector comparison of ZMP during maximum external force in the first IP (Head).

IOP Conf. Series: Journal of Physics: Conf. Series **1007** (2018) 012068 doi:10.1088/1742-6596/1007/1/012068

The Inverted Pendulum 2 is shown in figure11. In data without PID controller (open-loop), the standard deviation (SD) of data is about 15 and the maximal minimal angle of data is about (50 ~ 1) degree. In data with PID controller (closed-loop), the SD of data is about 9.738 and the maximal minimal angle of data is about (33 ~ -10) degree

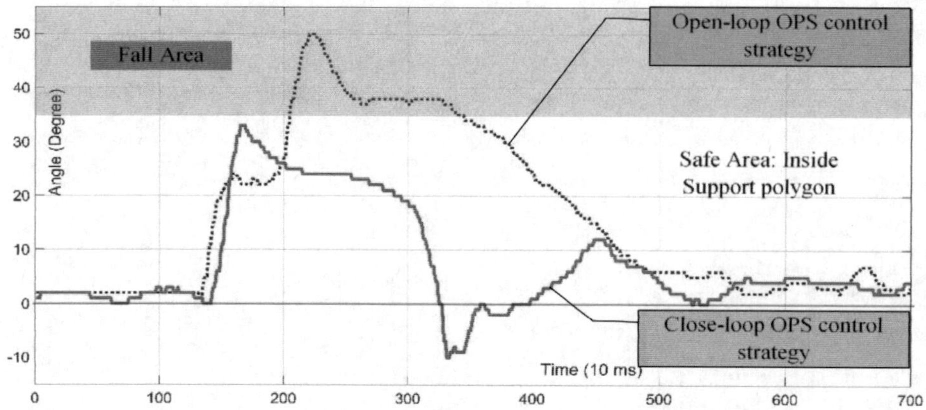

Figure 11. Position vector comparison of ZMP during maximum external force in the second IP (Trunk).

The Inverted Pendulum 3 is shown in figure 12. In data without PID controller (open-loop), the standard deviation (SD) of data is about 15.59 and the maximal minimal angle of data is about (51 ~ 1) degree. In data with PID controller (closed-loop), the SD of data is about 7.398 and the maximal minimal angle of data is about (25 ~ -7) degree.

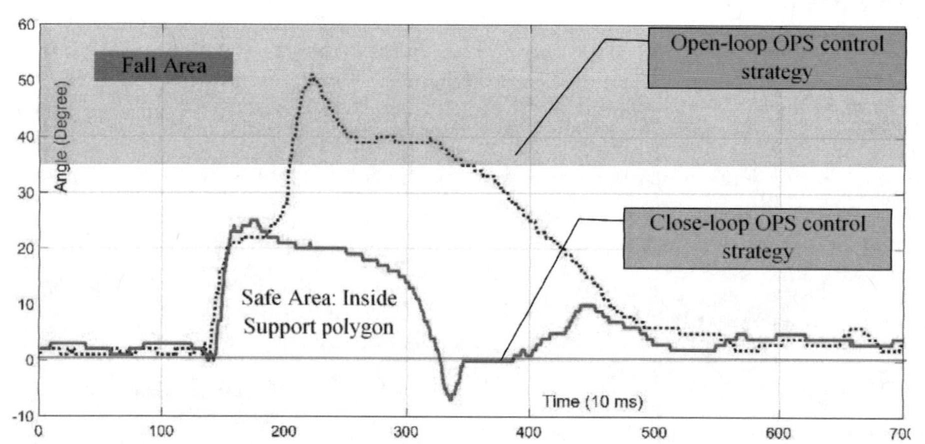

Figure 12. Position vector comparison of ZMP during maximum external force in the third IP (Legs).

7. Conclusion and Discussion

The main result of the experiment when receiving maximum external force at 30 degrees of pendulum angle is arobotwith open-loop control systemhas been falling and arobot with closed-loop control systemhas been successfully (withstanding) from external force and keeps standing.The difference in percentage of successful withstand is about 85.71% when using push recovery control strategy (closed-loop control system) and about 28.57% without using push recovery control strategy (open-loop control system).

But a new problem is appeared, that is, push recovery control strategy (closed-loop control system) cannot handle and damp the external force up to 30 degrees of pendulum angle. It isbecausethe position vector of ZMP in the critical condition (the position vector of ZMP almost outside SP). With this problem, the future research will be focusedon a strategy deals with the walking stepsoff whenreaching the critical condition and try touse another controller such as FUZZY, Neural Network, and etc to improve and compare the differences result from thepresentresearch.

References

[1] Technical Committee of RoboCup Humanoid League, 2017, *Robocup Soccer Humanoid League Laws of the Game 2016/2017*, pp. 58–63

[2] Luca B and Alberto L.D, 2008, Nonparametric decoupling of MIMO systems,*Proc.IEEE International Conference on Computer-Aided Control Systems, CACSD 2008,* pp. 1073–1078

[3] B. A. Ogunnaike, W. H. Ray, 1979, Multivariable controller design for linear systems having multiple time delays, *AIChE Journal, Vol. 25, No. 6*, pp. 1043–1057

[4] K.G. Eltohamy, 1997, Real time stabilization of a triple link inverted pendulum using single control input, *IEEE Proc. Control Theory and Application, Vol. 144, No. 5,* pp 498–504

[5] V.A. Tsachouridis and G.A. Medrano-Cerda, 1999, Discrete-time H∞ control of a triple inverted pendulum with single control input. *IEEE Proc. Control Theory and Application, Vol. 146, No. 6,* pp 567–577

[6] R. Dimas Pristovani, W. M. Rindo, B. Eko H, K. H. Achmad S, and Dadet P, 2016, Basic walking trajectory analysis in FLoW ROBOT, *Proc. 2016 Int. Electronics Symposium (IES),* pp 333–338

[7] K. Nuril E, D. Raden S, Dadet P, 2017, "FLoW" bipedal robot: squat and walk motions, *Int. J. Advanced Science Letters, Vol. 23, Issue 5*, pp. 3838-3842

[8] M. Vukobratovic, A. A. Frank, and D. Juricic,1970, On the stability ofbiped locomotion,*in: IEEE Transactions on Biomedical Engineering*, pp. 25–36

[9] Benjamin J.S, and Christopher G.A, 2010, Dynamic balance force control for compliant humanoid robots. *2010 IEEE/RSJ International Conference on Intelligent Robots and Systems (IROS),* pp 1248–1255.

[10] P. Derry, B. Eko H, A. Fernando, 2015, Movement control of two wheels balancing robot using cascaded PID controller. *2015 International Electronics Symposium (IES),* pp 100–105

[11] Anubha P, Apoorva P, and Somya G, 2016, Push recovery for humanoidrobot in dynamicenvironment and classifying the data using K-Mean,*Int. J. of Interactive Multimedia and AI, Vol. 4, No. 2*, pp. 29–34

[12] Y.F. Zheng, 1989, Acceleration compensation for biped robots to reject externaldisturbances,*In: IEEE Transactions on Systems, Manand Cybernetics, Vol. 19, No. 1*,pp.74–84

[13] Y. Wei, B. Gang, W. Zuwen, Balance recovery for humanoid robot in the presence of unknown external push, *In: ICMA 2009 Int. Conf. on Mechatronics and Automation*, pp.1928–1933

[14] Benjamin J. Stephens, 2007, Humanoid Push Recovery, In *Proc. IEEE-RAS Int. Conf. on Humanoid Robots (Humanoids)*, pp. 589 – 595

[15] Chao Li, Rong XIONG, Qiu-guo ZHU, Jun WU, Ya-liang WANG, Yi-ming HUANG, 2015, Push Recovery for The Standing Under-Actuated Bipedal Robot Using the Hip Strategy, In *Proc. Int. Journal: Frontiers of Information Technology & Electronic Engineering*, pp. 579 – 593

[16] Seung-Joon Yi, Byoung-Tak Zhang, Dennis Hong, Daniel D. Lee, 2012, Active Stabilization of a Humanoid Robot for Impact Motions with Unknown Reaction Forces, In *Proc. IEEE/RSJ Int. Conf. on Intelligent Robots and System*, pp. 4034 – 4039

[17] Sang-Ho Hyon, Joshua G. Hale, and Gordon Cheng, 2007, Full-Body Compliant Human–Humanoid Interaction: Balancing in the Presence of Unknown External Forces, In *Proc. Int, Journal: IEEE Transaction on Robotics*, Vol. 23, No. 5, pp. 884 – 898

[18] Awais Yasin, Qiang Huang, Qian Xu, M. Saad Sultan, 2012, Humanoid Robot Push Recovery through Foot Placement, In *Proc. IEEE Int. Conf. on Mechatronics and Automation*, pp. 59 – 63.

[19] Christian Ott, Maximo A. Roa, Gerd Hirzinger, 2011,Posture and Balance Control for Biped Robots based on Contact Force Optimization, In *Proc. IEEE-RAS Int. Conf. on Humanoid Robots (Humanoids)*, pp. 26 – 33

[20] Albertus Hendrawan A, Chee-Meng Chew, Weiwei Huang, Van Huan D, 2010, Humanoid Robot Push Recovery through Walking Phase Modification, In *Proc. IEEE Int. Conf. on Robotics Automation and Mechatronics (RAM)*, pp. 569 – 574

[21] Kenji Kaneko, Fumio Kanehiro, Mitsuharu Morisawa, Eiichi Yoshida, 2012, Disturbance Observer that Estimates External Force acting on Humanoid Robots, In *12th IEEE Int. Workshop on Advance Motion Control (AMC)*

[22] Ziegler. J.G and Nichols. N. B, 1942,Optimum setting for automatic controllers, In *Proc. IEEE Transactions of the ASME.* Vol. 64, pp. 759– 768

Acknowledgments
Gratefulness to Ministry of Research, Technology and Higher Education of the Republic of Indonesia for financial support and EEPIS Robotics Research Center (ER2C) laboratory.

IOP Conf. Series: Journal of Physics: Conf. Series **1007** (2018) 012069 doi:10.1088/1742-6596/1007/1/012069

Flex Sensor based Biofeedback Monitoring for Post-Stroke Fingers Myopathy Patients

Y R Garda[1], W Caesarendra[1], T Tjahjowidodo[2], A Turnip[3], S Wahyudati[4], L Nurhasanah[4] and D Sutopo[5]

[1]Mechanical Engineering Department, Diponegoro University, Semarang, Indonesia
[2]School of Mechanical and Aerospace Engineering, Nanyang Technological University, Singapore
[3]Technical Implementation Unit for Instrumentation Development, Indonesian Institute of Sciences, Bandung, Indonesia
[4]Medical Specialty Programs Physical Medicine and Rehabilitation, Medical Faculty, Diponegoro University, Semarang, Indonesia
[5]Electrical Engineering Department, Politeknik Negeri Batam, Batam, Indonesia

E-mail: yogaruki22@gmail.com

Abstract. Hands are one of the crucial parts of the human body in carrying out daily activities. Accidents on the hands decreasing in motor skills of the hand so that therapy is necessary to restore motor function of the hand. In addition to accidents, hand disabilities can be caused by certain diseases, e.g. stroke. Stroke is a partial destruction of the brain. It occurs if the arteries that drain blood to the brain are blocked, or if torn or leak. The purpose of this study to make biofeedback monitoring equipment for post-stroke hands myopathy patients. Biofeedback is an alternative method of treatment that involves measuring body functions measured subjects such as skin temperature, sweat activity, blood pressure, heart rate and hand paralysis due to stroke. In this study, the sensor used for biofeedback monitoring tool is flex sensor. Flex sensor is a passive resistive device that changes its resistance as the sensor is bent. Flex sensor converts the magnitude of the bend into electrical resistance, the greater the bend the greater the resistance value. The monitoring used in this biofeedback monitoring tool uses Graphical User Interface (GUI) in C# programming language. The motivation of the study is to monitor and record the progressive improvement of the hand therapy. Patients who experienced post-stroke can see the therapy progress quantitatively.

1. Introduction

Hands are one of the crucial parts of the human body in carrying out daily activities. However, hand injuries or hand accidents are the most common. As a result of this accident is a decrease in motor skills of the hand so that therapy is needed to restore motor function of the hand on a person. In addition to accidents, hand disabilities can be caused by certain diseases, one of which is a stroke. Stroke was the second leading cause of death after heart disease. Based on WHO data in 2012 there were 6.7 million people worldwide suffering from stroke [1]. Biofeedback is an alternative method of treatment that involves measuring body function measured subjects such as skin temperature, sweat activity, blood pressure, heart rate and hand paralysis due to stroke [2]. To date existed biofeedback-based monitoring system use sensors such as electromyograph (EMG), electrodemograph (EDG), electroencephalograph (EEG) and electrocardiogram (ECG). In hand therapy-based mechatronic systems, EMG sensors are widely used in decades due to its capability for measuring the muscle strength [3]. In addition to the muscle strength measurement, another indicator that showing the improvement of hand therapy program is necessary.

This indicator is the angle of finger motion. By monitoring the angle of finger motion and showing the result to the post-stroke patients, it will stimulate the patient's motivation for doing regular therapy. This idea is the main contribution of this paper. According to the capability of flex sensor that reported in the latest research papers [4, 5], we select flex sensor to measure and to monitor the angle of finger motion of myopathy patients. Flex sensor is a mechanical sensor where it works is to change the value of resistance when the sensor is curved. Flex sensor is designed with some resistive material so it can produce the resistance [3]. Biofeedback with flex sensor aims to monitor hand rehabilitation therapy of patients who are paralyzed due to stroke. This paper present preliminary study in developing a biofeedback-based monitoring system using Arduino and C# programming language. The C# programming language is a tool of programming (Rapid Application tool) created by Microsoft Corporation and can be used to create Grades-based programming using a C++-like programming language [4].

2. System

2.1. Wiring diagram of electronic parts and circuit

Wiring diagram is a visual representation of the connections and layout of electrical parts and systems. Wiring diagram show how the electrical components and cables are connected to build a system to assist the manufacturing process. In this paper, fritzing software was used to show the wiring diagram as presented in figure 1. This software can also display schematics for print circuit boards (PCB) and assist PCB manufacturing process. The schematic circuit diagram is presented in figure 2.

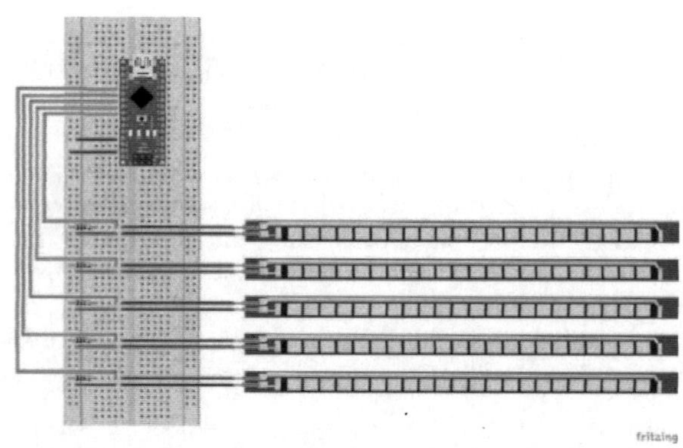

Figure 1. Wiring diagram of biofeedback system.

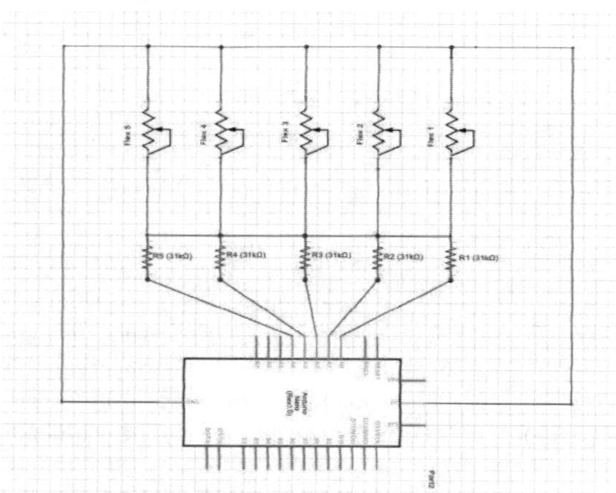

Figure 2. Schematic circuit diagram.

Table 1. Analog input configuration.

Analog input code	Description of fingers
A0	Thumb
A1	Index
A2	Middle
A3	Ring
A4	Little

2.2. Flex sensor attachment on the glove

Each flex sensor has a different resistance value. The resistance value increase proportionally to the bending angle of the flex sensor. Therefore, the flex sensor must be placed in a mixed condition instead of in a free movement. In this study, the flex sensor is attached on the glove with sewing method and glue as shown in figure 3. Sensors that have been placed on the gloves then read the size of the resulting resistance by means of integrating with microcontroller. Microcontroller used is Arduino Nano V3.0 ATM328 because compared with other Arduino, Arduino Nano type have smaller size than others, so can be more portable. Flex sensor has two cable connection parts, positive and negative. These are connected to PCB. The glove, Arduino and PCB connection of biofeedback monitoring tool can be seen in figure 4.

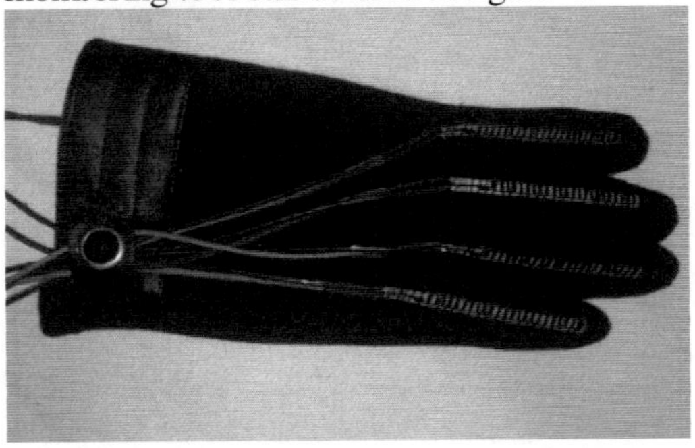

Figure 3. Glove with flex sensors.

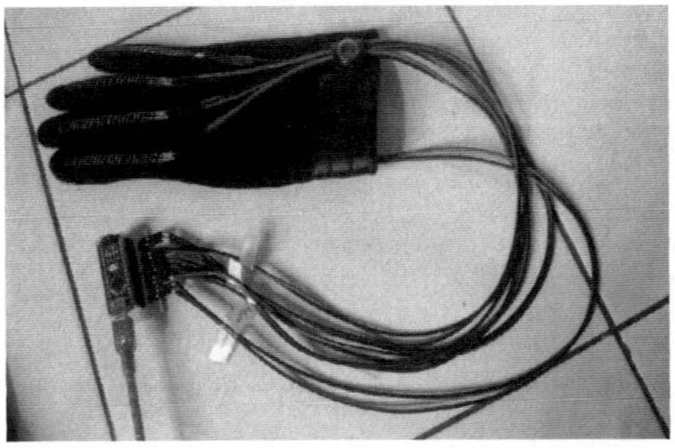

Figure 4. Glove, PCB and Arduino installation.

2.3. Flex sensor calibration

Flex sensor calibration is done by reading the ADC value generated when the sensor is in normal and bent condition. The purpose of this calibration is to align flex sensor motion with appropriate angular output results. Flex sensor calibration with gloves can be done by making the angle on the paper depicted by the measured lines of the angle by using the arc ruler shown in Fig. 5. Then the resultant angle that comes out on the arduino is equal to the angle measurement with the arc ruler.

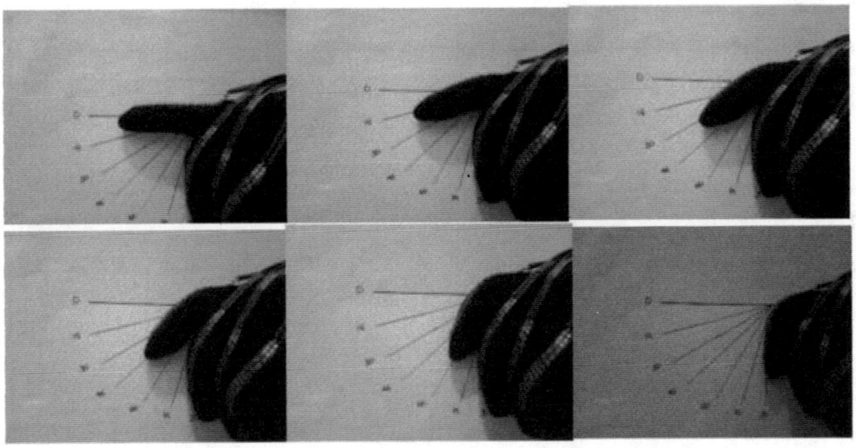

Figure 5. Manually-calibrated angle.

Arduino has an analog pin connected to a converter known as analog-to-digital converter (abbreviated ADC or A/D). This converter converts the analogue value of a Voltage signal into a digital form so that these analog values can be used more easily and applicable. In Arduino, this converter has a 10-bit resolution, meaning the conversion value ranges from 0 to 1023. At a voltage of 5 Volts (maximum), the Arduino will read a value of 1023, and when the voltage is 0 Volt (minimum), the read value is 0.

IOP Conf. Series: Journal of Physics: Conf. Series **1007** (2018) 012069 doi:10.1088/1742-6596/1007/1/012069

After the flex sensor can adapt to get the maximum and minimum values, the value is mapped to a value of 0-180. The resistance change can be measured through Arduino programming (analog pin reading) by putting a 31 kOhm resistor, which serves as a voltage divider that divides the 5 Volt voltage between the flex sensor and the resistor. By placing a 31 kOhm resistor, we will find an equation related to the resistance (partly due to flex) of the flex sensor to the voltage reading through which it passes. This resistance will be obtained by using a voltage divider as shown in quation (1).

$$V_{out} = \frac{legible\, value}{1023} x5 volt \tag{1}$$

where V_{out} is flex output voltage

The V_{out} value is obtained from the value shown on the Arduino serial monitor showing the voltage value of the analog input port. Since Arduino uses 10 bit analog inputs then to find the value of V_{out} using equation (2).

$$V_{out} = V_{in} \frac{R_2}{R_1 + R_2} \tag{2}$$

where V_{in} is an Arduino input voltage (volt), R_1 is flex sensor resistance (kOhm) and R_2 is resistor resistance (kOhm).

3. Results

Graphical User Interface (GUI) is an interface on an operating system that uses a graphical display, can be controlled using a variety of input devices, such as sensors. The GUI used for biofeedback monitoring uses Microsoft Visual Studio 2015 with the C# programming language (C sharp). The GUI made with C# programming language can be seen in figure 6. The generated GUI from the toolbox used is presented in figure 7 when the flex sensor gloves are paired on the fingers of the integrated Arduino Nano v3.0 Atm328 microcontroller.

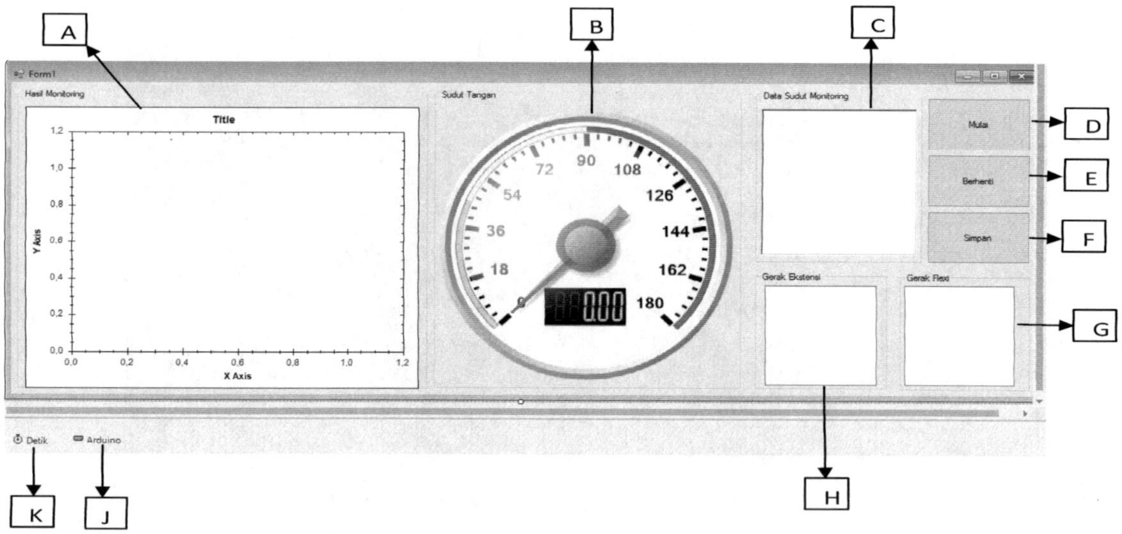

Figure 6. Biofeedback monitoring GUI.

Table 2. A description of biofeedback graphical user interface (GUI).

IOP Conf. Series: Journal of Physics: Conf. Series **1007** (2018) 012069 doi:10.1088/1742-6596/1007/1/012069

Code	Type of function on GUI	The purpose of the GUI function
A	Zedgraph	To create real-time degree monitoring of the fingers being monitored
B	Angular Gauge	To display angle animation in real-time
C	Richtextbox	To display the angle number of the finger being monitored
D	Start button	To start the monitoring process
E	Stop button	To stop the monitoring process
F	Save button	To store monitoring results in richtextbox
G	Textbox (extension motion)	For extension motion measurement of biofeedback monitoring user
H	Textbox (flexion motion)	For measurement of flexion motion user biofeedback monitoring
I	Digital monitoring	To display measurement result digitally
J	Arduino serial port	To create a relationship between Arduino IDE with C # programming language
K	Timer (second)	To enable the program operate in real-time

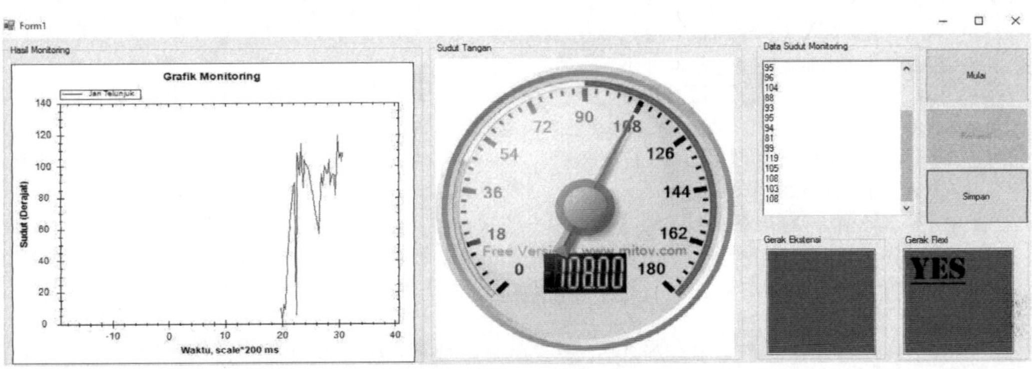

Figure 7. GUI display during the monitoring.

A biofeedback equipment that can be used and can be integrated with a GUI created in Microsoft Visual Studio 2015 with the C # programming language, here are the steps of measuring radius with a biofeedback equipment:

1. Hand to be measured, inserted into a glove that has been installed flex sensor as in figure 8.
2. Connect the integrated flex sensor gloves in the GUI in C # programming.
3. Start monitoring by looking at its interface on the GUI.
4. The monitoring results show that the extension box will be "Yes" when the angle is less than10^{o}, while the flexion box field will "Yes" when greater than 10°.
5. After the results of monitoring finger measurements have been obtained, then the results of monitoring angles can be stored in the form .xls file. And can be opened in Microsoft Office Excel.

IOP Conf. Series: Journal of Physics: Conf. Series **1007** (2018) 012069 doi:10.1088/1742-6596/1007/1/012069

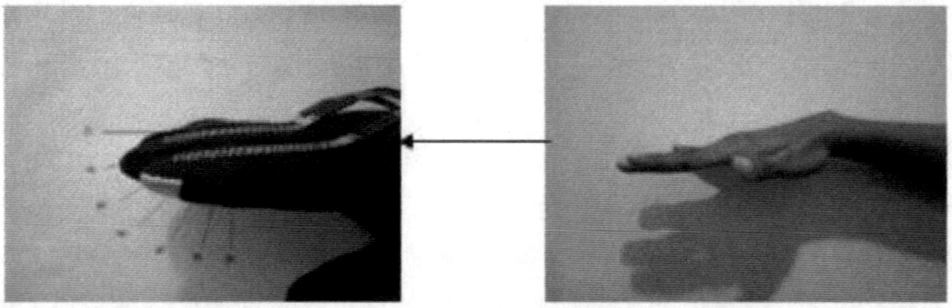

Figure 8. Before putting on biofeedback gloves and after wearing biofeedback gloves hand.

4. Conclusions

The biofeedback monitoring equipment successfully integrated with the flex sensor for flexion movement and finger extension, the equipment also managed to monitor the fingers, and can do the measurement for the motion of extension and flexion. The future works of present study is validate and test the biofeedback system to the myopathy patients in dr. Kariadi General Hospital, Semarang, Indonesia.

References

[1] World Health Organization, 2012. The 10 leading causes of death. Available at: http://www.who.int/mediacentre/factsheets/fs310/en/index.html [accessed on August 3, 2016]

[2] Giggins, Oonagh M; Persson, Ulrik MC; Caulfield, Brian . 2013. Journal of NeuroEngineering and Rehabilitation

[3] Vinstrup MSc, J., et.al. 2017 Hand strengthening exercises in chronic stroke patients: Dose-response evaluation using electromyography *Journal of Hand Therapy* (In Press)

[4] Saggio, G., 2014 A novel array of flex sensors for a goniometric glove, *Sensors and Actuators A: Physical* **205** pp 119-125

[5] Zheng, Y., Peng, Y., Wang, G., Liu, X., Dong, X., and Wang, J., 2016 Development and evaluation of a sensor glove for hand function assessment and preliminary attempts at assessing hand coordination *Measurement* **93** pp 1-12

[6] https://www.sparkfun.com, 2014. Flex Sensor SparkFun Electronics. Available at: https://www.sparkfun.com/products/10264 [accessed on August 30, 2016]

[7] Suharly , Rizky., 2015 . Design and implementation of finger robot hand wireless based on microcontroller. Bandung : Universitas Telkom.

Acknowledgement

Authors thank to Medical Specialty Programs Physical Medicine and Rehabilitation, Medical Faculty, Diponegoro University, Semarang, Indonesia for sharing information related to the hand therapy for myopathy patiens.

IOP Conf. Series: Journal of Physics: Conf. Series **1007** (2018) 012070 doi:10.1088/1742-6596/1007/1/012070

Generate an optimum lightweight legs structure design based on critical posture in A-FLoW Humanoid Robot

A Luthfi, K A Subhan, B Eko H, D R Sanggar, Pramadihanto D

EEPIS Robotics Research Center (ER2C), Politeknik Elektronika Negeri Surabaya (PENS), St. Raya ITS, Keputih, Sukolilo, Surabaya, 60111

E-mail: *luthfi.aminulloh@gmail.com, subhankh@pens.ac.id

Abstract. Lightweight construction and energy efficiency play an important role in humanoid robot development. The application of computer-aided engineering (CAE) in the development process is one of the possibilities to achieve the appropriate reduction of the weight. This paper describes a method to generate an optimum lightweight legs structure design based on critical posture during walking locomotion in A-FLoW Humanoid robot.The criticalposture can be obtained from the highest forces and moments in each joint of the robot body during walking locomotion. From the finite element analysis (FEA) result can be realized leg structure design of A-FLoW humanoid robot with a maximum displacement value of 0.05 mmand weight reduction about 0.598 Kg from the thigh structure and a maximum displacement value of 0,13 mmand weight reduction about 0.57 kg from the shin structure.

1. Introduction

Research in the field of robotics grows rapidly, one of them is ahumanoid robot. Humanoid robots are human-like robots, many researchers have successfully developed humanoid robots such as ATLAS [1], WABIAN [2], HUBO [3], LOLA [4] and ASIMO [5]. In previous research,EEPIS Robotic Research Centre (ER2C) laboratory has developed adult-sized humanoid robot named A-FloW[6]. This robot was developed using parallel link mechanism with 24 servo motors and 4 linear motors.Too many motor and mechanical parts increase the weight of robot to 25 kg in lower robot design. In order to reduce the weight, the robot will be redesigned using serial link mechanism. Computer-aided design (CAD) offers a wide variety of function that facilitates a design process. Traditional CAD systems primarily support modeling and analysis of conceptual solutions [7]. However, these systems rely on theuser to interpret their design problems and create conceptual solutions on their own. The systems do not actually explore and provide asolution for users [8]. Generative design method will be used to reduce the mechanical weight in this research. In this generative design method forces and moment are used as input in simulation to generate structure based on the distribution of the strain and stress [9,10]. There are three stages to design leg structure of A-FLoW humanoid robot as shown in figure 1.

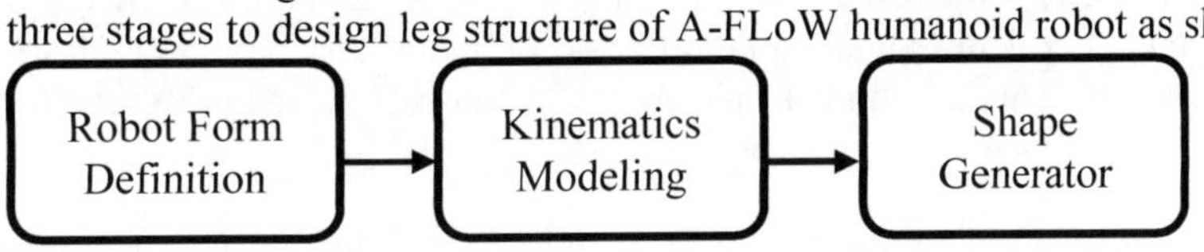

Figure 1. The design process of humanoid robot structure FLoW2.

IOP Conf. Series: Journal of Physics: Conf. Series **1007** (2018) 012070 doi:10.1088/1742-6596/1007/1/012070

The first stage is to define the humanoid robot form, the jointtype, and the mechanism. Then in the second stage, the kinematics of the robot is made. After that, a dynamic simulation is executed to find the value of force and moment that work on every joint. The highest value of force and moment in the quasi-dynamic simulation used as input in thegenerative design process at thefinal stage [11].

2. Kinematic Modelling

2.1. Inverse kinematics

This sectiondescribesthe model of inverse kinematic in A-FLoW humanoid robot. This work has been solved in previous research in T-FLoW humanoid robot [12]. So, can be obtained the inverse kinematic in Equation 1-3.

$$\theta_1 = Orientation \tag{1}$$

$$\theta_2 = \tan^{-1}\left(\frac{\sqrt{X^2+Y^2}*\sin(\tan^{-1}(Y/X)-\text{Orientation})}{Z}\right); \theta_6 = \theta_2 \tag{2}$$

$$\theta_3 = sin^{-1}\left(\frac{\sqrt{X^2+Y^2}}{\sqrt{{R_1}^2+aY^2}}\right) - cos^{-1}\left(\frac{{L_1}^2+{L_2}^2-{R_2}^2}{2\times {L_1}^2\times {L_2}^2}\right); \theta_4 = \pi - sin^{-1}\left(\frac{L_2*sin(\theta_{sc})}{\sqrt{{R_1}^2+aY^2}}\right); \theta_5 = \theta_3 + \theta_4 \tag{3}$$

Where $\theta_1, \theta_2, \theta_3$ are hip yaw, roll, pitch, θ_4 is knee pitch and θ_5, θ_6 are ankle pitch, roll.There are three planes to solve this inverse kinematic model. Where Equation 1is in transversal plane, Equation 2 is in frontal plane and Equation 3 is in sagittal plane. The angle calculation at each joint of the leg will be used as input parameters in quasi-dynamic simulation using Autodesk Inventor.

2.2. Walking Strategy

Walking strategyis needed to find critical posture and step conditions insingle support phase (SSP) and Double Support Phase (DSP). Walking strategy is used to organize trajectory on thecartesian coordinate system (x, y, z-axis). This walking strategy has been solved in the previous research [13]. Then can be obtained the walking strategy in Equation 4-7.

$$\Delta S_{(LT,RT),(x,y,z),(i)} = PP_{(LT,RT),(x,y,z),(i+1)} - PP_{(LT,RT),(x,y,z),(i)} \tag{4}$$

$$X_{(LT,RT),(T_T+T_{mn})} = \left((-\cos(T_T + T_{mn}) + 1) \times \Delta S_{(LT,RT),x,(i)}\right) + PP_{(LT,RT),x,(i)} \tag{5}$$

$$(Y,Z)_{(LT,RT),(1,2,4,5),(T_T+T_{mn})} = \left(\sin(T_T + T_{mn}) \times \Delta S_{(LT,RT),(y,z),(i)}\right) \times PP_{(LT,RT),(y,z),(i)} \tag{6}$$

$$(Y,Z)_{(LT,RT),(3,6),(T_T+T_{mn})} = \left((1 - \sin(T_T + T_{mn}) + 90) \times \Delta S_{(LT,RT),(y,z),(i)}\right) + PP_{(LT,RT),(y,z),(i)} \tag{7}$$

Where LT indicate left leg and RT indicate right leg. x, y, and z are walking trajectory parameter in cartesian. Then i is the number of steps while the robot walking.Equation 4 used to calculate the distance between apresent point and the last point parameters. Equation 5is used to obtain trajectory generator in thex-axis (step 1 to 6) and Equation 6 and Equation 7used to obtain trajectory generator in y,z-axis. Equation 6is in rising and transition step butEquation 10 is in landing step.

2.3. Critical Posture

Critical posture is a condition in which the movement of the foot structure of the robot holds the highest load that may occur when the robot is walking. Critical posture is derived from a simulation ofwalking motion on A-FloW robot.

IOP Conf. Series: Journal of Physics: Conf. Series **1007** (2018) 012070 doi:10.1088/1742-6596/1007/1/012070

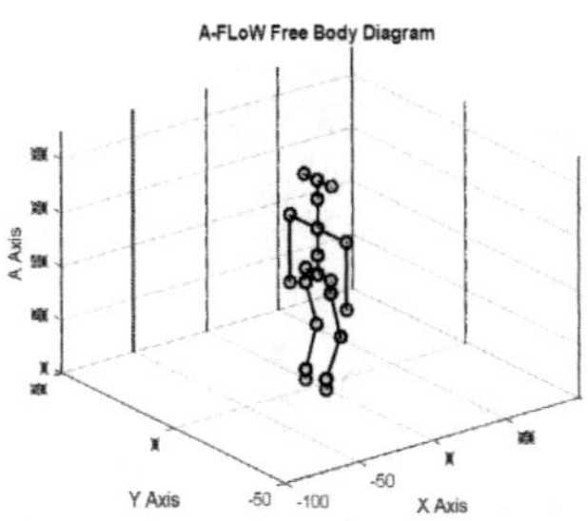

Figure 2. The Free-Body Diagram of A-FLoW in MATLAB simulation.

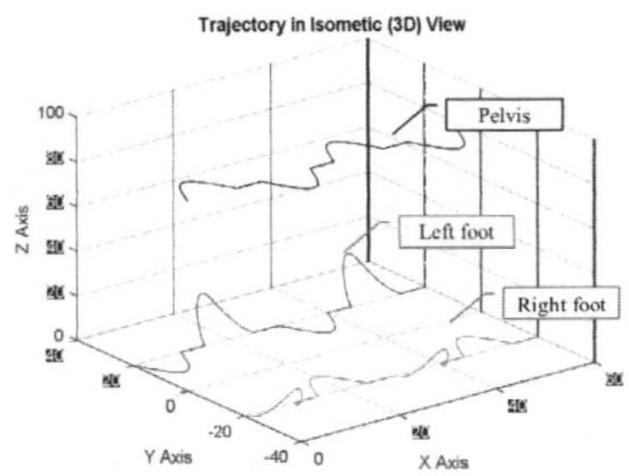

Figure 3. The 3D trajectory generated when the robot walking.

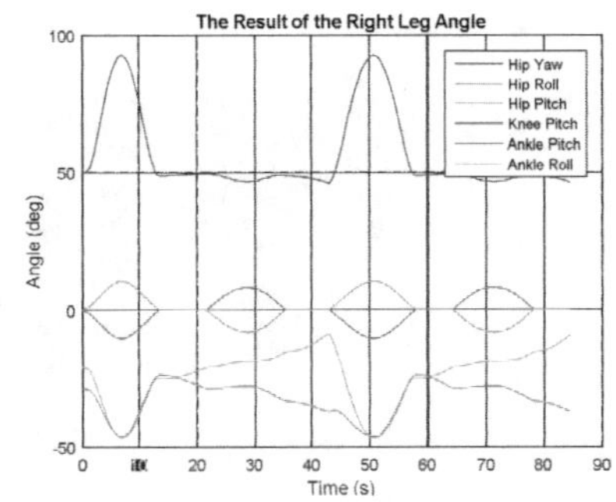

Figure 4. The angle values generated on each joint in left leg when robot walking.

When the walking motion implemented in the free-body diagram in Figure 2. Then trajectory of walking can be generated as shown in Figure 3. From the kinematic simulation, the angle in each joint of the legcan be obtained. Those resulting angle shown in Figure 4 are what will be used asinput parameters in thequasi-dynamic simulation.

3. Generative Design Implementation

3.1. Quasi-Dynamic Load Analysis

In the dynamic simulation, the balancing control on the robot is not applied. The input parameters for the robot motion in the dynamic simulation is generated from the inverse kinematic calculated on MATLAB. So, in the dynamic simulation, the robot can walk with a predetermined trajectory.

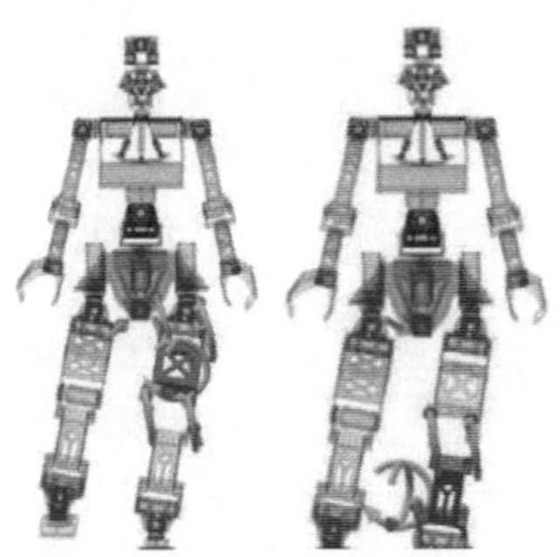

Figure 5. The moments and forces work on the leg, both in Thigh and shin.

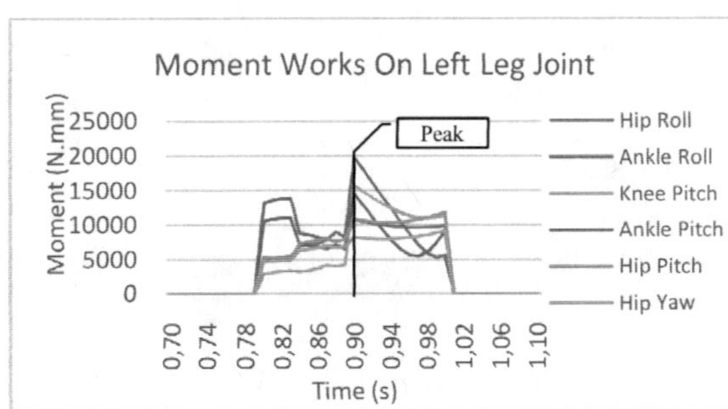

Figure 6. The moment works on leg joint, the simulation performs a critical posture on SSP and capture the moment 0.1s before and after.

Quasi-dynamic load analysis is used to simplify the dynamic model of the robot. The motion simulation in Autodesk Inventor is captured only in the peak value of SSP condition without losing its inertial effects as shown in Figure 5.So, the highest moments work on each leg joint will be obtained as displayed in Figure 6.

3.2. Shape Generator

Shape Generator provides an intelligent strategy for maximizing part stiffness based on the constraints that have been specified [9,10].Shape Generator produces a 3D mesh that can be used to guide the design refinement. The technology of shape generator integrated into some CAD software.The Autodesk fusion 360 used in this research make thedesign process more convenient to incorporate.

Figure 7. The distribution of stress and strain on thigh and shin structure.

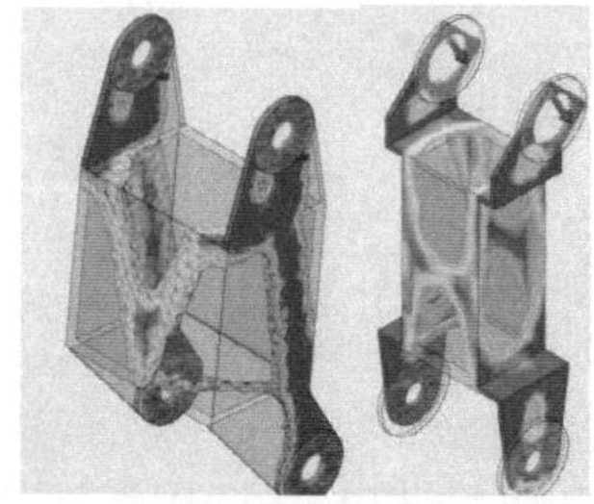

Figure 8. Weight reduction on the structure of shin based on the stress and strain distribution.

The distribution of stresson the thigh and shin structure after the load given to the structure are shown in Figure 7. The distribution of the most stress in the structure is indicated by the red area in the structure. So, the area is avoided to be removed in the load reduction process, Figure 8 shows the load reduction process by removing the blue area which gets the least stress and strain distribution.

4. Experimentand Analysis

4.1. Design Optimization Process

In the design optimization process of leg structure in A-FLoW humanoid robot,The conventional method is also presented using iterative design process as described by A. Albers in his research [8]. The conventional method will be compared to the proposed generative design method. So, it can be calculated the weight reduction from the conventional method to generative design method.

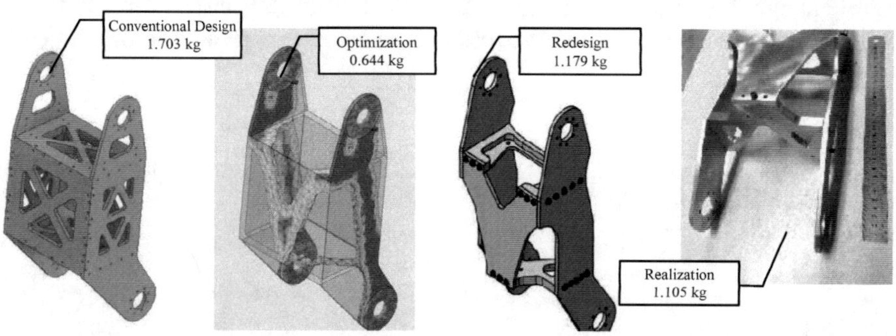

Figure 9. The comparison of weight in thigh design optimization process between conventional design proses, optimization result, redesign process and the manufactured part.

The design optimization process is displayed in figure 9.The weight from the conventional design of thigh structure is 1.703 kg, after redesign using generative design method the weight of the thigh structure reach 1.179 kg and the finalweight ofproduced part become 1.105 kg.

IOP Conf. Series: Journal of Physics: Conf. Series **1007** (2018) 012070 doi:10.1088/1742-6596/1007/1/012070

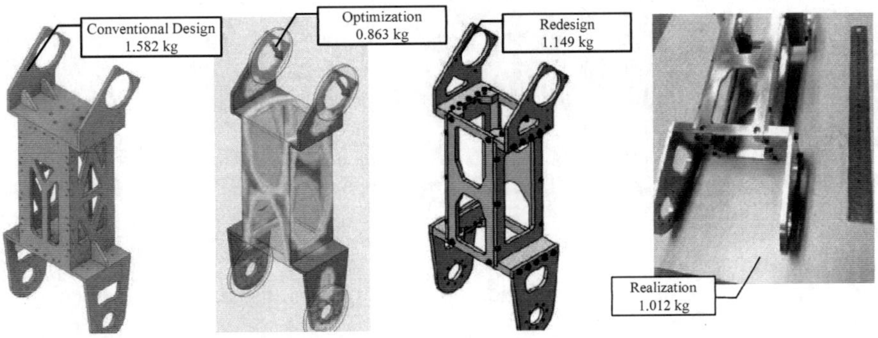

Figure 10. The comparison of weight in shin design optimization process between conventional design proses, optimization result, redesign process and the manufactured part.

The design optimization process is displayed in Figure 10. The weight of the conventional design of shin structure is 1.582 kg, after redesign using generative design method the weight of the thigh structure reach 1.149 kg and the final weight of produced part become 1.012 kg.

4.2. Design Analysis

In design analysis, Finite Element Analysis (FEA) method is presented. FEA shows whether a product will break, wear out, or work the way it was designed. EA works by breaking down a real object into a larger number of finite elements, such as little cubes. A computer then adds up all individual behaviors to predict the behavior of the actual object.

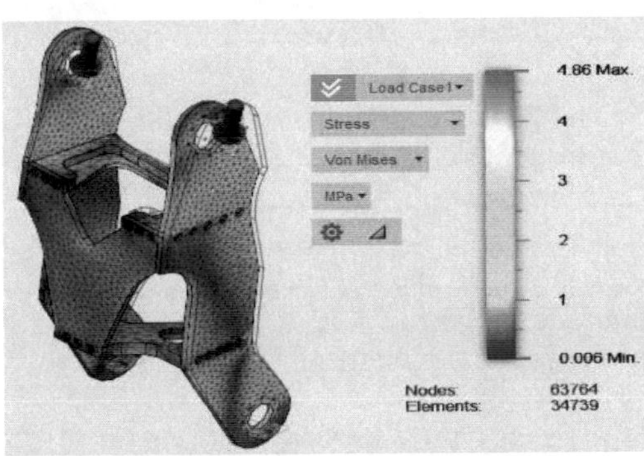

Figure 11. The FEA method is used in structural analysis on the design of the thigh.

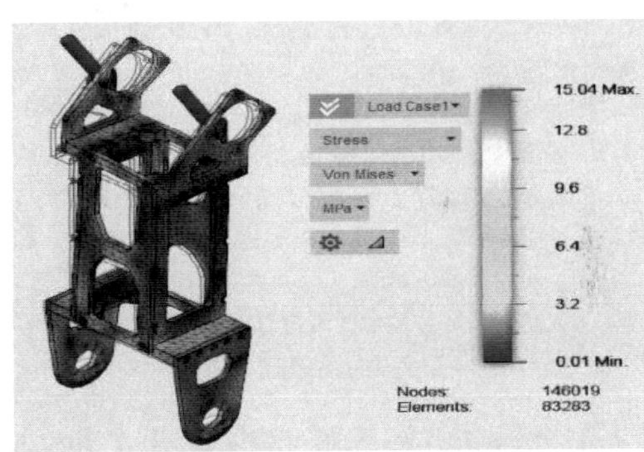

Figure 12. The FEA method is used in structural analysis on the design of the thigh.

The material used in the analysis is aluminum 7075. The analysis in Figure 11-12 shows the strain and stress spread evenly on the generated structure. The structure is analyzed by the loading of critical posture on the robot. The maximum stress and displacement in the thigh structure are 4.86 MPa and 0.05 mm. The maximum stress and displacement in the shin structure are 15.04 MPa and 0.13 mm.

4.3. Manufacturing Result

The manufactured part will be presented in this section. Designed structure in the previous stage will be disassembled into some part both in the thigh and shin structure. The raw material used to build each part is aluminum plat 7000 series. With the various thickness are 5 mm, 8 mm and 12 mm. Then the parts are manufactured using 3-axis CNC machine.

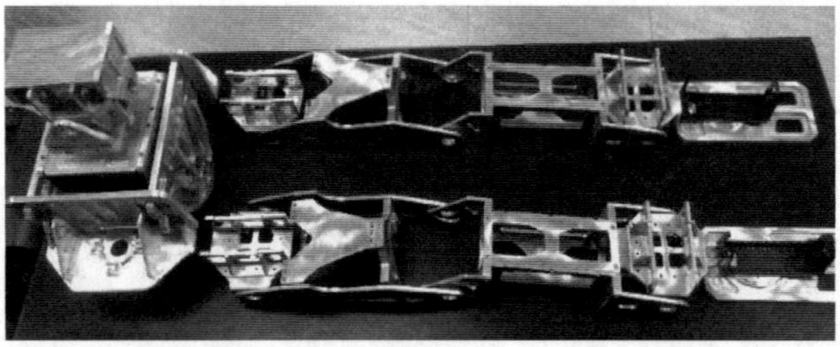

Figure 13. This is manufactured structureof the design provided by generative design method. These parts are ready to be assembled into A-FLoW Humanoid Robot with a height approximately 160 cm.

The result of machining process can be seen in figure 13. The weight of each leg structure is 3.57 kg and 7.14 for both leg structure. And after actuators are placed in each joint the weight reach 16.14 kg for both legs.

5. Conclusion

The main result of the experiment is areduction of leg structure design using generative design method compared to conventional method. The weight reduction is 0.598 kg in the thigh structure with the maximum stress and displacement from the thigh structure are 4.88 MPa and 0.05 mm. The weight reduction is 0.57 kgfrom the shin structure with the maximum stress and displacement are 15.04 MPa and 0.13 mm. The total weight is 16.14 kg in both of leg after the actuator has assembled, it finds out that the new leg design of A-FLoW humanoid robot more lightweight than its previous model.

But new problems appear in this research. The problem is thevalue of weight reduction is also based on the manufacturing constraints. So, the redesigned model can't fully meet the generated shape proposed by the generative design method. By the appeared of new problems, the next research will be afocus on manufacturing generative design using ametal 3D printer. So, the weight of the structure can be optimized more by having flexibility geometric shape and removing bolt connector.

References

[1] Banerjee, N., Long, X., Du, R., Polido, F., Feng, S., Atkeson, C.G., Gennert, M. and Padir, T., 2015, Human-supervised control of the ATLAS humanoid robot for traversing doors. *IEEE-RAS 15th International Conference on Robotics and Automation Society*, pp. 722-729

[2] Ogura, Y., Aikawa, H., Shimomura, K., Morishima, A., Lim, H. O., & Takanishi, A. 2006. Development ofa new humanoid robot WABIAN-2. *ICRA 2006. Proceedings 2006 IEEE International Conference on Robotics and Automation*, pp. 76-81

[3] Park, I.W., Kim, J.Y., Lee, J. and Oh, J.H., 2007. Mechanical design of the humanoid robot platform, HUBO. *Advanced Robotics, 21(11)*, pp.1305-1322

[4] Lohmeier, S., Buschmann, T., and Ulbrich, H., 2009, Humanoid robot LOLA. *IEEE International Conference on Robotics and Automation*, pp. 775-780

[5] Chestnutt, J., Lau, M., Cheung, G., Kuffner, J., Hodgins, J., & Kanade, T. 2005, April, Footstep planning for the hondaasimo humanoid. In *Robotics and Automation, 2005. ICRA 2005. Proceedings of the 2005 IEEE International Conference on* (pp. 629-634)

[6] K. Nuril E, D. Raden S, Dadet P, 2017, FLoW Bipedal Robot: Squat and Walk Motions, *Int. J. Advanced Science Letters*, Vol. 23, Issue 5, pp. 3838-3842

[7] Schwartz, M., Hwang, S., Lee, Y., Won, J., Kim, S. and Park, J., 2014, Aesthetic design and development of humanoid legged robot. In *Humanoid Robots (Humanoids), 2014 14th IEEE-RAS International Conference on* (pp. 13-19)

[8] Albers, A., Ottnad, J., Weiler, H., & Haeussler, P. 2007, Methods for lightweight design of mechanical components in humanoid robots, In Humanoid Robots, *7th IEEE-RAS International Conference on* (pp. 609-615)

[9] Krish, S., 2011. A practical generative design method. *Computer-Aided Design, 43(1)*, pp.88-100

[10] Zheng, B. and Gea, H.C., 2005. Structural topology optimization under design-dependent loads. In:*ASME Design Engineering Technical Conferences*, DETC (Vol. 85605, p. 2005)
[11] Hardee, E., Chang, K.H., Tu, J., Choi, K.K., Grindeanu, I. and Yu, X., 1999. A CAD-based design parameterization for shape optimization of elastic solids. *Advances in Engineering Software, 30(3)*, pp.185-199
[12] R. Dimas Pristovani, Ajir, B. Eko H, K. Achmad S, Dewanto. S, and Pramadihanto. D, 2017, Implementation of direct pass strategy during moving theball for "T-FLoW" humanoid robot, *in: 2nd International conferences on Information Technology, Information Systems and Electrical Engineering (ICITISEE)*
[13] R. Dimas Pristovani, B. Eko H, Dewanto. S, and Pramadihanto. D, 2017, Walking strategy model based on zero moment point with single inverted pendulum approach in "T-FLoW" humanoid robot, *in: 2nd International conferences on Information Technology, Information Systems and Electrical Engineering (ICITISEE)*

Acknowledgments
Gratefulness to Ministry of Research, Technology and Higher Education of the Republic of Indonesia for financial support and EEPIS Robotics Research Centre (ER2C) laboratory.

IOP Conf. Series: Journal of Physics: Conf. Series **1007** (2018) 012071 doi:10.1088/1742-6596/1007/1/012071

Variational iteration solution to the gravity wave-model equations

Sudi Mungkasi[1*], and Leo Hari Wiryanto[2]

[1]Department of Mathematics, Faculty of Science and Technology, Sanata Dharma University,Mrican, Tromol Pos 29, Yogyakarta 55002, Indonesia
[2]Department of Mathematics, Faculty of Mathematics and Natural Sciences, Bandung Institute of Technology,JalanGanesha 10, Bandung, Jawa Barat40132, Indonesia

E-mail: *sudi@usd.ac.id

Abstract. The gravity wave-model equations are considered. We solve these equations using the variational iteration method. The variational iteration solutions approximate the exact solution. The main advantage of using the variational iteration method is that we have an explicit function of the time and space variables as an approximate solution to the gravity wave-model problems.

1. Introduction

Shallow water waves occur in rivers and oceans. Water wave is considered shallow wave if the wave amplitude is smaller than the wave length. Shallow water waves can be governed by the shallow water wave equations. Furthermore, if the convective process is not significant, we can neglect the convective term in the shallow water wave equations and obtain the so called the gravity wave-model equations [1-3]. For a horizontal topography, the shallow water wave equations are

$$\frac{\partial}{\partial t}h(x,t)+\frac{\partial}{\partial x}(u(x,t)h(x,t))=0, \tag{1}$$

and

$$\frac{\partial}{\partial t}(u(x,t)h(x,t))+\frac{\partial}{\partial x}\left(u^2(x,t)h(x,t)+\frac{1}{2}gh^2(x,t)\right)=0. \tag{2}$$

Here h is the water depth;uh is the water discharge; gis the acceleration due to gravity; t is the time variable and x is the space variable. If we neglect the convective term u^2h, we obtain the gravity wave-model equations [1-3]

$$\frac{\partial}{\partial t}h(x,t)+\frac{\partial}{\partial x}(u(x,t)h(x,t))=0, \tag{3}$$

and

$$\frac{\partial}{\partial t}(u(x,t)h(x,t))+\frac{\partial}{\partial x}\left(\frac{1}{2}gh^2(x,t)\right)=0. \tag{4}$$

In this paper, we take the convention that all quantities have the SI units with the MKS system. In general, the exact solution to the gravity wave-model equations is not available until this paper is

IOP Conf. Series: Journal of Physics: Conf. Series **1007** (2018) 012071 doi:10.1088/1742-6596/1007/1/012071

written. Therefore, approximate solutions are usually desired. The aim of this study is to provide a way of approximating the exact solution to the gravity wave-model equations using the variational iteration method. The variational iteration method is chosen, because it is an analytical approach [4-7], so we do not need to discretise the time and space domains in order to solve the gravity wave-model problems.We note that some variants of variational iteration methods have been used to solve a number of different problems in physics [8-11], mathematics [12], computer science [13], and chemistry [14-15]. This paper is organised as follows. We provide the derivation of approximate solutions using the variational iteration method in Section 2. Results and discussion are given in Section 3. The paper is concluded in Section 4.

2. Variational iteration method

The variational iteration method works by first we construct the correction functionals for the gravity wave-model equations. Then we specify the value of the Lagrange multiplier of the correction functionals. Once we obtain the Lagrange multiplier, we substitute this Lagrange multiplier into the correction functionals to form the variational iteration formulas. The variational iteration formulas for the gravity wave-model equations are

$$h_{n+1}(x,t) = h_n(x,t) - \int_0^t \left(\frac{\partial h_n}{\partial \tau} - \frac{\partial (q_n)}{\partial x}\right) d\tau, \tag{5}$$

and

$$q_{n+1}(x,t) = q_n(x,t) - \int_0^t \left(\frac{\partial (q_n)}{\partial \tau} - \frac{\partial \left(\frac{1}{2} g h_n^2\right)}{\partial x}\right) d\tau. \tag{6}$$

Here $q_n = u_n h_n$ is the approximate water discharge function at the nth iteration level. The initialisation is set by taking the initial condition, that is, $h_0(x,t) = h(x,0)$ and $u_0(x,t) = u(x,0)$ so $q_0(x,t) = q(x,0) = u(x,0)h(x,0)$. When the variational iterations converge, they converge to the exact solution to the problem.

3. Results and discussion

In this section we provide our computational results and discussions.

3.1 Surface wave simulation

Surface wave is simulated using the initial water depth

$$h(x,0) = 0.1 \; + 0.25sech(x) \; + \; exp(-x^2)/(1 + exp(-x^2)) \tag{7}$$

for all x, as shown in Figure 1. We take the gravitational constant $g = 1$. The simulation is supposed to find the solution at any timet. The initial fluid velocity is $u(x,0) = 0$, for all x. Using the MAPLE software, the variational solutions are computed for $h_0(x,t)$, $q_0(x,t)$ and $h_1(x,t)$, $q_1(x,t)$, as well as $h_2(x,t)$, $q_2(x,t)$. Our results are as follows. Note that the star " * " sign is the multiplication operator in the MAPLE software. For the initialisation, we have

$$h_0(x,t) = 0.1 + 0.25 * sech(x) + \frac{exp(-x^2)}{1 + exp(-x^2)}$$

and

$$q_0(x,t) = 0\,.$$

The first levels of variationaliteration solutions are

$$h_1(x,t) = h_0(x,t)$$

and

IOP Conf. Series: Journal of Physics: Conf. Series **1007** (2018) 012071 doi:10.1088/1742-6596/1007/1/012071

$$q_1(x,t) = \left(0.1 + 0.25 * sech(x) + \frac{exp\,(-x^2)}{1+exp\,(-x^2)}\right) * \left(-0.25 * sech(x) * tanh(x) - 2 * x * \frac{exp\,(-x^2)}{1+exp\,(-x^2)} + 2 * \frac{exp\,(-x^2)^2}{(1+exp\,(-x^2))^2} * x\right) * t\,.$$

The second levels of variationaliteration solutions are

$$h_2(x,t) = 0.1 + 0.25 * sech(x) + \frac{exp(-x^2)}{1+exp(-x^2)} - 0.5 * \left(-1 * \left(-0.25 * sech(x) * tanh(x) - 2 * x * \frac{exp(-x^2)}{1+exp(-x^2)} + 2 * \frac{exp(-x^2)^2}{(1+exp(-x^2))^2} * x\right)^2 - 1 * \left(0.1 + 0.25 * sech(x) + \frac{exp(-x^2)}{1+exp(-x^2)}\right) * \left(0.25 * sech(x) * tanh(x)^2 - 0.25 * sech(x) * (1 - 1 * tanh(x)^2) - 2 * \frac{exp(-x^2)}{1+exp(-x^2)} + 4 * x^2 * \frac{exp(-x^2)}{1+exp(-x^2)} - 12 * x^2 * \frac{exp(-x^2)^2}{(1+exp(-x^2))^2} + 8 * \frac{exp(-x^2)^3}{(1+exp(-x^2))^3} * x^2 + 2 * \frac{exp(-x^2)^2}{(1+exp(-x^2))^2}\right)\right) * t^2$$

and

$$q_2(x,t) = \left(0.1 + 0.25 * sech(x) + \frac{exp\,(-x^2)}{1+exp\,(-x^2)}\right) * \left(-0.25 * sech(x) * tanh(x) - 2 * x * \frac{exp\,(-x^2)}{1+exp\,(-x^2)} + 2 * \frac{exp\,(-x^2)^2}{(1+exp\,(-x^2))^2} * x\right) * t\,.$$

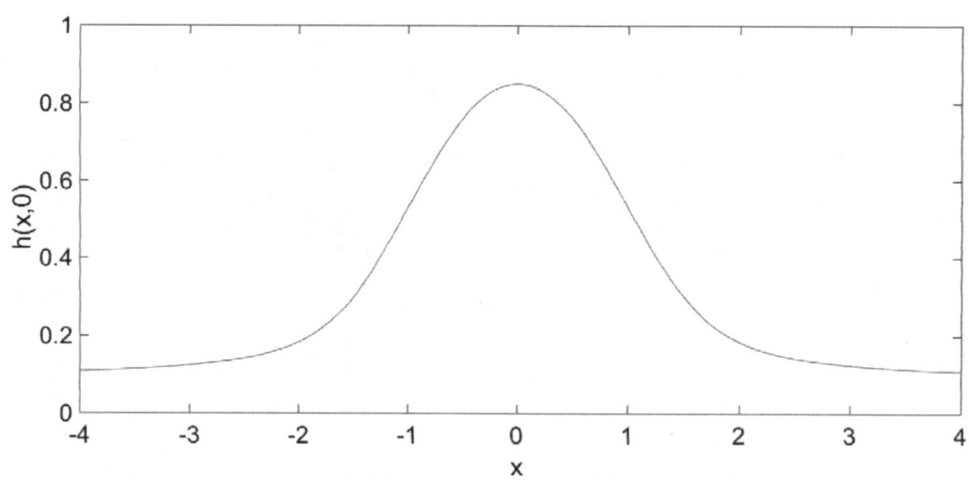

Figure 1. The initial water surface with the horizontal axis is for x variable.

3.2 Dam-break-like simulation

In this simulationwe solve a dam-break-like problem.The initial water depth is

$$h(x,0) = 2 - 1/(1 + exp(-1000x)) \quad (8)$$

for all x, as shown in figure 2. Again we take the gravitational constant$g = 1$. We want to find the solution at any time$t$. The initial fluid velocity is again set to be$u(x, 0) = 0$, for all$x$.

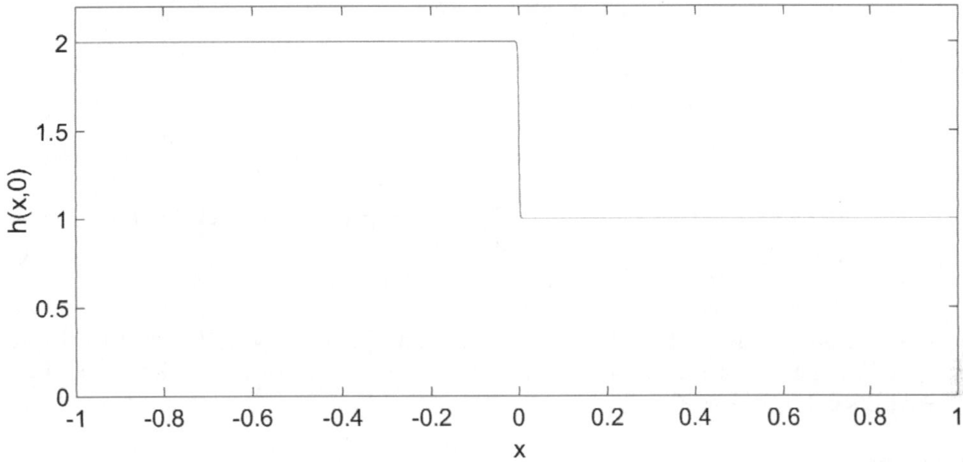

Figure 2. The initial water surface of the dam-break-like problem. The surface is actually smooth, but looks discontinuous at position x = 0.

Again using the aid of the MAPLE software, the variationaliterationsolutions are computed up to the second levels$h_2(x,t)$ and $q_2(x,t)$. The zeroth levels of variationaliterationsolutions are

$$h_0(x,t) = 2 - \frac{1}{1 + exp(-1000 * x)}$$

and

$$q_0(x,t) = 0\,.$$

The first levels of variational iteration solutions are

$$h_1(x,t) = h_0(x,t)$$

and

$$q_1(x,t) = -1000 * \frac{2 - \frac{1}{1+exp\,(-1000*x)}}{\left(1 + exp(-1000 * x)\right)^2} * exp(-1000 * x) * t\,.$$

The second levels of variationaliterationsolutions are

IOP Conf. Series: Journal of Physics: Conf. Series **1007** (2018) 012071 doi:10.1088/1742-6596/1007/1/012071

$$h_2(x,t) = 2 - \frac{1}{1+exp(-1000*x)} - 0.5 * \left(\frac{-1000000}{\left(1+exp(-1000*x)\right)^4} * exp(-1000*x)^2 + 2000000 * \frac{2 - \frac{1}{1+exp\,(-1000*x)}}{\left(1+exp(-1000*x)\right)^3} * exp(-1000*x)^2 - 1000000 * \frac{2 - \frac{1}{1+exp\,(-1000*x)}}{\left(1+exp(-1000*x)\right)^2} * exp(-1000*x) \right) * t^2$$

and

$$q_2(x,t) = -1000 * \frac{2 - \frac{1}{1+exp\,(-1000*x)}}{\left(1+exp(-1000*x)\right)^2} * exp(-1000*x) * t\,.$$

Obviously we can continue our calculation to any level of iteration. More number of iterations makes the approximate solutions too long to write in this paper. Therefore, we stop our iterations at the second level. We consider it enough to demonstrate how the variational iteration method works for the gravity wave-model equations.

4. Conclusion

We have demonstrated how to solve the gravity wave-model equations using the variational iteration method.Variational iteration solutions are successfully found as approximate solutions to the exact solution. The variational iteration solutions may need enhancement for approximating very large domain of time and space which could be a future direction to research.

References

[1] Apriani C M and Mungkasi S 2017 A staggered grid finite difference method for solving the gravity wave-model equationsJournal of Physics: Conference Series **909** 012046

[2] Martins R, Leandro J and Djordjević S 2016 Analytical solution of the classical dam-break problemfor the gravity wave-model equations Journal of Hydraulic Engineering **142**06016003

[3] Martins R, Leandro J and Djordjević S 2015 A well balanced Roe scheme for the local inertial equations with an unstructured meshAdvances in Water Resources **83** 351

[4] Mungkasi S and Wiryanto L H 2016 On the relevance of a variational iteration method for solving the shallow water equations AIP Conference Proceedings **1707** 050010

[5] Setianingrum P S and Mungkasi S 2016 Variational iteration method used to solve steady state problems of shallow water flows AIP Conference Proceedings **1746** 020057

[6] Setianingrum P S and Mungkasi S 2016 Variational iteration method used to solve the one-dimensional acoustics equations Journal of Physics: Conference Series **856**012010

[7] Wazwaz A M 2009 Partial Differential Equations and Solitary Waves Theory (New York: Springer)

[8] Ghorbani A and Bakherad M 2017 A variational iteration method for solving nonlinear Lane–Emden problemsNew Astronomy **54** 1

[9] Li Y and Yang Y 2017 Vibration analysis of conveying fluid pipe via He's variational iteration methodApplied Mathematical Modelling **43** 409

[10] Mohyud-Din S T, Sikander W, Khan U and Ahmed N 2017 Optimal variational iteration method using Adomian's polynomials for physical problems on finite and semi-infinite intervalsThe European Physical Journal Plus **132** 236
[11] Singh R, Das N and Kumar J 2017 The optimal modified variational iteration method for the Lane-Emden equations with Neumann and Robin boundary conditionsThe European Physical Journal Plus **132** 251
[12] Khuri S A and Sayfy A 2017 Generalizing the variational iteration method for BVPs: Proper setting of the correction functionalApplied Mathematics Letters **68** 68
[13] Armand A, Allahviranloo T, Abbasbandy S and Gouyandeh Z 2017 Fractional relaxation-oscillation differential equations via fuzzy variational iteration methodJournal of Intelligent & Fuzzy Systems **32** 363
[14] Wazwaz A-M 2017 Solving the non-isothermal reaction-diffusion model equations in a spherical catalyst by the variational iteration methodChemical Physics Letters **679** 132
[15] Wazwaz A-M 2017 The variational iteration method for solving systems of third-order Emden-Fowler type equationsJournal of Mathematical Chemistry **55** 799

Acknowledgements
This work was financially supported by a research grant from Direktorat Riset dan Pengabdian Masyarakat (DRPM) of the Ministry of Research, Technology and Higher Education of The Republic of Indonesia year 2017. Both authors are very grateful for the financial support.